A New Generation Material Graphene: Applications in Water Technology

Mu. Naushad
Editor

A New Generation Material Graphene: Applications in Water Technology

Springer

Editor
Mu. Naushad
Department of Chemistry, College
of Science
King Saud University
Riyadh
Saudi Arabia

ISBN 978-3-030-09244-3 ISBN 978-3-319-75484-0 (eBook)
https://doi.org/10.1007/978-3-319-75484-0

Softcover re-print of the Hardcover 1st edition 2019

Printed on acid-free paper

This Springer imprint is published by the registered company Springer International Publishing AG part of Springer Nature
The registered company address is: Gewerbestrasse 11, 6330 Cham, Switzerland

Preface

Water is the most essential substance for all life on earth and a precious resource for human civilization. The quantity of drinkable water is extremely limited. Out of all the water on earth, less than 1% water is available as groundwater and surface water, which is suitable for human uses such as drinking and cooking. Groundwater and surface water are continuously contaminating with various types of inorganic and organic pollutants. Current water and wastewater treatment technologies and infrastructure are reaching their limit for providing adequate water quality to meet human and environmental needs. Over the last few decades, nanotechnology is emerging as a rapidly growing sector of a knowledge-based economy due to unique physiochemical properties of nanomaterial. This technology gained a tremendous impetus due to its capability of reformulating the particle of metals into new nanosized form, with dimension less than 100 nm in size.

In recent years, various types of nanomaterials such as magnetic nanoparticles, layered double hydroxides, ion exchange nanocomposites, aluminosilicates, nanozeolites, quantum dots, carbon nanotubes and graphene-based nanocomposite materials have been widely explored for the remediation of wastewater. Among these materials, graphene is considered to be the most 'hot' material of recent years due to its numerous properties (mechanical, optical, environmental, etc.)

Therefore, it was thought worthwhile and opportune to prepare a reference book which describes the applications of graphene-based nanocomposite materials for water treatment. This book is the result of remarkable contribution from the experts of interdisciplinary fields of science with comprehensive, in-depth and up-to-date research and reviews.

Chapter 1 focuses on the main groups of the marine contaminants, their effects on mankind and methodologies to diminish the pollution of the ground and drinking water resources. This chapter will therefore help the readers to understand the issues and challenges that the humankind is facing in dealing with water quality issues.

Chapter 2 highlights the different water quality assessments, several sources of water pollution and the methods used for treating water for various purposes such as for drinking, industrial water supply, irrigation, water recreation or many other uses.

Overview regarding qualitative and quantitative measurements which are needed from time to time to constantly monitor the quality of water is also presented in this chapter.

Chapter 3 shows the current state of the art on the development and application of 3D graphene-based macroscopic assemblies for ultrafast and recyclable oils and organic solvents absorption. Furthermore, it distinguishes the fundamental knowledge gaps in the domain and lays out novel strategic research guidelines, all of which will promote further progress in this rapidly evolving cross-disciplinary field of current global interest.

In Chap. 4, different types of green and eco-friendly materials for the adsorption of various types of inorganic and organic pollutants are discussed.

Chapter 5 mainly focuses on the reduction of trihalomethanes (THMs) in drinking water in Puerto Rico. Three different nanostructured materials (graphene, mordenite and multiwalled carbon nanotubes) were used to reduce the THMs formation by adsorption in specific contact time. The results showed that graphene is the best nanomaterial to reduce THMs in drinking water.

Chapter 6 focuses on the recent advances in water treatment using chemically modified graphene/GO and aims to provide insights into the developments in water contamination removal technologies based on these novel nanomaterials. These GO-based composites have shown outstanding performance for the removal of water contaminants as well as act as good adsorbent of various types of inorganic pollutants like cadmium, chromium, arsenic, mercury, antimony, lead, fluoride, zinc, copper, etc. and organic dyes such as methylene blue (MB), methyl violet (MV), methyl orange (MO), rhodamine B (RB), etc.

Chapter 7 provides a detailed discussion on pharmaceuticals in general, their occurrence in water and their health consequences. It also delved into the photocatalytic degradation of these chemicals in water with emphasis on the use of graphene-based materials.

In Chap. 8, the pros and cons of the ozonation reaction catalyzed by graphene and their derivatives have been discussed tentatively. The unique properties of graphene that are of relevance to catalysis, with emphasis on the adsorption, electrostatic interaction, active sites that have been proposed to be responsible for the catalytic activity have been discussed. Moreover, some challenging issues of graphene-based carbocatalysts have been proposed to be resolved for the future development in the field.

Chapter 9 presents advances made in the synthesis of graphene oxides and their composites, and summarizes the application of these materials as a superior adsorbent for the removal of arsenic from water. The adsorption affinity in terms of contact time, pH and temperature has been discussed. Competitive ion effect and regeneration are included within the text. Moreover, the challenges for the commercial uses are discussed.

In Chap. 10, the formation of highly toxic bromate in the drinking water and its removal using graphene-based materials have been discussed. It is believed that graphene-based materials can meet water challenges in a sustainable way.

Chapter 11 focuses on the catalytic ozonation of aqueous aromatic pollutants using graphene/CNT and their derivative materials. The various sources of aqueous aromatic pollutants were demonstrated. Some compatible treatment methods of aromatic pollutants in water with their merits and demerits were debated.

Chapter 12 describes the promising application of graphene-based nanocomposites in adsorbing and removing nitrates and phosphates from the water bodies. Graphene nanocomposites can be obtained from its derivatives such as graphene oxide, chemically and thermally reduced graphene oxides. They act as nanosorbents and are widely used owing to their properties such as biocompatibility, stability, high surface area to volume ratio, good conductivity and low-cost synthesis.

Chapter 13 focuses on the applicability of graphene, its composites and its modified forms for the treatment of pesticides-containing effluents. The main aim of this chapter is to discuss the preparation, characterization and various types of application of graphene-based materials for the purification of pesticide-containing water.

Chapter 14 summarizes the various fabrication techniques used for the construction of sensors and biosensors based on graphene oxide and CNTs related materials for detection and removal of bisphenol A from different kinds of solutions and materials. This chapter also provides an overview of analytical performance for the application in clinical, environmental and food sciences research, and comments on future and interesting research trends in this field.

Chapter 15 presents the application of graphene oxide nanocomposites in the removal of toxic metal ions from water. A brief overview of recently developed/modified graphene oxide nanocomposites, their efficiency in heavy metal removal from water, adsorption kinetics and thermodynamic models, and their regeneration potential is presented in this chapter.

Chapter 16 presents the new analytical approaches for pharmaceutical wastewater treatment using graphene-based materials.

Chapter 17 focuses on solar light-driven renewable water treatment tool using less expensive semiconductor photocatalyst materials. In particular, it covers the recent development of graphene-based composite (metal, semiconductor and alloys) materials in organic dye pollutant degradation. The photocatalyst synthesis routes, comparative pollutant degradation analysis and required material property for effective water treatment are discussed in detail.

Chapter 18 presents a short introduction to adsorption principles and adsorption isotherms. It explains the synthesis and use of nano-graphene materials for the remediation of dyes. It also consolidates the recent literature available for dye adsorption using graphene materials and its mechanism.

Riyadh, Saudi Arabia Dr. Mu. Naushad

Acknowledgements

First and foremost I would like to thank God for giving me the skills, knowledge and abilities to achieve anything that set my mind out to do.

This book is the outcome of the remarkable contribution of experts from various interdisciplinary fields of science and covers the most comprehensive, in-depth and up-to-date research and reviews. I am thankful to all the contributing authors and their co-authors for their esteemed work. We would like to express my gratitude to all the authors, publishers and other researchers for granting us the copyright permissions to use their illustrations, although every effort has been made to obtain the copyright permissions from the respective owners to include citation with the reproduced materials. I would still like to offer my deep apologies to any copyright holder if unknowingly their right is being infringed.

I would like to thank the Springer Publication house for giving me this opportunity to edit this book. Special thanks to Dr. Nabil Khelifi and Reyhaneh Majidi, who provide me the editorial assistant and support. Without their support, this book would not have been possible.

I express my deep gratitude to the Chairman, Department of Chemistry, College of Science, King Saud University, Saudi Arabia for his valuable suggestions, guidance and constant inspiration. I am feeling short of words to express my thanks to all my friends and colleagues for their timely help, good wishes, encouragement and affections.

Above all, I want to thank my parents and all my family members who supported and encouraged me in spite of all the time it took me away from them. It was a long and difficult journey for them. Last and not least, I beg forgiveness of all those who have been with me over the course of the years and whose names I have failed to mention.

Finally, I would extend my appreciation to the Deanship of Scientific Research at King Saud University for the support.

Dr. Mu. Naushad

Contents

Editor and Contributors

About the Editor

Dr. Mu. Naushad is presently working as Associate Professor in the Department of Chemistry, College of Science, King Saud University (KSU), Riyadh, Kingdom of Saudi Arabia. He obtained his M.Sc. and Ph.D. degrees in Analytical Chemistry from Aligarh Muslim University, Aligarh, India in 2002 and 2007, respectively. He has vast research experience in the multidisciplinary fields of Analytical Chemistry, Materials Chemistry and Environmental Science. He holds several U.S. patents, over 200 publications in the international journals of repute, 15 book chapters and four books published by renowned international publishers. He has >4400 citations with a Google Scholar H-Index of >40. He has successfully run several research projects funded by National Plan for Science and Technology (NPST) and King Abdulaziz City for Science and Technology (KACST), Kingdom of Saudi Arabia. He is the editor/editorial member of several reputed journals like Scientific Report (Nature); Process Safety & Environmental Protection (Elsevier); Journal of Water Process Engineering (Elsevier) and International Journal of Environmental Research & Public Health (MDPI). He is also the associate editor for Environmental Chemistry Letters (Springer) and Nanotechnology for Environmental Engineering Journal (Springer). He has been awarded the Scientist of the Year Award-2015 from National Environmental Science Academy, Delhi, India and Almarai Award-2017, Saudi Arabia.

Contributors

P. Ajmal Koya Department of Chemistry, National Institute of Technology Mizoram, Aizawal, Mizoram, India

Basma Al-Najar Department of Physics, College of Science, University of Bahrain, Sakheer, Bahrain

William W. Anku Department of Applied Chemistry, University of Johannesburg, Johannesburg, South Africa

Tanvir Arfin Environmental Materials Division, CSIR-National Environmental Engineering Research Institute (CSIR-NEERI), Nagpur, India

Rajasekhar Balasubramanian Department of Civil and Environmental Engineering, National University of Singapore, Singapore, Singapore

Qi Bao Hong Kong Applied Science and Technology Research Institute Company Limited (ASTRI), Shatin, Hong Kong SAR, China

Rajesh Babu Bitra Department of Physics, G.V.P. College of Engineering for Women, Visakhapatnam, Andhra Pradesh, India

Mohamed Bououdina Department of Physics, College of Science, University of Bahrain, Sakheer, Bahrain

Jorge L. Hernández Bourdón School of Science and Technology, Universidad Del Turabo, Gurabo, PR, USA

A. Carmalin Sophia National Environmental Engineering Research Institute (NEERI), Chennai Zonal Laboratory, Taramani, Chennai, India

Saif Ali Chaudhry Department of Chemistry, Jamia Millia Islamia University, New Delhi, India

Shamik Chowdhury Centre for Advanced 2D Materials, National University of Singapore, Singapore, Singapore

Papita Das Department of Chemical Engineering, Jadavpur University, Kolkata, India

Akira Fujishima Photocatalysis International Research Center, Tokyo University of Science, Noda, Chiba, Japan

Pallavi Goel Amity Institute of Applied Sciences, Amity University, Noida, Uttar Pradesh, India

Penny P. Govender Department of Applied Chemistry, University of Johannesburg, Johannesburg, South Africa

Mirna Habuda-Stanic Faculty of Food Technology Osijek, Josip Juraj Strossmayer University of Osijek, Osijek, Croatia

Girish M. Joshi Polymer Nanocomposite Laboratory, Center for Crystal Growth, VIT University, Vellore, Tamilnadu, India

Parul Khurana Department of Chemistry, G. N. Khalsa College of Arts, Science and Commerce, Mumbai University, Mumbai, India

Ephraim M. Kiarii Department of Applied Chemistry, University of Johannesburg, Johannesburg, South Africa

Venkata Krishnan School of Basic Sciences and Advanced Materials Research Center, Indian Institute of Technology Mandi, Mandi, Himachal Pradesh, India

Dinesh Kumar School of Chemical Science, Central University of Gujarat, Gandhinagar, India; Department of Chemistry, Central University of Gujarat, Gandhinagar, India

Ilango Aswin Kumar Department of Chemistry, Anna University, University College of Engineering Dindigul, Dindigul, Tamil Nadu, India

Rajesh Kumar Centre for Semiconductor Components and Nanotechnology (CCS Nano), University of Campinas (UNICAMP), Campinas, Sao Paulo, Brazil

Suneel Kumar School of Basic Sciences and Advanced Materials Research Center, Indian Institute of Technology Mandi, Mandi, Himachal Pradesh, India

Vinod Kumar Department of Materials Engineering, Ben Gurion University of the Negev, Beer-Sheva, Israel

Eder C. Lima Institute of Chemistry, Federal University of Rio Grande do Sul (UFRGS), Porto Alegre, RS, Brazil

Stanislav A. Moshkalev Centre for Semiconductor Components and Nanotechnology (CCS Nano), University of Campinas (UNICAMP), Campinas, Sao Paulo, Brazil

S. Muthusaravanan Department of Biotechnology, Kumaraguru College of Technology, Coimbatore, TN, India

Mu. Naushad Department of Chemistry, College of Science, King Saud University, Riyadh, Saudi Arabia

Marija Nujic Faculty of Food Technology Osijek, Josip Juraj Strossmayer University of Osijek, Osijek, Croatia

Ritu Painuli Department of Chemistry, Banasthali University, Vanasthali, Rajasthan, India

Sharadwata Pan School of Life Sciences Weihenstephan, Technical University of Munich, Freising, Germany

T. Paramasivan Department of Biotechnology, Kumaraguru College of Technology, Coimbatore, TN, India

Soodamani Periyasamy Department of Chemistry, Anna University, University College of Engineering Dindigul, Dindigul, Tamil Nadu, India

Sudhagar Pitchaimuthu Multi-functional Photocatalyst & Coatings Group, SPECIFIC, College of Engineering, Swansea University (Bay Campus), Swansea, Wales, UK

J. Prakashmaran Department of Food Science and Nutrion, Periyar University, Salem, TN, India

Sapna Raghav Department of Chemistry, Banasthali University, Vanasthali, Rajasthan, India

S. Ramalingam Department of Chemical Engineering, University of Louisiana at Lafayette, Lafayette, LA, USA

Rangnath Ravi Department of Chemistry, Jamia Millia Islamia University, New Delhi, India

A. Saravanan Department of Biotechnology, Vel Tech High Tech Dr. Rangarajan Dr. Sakunthala Engineering College, Chennai, India

P. Senthil Kumar Department of Chemical Engineering, SSN College of Engineering, Chennai, India

Rama Sharma Department Agriculture Science, AKS University, Satna, Madhya Pradesh, India

Sudheesh K. Shukla Department of Applied Chemistry, University of Johannesburg, Johannesburg, South Africa; Department of Biomedical Engineering, Ben-Gurion University of the Negev, Beer Sheva, Israel; Vinoba Bhave Research Institute, Saidabad, Allahabad, Uttar Pradesh, India

Sharf Ilahi Siddiqui Department of Chemistry, Jamia Millia Islamia University, New Delhi, India

Rajesh Kumar Singh School of Physical & Material Sciences, Central University of Himachal Pradesh (CUHP), Kangra, Dharamshala, India

S. Sivamani Chemical and Petrochemical Engineering Section, Engineering Department, Salalah College of Technology, Sultanate, Oman

N. Sivarajasekar Department of Biotechnology, Kumaraguru College of Technology, Coimbatore, TN, India

R. Subashini Department of Biotechnology, Kumaraguru College of Technology, Coimbatore, TN, India

S. Suganya Department of Chemical Engineering, SSN College of Engineering, Chennai, India

Tetiana Tatarchuk Department of Pure and Applied Chemistry, Faculty of Natural Science, Vasyl Stefanyk Precarpathian National University, Ivano-Frankivsk, Ukraine

Chiaki Terashima Photocatalysis International Research Center, Tokyo University of Science, Noda, Chiba, Japan

Sheenam Thatai Amity Institute of Applied Sciences, Amity University, Noida, Uttar Pradesh, India

Rohit Verma Amity Institute of Applied Sciences, Amity University, Noida, Uttar Pradesh, India

Natrayasamy Viswanathan Department of Chemistry, Anna University, University College of Engineering Dindigul, Dindigul, Tamil Nadu, India

P. R. Yaashikaa Department of Chemical Engineering, SSN College of Engineering, Chennai, India

Chapter 1
Threats to Water: Issues and Challenges Related to Ground Water and Drinking Water

Sapna Raghav, Ritu Painuli and Dinesh Kumar

Abstract The extreme burden of the mankind on the earth is instigating variety of environmental changes worldwide, which in turn directly affects the safe and protected water for the lives in the world. In this chapter, we focused on the main groups of the marine contaminants, their effects on mankind and methodologies to diminish the pollution of the ground and drinking water resources. The pollution caused by the heavy metal and metalloids is also highlighted as they pose a severe threat to all life forms in the environment owing to its lethal effects. Some aspects of waterborne diseases and the basic requirements for the enriched sanitation in developing countries are also discussed. The chapter also reports the current scientific improvements to deal with the variety of pollutants. This chapter will therefore help the readers to understand the issues and challenges that the human kind is facing in dealing with water quality issues.

Keywords Ground and surface water · Water pollution · Issues Challenges

Abbreviations

DDT	Dichloro diphenyl trichloroethane
PAHs	Polycyclic aromatic hydrocarbons
PBDEs	Polybrominated diphenyl ethers
PCBs	Polychlorinated biphenyl
PCDD	Polychlorinated dibenzo-p-dioxins
PCDFS	Polychlorinated dibenzo-p-furans

S. Raghav · R. Painuli
Department of Chemistry, Banasthali University, Vanasthali 304022, Rajasthan, India
e-mail: sapnaraghav04@gmail.com

R. Painuli
e-mail: ritsjune8.h@gmail.com

D. Kumar (✉)
School of Chemical Science, Central University of Gujarat, Gandhinagar 382030, India
e-mail: dinesh.kumar@cug.ac.in

M. Naushad (ed.), *A New Generation Material Graphene: Applications in Water Technology*, https://doi.org/10.1007/978-3-319-75484-0_1

1 Introduction

The majority of issues that humankind is facing in this area are interlinked to the water quality or quantity concerns (UN Educ. Sci. Cult. Organ. (UNESCO) 2009). In the related future, these grievous problems are going to be more intensified via the climatic changes. This will result in the melting of the glaciers, enhanced water temperature, and an increase in the water cycles with potentially more droughts and floods (Huntington 2006). Owing to these water problems the severe impact on the human health includes lack of better sanitation, exposure to the pathogens via the recreation or food chain etc. (Fenwick 2006). Currently, these problems are affecting more than a third of the population in the world. The reachable and the renewable freshwater of our earth is consumptively utilized for the industrial, agricultural and for the domestic purpose (Schwarzenbach et al. 2006). These activities lead to the water contamination by discharging the miscellaneous synthetic and the organic chemicals. Because of all these human activities, the chemical contamination of the natural water resources has become the main issue in all over the globe (Alqadami et al. 2017; Naushad et al. 2018). The Gallup poll taken up in 2009 publicized that the pollution of drinking water is the primary U.S. environmental concern (Saad 2009). The water pollutants can be grouped into the two classes, the relatively small number of the macro pollutants, which usually occur at mg/g level for instances, as nitrogen (Gruber and Galloway 2008) and phosphorous species (Filippelli 2008) as well as natural organic constituents (Jorgenson 2009). The sources as well as the problems caused by these pollutants are well known, but planning some sustainable treatment machineries for them still remains a technical challenge (Larsen et al. 2007). For instance, enhanced nutrient loads enhance the production of the biomass, O_2 reduction as well as the lethal algal blooms (Lohse et al. 2009; Heisler et al. 2008). An increased salt load entering the surface water results in an additional long-time difficulty (Kaushal et al. 2005). The increased concentration of salt prevents the crop growth in agriculture as well as the utilization of the drinking water.

The problem is highlighted in several coastline areas, for instances, India, and China by marine salt intrusion into groundwater owing to overexploitation of aquifers and sea level rise (Post 2005). However, there are numerous political and the technical strategies to handle these problems which have been discussed in the literature (An et al. 2009; Gray 2005).

In this chapter, we will explore a variety of natural and synthetic trace pollutants which are existing in water resources. Most of these pollutants possess noxious effects, even at the low concentrations (Gaur et al. 2014; Tak et al. 2013). Hence, by considering the efforts in assessing the micro pollutants effects on the human life and the aquatic life, the inexpensive water treatment approaches for their efficient exclusion should be accomplished. The major efforts for instances, regulated utilization, and exchange or oxidation treatment have to be anticipated in order to inhibit these pollutants from getting into the natural water. Moreover, it is well known that these tasks imply a difficult challenge not only from methodological but also from communal, commercial and the governmental standpoints. It is assumed that globally about 30% of the available renewable freshwater is utilized by the municipalities and industries (Cosgrove and Rijsberman 2000). This results in producing altogether a massive amount of waste water possessing several chemicals in variable concentrations. In nations like China, these wastewaters are still under treatment or suffer from the treatment that is insufficient in eliminating the majority of the micro contaminants (Shao et al. 2006). Inputs from the agriculture are the other imperative sources of the micro pollutants (Bockstaller et al. 2009), which results in an enormous amount of pesticides every year; from oil and gasoline spills (Eliopoulou and Papanikolaou 2007); and from noxious chemicals, i.e. toxic metal ions (Watson 2004) etc. A large number of municipal and precarious wastes areas from where toxic chemicals enter into the natural water, particularly into the groundwater. By considering a variety of micro pollutants from diverse sources, we put an effort to give a typical representation of the water pollution worldwide. As an overview of these issues, we will discuss certain common problems and challenges in assessing micro pollutants in water resources.

2 Effect of Heavy Metal Ions on the Water Quality

Soils have an excessive adsorption capacity for the naturally occurring heavy metals and hence they are not freely accessible for living beings. The cohesive energy of heavy metal for the soil is very much higher in comparison to the other anthropogenic resources (Fulekar et al. 2009). The natural phenomenon is responsible for the occurrence of heavy metals in the environment, processes like weathering of minerals, comets, erosion, and volcanic eruptions. Heavy metals which are available from anthropogenic sources characteristically have a very high bioavailability because of its soluble nature (Bolan et al. 2014). These anthropogenic sources are chemical fertilizers, leather tanning, pesticides, atmospheric deposition, alloy manufacture industries, battery manufacturing, bio solids, coating, manufacturing of explosives, mining, photographic substances, pigments used in printing, sewage irrigation, steel casting, and trades for electroplating etc. (AL-Othman et al. 2012; Sharma et al. 2014) (Table 1). The heavy metals in higher concentration in the atmosphere cause danger to all living organisms (D'amore et al. 2005). The way to entrance for these heavy metals into the atmosphere generally contains the weathering of parent materials, soil ingestion, the change of the geochemical cycle by humans, the release of high concentrations from the industries, and the transfer from mines from one to another place. The major factor of increasing the concentration of heavy metal in the soil is mining and ore processing, and the repossession of environments from this mining works could take several years. These mining activities are responsible for the large number of dumps and stockpiles. Abandoned mines pollute water life by chemical surplus overflow

Table 1 Effect of water pollutant on human, plant and microorganism

Effect on human	Effect on plants	Effect on microorganisms	Reference
Cancer, dermatitis, liver sicknesses, and nasal ulceration	Decreases synthesis of metabolites, and constrain chlorophyll synthesis	Constrain enzyme activities, and decrease growth rate	Blais et al. (2008), An and Kim (2009)
Brain injury, conjunctivitis, skin tumor, cardiovascular and respiratory syndrome	Impairment of cell membrane, loss of fertility, yield and fruit production, and oxidative stress, growth inhibition, constrained roots extension, proliferation	Deactivation of enzyme	Abdul-Wahab and Marikar (2012), Finnegan and Chen (2012), Bissen and Frimmel (2003)
Hypersensitive reactions, heart and lung diseases	Seed incubation inhibited	Chromosomal abnormality and change	Gordon and Bowser (2003)
Bone ailment, coughing, hypertension, headache, kidney diseases, and anemia	Chlorosis, inhibits growth, decrease the nutrient ingredients of plants, germination of seed reduces	Impairment of nucleic acid, protein denature, and cell division prevents	Fashola et al. (2016), Chibuike and Obiora (2014), Sebogodi and Babalola (2011), Sankarammal et al. (2014)
Chronic bronchitis, headache, skin irritation, nausea, renal failure, respiratory tract itching, and liver diseases	Biochemical lesions, chlorosis, wilting, delayed senescence, stunted growth, decreases germination biosynthesis, and oxidative stress	Growth inhibition, and oxygen uptake inhibition	Barakat (2011), Mohanty et al. (2012), Cervantes et al. (2001)
Abdominal pain, anemia, diarrhea, headache, liver and kidney damage, metabolic disorders, nausea, and vomiting	Chlorosis, oxidative stress, and delay growth	Disrupt cellular function, and prevent enzyme activities	Salem et al. (2000), Nagajyoti et al. (2010)
High BP, anorexia, impairment to neurons, hyperactivity, chronic nephropathy, insomnia	Disturbs photosynthesis and evolution, decreases seed propagation, chlorosis, constrained the activity of enzyme and, oxidative stress	Denatures proteins and also nucleic acid and transcription	Wuana and Okieimen (2011), Mupa (2013)

(continued)

Table 1 (continued)

Effect on human	Effect on plants	Effect on microorganisms	Reference
Blindness, gastrointestinal irritation, decrease immunity, deafness, loss of memory, reduction in rate of fertility, dementia, kidney problems, dizziness, sclerosis, dysphasia, and gingivitis	Boost lipid peroxidation disturbs the antioxidative system, disturbs photosynthesis process, prevent growth of the plant, stimulated genotoxic effect, nutrient acceptance, homeostasis, and oxidative stress	Decrease population size, denature protein, cell membrane, prevents enzyme function	Wang et al. (2012), Ali et al. (2013)
Nausea, cardiovascular diseases, nasal cancer, chest pain, lung cancer, dermatitis, kidney problems, dizziness, headache, breathing problems, and dry cough	Decrease in the chlorophyll content, prevent functions of enzyme and its evolution, reduced nutrient uptake	Disorder cell membrane	Malik (2004)
Dysfunction of the endocrine system, gastrointestinal disturbances, impairment of natural killer cells activity, and liver damage	Alteration of protein properties, reduction of plant biomass	Inhibit growth rate	Germ et al. (2007), Dixit et al. (2015)
Cytopathological effects in fibroblast, Argyria and argyrias, bronchitis, and keratinocytes	Affects homeostasis, reduction of chlorophyll content, prevents growth	Cell lysis, inhibit cell transduction	Prabhu and Poulose (2012), Qian et al. (2013)
Nausea, alopecia, insomnia ataxia, hypotension, burning feet syndrome, hair fall, coma, gastroenteritis, convulsions, fatigue and delirium	Prevents activities of enzyme, reduce growth rate	Damage DNA, prevents enzyme activities and growth	Babula et al. (2008)

and particulates that collect in water resources. So, there is a need for the treatment of wastewater which is polluted with the heavy metals, before release into the atmosphere occurs (Alder et al. 2007).

3 Chemicals and Water Pollution

Chemical water pollution includes the various types of pollutant which severe water pollution or hazardous to ground and drinking water. Table 2 gives an overview about different features of worldwide water pollution, together with important kinds of pollutant sources and pollutants.

4 Agriculture and Water Pollution

For the agricultural productions to preserve and enhanced crop yields many million tons of chemicals are utilized for the controlling of insects, weeds, fungi and other pests. Chemical utilized in the pesticides and agrochemicals contains hundreds of active chemical components which are available commercially on the market. Due to the toxicity of chemicals used in agriculture for the both humans and biota and their intentional release into the surrounding atmosphere, the utilization of advanced agrochemicals products in the agriculture. Country-specific registering and risk assessment procedures, objective of defending not only soil and water resources but also consumers and farmers (UN Food Agric. Organ. (FAO) 2008; US Environ. Prot. Agency (EPA) 2008a, b; Galt 2008; Wiley-VCH, ed 2007; Reus et al. 2002; Eur. Comm. 1991). Pollution of water capitals in catchment areas of farming land and constant exposure of biota and humans to biologically active substances is of much concern. The highest concentrations of insecticides and pesticides and their alteration products, such as the commonly detected chloroacetanilides or triazanes in U.S. rivers, can exceed ecotoxic levels for aquatic systems and cooperation in the utilization of surface and groundwater for drinking water supplies (Gilliom 2007). Pesticides can infiltrate into the water resources, aquatic systems and the soil via sewage treatment plants reliant on connections to sewer systems. High-concentration level in the outlet of a catchment area can be caused by point sources, but they do not necessarily constitute a share of the mass input (Leu et al. 2004a, b). Instead, diffuse losses, as well as field overflow, spray drift or drainage/leakage into the subsurface, are of much greater concern, and a wide diversity of mitigation measures has been assessed to decrease their effect on water resources. Losses of pesticide from the overflow is determined by the hydraulic properties of soils such as water flow patterns, and the meteorological and topography conditions of the soil, while compound-specific properties are less important (Leu et al. 2004a, b). Water pollution also rises in the sewers and drainage systems from the applications of pesticides in urban or non-agriculture areas through enhanced overflow of pesticide-containing rainwater over sealed surfaces, such as roads and roofs (Kristoffersen et al. 2008). Finally concluded that from the direct contact of pesticides, there is acute poisoning, this is considerably hazardous for the farmers. While the influence of this exposure pathway is disputed in Europe and North America (Calvert et al. 2008; Steerenberg et al. 2008), accidental revelation

Table 2 An overview about different features of worldwide water pollution, together with important kinds of pollutant sources and pollutants

Pollutant source	Source type	Pollutant types	Examples	Water quality problems	Major challenges
Multiple (waste sites, spills)	Globally distributed point and diffuse	Persistent organic pollutants (POPs)	PCBs, PBDEs, DDT, PAHs, PCDD, PCDFs	Bio magnification in food chain, diverse health effects	Phase out existing POPs, confine existing sources, prevent use of new POPs
Agriculture	Diffuse	Pesticides	Triazines, chloroacetanilides, DDT, lindane	Pollution of ground and surface water with biologically active chemicals, accidental poisoning	Control of pesticides runoff from agricultural land, pesticides misuse
Natural contaminants	Diffuse	Inorganic contaminants, and cyanotoxins, taste	As, F, Se, U, microcystins, geosmin	Cancer, fluorosis, human health, aesthetic	Development of effective house hold treatment systems, control eutrophication
Mining	Mostly points	Acids, leaching agents, heavy metals	Sulfuric acid, cyanide, mercury, copper	Metal remobilization, acute toxicity, chronic neurotoxicity	Acid neutralization, metal removal, introducing effective nontoxic reagent
Hazardous waste	Point	Diverse	U, Cr, chlorinated solvents, nitro aromatic explosives	Long term contamination of drinking water resources	Containment of pollutants, monitoring of mitigation process
Urban wastewater in industrialized countries	Point	Pharmaceuticals hormones	Diclophenac, 17 α-ethinyl estradiol	Ecotoxicology effects in rivers, feminization of fish	Decrease of micro pollutants loads from wastewater by polishing treatment
Urban wastewater in developing and emerging countries	Point	Microorganism	Diarrhea, hepatitis A schistosomiasis, dengue	Human health, child mortality, malnutrition	Improving sanitation and hygiene, safe drinking water

and misuse of agrochemicals appears to more frequently used in developing countries (Oluwole and Cheke 2009; Sarkar et al. 2008; Williamson et al. 2008), which resulting in an expectable poisoning of three million people (UN Environ. Program. (UNEP) 2007). Because of both urbanization and as well as industrialization, these agricultural advances are producing water quality issues (Atapattu and Kodituwakku 2009). Pesticide utilized for the cropland per hectare enhanced in the current years (Khan et al. 2009). Capitals and abilities for regulating pesticide amount in aquatic systems and evaluating the danger for humans and the environment are often limited in developing countries (Menezes and Heller 2008). Monitoring or regulating plans for the pesticide occurrence and distribution demonstrates that the range of active ingredients can still different from those utilized in the developed countries. Particularly, the derivatives of organochlorine pesticides like DDT, HCHs are applied broadly in the agriculture as well as sanitation purposes also, since these pesticides are low-priced and effective (Agrawal 1999).

5 Groundwater Pollution by Hazardous Waste Sites and Spills

Pollution of ground and surface water from the municipal waste, harmful waste sites, abandoned production facilities, and accidental spills are the main cause of ground water pollution. A number of waste sites throughout in the world, in which hundred million tons of waste are needed to be discarded. In these waste sites, a number of sites contain radioactive hazardous materials in huge amount (Giusti 2009; Eur. Environ. Agency (EEA) 2000; US Environ. Prot. Agency (EPA) 2008a, b). But, the point detected are even much higher in number as we known, groundwater-contaminating landfills. The number of official contaminated sites are under controlled observation, mostly of them are releasing chemicals into the atmosphere of nearby areas. Including all these a thousand of gasoline, oil and a number of chemical spills are made every year on land and water, by which different of incidents happened, in their transportation and facility releases. To detect these numbers and fluxes of noxious chemicals from such polluted area to the groundwater and land is very much difficult (Baun and Christensen 2004; Christensen et al. 2001). The primary contaminants in the cases of spills, abandoned facilities, and waste disposal sites are known: chlorinated ethenes, fuel hydrocarbons, methylmercury content in wastewater, nitroaromatic explosives from ammunition plants, PCBs and PCDDs in the pesticide which is manufacturing from the waste, radionuclides and radioactive waste, etc. (Farhadian et al. 2008; Smidt and de Vos 2004; Engelhaupt 2008; Selin 2009; Kersting et al. 1999; Ewing 2006; Spain et al. 2000). Discarded materials are, however, often not well characterized and heterogeneous (Eighmy et al. 1995). The utilization of groundwater as a drinking water source is extensive and the persistence of contaminations for long period, valuation of human health risks from the exposure of the huge amount of chemicals and need to execution of suitable and appropriate, cost-effective remediation strategies are necessary (Bayer-Raich et al. 2006; Vrijheid 2000). One of the major scientific challenges and prerequisites for a thorough assessment of groundwater pollution by spills and hazardous waste sites is thus to quantify the site-specific, relevant processes that determine the transport and transformation behavior of a given pollutant and its transformation products. One promising analytical tool to obtain such information is compound-specific stable-isotope analysis (Hofstetter et al. 2008).

6 Global Health Problems Related to Sanitation and Drinking Water

In industrialized and developing countries, issues related to hygiene, drinking water and sanitation, differs fundamentally. During the next 20–30 years, in high-income countries, maintenance and replacement of the installed water supply substructure and sanitation are the major responsibilities. In areas, where almost of the sewage is discharged by lacking of any treatment, the enhancement of sanitation and source of safe drinking water are of prime importance (UN Educ. Sci. Cult. Organ. (UNESCO) 2009). Most of the population of the countries increase in urban areas of developing countries, and from the current studies about 67% of the world's population will still not be linked to public sewerage systems. Presently, 1.1 billion peoples lack access to safe drinking water, and approximately 2.6 billion people do not even have appropriate sanitation, in developing countries, and this inequality occurs between urban and rural areas for both safe drinking water supply and improved sanitation. Out of the total population of the world four out of five inhabitants with no access to safe sources of drinking water which live in a rural area (World Health Organ. (WHO)/UN Child. Fund (UNICEF) 2008). On a worldwide scale, 1.6 million deaths per year due to the controlled access to safe water (World Health Organ. 2009) and more than 99% thereof occur in the developing world. Fifty percent of children deaths occur in sub-Saharan Africa country due to unsafe drinking water and nine out of ten happenings occurs with the children (World Health Organ./UN Child. Fund (UNICEF) 2006). From the recent studies, 15–30% of gastrointestinal diseases due to the unsafe drinking water and almost 6.11% health related problems are due to consumption of unsafe drinking water. The easily preventable diarrheal diseases caused by unsafe water and lack of sanitation and hygiene. The foremost acute disease risk in developing and transition countries in the drinking water is because of bacteria, protozoa, and virus which spread via the fecal route in the groundwater (Ashbolt 2004). According to WHO records of infectious disease, waterborne diseases are top of the list outbreaks in 132 countries, and cholera as the next most frequent disease which also caused due to unsafe water consumptions, followed by acute diarrhea, legionellosis, and typhoid fever (World Health Organ. 2002). Due to wastewater and drinking water supply in many of the health problems associated which are also intimately linked to sanitation. For the projects related to the safe drinking water 82% of the funds released by the Organization for Economic Co-operation and Development (OECD) (World Health Organ. 2008). This preference contrasts with strong epidemiological evidence, which suggests that better sanitation and safe drinking water would extremely decrease the problem of infectious diseases and, linked to this, also malnutrition. To decrease the human health problems due to bad water quality, World Health Organization and the United Nations Children's Fund have launched as a millennium development goal to half the population lacking access to safe drinking water by 2015 (World Health Organ. (WHO)/UN Child. Fund (UNICEF) 2006). Also, hepatitis A and E viruses, parasitic protozoa and rotaviruses, are due to inadequate water supply and hygiene. A study in Bangladesh reported that 75% of diarrheal and 44% of the children were diseased with unsafe drinking water having number microbes i.e. enterotoxigenic and enteropathogenic like *Escherichia coli* (Nelson and Murray 2008). Outbreaks of typhoid fever happen periodically. Eruptions caused by pathogenic *E. coli* and cryptosporidiosis are also reported in high-income countries, and *Legionella pneumophila* is increasingly scattered in warm water supplies and air conditioning systems of large buildings, such as hospitals, companies etc.

7 Improved Sanitation and Safe Drinking Water Supply

As many waterborne pathogens spread primarily via feces-contaminated water, by using the methods for the clear separation between drinking water and wastewater systems is the fruitful way for water management. To decrease the concentration of viruses and pathogenic microbes into ground water from wastewater, a multiple method for treatment are accessible, and feasible in rural areas. Most of the methods based on physical elimination of the pathogens by coagulation, filtration, and sedimentation. Now a day, UVC irradiation or chemicals are utilized for the disinfection of treated wastewater in some countries. By utilizing the multiple barriers, removal of pathogenic microbes from contaminated water. These barriers encompass riverbank filtration, sand filtration, filtration by soil aquifer treatment, or membrane systems and also disinfection steps, like boiling, chemical disinfection, or UV light etc. For disinfecting drinking water, chlorination is still the most widely utilized method because it is very much economical and effective and also the byproduct of chlorinated water is now a day considered as insignificant when compared to the health benefits from the inactivation of pathogens (Albert et al. 1999). From the previous time, for the municipal water treatment membrane based methods became cost-effective and are gradually utilized as enhancing steps to eliminate microbes and viruses from pretreated water (Peter-Varbanets et al. 2010). A current effort recommends that in low-income countries gravity based low-flow ultrafiltration is a good option for producing drinking water from low-quality source of water. The efficiency of the low flow ultrafiltration process highly depends on their implementation as centralized versus decentralized solutions. In much populated areas, the centralized drinking water distribution and production technology are much economically favorable and, hence, the common case in industrialized areas. While in the large area or large cities of the low-income countries centralized systems is no good for the distribution of safe drinking water to their mankind (Peter-Varbanets et al. 2009). Because of insufficient maintenance owing to lack of finances or proficiency, as well as to pressure failure, illegal tapping, etc. So, in low-income countries, for safe drinking water treatment at the household level is essential not only in rural areas but also in the centralized system area (Hunter 2009). The consistency of these methods is of prime importance because consumption of unsafe water results in an enhanced health risk, mainly for children (Hunter 2009).

8 Issues in Tackling Groundwater Contamination and Pollution

In order to apprehend the main cause/origin, type and extent of pollution, the first step evolves in making reliable and precise information via the water quality management. But they are very few observatories in the nation that cover all the necessary information about the water quality; therefore, the obtained data are not decisive on the status of the water quality. Additionally, the water quality management involves costly as well as superior equipment's that are very hard to control and maintain and also required skilled manpower in assembling and investigating the obtained data. To determine the diverse sources of water pollution, the existing methodology for the water quality management is very insufficient. Combining the data on water quality with data on water supplies, which is very significant for assessing water obtainability for meeting several societal, financial and ecological objectives, is barely completed. Lastly, without any strict norms on water quality testing, and depending upon the period of testing, testing procedure and instruments, the results can be changed all across the agencies. Let us observe the technical issues in extenuating contamination. For seawater intrusion, artificial recharge methods are present in India for diverse geo-hydrological settings. Artificial recharge could push seawater-freshwater interface seawards. On the basis of dilution principle, these techniques can be utilizing to decrease the fluoride, arsenic levels or salinity in aquifer waters. Then again, the problem is of accessibility of healthy water for revitalizing in arid and semi-arid regions given the large aerial extent of contaminated aquifers.

For the industrial pollution, there are 3 types of issues for instances; propelling out polluted water from the aquifer; treatment of this water to non- toxic boundaries; and replacing the depleted aquifer with freshwater. But finding sufficient freshwater for replacement was also problematic there. In the context of our country, it is not economically practical to clean the aquifers.

To remove the various chlorinated solvents in groundwater, zerovalent iron permeable reactive barriers are utilized in situ, in the United States. These techniques are also utilized for the removal of the toxic arsenic. Our country is too poor to pay for some of the technologies as they are very expensive. In India, groundwater quality monitoring is primarily the concern of the Central Ground Water Board and state groundwater agencies.

Here is the list of issues regarding the appropriateness of scientific data obtainable from them:

1. The monitoring stations network is inadequate.
2. Water quality analysis eliminates serious parameters that assist to identify pollution by pesticide and fertilizer, heavy metals and other lethal discharges.
3. The Central Pollution Control Board (CPCB) and the State Pollution Control Boards (SPCBs) are the pollution supervisory body in India? Although observing the quality of the groundwater and water quality of rivers has come under their purview only recently. To perform its functions the GPCB also suffers from the shortages of the working staff. There are numerous problems with the design of the institution itself. The two-function executed by the SPCBs includes the monitoring and administering pollution control standards. The agency lacks legal teeth and managerial apparatus to penalize contaminators. This decreases the efficiency of the organization in administering pollution controller norms.

The geo-hydrological chemical process gets activated by the pumping and is the foremost origin of the ground water contamination. It is very difficult to ban pumping as in our country the masses of rural families rely upon the groundwater for supporting irrigated agriculture and livelihoods. Any legal/regulatory contributions for the ban of pumping would result in depriving societies from their rights. The nitrate pollution can be significantly organized by the crop rotation, utilization of the organic manure, controlled dosage of the fertilizers, regulated timing of fertilizer application, but there are no official administrations governing the utilization of the fertilizer and dumping of animal waste.

9 Emerging Challenges

The existing water treatment plants works on the principle based on chemistry and physics. Therefore, the efficacy of these systems relies upon sustaining certain definite operating conditions. This in turn depends upon the skilled manpower, regular operations as well as the maintenance, which are mostly not present. Since 1989, approximately 28 desalinations systems were set up by the Gujarat Water Supply and Sewerage Board. All of these desalination systems became dysfunctional in very less period. Most of the desalinations plants commissioned by the government agencies in 8 states become non-functional owing to the lack of technical assistance and improper membrane selection. It has been tried out that the majority of the treatment plants for the drinking water at the public level should be inexpensive and reasonable. Hence, constructing and operating the water treatment systems on the basis of full cost recovery will be more applicable. The supply of water has to be reasonable for the all class of society as the water is vital for the existence. Therefore, unit cost of production should be decreased for commercial viability. By creating the sufficient demand i.e. by running the plant at peak capacity, the unit cost of production can be brought down. The factors which will help in bringing down the cost of production include selection of proper membrane, ideal plant design, generation of satisfactory demand, etc. New techno institutional models need to be developed to manage these system in order to make them self-sustaining. For providing the safe and clean drinking as well as ground water the involvement of the private sector would be a key step towards achieving this.

In order to respond to the water quality problems civil society as well as the institutions are needed to be reinforced. This can be achieved via the awareness and information about the nature of ground water pollution, the ill-effects of using contaminated water, potential sources of threats to groundwater quality in their region, and the possible preventive measures. As the variations of the ground water quality in nature are often infrequent, and it is very problematic for the monitoring agencies to create a large setup of water quality management stations because of the extraordinary prices and technical power involved. Therefore, it is imperative to strengthen the civil society and the institutions. The incomplete information about the water quality in numerous sources, it is impossible for various line agencies to determine the suitable treatment measures. Also, the enthusiasm of the societies to pay for water is directly linked to their information and alertness about bad-effects of drinking polluted/contaminated water. Reliable and technically knowledgeable government agencies will also play a big role in reinforcing the general public, by producing the vital database on groundwater quality.

10 Conclusions and Policy Inferences

For handling the water pollution worldwide an effectual group of policies, scientific advancements and technologies on diverse scales is required. The volatile chemicals that cannot undergo the bioremediation process but is having the capacity to accumulate in the food chain, should be controlled in their utilizations and in applications. In order to enhance the agricultural production yields as well as to protect the ecology and to maintain the food chain against contamination, the global agriculture faces various challenges. For the enhancement of water quality in agricultural field more integrated methodologies are required. In the developing countries, rural populations rely upon the contaminated ground water wells. For these issues, determining the alternate water resources or employing easy, trustworthy domestic water treatment machineries are required. For the cleaning of huge scale of water pollution from the miming activities and the pollution of ground water from the waste sites the science based decisions are required, by taking into considerations the specific hydrological conditions, the microbial and geochemical transformation pathways, and remediation technologies. Hence, the are challenges that the water utilities would face for instances, constructing practical and administrative skills to design, install, control and accomplish water treatment systems, making mankind to knowledge and awareness among communities about groundwater and drinking water quality issues and treatment measures. At last, the policies need to be motivated on constructing the scientific abilities of line agencies related with water quality management, pollution control, water supplies, and execution of pollution control norms efficiently and to empower them implement environmental management projects.

Acknowledgements We gratefully acknowledge support from the Ministry of Science and Technology and Department of Science and Technology, Government of India under the scheme of Establishment of Women Technology Park, for providing the necessary financial support to carry out this study vide letter No, F. No SEED/WTP/063/2014.

Conflict of Interest The authors declare no conflict of interest.

References

Abdul-Wahab S, Marikar F (2012) The environmental impact of gold mines: pollution by heavy metals. Open Eng 2:304–313. https://doi.org/10.2478/s13531-011-0052-3

Adler RA, et al. (2007) Water, mining and waste: an historical and economic perspective on conflict management in South Africa. Econ Peace Secur J 2:32–41

Agrawal GD (1999) Diffuse agricultural water pollution in India. Water Sci Technol 39:33–47

Albert MJ, et al. (1999) Case-control study of enteropathogens associated with childhood diarrhea in Dhaka, Bangladesh. J Clin Microbiol 137:3458–3464

Ali H, et al. (2013) Phytoremediation of heavy metals—concepts and applications. Chemosphere 91:869–881. https://doi.org/10.1016/j.chemosphere.2013.01.075

AL-Othman ZA, et al. (2012) Hexavalent chromium removal from aqueous medium by activated carbon prepared from peanut shell: adsorption kinetics, equilibrium and thermodynamic studies. Chem Eng J 184:238–247. https://doi.org/10.1016/j.cej.2012.01.048

Alqadami AA, et al. (2017) Novel metal-organic framework (MOF) based composite material for the sequestration of U(VI) and Th(IV) metal ions from aqueous environment. ACS Appl Mater Interfaces 9:36026–36037. https://doi.org/10.1021/acsami.7b10768

An YJ, Kim M (2009) Effect of antimony on the microbial growth and the activities of soil enzymes. Chemosphere 74:654–659. https://doi.org/10.1016/j.chemosphere.2008.10.023

An Q, et al. (2009) Influence of the three gorges project on saltwater intrusion in the Yangtze River Estuary. Environ Geol 56:1679–1686. https://doi.org/10.1007/s00254-008-1266-4

Ashbolt NJ (2004) Microbial contamination of drinking water and disease outcomes in developing regions. Toxicology 198:229–238. https://doi.org/10.1016/j.tox.2004.01.030

Atapattu SS, Kodituwakku DC (2009) Agriculture in South Asia and its implications on downstream health and sustainability: a review. Agric Water Manage 96:361–373. https://doi.org/10.1016/j.agwat.2008.09.028
Babula P, et al. (2008) Uncommon heavy metals, metalloids and their plant toxicity: a review. Environ Chem Lett 6:189–213. https://doi.org/10.1007/s10311-008-0159-9
Barakat M (2011) New trends in removing heavy metals from industrial wastewater. Arab J Chem 4:361–377. https://doi.org/10.1016/j.arabjc.2010.07.019
Baun DL, Christensen TH (2004) Speciation of heavy metals in landfill leachate: a review. Waste Manage Res 22:3–23
Bayer-Raich M, et al. (2006) Integral pumping test analyses of linearly sorbed groundwater contaminants using multiple wells: inferring mass flows and natural attenuation rates. Water Resour Res 42:W08411. https://doi.org/10.1029/2005wr004244
Bissen M, Frimmel FH (2003) Arsenic—a review. Part I: occurrence, toxicity, speciation, mobility. Acta Hydrochim Hydrobiol 31:9–18. https://doi.org/10.1002/aheh.200390025
Blais J, et al. (2008) Metals precipitation from effluents: review. Pract Period Hazard Toxic Radioact Waste Manage 12:135–149
Bockstaller C, et al. (2009) Comparison of methods to assess the sustainability of agricultural systems. A review. Agron Sustain Dev 29:223–235. https://link.springer.com/chapter/10.1007%2F978-90-481-2666-8_47
Bolan N, et al. (2014) Remediation of heavy metal(loid)s contaminated soils—to mobilize or to immobilize? J Hazard Mater 266:141–166. https://doi.org/10.1016/j.jhazmat.2013.12.018
Calvert GM, et al. (2008) Acute pesticide poisoning among agricultural workers in the United States, 1998–2005. Am J Ind Med 51:883–898. https://doi.org/10.1002/ajim.20623
Cervantes C, et al. (2001) Interactions of chromium with microorganisms and plants. FEMS Microbiol Rev 25:335–347. https://doi.org/10.1111/j.1574-6976.2001.tb00581.x
Chibuike G, Obiora S (2014) Heavy metal polluted soils: effect on plants and bioremediation methods. Appl Environ Soil Sci 2014:1–12. https://doi.org/10.1155/2014/752708
Christensen TH, et al. (2001) Biogeochemistry of landfill leachate plumes. Appl Geochem 16:659–718. https://doi.org/10.1016/S0883-2927(00)00082-2
Cosgrove WJ, Rijsberman FR (2000) World water vision: making water everybody's business. World Water Counc, London
Damore J, et al. (2005) Methods for speciation of metals in soils. J Environ Qual 34:1707–1745. https://doi.org/10.2134/jeq2004.0014
Dixit R, et al. (2015) Bioremediation of heavy metals from soil and aquatic environment: an overview of principles and criteria of fundamental processes. Sustainability 7:2189–2212. https://doi.org/10.3390/su70221890
Eighmy TT, et al. (1995) Comprehensive approach toward understanding element speciation and leaching behavior in municipal solid-waste incineration electrostatic precipitator ash. Environ Sci Technol 29:629–646. https://doi.org/10.1146/annurev.environ.051308.084314
Eliopoulou E, Papanikolaou A (2007) Casualty analysis of large tankers. J Mar Sci Technol 12:240–250. https://doi.org/10.1007/s00773-007-0255-8
Engelhaupt E (2008) Happy birthday, love canal. Environ Sci Technol 42:8179–8186
Eur. Comm. (1991) Plant protection products—directive 91/414/EEC. http://ec.europa.eu/food/plant/protection/evaluation/dir91-414eec_en.htm
Eur. Environ. Agency (EEA) (2000) Management of contaminated sites in Western Europe: topic report No. 13/1999. EEA, Copenhagen. http://www.eea.europa.eu/publications/Topic_report_No_131999
Ewing RC (2006) The nuclear fuel cycle: a role for mineralogy and geochemistry. Elements 2:331–334
Farhadian M, et al. (2008) In situ bioremediation of monoaromatic pollutants in groundwater: a review. Bioresour Technol 99:5296–5308. https://doi.org/10.1016/j.biortech.2007.10.025
Fashola M, et al. (2016) Heavy metal pollution from gold mines: environmental effects and bacterial strategies for resistance. Int J Environ Res Public Health 13:1047

Fenwick A (2006) Waterborne infectious diseases—could they be consigned to history? Science 313:1077–1081. https://doi.org/10.1126/science.1127184

Filippelli GM (2008) The global phosphorus cycle: past, present, and future. Elements 4:89–95. https://doi.org/10.2113/GSELEMENTS.4.2.89

Finnegan P, Chen W (2012) Arsenic toxicity: the effects on plant metabolism. Front Physiol 3:182

Fulekar M, et al. (2009) Genetic engineering strategies for enhancing phytoremediation of heavy metals. Afr J Biotechnol 8:529–535. https://www.ajol.info/index.php/ajb/article/view/59858/48132

Galt RE (2008) Beyond the circle of poison: significant shifts in the global pesticide complex, 1976–2008. Glob Environ Change 18:786–799. https://doi.org/10.1016/j.gloenvcha.2008.07.003

Gaur N, et al. (2014) A review with recent advancements on bioremediation-based abolition of heavy metals. Environ Sci Process Impacts 16:180–193. https://doi.org/10.1039/c3em00491k

Germ M, et al. (2007) Combined effects of selenium and drought on photosynthesis and mitochondrial respiration in potato. Plant Physiol Biochem 45:162–167. https://doi.org/10.1016/j.plaphy.2007.01.009

Gilliom RJ (2007) Pesticides in U.S. streams and groundwater. Environ Sci Technol 41:3407–3413. https://doi.org/10.1021/es072531u

Giusti L (2009) A review of waste management practices and their impact on human health. Waste Manage 29:2227–2239. https://doi.org/10.1016/j.wasman.2009.03.028

Gordon T, Bowser D (2003) Beryllium: genotoxicity and carcinogenicity. Mutat Res, Fundam Mol Mech Mutagen 533:99–105. https://doi.org/10.1016/j.molliq.2015.09.026

Gray NF (2005) Water technology: an introduction for environmental scientists and engineers. Elsevier-Butterworth-Heinemann, Oxford

Gruber N, Galloway JN (2008) An Earth-system perspective of the global nitrogen cycle. Nature 451:293–296. https://doi.org/10.1038/nature06592

Gumpu MB, et al. (2015) A review on detection of heavy metal ions in water—an electrochemical approach. Sens Actuators, B Chem 213:515–533. https://doi.org/10.1016/j.snb.2015.02.122

Heisler J, et al. (2008) Eutrophication and harmful algal blooms: a scientific consensus. Harmful Algae 8:3–13. https://doi.org/10.1016/j.hal.2008.08.006

Hofstetter TB, et al. (2008) Assessing transformation processes of organic compounds using stable isotope fractionation. Environ Sci Technol 42:7737–7743. https://doi.org/10.1021/es801384j

Hunter PR (2009) Household water treatment in developing countries: comparing different intervention types using meta-regression. Environ Sci Technol 43:8991–8997. https://doi.org/10.1021/es9028217

Hunter PR, et al. (2009) Estimating the impact on health of poor reliability of drinking water interventions in developing countries. Sci Total Environ 407:2621–2624. https://doi.org/10.1016/j.scitotenv.2009.01.018

Huntington TG (2006) Evidence for intensification of the global water cycle: review and synthesis. J Hydrol 319:83–95. https://doi.org/10.1016/j.jhydrol.2005.07.003

Jorgenson AK (2009) Political-economic integration, industrial pollution and human health: a panel study of less-developed countries, 1980–2000. Int Soc 24:115–143. http://journals.sagepub.com/doi/pdf/10.1177/0268580908099156

Kaushal SS, et al. (2005) Increased salinization of fresh water in the northeastern United States. Proc Natl Acad Sci USA 102:13517–13520. https://doi.org/10.1073/pnas.0506414102

Kersting AB, et al. (1999) Migration of plutonium in ground water at the Nevada Test Site. Nature 397:56–59. https://doi.org/10.1038/16231

Khan S, et al. (2009) Water management and crop production for food security in China: a review. Agric Water Manage 96:349–360. https://doi.org/10.1016/j.agwat.2008.09.022

Kristoffersen P, et al. (2008) A review of pesticide policies and regulations for urban amenity areas in seven European countries. Weed Res 48:201–214. https://doi.org/10.1111/j.1365-3180.2008.00619.x

Larsen TA, et al. (2007) Nutrient cycles and resource management: implications for the choice of wastewater treatment technology. Water Sci Technol 56:229–237. https://doi.org/10.2166/wst.2007.576

Leu C, et al. (2004a) Simultaneous assessment of sources, processes, and factors influencing herbicide losses to surface waters in a small agricultural catchment. Environ Sci Technol 38:3827–3834. https://doi.org/10.1021/es0499602

Leu C, et al. (2004b) Variability of herbicide losses from 13 fields to surface water within a small catchment after a controlled herbicide application. Environ Sci Technol 38:3835–3841. https://doi.org/10.1021/es0499593

Lohse KA, et al. (2009) Interactions between biogeochemistry and hydrologic systems. Annu Rev Environ Resour 34:65–96. https://doi.org/10.1146/annurev.environ.33.031207.111141

Malik A (2004) Metal bioremediation through growing cells. Environ Int 30:261–278

Menezes CT, Heller L (2008) A method for prioritization of areas for pesticides surveillance on surface waters: a study in Minas Gerais, Brazil. Water Sci Technol 57:1693–1698. https://doi.org/10.2166/wst.2008.277

Mohanty M, et al. (2012) Bio-concentration of chromium—an in-situ phytoremediation study at South Kaliapani chromite mining area of Orissa, India. Environ Monit Assess 184:1015–1024

Mupa M (2013) Lead content of lichens in metropolitan Harare, Zimbabwe: air quality and health risk implications. Greener J Environ Manage Public Saf 2:75–82

Nagajyoti P, et al. (2010) Heavy metals, occurrence and toxicity for plants: a review. Environ Chem Lett 8:199–216. https://doi.org/10.1007/s10311-010-0297-8

Naushad M, et al. (2018) Efficient removal of toxic phosphate anions from aqueous environment using pectin based quaternary amino anion exchanger. Int J Biol Macromol 106:1–10. https://doi.org/10.1016/j.ijbiomac.2017.07.169

Nelson KL, Murray A (2008) Sanitation for unserved populations: technologies, implementation challenges, and opportunities. Annu Rev Environ Resour 33:119–151. https://doi.org/10.1146/annurev.environ.33.022007.145142

Oki T, Kanae S (2006) Global hydrological cycles and world water resources. Science 313:1068–1072. https://doi.org/10.1126/science.1128845

Oluwole O, Cheke RA (2009) Health and environmental impacts of pesticide use practices: a case study of farmers in Ekiti State, Nigeria. Int J Agric Sustain 7:153–163

Peter-Varbanets M, et al. (2009) Decentralized systems for potable water and the potential of membrane technology. Water Res 43:245–265. https://doi.org/10.1016/j.watres.2008.10.030

Peter-Varbanets M, et al. (2010) Stabilization of flux during dead-end ultra-low pressure ultrafiltration. Water Res 44:3607–3616. https://doi.org/10.1016/j.watres.2010.04.020

Post VEA (2005) Fresh and saline groundwater interaction in coastal aquifers: is our technology ready for the problems ahead? Hydrogeol J 13:120–123. https://doi.org/10.1007/s10040-004-0417-2

Prabhu S, Poulose EK (2012) Silver nanoparticles: mechanism of antimicrobial action, synthesis, medical applications, and toxicity effects. Int Nano Lett 2:1–10. https://doi.org/10.1186/2228-5326-2-32

Qian H, et al. (2013) Comparison of the toxicity of silver nanoparticles and silver ions on the growth of terrestrial plant model *Arabidopsis thaliana*. J Environ Sci 25:1947–1956. https://doi.org/10.1016/S1001-0742(12)60301-5

Reus J, et al. (2002) Comparison and evaluation of eight pesticide environmental risk indicators developed in Europe and recommendations for future use. Agric Ecosyst Environ 90:177–187. https://doi.org/10.1016/S0167-8809(01)00197-9

Saad L (2009) Water pollution Americans' top green concern. http://www.gallup.com/poll/117079/waterpollution-americans-top-greenconcern.aspx

Salem HM, et al. (2000) Heavy metals in drinking water & their environment impact on human health. In: Proceedings of the international conference for environmental hazards mitigation ICEHM 2000, Cairo University, Giza, Egypt, 9–12 September 2000, pp 542–556

Sankarammal M, et al. (2014) Bioremoval of cadmium using pseudomonas fluorescens. Open J Water Pollut Treat 1:92–100

Sarkar SK, et al. (2008) Occurrence, distribution and possible sources of organochlorine pesticide residues in tropical coastal environment of India: an overview. Environ Int 34:1062–1071. https://doi.org/10.1016/j.envint.2008.02.010

Schwarzenbach RP, et al. (2006) The challenge of micropollutants in aquatic systems. Science 313:1072–1077. https://doi.org/10.1126/science.1127291

Sebogodi KM, Babalola OO (2011) Identification of soil bacteria from mining environments in Rustenburg, South Africa. Life Sci J 8:25–32

Selin NE (2009) Global biogeochemical cycling of mercury: a review. Annu Rev Environ Resour 34:43–63

Shao M, et al. (2006) City clusters in China: air and surface water pollution. Front Ecol Environ 4:353–361. http://www.frontiersinecology.org/specialissue/ESA_Sept06_ONLINE-04.pdf

Sharma G, et al. (2014) Preparation, characterization and antimicrobial activity of biopolymer based nanocomposite ion exchanger pectin zirconium (IV) selenotungstophosphate: application for removal of toxic metals. J Ind Eng Chem 20:4482–4490. https://doi.org/10.1016/j.jiec.2014.02.020

Smidt H, de Vos WM (2004) Anaerobic microbial dehalogenation. Annu Rev Microbiol 58:43–73. https://doi.org/10.1146/annurev.micro.58.030603.123600

Spain JC, et al. (2000) Biodegradation of nitroaromatic compounds and explosives. Lewis, Boca Raton, p 434

Steerenberg P, et al. (2008) Toxicological evaluation of the immune function of pesticide workers, a European wide assessment. Hum Exp Toxicol 27:701–707. http://journals.sagepub.com/doi/abs/10.1177/0960327108095993

Tak HI, et al. (2013) Advances in the application of plant growth-promoting rhizobacteria in phytoremediation of heavy metals. In: Reviews of environmental contamination and toxicology. Springer, New York, pp 33–52. https://doi.org/10.1007/978-1-4614-5577-6

UN Educ. Sci. Cult. Organ. (UNESCO) (2009) The United Nations World Water Development Report 3: water in a changing world. UNESCO/Berghahn Books, Paris/New York

UN Environ. Program. (UNEP) (2007) Global environment outlook GEO4: environment for development (e-book). http://www.eoearth.org/article/Global_Environment_Outlook_%28GEO-

UN Food Agric. Organ. (FAO) (2008) FAOSTAT statistical database. http://faostat.fao.org/site/424/default.aspx

US Environ. Prot. Agency (EPA) (2008a) 2008 Report on the environment. EPA/600/R-07/045F, Natl. Cent. Environ. Assess., Washington, DC. http://www.epa.gov/roe/

US Environ. Prot. Agency (EPA) (2008b) Pesticides. http://www.epa.gov/pesticides/

Vrijheid M (2000) Health effects of residence near hazardous waste landfill sites: a review of epidemiologic literature. Environ Health Perspect 108:101–112

Wang J, et al. (2012) Remediation of mercury contaminated sites—a review. J Hazard Mater 221:1–18. https://doi.org/10.1016/j.jhazmat.2012.04.035

Watson SB (2004) Aquatic taste and odor: a primary signal of drinking-water integrity. J Toxicol Environ Health 67:1779–1795. https://doi.org/10.1080/15287390490492377

Wiley-VCH (ed) (2007) Ullmann's agrochemicals, vol 2. In: Plant growth and crop protection. Wiley-VCH, Weinheim. https://doi.org/10.1039/b606377b

Williamson S, et al. (2008) Trends in pesticide use and drivers for safer pest management in four African countries. Crop Prot 27:1327–1334. https://doi.org/10.1016/j.cropro.2008.04.006

World Health Organ. (2008) UN-Water GLAAS Pilot Report. WHO, Geneva

World Health Organ. (2009) Global health risks: mortality and burden of disease attributable to selected major risks. WHO, Geneva

World Health Organ. (WHO) (2002) Global defense against the infectious disease threat. In: Kindhauser MK (ed) Emerging and epidemic-prone diseases. WHO, Geneva, pp 56–103

World Health Organ. (WHO)/UN Child. Fund (UNICEF) (2006) Meeting the MDG drinking-water and sanitation target: the urban and rural challenge of the decade. WHO/UNICEF, Geneva/New York

World Health Organ. (WHO)/UN Child. Fund (UNICEF) (2008) Progress on drinking-water and sanitation: special focus on sanitation. WHO/UNICEF, Geneva/New York, 58 pp

Wuana RA, Okieimen FE (2011) Heavy metals in contaminated soils: a review of sources, chemistry, risks and best available strategies for remediation. ISRN Ecol 2011:1–20. https://doi.org/10.5402/2011/402647

Chapter 2
Water Quality Standards, Its Pollution and Treatment Methods

Sheenam Thatai, Rohit Verma, Parul Khurana, Pallavi Goel and Dinesh Kumar

Abstract Water is not only the most essential source of our day-to-day life, but the development of this natural resource also plays a crucial role in economic and social development processes. Waste disposal has become a worldwide problem for increased environmental awareness, for more rigorous environmental standards and dewatering challenges. Therefore, water quality management is a great work controlled by monitoring of discharge and various effluents. World health organization has issued guidelines for drinking water quality, its contaminants and how to handle water supplies in small rural communities. Qualitative and quantitative measurements are needed from time to time to constantly monitor the quality of water from the various sources of supply. This chapter highlights different water quality assessments, several sources of water pollutions and the methods used for treating water for various purposes such as for drinking, industrial water supply, irrigation, water recreation or many other uses.

Keywords Water · Quality · Pollution · Standard · Treatment

S. Thatai (✉) · R. Verma (✉) · P. Goel
Amity Institute of Applied Sciences, Amity University, Noida, Uttar Pradesh, India
e-mail: messageforsheenam@gmail.com

R. Verma
e-mail: rverma85@amity.edu

P. Goel
e-mail: pallavigoel85@gmail.com

P. Khurana
Department of Chemistry, G. N. Khalsa College of Arts, Science and Commerce, Mumbai University, Mumbai, India
e-mail: messageforparul@gmail.com

D. Kumar
Department of Chemistry, Central University of Gujarat, Gandhinagar, India
e-mail: dsbchoudhary2002@gmail.com

M. Naushad (ed.), *A New Generation Material Graphene: Applications in Water Technology*, https://doi.org/10.1007/978-3-319-75484-0_2

1 Introduction

Water is one of the prime elements responsible for life on earth. Water is used every day for various purposes and involved in all the bodily purposes. An average body consists of 42 l of water. Safe water availability is important for public health. UN general assembly recognized the right of humans to clean water and sanitation. Since 1990, 2.6 billion people have gained access water quality standard to improved drinking water (Pathak 2013). 97% of water available on earth is saline and only 3% of water is freshwater. Of the 3% freshwater, 68.7% is available in glaciers and icecaps and 30.1% is available as ground water. And the remaining 0.3% is available as surface water. Out of this 0.3% of surface water, 87% is present in lakes, 11% in swamps and 2% in rivers (Pontius 1990).

India is ranked among the top ten water rich countries with 4% of world's freshwater resources. Surface water sources in India include rivers, lakes or fresh water wetlands. India has 12 major river systems. The perennial Himalayan Rivers in the north include Ganga, Yamuna, Indus and Brahmaputra. The south has rivers Krishna, Godavari and Cauvery while rivers Narmada, Tapti and Mahanadi drain central India. The Bay of Bengal receives 70% of total drainage while the Arabian Sea receives 20% of the total drainage. Rivers are the lifeline of growth. They provide drinking as well as raw water for industrial use. Lakes also serve as a source of water for agriculture, drinking and industries. They also act as recharge zones for groundwater. Lakes are a source of livelihood for many people. Indian cities such as, Ahmedabad, Bangalore and Hyderabad have a number of lakes. Some of the lakes are Dal Lake, Loktak Lake, Ropar Lake, etc. Water sources within aquifers are known as ground water resources. The assessment of ground water resources of the country is carried out by Central Ground Water Board. Ground water contributes to 85% of drinking water, 58% of irrigation water and about 50% of industrial requirements (Durfor and Becker 1964; Kumar et al. 2017).

A World Health Organization (WHO) report also suggests that about 1.8 billion people rely on sources of drinking water which are contaminated. Water pollution is the most serious ecological threat that the world faces today. Water pollution is defined as the condition in which one or more toxic substances accumulate in water bodies and degrade the quality of water. The Economist report in 2008 stated that in India each day over 1000 children die of diarrheal sickness. Water is polluted by both natural and manmade activities. Let, us now discuss the major causes that lead to the pollution of water.

2 Causes of Water Pollution

There are several causes of water pollution. The main causes are as follows.

2.1 Urbanization

The population of urban dwellers has risen to 285 million from 25.8 million people in 1901. With population growth the demand for housing, food and cloth has also increased. This rapid urbanization has led to various issues such as inadequate water supply, production of wastewater in large amount, its disposal and treatment. Water is supplied from rivers, lakes, ponds for domestic and industrial use. After the work is done, this water is released out as wastewater, which is left untreated most of the times. Thus, causing large scale surface water pollution. Due to rapid population growth, the use of water for domestic purposes and use of soaps and detergents going to sink has also increased. But the facility of sewerage lags far behind. Only 22% of the waste water from class 1 cities is collected through sewerage. Thus wastewater in large amount is left uncollected. According to the CPCB 2003 report, wastewater about 22,900 million liter per day (mld) was generated. Out of which, only 5900 mld which accounts to only 26% is treated while the remaining wastewater, which is about 17,100 mld is left untreated (http://greencleanguide.com/earths-water-distribution-and-indian-scenario/). It has been found that permanent treatment facility is available only in 27 cities while the rest forty-nine cities have access to only primary and secondary facility for treatment of water. Table 1 explains the trend of population, water supplied, wastewater generated and the amount of wastewater treated and untreated in class I cities and class II towns.

Table 1 Water supply trend and sanitation status in class I cities and class II towns

Parameters	Class I cities			Class II towns		
Year	1978–79	1989–90	1994–95	1978–79	1989–90	1994–95
Number	142	212	299	190	241	345
Population (millions)	60	102	128	12.8	241	23.6
Water supply (mld)	8638	15,191	20,607	1533	1622	1936
Wastewater generated (mld)	7007	12,145	16,662	1226	1280	1650
Wastewater treated (mld)	2756 (39%)	2485 (20.5%)	4037 (24%)	67 (5.44%)	27 (2.12%)	62 (3.73%)
Wastewater untreated (mld)	4251 (61%)	9660 (79.5%)	12,625 (76%)	1160 (94.56%)	1252 (97.88%)	1588 (96.27%)

The numbers in round brackets show the percentage of the treated and untreated water (mld stands for million liters per day)

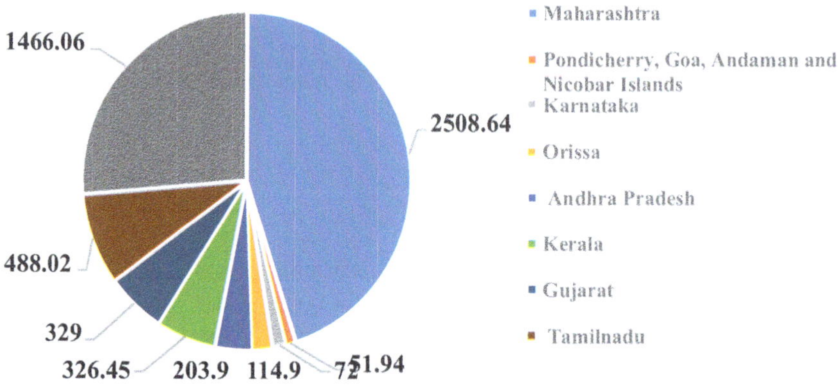

Chart 1 Municipal wastewater generated in coastal cities of India

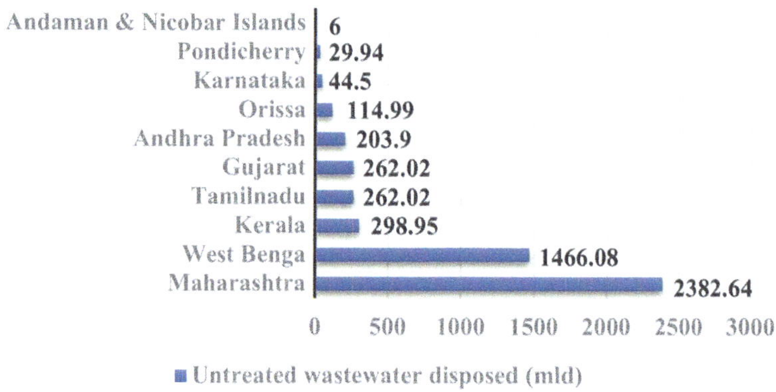

Chart 2 Untreated municipal wastewater disposed in the coastal cities of India

From the above table, it is evident that the amount of wastewater treated in comparison to the amount generated is very less. Thus, large amount of wastewater is left untreated and released into the water bodies leading to pollution of water and causing harmful effects on marine life, plants, and humans who consume this polluted water.

Large amount of wastewater is generated and left untreated in the coastal cities. India has an 8118 km long coastline. Cities near the coastline generate 5560.99 mld of wastewater and 90.62% of the total wastewater generated is released into the coastal water untreated. Chart 1 shows the amount of wastewater generated in different coastal cities. And it has been found that Maharashtra generates the highest amount of wastewater, followed by West Bengal (Aral 2009).

While Chart 1 depicts the amount of wastewater generated and Chart 2 depicts the amount of wastewater which is left untreated. Maharashtra, which generates the largest amount of wastewater also disposes most of its wastewater untreated, followed by West Bengal and Kerala.

Another survey shows that during 2015, 61,754 mld wastewater was generated in India and about 38,791 mld of untreated sewage was discharged directly into water bodies (http://cpcb.nic.in/).

2.2 *Industries*

Most of the fresh water sources like rivers are contaminated by toxic wastes produced by industries. A water pollution control programme was launched by the Central Pollution Control Board (CPCB) in 1992 for industries. This program identified 1551 industries which did not comply with the pollution standards. CPCB gave them a time period for compliance with prescribed standards. It has been estimated that the major industrial sources generate around 83,048 mld of waste water. The major waste water generating industries in terms of volume are engineering industries and electroplating units. Steel plants, paper mills, textile and sugar industries also contribute significantly to waste water. Table 2 shows the wastewater generated by different industrial sectors (http://www.sulabhenvis.nic.in).

Both large and small scale industries contribute to water pollution. In 1995, only 59% of large and medium industries had adequate effluent treatment plants. There are about 3 million small scale industries in India which cannot afford ETP's of their own. Such small scale industries contribute to 40% of industrial water pollution (Murty and Kumar 2011).

2.3 *Agriculture*

Agriculture is the largest user of freshwater resources. It is also a major cause of surface and ground pollution of water. Very large parts of total land area are under

Table 2 Wastewater generated by different industrial sectors

Industrial sector	Annual waste water discharge (million cubic meters) (%)
Thermal power plants	27,000.9
Engineering	1551.3
Pulp and paper	695.7
Textiles	637.3
Steel	396.8
Sugar	149.7
Others	241.3
Total	30,729.2

agriculture. Agricultural runoffs contain pesticides and fertilizers which pollute the ground and surface water resources. A 40 km long drain in Haryana, pours 250,000 kg/day of chlorides into Yamuna. Thus, raising the concentration of chlorides in Yamuna to 150 mg/l. Majority of chlorides are from agricultural return flow. Nutrients which are derived from farming lead to enrichment of water. These principally include nitrogen and phosphorous and their various forms, which often lead to eutrophication. Also, the transfer of soil from agricultural land into water resources affects the amount of light entering the water and fish spawning. Irrigation is also a major cause of surface and ground pollution of water. Runoff of salts leads to salinization of surface water and runoff of fertilizers and pesticides to surface water leads to bioaccumulation in aquatic organisms. Other agricultural activities such as aquaculture and silviculture also deteriorate the quality of surface and ground water (Memon and Schröder 2009; Agarwal et al. 1997).

These were the major causes of water pollution. From the above discussion, it can be concluded that large amount of wastewater is disposed off in water bodies untreated. For treatment of this wastewater, techniques have been designed in accordance with the pollutants present in wastewater. In order to determine the degree of treatment required, water is tested for various characteristics which are discussed below.

3 Characteristics of Water Tested for Water Quality Assessment

Water contamination is caused due to chemicals, pathogens, heavy metals, pesticides, fertilizers, etc. which are a product of agricultural and industrial activities (Arora et al. 2011). Such activities have become a hazard to human health. It is necessary to characterize the various characteristics of water in order to develop treatment plans.

The following characteristics are tested for water quality.

1. **pH**: pH is defined as the measure of acidic or basic nature of a solution. It is determined by the concentration of hydrogen ion (H^+) activity in a solution. Mathematically,

$$\mathrm{pH} = -\log(\mathrm{H}^+)$$

2. **Temperature** The metabolism of the aquatic system is regulated by the temperature of the water. Under high water temperature the ability of water to hold dissolved gases such as oxygen reduces. Thus, leading to killing of fishes due to reduced availability of oxygen in water.

3. **Hardness** Hardness is the defined as the measure of the water capacity to precipitate soap. It is also defined as the sum of concentration of calcium and magnesium (Memon and Schröder 2009).
4. **Nitrate and Nitrite** Nitrates and nitrites are a part of the nitrogen cycle. Excessive use of fertilizers, pesticides, sewage disposal, discharge from industries, domestic effluent, water from decayed vegetables and atmospheric precipitation, have led to increase in concentration of nitrate in drinking water. Excessive concentration of nitrate causes heart and lung diseases. Excessive nitrites leads to production of methemoglobin in warm blooded animals leading to the condition of methemoglobinemia (Memon and Schröder 2009).
5. **Chlorides** Combination of chlorine (Cl) gas with a metal results in formation of inorganic compounds known as chlorides such as sodium chloride (NaCl). Chlorine is used as a disinfectant and is highly toxic in nature. Surface water is contaminated by chlorides through several sources such as agricultural run-off, wastewater from industries, and effluent waste from water treatment plants and also from rocks containing chlorides. High level of chlorides in freshwater and lakes affects the health of aquatic organisms. Chlorides also affect the taste of food products and corrode metals (Memon and Schröder 2009; Agarwal et al. 1997).
6. **Fluoride** In 20 states of India, the fluoride levels in ground water are higher than permissible limits (Memon and Schröder 2009). This means more than 60 million people consume water which has fluoride greater than 1 mg/l. High concentration of fluoride in drinking water leads to severe conditions such as dental and skeletal fluorosis. Consumption of high levels of fluoride in drinking water also reduces the absorption of iron which further leads to iron deficiency and related problems. The fluoride endemic states in India are Andhra Pradesh, Kerala, Maharashtra, Gujarat, Karnataka, Punjab, Tamil Nadu, Rajasthan, Haryana, Jammu and Kashmir and Delhi (Memon and Schröder 2009).
7. **Arsenic** Arsenic is an odourless metalloid. The most toxic form of arsenic is Arsenite. Ground water is contaminated by arsenic during weathering of rocks and minerals. Exposure sources of arsenic also include anthropogenic sources such as agrochemicals, burning of fossil fuels, industrial sources, wood preservatives, etc. Ground water arsenic contamination was first observed in West Bengal in the year 1983 (Memon and Schröder 2009; Arora et al. 2011). High levels of arsenic leads to the condition called arsenicosis which refers to arsenic poisoning. Arsenic causes skin damage, skin cancer, and internal cancers and also damages the vascular system. The crops grown using arsenic contaminated water are transported to other unaffected places. Thus, affecting the habitants of the unaffected region and posing new dangers (Memon and Schröder 2009; Arora et al. 2011).
8. **Lead** Natural sources contribute to presence of lead in tap water to some extent but primarily household plumbing systems contain lead and thus contaminate tap water. Lead compounds present in PVC pipes can be leached from them and contribute to high concentration of lead in water. Lead poisoning has several health hazards such as anemia, lower IQ and hyperactivity in children, stunted

growth of the fetus in pregnant women, reproductive problems, and cardiovascular effects and reduced kidney function in adults (Memon and Schröder 2009; Bushra et al. 2015; Mittal et al. 2016; Nabi et al. 2009).

9. **Phosphorous** Phosphorous is one of the most important elements for growth of biotic things. Phosphate a compound formed from phosphorous exists in three forms, orthophosphate, metaphosphate, and organically bound phosphate. Excessive growth of phosphate leads to rapid growth of algae and aquatic plants. Thus, choking up the water and reducing the availability of water, leading to the condition of eutrophication. Rapid growth of aquatic vegetation causes death and decay of vegetation and deteriorates the quality of aquatic life by reducing the availability of dissolved oxygen in water (Atwater and Polman 2010).
10. **Iron** Iron is the second most abundant metal and it is most commonly found in form of its oxides. According to WHO, the minimum requirement of iron ranges between 10 and 50 mg/day. Iron is a vital element required in the oxygen transport mechanism in the blood. Consumption of iron doses as low as 40 mg/kg of body weight have resulted in deaths. Increased iron absorption in the body results in chronic iron overdose. Iron present in domestic water supply leads to staining of laundry and porcelain. Excessive iron in drinking water gives it a bitter and astringent taste (Metcalf et al. 2003).
11. **Dissolved Oxygen** Dissolved oxygen refers to the amount of free oxygen dissolved in an aqueous solution. Dissolved oxygen has influence on aquatic organisms and thus is a very important parameter to access the quality of water. Dissolved oxygen enters water through air by diffusion and as a byproduct of photosynthesis. The amount of dissolved oxygen depends on temperature, pressure and salinity. As temperature decreases the solubility of oxygen decreases. Dissolved oxygen decreases with increase in salinity and increases with increase in pressure. With the increase in concentration of dissolved oxygen the fish mortality rate increases. High level of dissolved oxygen can result in gas bubble disease in fishes (http://nptel.ac.in/courses/105105048/M13L16.pdf).
12. **Biological Oxygen Demand (BOD)** BOD is defined as the amount of oxygen required for decomposing the organic matter by microorganisms. It also measures the chemical oxidation of organic matter. The amount of dissolved oxygen in water sources is directly affected by biological oxygen demand. Greater the value of BOD, more rapid will be the depletion of oxygen in water sources. Thus reducing the amount of oxygen available for aquatic organisms in the water. For raw sewage the general range of BOD is 100–400 mg/l. Dead plants and animals, animal manure, water treatment and food processing plants and effluents from paper mills are the major sources of biological oxygen demand.

    The water quality monitoring results obtained during 1995 and 2011 with respect to biological oxygen demand show that the quality of water has degraded gradually. Chart 3 shows the number of observations having

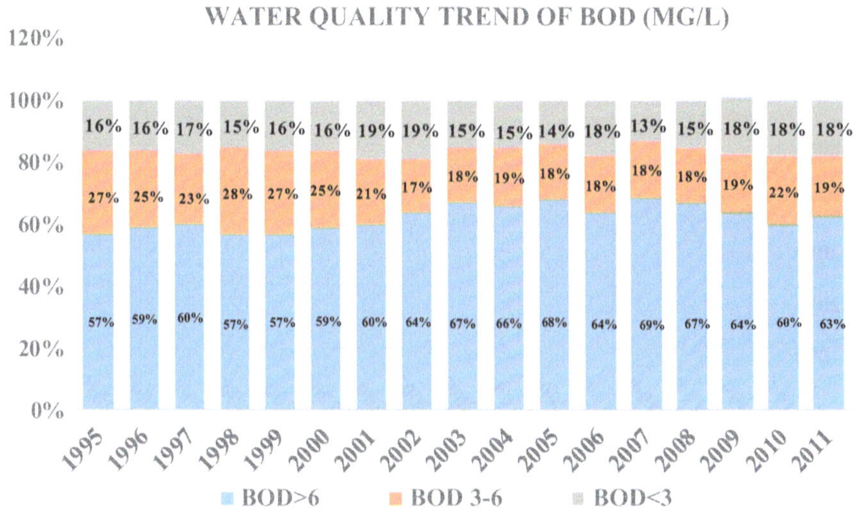

Chart 3 Water quality trend of biological oxygen demand (BOD)

biological oxygen demand in different ranges for the period 1995–2011. A gradual decrease was observed in the number of observations having BOD < 3 and the highest value of 69% was observed in the year 2007. Similarly there was a gradual decrease in the number of observations having BOD between 3 and 6 mg and the highest value of 28% was observed in the year 1998. But the number of observations having BOD > 6 was maximum in the year 2001 and 2002 and then decreased gradually. While in the year the 2011 maximum value of 18% was observed (http://nptel.ac.in/courses/105105048/M13L16.pdf; Drinan 2001).

13. **Fecal Coliform Bacteria** Fecal coliform serves as indicators of fecal pollution in the water. Fecal coliform present in water suggests that the water has been polluted by the fecal material. Absence of coliform organism indicates that the water is free from disease spreading organisms. Coliform bacteria reside in the intestine of human beings. *Escherichia coli* is the most common member of this group. Typhoid fever, hepatitis A and viral are some of the most common pathogenic diseases. Chart 4 shows the water quality monitoring results obtained during the period of 1995–2011 with respect to the indicator of fecal coliform bacteria. During the year 1995–2011 the numbers of observed fecal coliform values of 500 MPN/100 ml was found to be between 48 and 70% and that of fecal coliform values of the ranges 500–5000 MPN/100 ml was between 20 and 35%. In the former case maximum value of 70% was observed in the year 2009 while in the latter case the maximum value of 35% was observed in the year 1999. For the third case, i.e. for fecal coliform values >5000 MPN/100 ml, the numbers of observed values was between 7 and 21% in the year 1995–2011 and the maximum value of 21% was observed in the year 2006 (http://cpcb.nic.in/water.php).

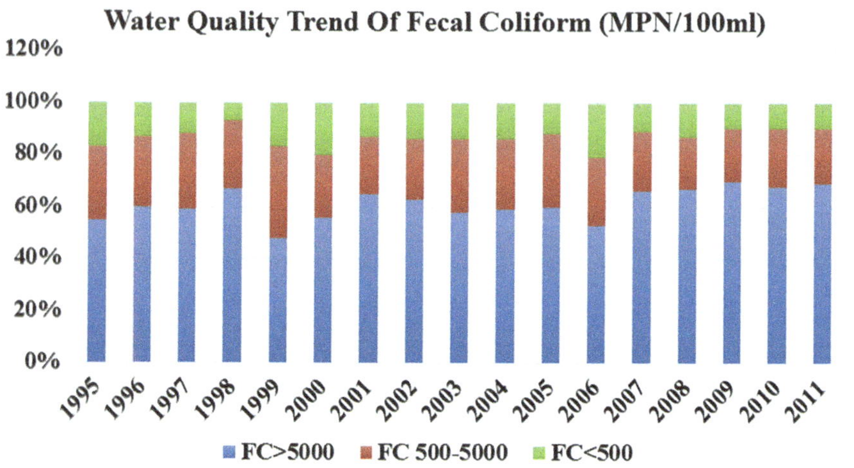

Chart 4 Water quality trend of fecal coliform

In the above section of the chapter, we have discussed the causes of water pollution, major trends associated with it and the important characteristics which are tested for treatment of contaminated water. Now, in the upcoming section, we will describe the harmful effects of polluted water on human health. Contaminated water leads to various chronic diseases, killing millions of people each year. The various water related diseases are discussed below.

4 Water-Related Diseases

Water related diseases lead to death of millions of people each year and a large proportion of about 2.3 million people suffer from water related diseases (http://cpcbenvis.nic.in). Different water related diseases have different nature, transmission methods and harmful effects.

4.1 Water-Borne Diseases

Diseases are caused by water which is contaminated by animals, humans and chemical wastes. Typhoid, polio, hepatitis A, hepatitis E, cholera and meningitis come under the water-borne diseases. Improper sanitation facilities lead to various water borne diseases majorly diarrhea. About 4 billion cases of diarrhea are reported every year. Presence of toxic substances such as fertilizers, chemicals, pesticides, industrial wastes, etc. can give rise to chronic diseases such as cancers. Presence of pesticides such as DDT also has harmful effects on human health. They lead to cancers and are also responsible for reduced sperm count and neurological diseases.

4.2 Water-Based Diseases

These diseases are caused by aquatic organisms that live both in water and as parasites of animals. They can reside in polluted as well as unpolluted water. Various organisms such as flukes, worms, helminths, etc. cause these diseases. Some examples of water based diseases are bilharzia, paragonimiasis and guinea worm.

4.3 Water-Related Vector Diseases

Vectors or insects are those which breed near polluted and unpolluted water, transmit infections and cause diseases such as malaria, dengue fever, filariasis, sleeping sickness, etc. A rapid increase has been observed in water related vector diseases. Mosquitoes are developing resistance to insecticides such as DDT. Other factors such as climate change, migration, creation of new breeding sites, etc. has also contributed in the increase of such diseases (http://cpcbenvis.nic.in). Table 3 shows the water related diseases along with the respective causative organism (http://cpcbenvis.nic.in).

Now, that we have discussed the harmful effects of contaminated water on human health, we must explain about the steps involved to reduce water

Table 3 Water related diseases and their causative organism

Disease	Causative organism
Water-borne diseases	
Bacterial	
Typhoid	*Salmonella typhi*
Cholera	*Vibrio cholera*
Paratyphoid	*Simonella parayphi*
Bacterial dysentery	*E. coli*
Viral	
Hepatitis	Hepatitis-A
Diarrheal diseases	Rota-virus
Water-washed diseases	
Scabies	Skin-fungus species
Bacillary dysentery	*E. coli*
Water-based diseases	
Guinea worm	Guinea worm
Schistosomiasis	Schistosoma sp.
Infection caused due to water related vectors	
Malaria	*Plasmodium*
Sleeping sickness	*Trypanosoma*

contamination. One of the most important steps in this regard is standardization of water quality. The water quality standards set for various purposes are discussed below.

5 Water Quality Standards for Various Purposes

Drinking water is used for various purposes such as drinking, cooking, etc. Water is a basic human need and a prime resource. Therefore, the provision of safe drinking water is a matter of high priority. According to the Bureau of Indian Standards (BIS)-10500, drinking water shall comply with the requirements given in the Table 4 (http://cpcbenvis.nic.in):

According to the Central Pollution Control Board (CPCB), under the National Water Quality Monitoring Programme (NWMP), the classification of surface waters in India on the basis of their usage and the respective water quality criteria is discussed in Table 5 (http://cpcbenvis.nic.in):

According to the World Health Organization (WHO) guidelines, drinking water shall comply with the requirements discussed below in the Table 6 (http://cpcbenvis.nic.in/water_pollution_main.html).

Now, as we are aware of the water quality standards for various purposes, let us discuss about the treatment methods involved. Contaminated water has to be treated in order to meet the quality standards. Treatment of wastewater involves various techniques and processes which are used according to the type of pollutants present in the wastewater.

6 Treatment of Wastewater

Various treatment techniques are used for removal of pollutants from water. We have already discussed the various characteristics for which water is tested. Let's talk about the techniques involved in testing of the discussed parameters.

6.1 Techniques for Testing of Various Parameters in Wastewater

Drinking water is tested for various parameters (http://cpcbenvis.nic.in; http://cpcbenvis.nic.in/water_pollution_main.html). The methodology for measurement of various parameters is as follows.

Table 4 Water quality standards

S. No.	S. No.	Parameter	BIS, Indian Standards (IS 10500:1991)	BIS, Indian Standards (IS 10500:1991)	Risk or effects	Sources
		Desirable unit	**Permissible unit**			
1.	**Color**	5 Hazen Units	25 Hazen Units	Visible tint, acceptance decreases	Tannins, Iron, Copper, Manganese, natural deposits	**Color**
2.	**Odour**	Unobjectionable	Unobjectionable	Rotten egg, Musty, Chemical	Chlorine, Hydrogen Sulphide, Organic matter, Methane gas	**Odour**
3.	**pH**	6.5–8.5	No relaxation	Low pH: corrosion, metallic taste High pH: bitter/soda taste, deposits	Natural	**pH**
4.	**Total Dissolved Solids**	500 mg/l	2000 mg/l	Hardness, staining, salty or bitter taste, corrosion of pipes	Livestock waste, septic system, landfills, nature of soil, dissolved minerals, iron and manganese	**Total Dissolved Solids**
5.	**Hardness**	300 mg/l	600 mg/l	Soap scums, scale in utensils and hot water system	Calcium and magnesium from soil	**Hardness**
6.	**Alkalinity**	200 mg/l	600 mg/l	Low alkalinity leads to deterioration of plumbing	Plumbing fixtures, pipes, landfills	**Alkalinity**
7.	**Iron, Fe**	0.3 mg/l	1.0 mg/l	Discoloration of beverages, brackish color and bitter and metallic taste	Leaching of pipes in water distribution system	**Iron, Fe**
8.	**Manganese, Mn**	0.1 mg/l	0.3 mg/l	Black stains on laundry, bitter taste and brownish color of water	Deposits in rocks and soil	**Manganese, Mn**
9.	**Sulphate, SO_4**	200 mg/l	400 mg/l	Rotten—egg odor, bitter and medicinal taste	Sewage, industrial waste and natural deposits	**Sulphate, SO_4**

(continued)

Table 4 (continued)

S. No.	S. No.	Parameter	BIS, Indian Standards (IS 10500:1991)	BIS, Indian Standards (IS 10500:1991)	Risk or effects	Sources
10.	**Nitrate, NO_3^-**	45 mg/l	100 mg/l	Blue baby disease in infants	Fertilizers, waste water from households, manure lagoons	**Nitrate, NO_{3}-**
11.	**Chloride, Cl**	250 mg/l	1000 mg/l	High blood pressure, corrosion of pipes and appliances, blackening of stainless steel	Fertilizers, minerals and waste from industries	**Chloride, Cl**
12.	**Fluoride, F**	1.0 mg/l	1.5 mg/l	Coloration of teeth and damage to the bones	Industrial waste	**Fluoride, F**
13.	**Arsenic, As**	0.05 mg/l	No Relaxation	Depression, skin and nervous system toxicity	Improper waste disposal, rocks, pesticides	**Arsenic, As**
14.	**Chromium, Cr**	0.05 mg/l	No Relaxation	Nasal ulcer, lung tumors, Damage to nervous and circulatory system	Industrial discharge and mining	**Chromium, Cr**
15.	**Copper, Cu**	0.05 mg/l	1.5 mg/l	Anemia, damage to liver and kidney, staining of plumbing fixtures		**Copper, Cu**
16.	**Cyanide**	0.05 mg/l	No Relaxation	Thyroid, Damage to nervous system	Fertilizers	**Cyanide**
17.	**Lead, Pb**	0.05 mg/l	No Relaxation	Mental retardation, hearing loss, hypertension and blood disorders	Paint, discarded batteries, leaded gasoline	**Lead, Pb**
18.	**Mercury, Hg**	0.001 mg/l	No Relaxation	Vision and hearing loss, nervous system disorders	Electrical equipment, fungicides	**Mercury, Hg**
19.	**Zinc, Zn**	5 mg/l	15 mg/l	Metallic taste	Paints, dyes and leaching of galvanized pipes	**Zinc, Zn**

1. **Temperature** Temperature is measured by dipping the bulb of a thermometer in the river for about 2–3 min. Then the mercury level in the thermometer is read.
2. **pH** pH is measured with the help of a pH meter at the Divisional Laboratory (Level-2). The instrument is standardized before measuring the pH value.
3. **Electrical Conductivity** Electricity conductivity (EC) is measured by an electrical conductivity meter. An EC meter is used to measure the resistance of water between two platinized electrodes.
4. **Magnesium** It is measured by complex metric titration using the standard solution of EDTA under the buffer solution which is prepared from ammonium chloride and ammonium hydroxide of pH 10.0. Value of the EDTA solution gives the sum of calcium and magnesium concentration.
5. **Sodium** A flame photometer is used to measure sodium. Sample emits radiation which is measured through yellow filters.
6. **Potassium** A flame photometer is used to measure potassium. Potassium solution is used to standardize the instrument. A flame photometer work on the principle of measuring the radiation emitted by the atoms of potassium.
7. **Carbonate** Presence of carbonate is indicated when pH touches 8.30. It is measured with the help of process of titration with HCl and phenolphthalein as an indicator. As the value of pH exceeds 8.3 phenolphthalein becomes pink.
8. **Sulphate** Sulphate is measured with the help of nephelometric method. In this method the concentration of turbidity is measured with respect to the known sulphate solution.
9. **Chloride** For measuring chloride, a sample is titrated against silver nitrate solution. A potassium chromate solution in water is used as an indicator.
10. **Dissolved Oxygen** Dissolved oxygen is measured using the Winkler's method. It must be measured at the river site as it changes with time.
11. **Biochemical Oxygen Demand** Winkler's method is used to measure dissolved oxygen which is left in the sample. The difference between the initial and final values of dissolved oxygen gives the amount of oxygen consumed by bacteria during the incubation period. The value is calculated for one liter of sample.
12. **Total coliform** The bacteria which are present in water grow rapidly when placed in a nutritive medium. The movement of the bacteria growing under these special conditions is restricted and thus the number of bacteria is confined to the sample. Various colonies of bacteria together give the total number of coliform in the sample (http://cpcbenvis.nic.in).

Now, let us discuss a wastewater treatment plant and the different methods which are involved.

6.2 Methods Involved in a Wastewater Treatment Plant

A wastewater treatment plant involves use of different treatment methods (http://cpcbenvis.nic.in; http://cpcbenvis.nic.in/water_pollution_main.html; Camp 1963). These methods are classified into the following categories.

6.2.1 Preliminary Treatment

Preliminary treatment involves removal of floating material and settleable inorganic and organic solids from raw water. Preliminary treatment involves processes such as screening and grit removal.

Screening is the first operation in a wastewater treatment plant. It is used for removal of large floating objects such as plastics, clothes, paper, dead animals, etc. A set of inclined parallel bars are fixed at a certain distance in a channel. Screens are designated as hand cleaned and mechanically cleaned based on the method of cleaning. On the basis of the size of the opening, screens are differentiated into coarse and fine screens. Coarse screens consist of parallel bars with bar spacing of 50–150 mm.

Fine screens are mechanically cleaned and the opening size ranges from 0.035 to 6 mm. They are specifically used for pretreatment of industrial wastewater. They are also used for removing solids from primary effluents to prevent clogging of filters (http://cpcbenvis.nic.in).

Grit removal technique is used for removing heavy inorganic materials such as sand and ash. Working of grit chambers is based on the process of sedimentation due to gravitational forces. Grit chambers are very similar to sedimentation tanks, as they are also used for separation of heavier materials and to pass on the lighter materials. The velocity of water through the grit chamber is maintained very high in order to prevent the organic solids from settling. This velocity is known as the "differential sedimentation and differential scouring velocity". The critical velocity of flow "V_c" should always be less than the scouring velocity of grit. Because beyond the critical velocity of flow, already settled particles can be reintroduced in the flow (http://cpcbenvis.nic.in/water_pollution_main.html).

6.2.2 Primary Treatment

Primary treatment involves removal of organic and inorganic solids by physical processes such as sedimentation and floatation. In primary treatment a quiescent condition is maintained and the velocity of flow is reduced such that the denser material settles down and the lighter one floats to the surface. During primary treatment, about 50–70% of total suspended solids and 65% of oil and grease is removed. Chemicals that are used to precipitate insoluble substances are added in water in order to promote sedimentation. These chemicals are known as flocculants.

Flocculation leads to aggregation of small pollutants. Thus, forming large floc so that they can settle down fast. The solids that settle down are known as primary or raw sludge.

A sedimentation tank usually has 4 zones; inlet zone, settling zone, sludge zone and outlet zone (Camp 1963).

Based on shape the primary sedimentation tank is divided into three types.

- Rectangular tank: These are the most widely used tanks. Cost of maintenance is very low in case of rectangular tanks. Also, these are suitable for large capacity. Rectangular basins are least likely to short circuit. In a rectangular tank, the flow takes place lengthwise.
- Circular tank: These type of sedimentation tanks are preferred for continuous vertical flow type of sedimentation tanks. Circular tanks have high clarification efficiency but they are uneconomical as compared to rectangular tanks.
- Hopper bottom tank: In this type of tank, a deflector box which is placed at the top deflects the incoming influent downwards with the help of a central pipe. The collected sludge is disposed using a sludge pump (http://cpcbenvis.nic.in/water_pollution_main.html).

6.2.3 Secondary Treatment

After primary treatment, the wastewater is directed towards secondary treatment, which degrades the biological content of sewage by removing nutrients and remaining solids. Secondary treatment involves use of biological processes. Aerobic microorganisms perform biological treatment in presence of oxygen. These microorganisms consume the organic matter and produce more microorganisms and inorganic end-products. Secondary treatment involves use of sedimentation tanks known as secondary clarifiers. During secondary sedimentation, biological solids called as secondary or biological sludge are removed. Processes involved in secondary treatment include activated sludge process, trickling filters and aerated lagoons and oxidation ponds (http://cpcbenvis.nic.in; http://cpcbenvis.nic.in/water_pollution_main.html; Camp 1963).

Activated Sludge Process

Activated sludge plant involves the following steps

- Wastewater aeration in presence of microorganisms.
- Separation of solids and liquids.
- Discharge of clarified effluent.
- Wastage of extra biomass.
- Remaining biomass is returned to the aeration tank.

Table 5 Classification of surface water and respective water quality standard

Beneficial use	Criteria
Drinking water source without conventional treatment	1. Dissolved Oxygen 6 mg/l 2. BOD (5 days 20 °C) 2 mg/l 3. pH between 6.5 and 8.5 4. Total coliform organism MPN/100 ml must be 50 or less
Outdoor bathing	1. Dissolved Oxygen 5 mg/l 2. BOD (5 days 20 °C) 3 mg/l 3. pH between 6.5 and 8.5 4. Total coliform organism MPN/100 ml should be 500 or less
Source of drinking water after conventional treatment	1. Dissolved Oxygen 4 mg/l 2. BOD (5 days 20 °C) 3 mg/l or less 3. pH between 6 and 9 4. Total coliform organism MPN/100 ml should be 5000 or less
Propagation of wildlife and fisheries	1. Dissolved Oxygen 4 mg/l 2. pH between 6.5 and 8.5
Irrigation and waste disposal	1. pH between 6.0 and 8.5 2. Maximum boron 2 mg/l 3. Maximum sodium absorption ratio 26

In this process, wastewater is aerated in an aeration basin. Microorganisms are added in the waste water, which metabolize the organic matter present in wastewater and form floc. Also, these microorganisms multiply to produce new microorganisms. Wastewater from the aeration tank is transferred to the settling tank, where these stable solids (floc) settle down. This settled organic matter consists of numerous activated microorganisms and is called activated sludge. A part of this settled activated sludge is returned to the aeration tank and the remaining waste activated sludge is removed from the tank and fed to sludge digestor (Camp 1963).

Trickling Filter

Trickling filter is an aerobic treatment process. In this process, the microorganisms which are used for treatment are attached to a medium in order to remove organic matter form wastewater. Microorganisms are attached to material such as rock, sand, redwood and various synthetic materials. Microbes such as bacteria, fungi and algae grow on the filtering medium and are called biological slime. During the process, wastewater is sprayed in the air and is allowed to trickle through the medium. During this the organic matter which is present in the wastewater is metabolized by the biological slime attached to the medium. Thickness of the biological slime increases as the organic matter present in wastewater is synthesized into new cellular material. Eventually, the biological slime layer becomes very thick

Table 6 WHO water quality standards

S. No.	Parameter	WHO guideline (maximum allowable concentration)
1.	Color	15 True Color Units
2.	Turbidity	5.0 NTU
3.	pH	6.5–8.5
4.	Total hardness	500 mg/l
5.	Chlorides	250 mg/l
6.	Dissolved solids	1000 mg/l
7.	Sulphate	400 mg/l
8.	Nitrate	10 mg/l
9.	Fluoride	1.5 mg/l
	Micro pollutants	(Heavy metals and pesticides) (mg/l)
10.	Zinc	5.0
11.	Iron	0.3
12.	Manganese	0.1
13.	Copper	1.0
14.	Arsenic	0.05
15.	Cyanide	0.1
16.	Lead	0.05
17.	Chromium	0.05
18.	Aluminium	0.2
19.	Cadmium	0.005
20.	Selenium	0.01
21.	Mercury	0.001
22.	Sodium	200
23.	Aldrin and dieldrin	0.03
24.	DDT	1.0
25.	Lindane	3.0
26.	Methoxychlor	30.0
27.	Benzene	10.0
28.	Hexachlorobenzene	0.01
29.	Pentachlorophenol	10.0

and gets detached from the surface. This phenomenon of detachment of slime layer is known as sloughing. This detached slime layer and treated wastewater is collected through underdrainage and is passed to a settling tank for separation of solids and liquids (http://cpcbenvis.nic.in/water_pollution_main.html).

6.2.4 Tertiary Treatment

Tertiary treatment is also known as the final or advanced treatment. Tertiary treatment involves removal of organic matter left after secondary treatment. Tertiary treatment involves removal of different types of pollutants such as suspended solids, nutrients, pathogens, organic matter and heavy metals which are not removed by secondary treatment. In tertiary treatment, the methods/processes used depend on the characteristics of the secondary effluent.

Removal of Nitrogen

The most reduced nitrogen compound found in wastewater is ammonia. Under aerobic conditions, ammonia nitrogen is oxidized biologically to nitrate. Air stripping and biological nitrification and denitrification are the two most common processes used for removal of ammonia.

Air Stripping

This process involves conversion of ammonia to gaseous form and then dispersion of the liquid in air. Thus, leading to transfer of ammonia from wastewater to air. This process is well suited for wastewater with ammonia levels between 10 and 100 mg/l. It depends on the temperature of air and pH of the wastewater. This is a controlled process and produces no backwash. Also, the process is unaffected by toxic compounds (Camp 1963).

Biological Nitrification and Denitrification

Nitrogen is removed biologically by bacteria in a two-step process involving nitrification followed by denitrification.

Nitrification
It is defined as the biological conversion of ammonium to nitrate nitrogen. Nitrification involves two steps. In the first step, bacteria *Nitrosomonas* converts ammonia to nitrite, which in the second step is converted into nitrate by another bacteria called *Nitrobacter*. Nitrification requires long retention rate, sufficient alkalinity and low food to microorganism ratio. The process is also affected by the wastewater temperature and pH.

Denitrification
This process is carried out by heterotrophic bacteria under anaerobic conditions. These bacteria use dissolved oxygen or nitrogen as an oxygen source for oxidation of organic matter. For reduction of nitrogen to occur, the level of dissolved oxygen must be near zero and enough carbon supply must be available to the bacteria. The process is conducted in an aerobic filter.

Removal of Suspended Solids

Removal of suspended solids can be achieved using two techniques, namely micro straining and coagulation and flocculation.

Micro Straining

In a micro strainer, a very fine screen made of stainless steel or plastic is supported by a rotating drum. The wastewater fed into the drum is filtered outwards through the screen. And solids accumulate on the screen inside the drum. These accumulated solids are flushed in a removal trough by a backwash system (http://cpcbenvis.nic.in; http://cpcbenvis.nic.in/water_pollution_main.html; Camp 1963).

Coagulation and Flocculation

Suspended particles vary in size, source, charge and shape. Coagulation process aims at altering the particles in such a way that they adhere to each other. A coagulant neutralizes the charge of suspended particles allowing them to combine together to form large particles. The commonly used coagulant is alum. Once the floc formed has gained appropriate strength and size, the solution is transferred to a settling tank for the process of sedimentation (Camp 1963).

Acknowledgements The authors acknowledge Prof. Sunita Rattan, Head of Institute, Amity Institute of Applied Sciences (AIAS), Amity University Uttar Pradesh, Noida. ST and RV acknowledge Prof. Sangeeta Tiwari and Prof. R. S. Pandey, Head, Dept. of Chemistry and Physics, AIAS, Amity University Uttar Pradesh, Noida. ST would like to thanks Department of Science and Technology, Government of India, New Delhi, India.

References

Agarwal V, Vaish AK, Vaish P (1997) Groundwater quality: focus on fluoride and fluorosis in Rajasthan. Curr Sci 9:743–746

Aral MM (2009) Water quality, exposure and health: purpose and goals. Water Qual Expo Health 1:1–4

Arora P, Sindhu A, Dilbaghi N, Chaudhury A (2011) Biosensors as innovative tools for the detection of food borne pathogens. Biosens Bioelectron 1:1–12

Atwater H, Polman A (2010) Plasmonics for improved photovoltaic devices. Nat Mater 9:205–213

Bushra R, Naushad M, Adnan R et al (2015) Polyaniline supported nanocomposite cation exchanger: synthesis, characterization and applications for the efficient removal of Pb^{2+} ion from aqueous medium. J Ind Eng Chem 21:1112–1118. https://doi.org/10.1016/j.jiec.2014.05.022

Camp TR (1963) Water and its impurities. Reinhold, New York, p 355

Drinking water—specification, Bureau of Indian Standards (Second Revision of IS 10500). www.mdws.gov.in/sites/default/files/Drinking_water_quality_standards

Drinan JE (2001) Water and wastewater treatment: a guide for the non engineering professionals. CRC Press, US
Durfor CN, Becker E (1964) Public water supplies of the 100 largest cities in the United States. US Geog. Sur. water supply paper 1812, p 364
http://cpcb.nic.in/
http://cpcb.nic.in/water.php
http://cpcbenvis.nic.in
http://cpcbenvis.nic.in/water_pollution_main.html
http://greencleanguide.com/earths-water-distribution-and-indian-scenario/
http://nptel.ac.in/courses/105105048/M13L16.pdf
http://www.sulabhenvis.nic.in
Kumar S, Meena HM, Verma K (2017) Water pollution in India: it's impact on the human health: causes and remedies. Int J Appl Environ Sci 12:275–279
Memon AR, Schröder P (2009) Implications of metal accumulation mechanisms to phytoremediation. Environ Sci Pollut Res 16:162–175
Metcalf and Eddy, Tchobanoglous G, Burton FL, Stensel HD (2003) Wastewater engineering: treatment and reuse, 4th edn. Metcalf & Eddy, Inc. Tata McGraw-Hill
Mittal A, Naushad M, Sharma G et al (2016) Fabrication of MWCNTs/ThO_2 nanocomposite and its adsorption behavior for the removal of Pb(II) metal from aqueous medium. Desalin Water Treat 57:21863–21869. https://doi.org/10.1080/19443994.2015.1125805
Murty M, Kumar S (2011) Water pollution in India—an economic appraisal. India Infrastructure Report, pp 286–298
Nabi SA, Naushad M, Bushra R (2009) Synthesis and characterization of a new organic—inorganic Pb^{2+} selective composite cation exchanger acrylonitrile stannic (IV) tungstate and its analytical applications. NANO 152:80–87. https://doi.org/10.1016/j.cej.2009.03.033
Pathak AK (2013) Water pollution and treatment. Int J Environ Eng Manage 4:191–198
Pontius FW (1990) Water quality and treatment, 4th edn. McGraw-Hill Inc., New York

Chapter 3
Three-Dimensional Graphene-Based Macroscopic Assemblies as Super-Absorbents for Oils and Organic Solvents

Shamik Chowdhury, Sharadwata Pan, Rajasekhar Balasubramanian and Papita Das

Abstract With frequent oil spill incidents and industrial discharge of organic solvents, the development of highly efficient and environment friendly absorbents with both hydrophobic and oleophilic properties have become a top priority. Attributing to exceptionally large specific surface area, intrinsic hydrophobicity, outstanding electrochemical stability and superior mechanical properties, two-dimensional (2D) graphene holds significant promise as advanced absorbents for oil spill response and restoration. However, just as any other carbon allotrope, graphene as a bulk material tends to form irretrievable agglomerates due to strong π–π interactions between the individual graphene sheets. This leads to incompetent utilization of isolated graphene layers for environmental remediation applications. In order to overcome this restacking issue, the integration of 2D graphene macromolecule sheets into 3D macrostructures, and ultimately into a functional system, has materialized as a progressively critical approach in recent years. Consequentially, a wide array of exotic 3D graphene-based macroscopic assemblies (GMAs), such as aerogels, hydrogels, sponges, foams, etc., have been intensively

S. Chowdhury (✉)
Centre for Advanced 2D Materials, National University of Singapore, Singapore 117546, Singapore
e-mail: c2dsham@nus.edu.sg

S. Pan
School of Life Sciences Weihenstephan, Technical University of Munich, 85354 Freising, Germany
e-mail: sharadwata.pan@tum.de

R. Balasubramanian
Department of Civil and Environmental Engineering, National University of Singapore, Singapore 117576, Singapore
e-mail: ceerbala@nus.edu.sg

P. Das
Department of Chemical Engineering, Jadavpur University, Kolkata 700032, India
e-mail: papita.das@jadavpuruniversity.in

M. Naushad (ed.), *A New Generation Material Graphene: Applications in Water Technology*, https://doi.org/10.1007/978-3-319-75484-0_3

developed during the past five years. Owing to their well-defined and physically interconnected porous networks, these rationally designed macroscopic graphene architectures can support rapid mass transfer in 3D and provide adequate accessible surfaces for molecular absorption, thereby outspreading their application potential. This chapter aims at collating the current state-of-the-art on the development and application of 3D GMAs for ultrafast and recyclable oils and organic solvents absorption. Furthermore, it distinguishes the fundamental knowledge gaps in the domain, and lays out novel strategic research guidelines, all of which will promote further progress in this rapidly evolving cross-disciplinary field of current global interest.

Keywords Graphene macroassemblies · Absorption · Absorbents
Oils · Organic solvents

1 Introduction

The upsurge of extraordinary industrial development and consequential global oil exploration, have brought with them the archenemy of oil spillage (Aguilera et al. 2010; Al-Majed et al. 2012; Wahi et al. 2013; Beyer et al. 2016; Kumari and Singh 2016), as evidenced by the recent explosion and sinking of the Deepwater Horizon drilling rig in the northern Gulf of Mexico. The aftermaths of such blowouts are catastrophic as well as comprehensive, exceeding the domains of environment and ecosystems, human health, and general flora and fauna (Aguilera et al. 2010; Beyer et al. 2016). Numerous tactics have been contemplated to successfully combat the problem of oil spillage, ranging from suppression methods to elimination of spills employing booms (Muttin 2008), skimmers (Broje and Keller 2006), sorbents (Wahi et al. 2013), dispersants (Kujawinski et al. 2011), solidifiers (Vidyasagar et al. 2011), *in situ* burning (Buist et al. 1999) and bioremediation (Ron and Rosenberg 2014). Out of all these, absorbents with both oleophilic (oil-attracting) and hydrophobic (water-repellent) properties have garnered the most attention from the scientific community due to their enhanced viability, high oil/water separation efficiency, enviable specificity and self-adaption capability under harsh sea conditions (Wahi et al. 2013; Ge et al. 2017). Mainly three different types of materials have been utilized till date for oil spill cleanup: inorganic minerals (e.g., zeolite, silica gel and activated carbon), synthetic organics (e.g., polyurethane foam and polypropylene) and natural organics (e.g., wool fiber, saw dust and rice husk) (Al-Majed et al. 2012). Despite the many potential benefits, all these materials suffer from multiple disadvantages, including high density, daunting recovery, poor regeneration, pore clogging, non-biodegradability, exorbitant costs and limited applicability, thus precluding their practical implementation (Al-Majed et al. 2012). There is, therefore, a pressing need for better absorbents that can overcome these shortcomings, and are yet economical.

Since the last decade, two-dimensional (2D) graphene-based materials have proved to be highly beneficial for a multitude of environmental applications (Chowdhury and Balasubramanian 2014a, b; Balasubramanian and Chowdhury 2015), including oil and organic solvent absorption from water (Gupta and Tai 2016); all thanks to their inherent hydrophobicity, enormous surface area and many superlative properties (Dong et al. 2012; Cong et al. 2014). However, 2D graphene nanosheets have a tendency to agglomerate irreversibly due to their strong hydrophobicity, or even restack into graphite through strong van der Waals interactions (Chowdhury and Balasubramanian 2016), thus decreasing their competence for practical use. Several noteworthy efforts have been directed to tackle this challenge, among which the self-assembly of 2D graphene macromolecule sheets into 3D macrostructures, and ultimately into a functional system, has recently materialized as a progressively critical approach (Yin et al. 2012; Cong et al. 2014; Fan et al. 2015; Chowdhury and Balasubramanian 2017). The integration of nanoscale graphene into monolithic macroscopic materials not only prevents their restacking, but also largely translates the intriguing characteristics of individual graphene sheet into the resulting macrostructure and simplifies the processing of graphene materials (Shen et al. 2015; Xu et al. 2015a). Accordingly, a broad spectrum of 3D graphene-based macroscopic assemblies (GMAs), such as aerogels, hydrogels, sponges, foams, spheres, crumpled balls, frameworks with periodic structures, honeycomb-like structures, porous films and vertical sheets, have been intensively developed in the last five years (Fig. 1) (Chowdhury and Balasubramanian 2017). Owing to their highly interconnected graphene networks, these rationally designed macroarchitectures possess ultralow density, exceptionally high surface area, intense porosity, extraordinary flexibility and superior durability, leading to versatile and recyclable absorbents with extraordinary uptake capacities and ultrafast removal capabilities (Wan et al 2016).

In view of these exciting developments, the current chapter offers a comprehensive overview of the recent progress in the design, synthesis and utilization of various 3D GMAs (including sponges, foams and aerogels) for the selective absorption of a broad variety of oils and organic solvents. Additionally, it attempts to isolate the basic lacunae in the specific domain knowledge and outlines some original future research plans. We believe that these will attract further scientific

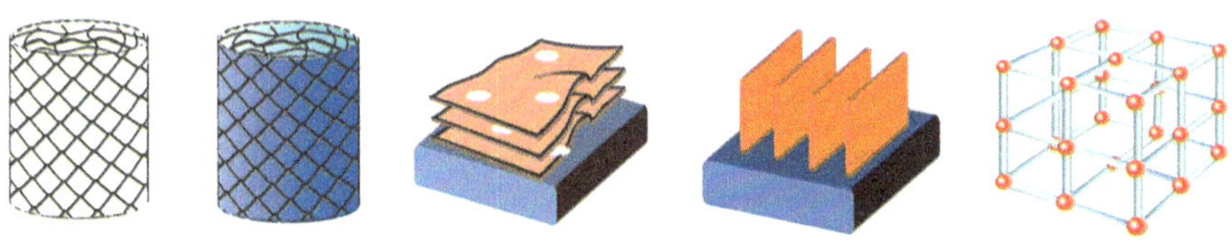

Fig. 1 Various types of 3D GMAs developed in recent years. Reproduced from El-Kady et al. (2016), Copyright 2016, with permission of Nature Publishing Group

Fig. 2 **a–c** Digital photographs of a GA, obtained through hydrothermal reduction and supercritical ethanol drying, supporting a 2 kg counterpoise, more than 26,000 times its own weight of 75 mg without any deformation. Reproduced from Cheng et al. (2017), Copyright 2017, with permission of Nature Publishing Group. **d** Digital image of GAs prepared by chemical reduction-induced self-assembly of GO sheets propped up on a green bristlegrass. Reproduced from Li et al. (2014a), Copyright 2013, with permission of the Royal Society of Chemistry. **e** Digital photo of a homogenous GO aqueous dispersion and the resulting GH after hydrothermal reduction. **f** Scanning electron microscopy (SEM) image revealing the unique 3D interconnected macroporous structure of the GH. Digital picture of GHs synthesized *via* hydrothermal reduction of aqueous GO dispersions at 180 °C with **g** different GO concentration (CGO) and **h** different reaction times. Reproduced from Xu et al. (2010), Copyright 2010, with permission of the American Chemical Society

interest and help to endorse more significant advances in this already fast evolving interdisciplinary research domain.

2 3D GMAs for Oil and Organic Solvent Absorption

2.1 Aerogels

Graphene aerogels (GAs) are an emerging class of mechanically robust, highly porous (>90% v/v), low density (typically <200 mg cm^{-3}) carbon materials (Qian et al. 2013), formed by interlocking of 2D graphene nanosheets in 3D space through non-covalent bonds, such as hydrogen bonds, van der Waals forces, electrostatic forces, hydrophobic–hydrophilic interactions and π–π stacking interactions (Yan et al. 2015) (Fig. 2a–d). They are usually produced *via* sol–gel chemistry, which involves self-assembly of graphene oxide (GO) sheets through chemical, electrochemical or hydrothermal reduction, to form a highly cross-linked 3D structure (Xu et al. 2015a), followed by drying to vent out the absorbed water (Ma and Chen 2015). The latter frequently involves non-evaporative techniques to eliminate capillary forces that could otherwise collapse the 3D architecture.

The hydrothermal or chemical reduction of GO significantly curtails the oxygenated surface functionalities and increases the van der Waals forces between graphene basal planes. The π–π stacking interactions and concurrent hydrophobic effects enforce the flexible, reduced GO sheets to partially overlap and interlock with each other to generate adequate physical cross-linking sites to facilitate the formation of a 3D porous scaffold with entrapped water, commonly referred to as graphene hydrogel (GH) (Yao and Zhao 2017) (Fig. 2e). A typical GH consists of well-defined and highly interconnected 3D porous graphene networks ($\sim$2 wt%) filled with water ($\sim$98 wt%) (Xu et al. 2015a) (Fig. 2f). The pore sizes are usually in the sub-micron to micron range, with pore walls comprising of thin layers of stacked graphene sheets (Xu et al. 2015a; Gorgolis and Galiotis 2017). In addition, the size and shape of the hydrogel can be easily adjusted by modulating the concentration of GO, treatment time and temperature, as well as the shape of the reactor (Fig. 2g, h) (Bi et al. 2012a; Wu et al. 2012). After subsequent freeze-drying or supercritical drying, the ensuing monolithic GA without substantial volume reduction or network compaction, exhibits hierarchical meso-/macroporosity with large specific surface area (100–700 m^2 g^{-1}), increased surface hydrophobicity, and superior mechanical strength (storage modulus of 450–490 kPa) (Xu et al. 2015a; Gorgolis and Galiotis 2017). Attributing to these beneficial structural features, GAs have outstanding absorption efficiencies and can absorb hydrocarbons and different types of organic solvents up to several times their own weight (Wang et al. 2012; Xu et al. 2015a; Liu et al. 2016; Zhang et al. 2016a; Ren et al. 2017), which poses them amongst the superlative absorbing materials.

A typical instance is that of Wang et al. (2012), who developed superhydrophobic, ultralight (10–20 mg cm^{-3}) GAs with abundant pore space through chemical reduction-induced self-assembly of GO sheets in the presence of a wide variety of phenolic acids, such as trihydroxybenzoic acid (gallic acid), dihydroxybenzoic acid (gentisic acid, protocatechuic acid) and monohydroxybenzoic acid (vanillic acid, ferulic acid). Regardless of the type of phenolic reductant, the as-prepared GAs could readily absorb various oils and organic solvents with high efficiency, including vegetable oil, paraffin oil, kerosene, pump oil, dimethylformamide (DMF), cyclohexane, toluene, hexane, ethanol, chlorobenzene and phenixin. The maximum absorption uptake was approximately 25–60 times the weight of the aerogel . Similarly, using L-phenylalanine as a reducing agent, Xu et al. (2015b) prepared a superoleophilic GA, with great absorption capacity and amazing recyclability, by *in situ* chemical reduction and self-assembly of GO sheets (Fig. 3a–e). The highly water repellent aerogel featured an outstanding absorption capacity of over 100 g g^{-1} for all common organic solvents (Fig. 3f–h), and hence merits further consideration in oil–water separation applications. More recently, Zhang et al. (2016a) described a facile and versatile strategy to efficiently synthesize GAs with cellular-like pores that involves chemical reduction of GO assemblies at oil–water interface under mild conditions (Fig. 3i). The as-synthesized GA was ultralight (with density <3 mg cm^{-3}), and yet mechanically resilient, making it a promising candidate for absorbing and desorbing organic liquids. As depicted in Fig. 3j, the GA saturated with n-hexane could be compressed more than 90% of its

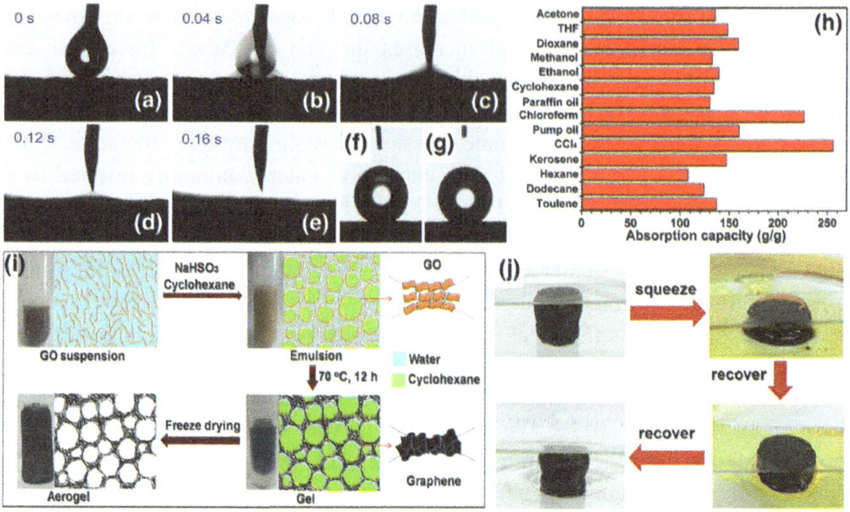

Fig. 3 **a–e** Video snapshots of the absorption process of a drop of hexadecane on the surface of GA, fabricated by facile chemical reduction, demonstrating the superoleophilicity of the material. Optical photo of a water droplet on the **f** surface and **g** cross-section of the as-produced GA. The water contact angle in **f** and **g** are 154° and 151°, respectively. **h** Absorption capacity of the bulk GA for various oils and organic solvents. Reproduced from Xu et al. (2015b), Copyright 2015, with permission of the Royal Society of Chemistry. **i** Major steps involved in the synthesis of GA comprising cellular-like pores. **j** Absorption-release cycle showing excellent recyclability of the cellular GA toward n-hexane (dyed with Sudan orange G). Reproduced from Zhang et al. (2016a), Copyright 2016, with permission of Nature Publishing Group

original height to squeeze-out the absorbed fluid. Further, the shape and size of the monolith was completely restored after reabsorbing the extruded oil. More importantly, this absorption/release process is highly repeatable without any change in liquid absorption capacity, validating the high structural stability of the aerogel.

Impressively, Liu et al. (2016) fabricated highly compressible GAs with ultralow density and anisotopic porosity by directional freezing of GH using anisotropically grown ice crystals as templates, followed by freeze-drying (Fig. 4a–c). Owing to the anisotropic structure, the resultant GA displayed high compressive strength in the axial (freezing) direction and decent compressibility in both axial and radial directions. When evaluated as absorbents, these novel hydrophobic GAs demonstrated exceptional absorption capacity toward various organic liquids (such as vegetable oil, pump oil, white oil, acetone, ethanol, n-hexane and n-heptane) under absorbing-burning, absorbing-distilling and absorbing-squeezing cycles with good recyclability (Fig. 4d–g), due to their large porosity, immense flexibility (in both liquid and gaseous phases) and satisfactory fire-retardancy (Fig. 4h).

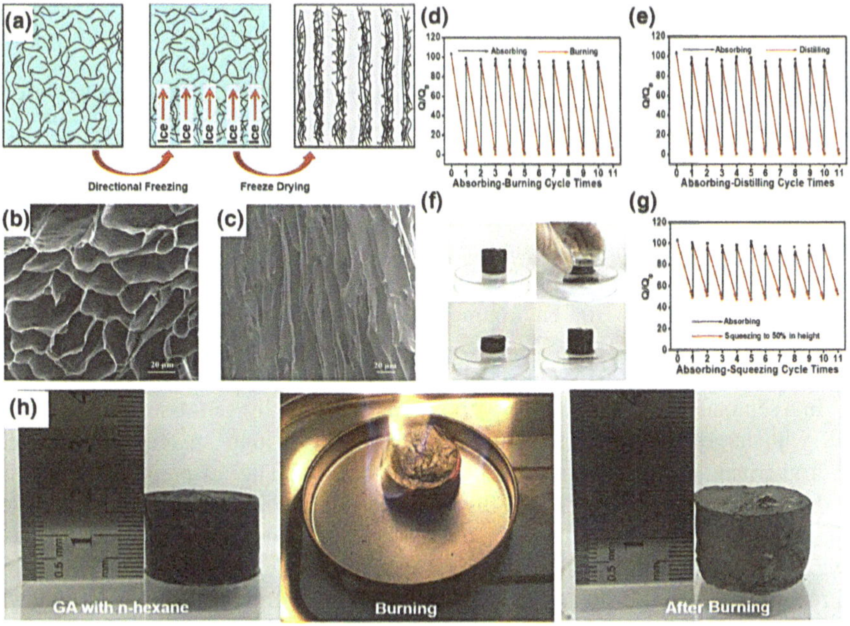

Fig. 4 **a** Schematic of the preparation of anisotropic GA (AGA) by directional-freezing of GH and subsequent freeze-drying. **b** Top-view and **c** side-view SEM images of the AGA. Recyclability of the AGA for absorption of n-hexane under **d** absorbing-burning and **e** absorbing-distilling cycles. **f** Digital images showing the exceptional compression-recovery performance of the AGA after absorbing n-heptane. **g** Recyclability of the AGA for absorption of n-heptane under absorbing-squeezing cycles. **h** Digital photographs illustrating the burning of AGA saturated with n-hexane. Even after the absorbed fluid burns up, the AGA maintains its original appearance except for change in color from black to gray. Reproduced from Liu et al. (2016), Copyright 2016, with permission of Elsevier

Very recently, Ren et al. (2017) conceived a superhydrophobic (water contact angle, WCA of 151°), robust reduced GA (rGA) *via* resorcinol-formaldehyde (RF) sol-gel chemistry, which presented incredible absorbance for oils and organic solvents (up to 19–26 times of its own weight), as well as remarkable cycling stability. Additionally, the high absorbing ability was retained even under extreme temperature conditions (−40 to 240 °C), which was ascribed to the exceptional thermal stability of the nanoscale graphene building blocks and good heat transfer characteristics of the 3D interconnected network structure. These findings clearly showcase the efficacy and versatility of the synthesized rGA for routine and emergency oil spill cleanup in oceanography, environmental protection and industrial production.

2.2 Sponges

Graphene sponges (GSs) represent novel, shape-mouldable, 3D graphene-based, porous monolithic macroscopic materials, which demonstrate both superelastic performance like rubber and near-zero Poisson's ratio in all directions like cork (Wu et al. 2015). They have densities comparable to air, and exhibit highly repeatable compression and complete recovery across a wide temperature range in both air and liquid without any substantial degradation (Wu et al. 2015). Catering to these exclusive features, GS has been the subject of intense research and development (R&D) efforts for environmental reclamation/restoration applications in recent years (Bi et al. 2012b, 2014; Zhao et al. 2012a; Xie et al. 2013; Zhang et al. 2014; Wu et al. 2015, 2016; Guo et al. 2016; Ge et al. 2017; Zhang et al. 2017a, b).

In a seminal study, Sun and co-workers (Bi et al. 2012b) self-assembled GO into GS with large specific surface area (423 m^2 g^{-1}) through an ammonia-assisted hydrothermal method, followed by freeze-drying. The as-prepared GS was hydrophobic (WCA = 114°), and displayed extremely high absorbing abilities for not only commercial petroleum products (viz., kerosene and pump oil) and fats (e.g., vegetable oil), but also for toxic solvents (e.g., toluene, chloroform, nitrobenzene, etc.). For example, the GS could absorb toluene at 55 times its own weight, which was orders of magnitude higher than contemporary absorbers, such as vaseline-loaded expanded graphite (<10 times its own weight). Furthermore, the GS could be regenerated and recycled without affecting its performance when heated close to the boiling point of the absorbate. Meanwhile, the emanating organic compound could be evaporated, condensated and recollected elsewhere for reuse. In order to enrich the absorption capacity of their GS for oils and organic solvents, Sun and colleagues (Xi et al. 2013) subsequently tailored the porosity of the GS by modulating the freezing temperature through a custom-prepared freezing temperature control unit (Fig. 5a–f). When the freezing temperature was lowered from −170 to −10 °C, both the pore diameter and wall thickness increased several folds. In addition, significant differences in pore morphology were also observed. As illustrated in Fig. 5g, a low ice solidification rate (i.e., high freezing temperature) created large anisotropic, lamellar ice crystals, resulting in larger pores with thicker pore walls, thereby causing the sponge to be water repellant (Fig. 5h). In contrast, at low freezing temperatures, the process was mainly nucleation dominated, producing sponges with smaller spacing and thin walls that appeared to be water permeable (Fig. 5h). Later, a soot-based treatment technology was recommended by the authors to further augment the surface hydrophobicity and hence the absorption capacity of the GS (Bi et al. 2014). In comparison to the pristine GS, the absorbance of the treated GS was approximately 8 times higher for chloroform.

In another notable contribution, Wu et al. (2014) adopted a novel solvothermal-induced self-assembly approach to construct shape-adjustable GSs with over 99.9% porosity and strong hydrophobicity (WCA = 135°), using GO as precursor (Fig. 6a). The homogenous and nearly isotropic GS could absorb remarkably large volumes of organic liquids (such as acetone, ethanol, DMF,

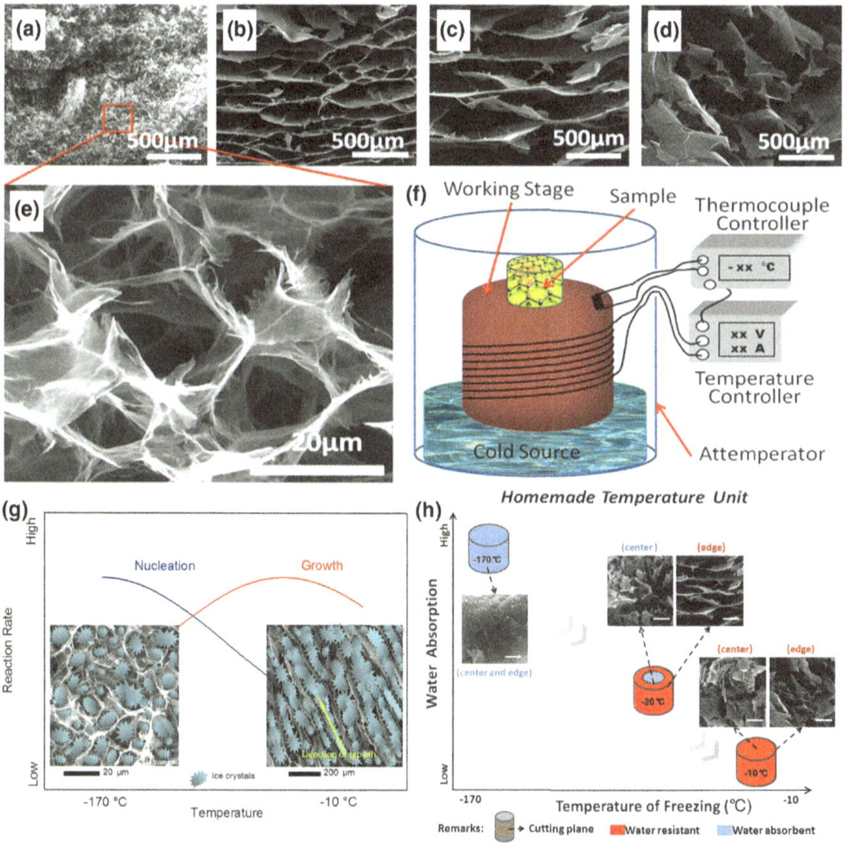

Fig. 5 SEM images of GS fabricated at different freezing temperatures: **a** −170 °C, **b** −40 °C, **c** −20 °C and **d** −0 °C. The mean pore size rose from 10 to 700 mm, while the wall thickness increased from 20 nm to 80 mm, upon lowering the freezing temperature from −170 to −10 °C. **e** High-magnification SEM image of the pore walls of the sponge in (**a**). Clearly, the pore walls were composed of graphene nanosheets. **f** Working principle of the homemade freezing temperature unit. **g** Qualitative schematic of the relationship between nucleation and crystal growth as a function of freezing temperature during ice solidification. The size of ice crystals is not drawn proportionally to that of the GS. Ice crystals are demonstrated artificially on the background, while the factual SEM images of the microstructures are provided. **h** Water absorption/resistance performance of the cross-section of the GS samples. Pores with sizes ranging from 250 to 700 mm showed water resistance, whereas pores smaller than 250 mm allowed water absorption. A cross-section may also possess different water absorption properties in the center and near the edge at transitive freezing temperatures, e.g., −20 °C. The SEM images represent the corresponding microstructures of the surface (scale bars = 500 mm). Reproduced from Xie et al. (2013), Copyright 2013, with permission of Nature Publishing Group

silicon oil and paraffin oil), and then reversibly and repeatedly discharge the liquid in seconds upon compression (Fig. 6b, c). For example, the GS could reversibly absorb and extrude up to 1010 times its weight of pump oil, which is 3 times higher

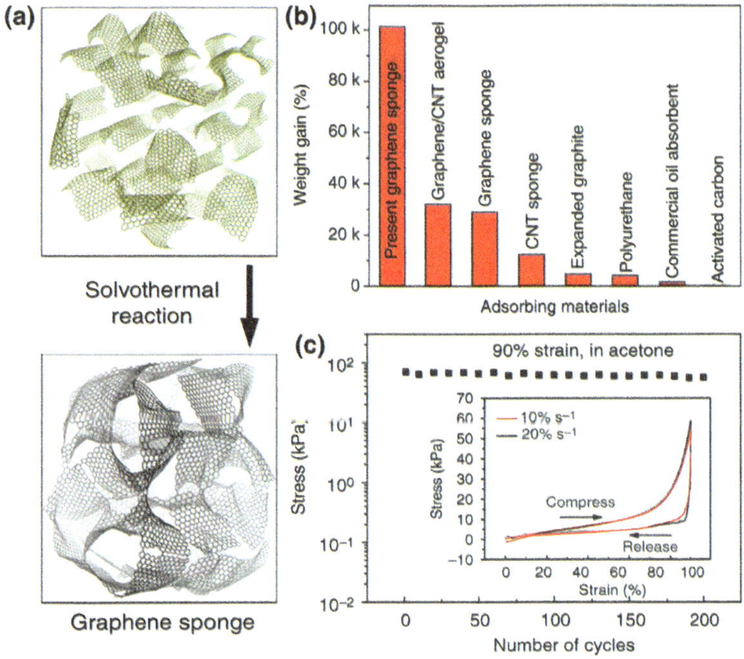

Fig. 6 **a** Illustration of the solvothermal conversion of ethanol-dispersed GO platelets into GS. **b** Liquid absorption capacity (in percent weight gain during liquid absorption) of the GS for pump oil, compared with that of other absorbents for either pump or engine oil. **c** Compressive stress at an engineering strain rate of 20% strain s^{-1} plotted as a function of cycle number for 200 cycles. The inset shows stress–strain cycles in acetone up to 90% strain for engineering strain rates of 10% strain s^{-1} (red line) and 20% strain s^{-1} (black line). Reproduced from Wu et al. (2014), Copyright 2014, with permission of Nature Publishing Group

than the best performing solid absorbent previously reported for pump/engine oil (Fig. 6b). Moreover, the long cycle life (Fig. 6c), the short time periods to achieve 90% liquid extrusion and the complete reabsorption for even viscous liquids (e.g., 4.5 s for DMF) suggest the potential application of the monolithic GS in environmental remediation.

Meanwhile, employing soap bubbles as soft templates, Zhang et al. (2014) built a GS with trimodal pore size distribution spanning across the micro-, meso- and macropore range. Compared to many competing absorbent materials, the GS showed excellent absorbance for a multitude of oils (like diesel oil and vegetable oil) and organic solvents (e.g., acetone, ethanol, methanol, ethylene glycol, etc.). Likewise, different types of hard/soft template-based synthesis strategies have been recently reported to fabricate high performance GS-based absorbents for oil spill control (Zhang et al. 2017a, b). In such 3D array of interconnected voids, the larger pores have a greater absorption capacity and faster desorption rate than the smaller ones, which is due to the higher pore volume and weaker interactions with the absorbed molecules (Guo et al. 2016).

Since heavy crude oil constitute 40% of the world's oil reserve, improving the diffusion of viscous crude oil, and thus the oil absorption rate of oil absorbents is critically important from a practical viewpoint. To this end, Yu and colleagues (Ge et al. 2017) explicitly designed a novel Joule-heated graphene wrapped sponge (GWS) (Fig. 7a), which benefiting from both the hydrophobic nature and giant electron mobility (200,000 $cm^2\ V^{-1}\ s^{-1}$) of 2D graphene could cleanup viscous crude oil spills at a high sorption speed. While the intrinsic hydrophobicity of graphene ensured a high oil/water separation efficiency, the colossal electrical conductivity allowed to heat up the sponge through the application of electric current. As a result, the Joule-heated GWS could rapidly absorb crude oil by heating up the liquid and decreasing its viscosity (Fig. 7b). Indeed, compared with that of non-heated GWS, the oil absorption time was reduced by ~95% utilizing the heated GWS (Fig. 7c). Additionally, because of its low viscosity, the absorbed hot oil could be squeezed out 7 times faster than natural crude oil. Furthermore, the increase in oil temperature lowered the mechanical pressure needed to compress the loaded absorbent. More interestingly, owing to the reduced viscosity, the heavy crude oil could be directly collected from above the surface of seawater by *in situ* pumping through the GWS (Fig. 7d). This important conceptual breakthrough in the design of 3D graphene-based oil absorbents is expected to stimulate further advances in developing a new generation of graphene-based 3D monolithic absorbents that can quickly remove spilled crude oil from our oceans in a highly efficient, economical and environmentally benign manner.

2.3 Foams

Graphene foams (GFs) are an important variant of 3D GMAs with foam-like morphology created primarily through a templating strategy (Yao and Zhao 2017). The concept of GF is to grow seamlessly interconnected 3D porous graphene networks with high purity and excellent quality, which is different from the approach to assemble 2D graphene building blocks into 3D macroscopic assemblies (Chen et al. 2011). This makes them extremely appealing as absorbents for removal of oils and oil-derived organic solvents (Niu et al. 2012; He et al. 2013; Shi et al. 2016; Yang et al. 2016a; Zhang et al. 2016b). For example, applying naturally abundant sea scallops as template, Shi et al. (2016) demonstrated the chemical vapor deposition (CVD) mediated growth of GFs with negligible impurity, ultralow density (~3 mg cm^{-3}) and exceptional mechanical flexibility (Fig. 8a). The self-supporting hydrophobic GF with continuous, open-pore geometry showed marvelous absorption potential and rapid uptake kinetics toward a wide range of organics, including ethanol, dimethyl sulfoxide, n-hexane, dichloromethane, pump oil, bean oil, petrol and motor oil (Fig. 8b–d). Remarkably, the biomass-derived GF was capable of absorbing over 250 times its own weight. This is much higher than

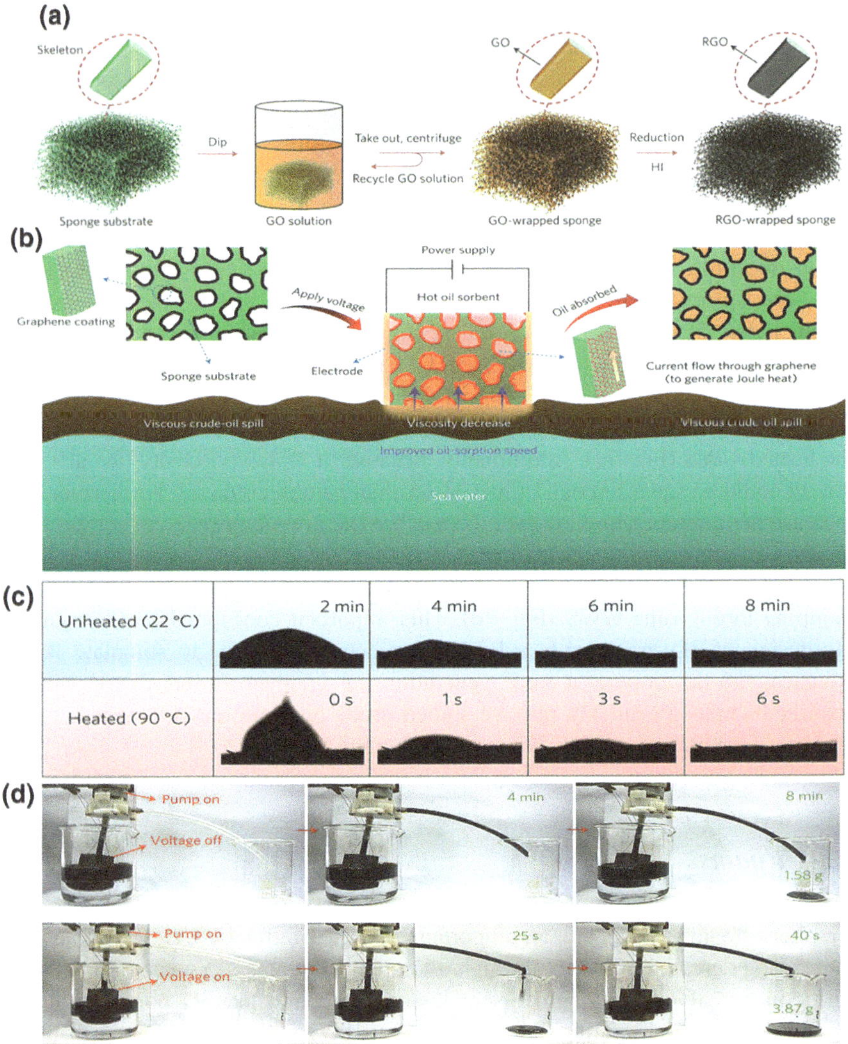

Fig. 7 **a** Fabrication process of the GWS. **b** Schematic of the Joule-heating effect of the GWS to cleanup a viscous crude oil spill. The red color applied to the graphene coating represents the increase in temperature. The dark brown color of the oil signifies the original highly viscous crude oil, while the light brown color of the oil denotes the crude oil with a decreased viscosity. **c** Permeating behavior of an oil droplet on the surface of the GWS with (lower row) and without (upper row) an applied voltage. **d** Continuous collection of viscous crude oil from the surface of sea water by *in situ* pumping through the GWS under applied voltages of 0 (upper row) and 17 V (lower row), respectively. Reproduced from Ge et al. (2017), Copyright 2017, with permission of Nature Publishing Group

other typical carbonaceous absorbents previously reported for pump oil (Fig. 8e). Similarly, Yang et al. (2016a) introduced an autoclaved leavening route that relies on utilization of polystyrene particles as sacrificial templates, to synthesize GFs. The developed GF was further modified through physical activation using carbon dioxide as the oxidizing agent. Owing to its well-defined trimodal pore system, impressive thermal stability and superior hydrophobic/oleophilic properties, the hierarchical porous GF (HGF) delivered excellent absorption uptake, outstanding recyclability and high selectivity when used to separate a range of oil/water mixtures compared to regular GF.

2.4 Surface Modification and Performance Enhancement

Clearly, a wide array of highly efficient and exceptionally durable 3D graphene-based macrostructures have been designed and developed to reduce the environmental impacts of oil spills on the ecosystems and prevent accumulation of organic solvents in water bodies. In spite of this extraordinary progress, their overall performance is yet to achieve a satisfactory crest from a practical deployment viewpoint. Therefore, tailoring the microstructural features and improving the separation abilities of 3D GMAs have attracted considerable research attention of late. In particular, doping heteroatoms (e.g., nitrogen (N), sulfur (S), boron (B), phosphorous (P), etc.) into the graphitic lattice is actively being pursued to manipulate the surface chemistry and elevate the absorption propensity of 3D GMAs (Zhao et al. 2012b; Du et al. 2016; Ren et al. 2016; Yang et al. 2016b; Liu et al. 2017). This is because the dopant atom generates an activation center on the 2D graphene building blocks by modulating the spin density and charge distribution of the neighboring carbon atoms. The electrochemically active sites serve as additional molecular traps, ultimately rendering the bulk material a superior absorption capacity. It is for this reason that Zhao et al. (2012b) doped GAs with highly electronegative N atoms by a facile, one-pot hydrothermal method without any post-synthesis modification. Essentially, because of their highly porous and mechanically stable graphene skeleton, the as-prepared N-GAs could quickly absorb a wide range of oils and organic solvents. For instance, 300 mg of gasoline was absorbed by a 1.2 mg of N-GA block within 6 s, with an average absorption rate (41.7 g g^{-1} s^{-1}) many folds higher than conventional GA. Moreover, the expended N-GAs could be easily regenerated by direct combustion in air for repeated usage. In addition to single-element doping, codoping with two different elements may induce synergistic effects and generate even more powerful active regions on the graphene plane, bringing about further improvements in the absorption behavior. As outlined in Fig. 9, a B,N-codoped GA (BN-GA), prepared by a combined solvothermal-induced self-assembly and freeze-drying process, showed an excellent absorption capacity of up to 23 g g^{-1} for different classes of oils and organic solvents, and therefore merits widespread applications in prevailing oil pollution control technologies (Liu et al. 2017).

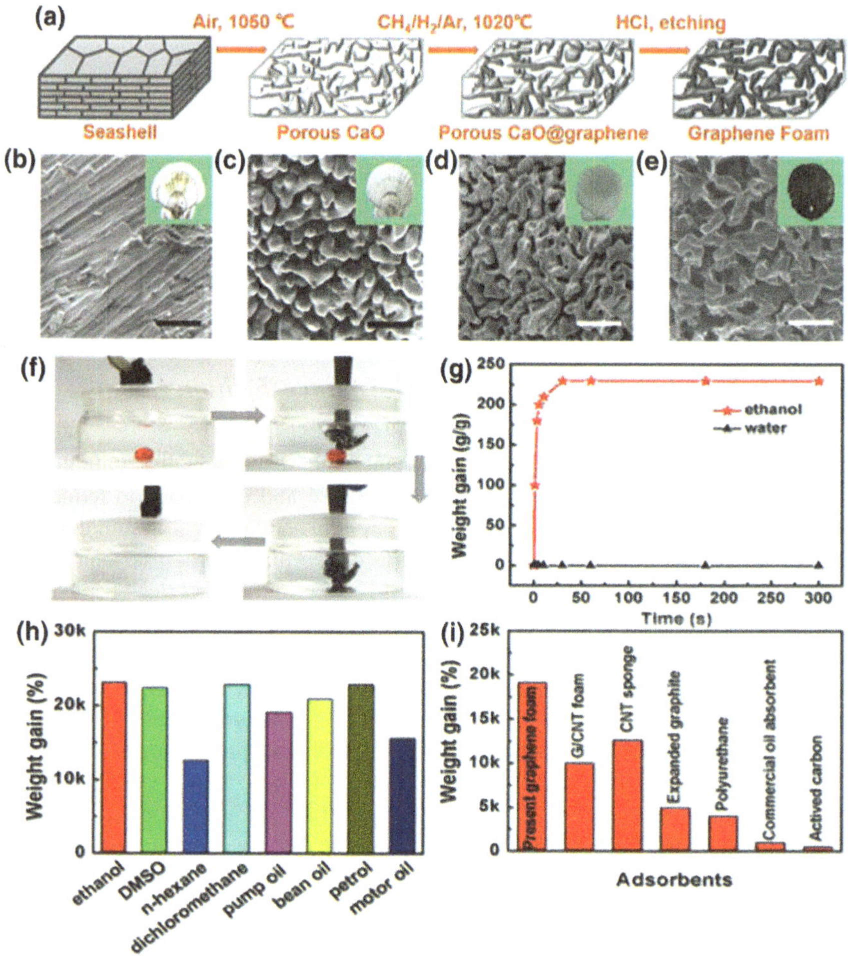

Fig. 8 **a** Scallop-based CVD mediated production of GF. SEM images of **b** pristine scallop , **c** scallop calcined at 1050 °C for 30 min in air, **d** graphene-coated calcined scallop after CVD growth at 1020 °C for 30 min and **e** GF after etching out of the calcium oxide framework. The scale bars represent 5 μm. Insets in (**b**)–(**e**) are the corresponding digital photographs of the samples. **f** Schematic of methylene chloride (dyed with Sudan III) absorption from water by the biomass-derived GF. Within 2 s, 133 mg of methylene chloride was taken up by the GF. **g** Ethanol and water absorption kinetics of the GF. **h** Organic solvents and oil absorption capacity of the GF. **i** Comparison of the pump oil absorption capacity of the GF with that of other competing carbon-based materials . Reproduced from Shi et al. (2016), Copyright 2016, with permission of the American Chemical Society

Alternatively, decoration with inorganic nanostructures and compounds, such as Cu_2O (Wu et al. 2013), Fe_2O_3 (Ye and Feng 2014; Qiu et al. 2015), Fe_3O_4 (Cong et al. 2012; Zhou et al. 2015; Subrati et al. 2017), FeOOH (Cong et al. 2012), etc., also appear to extend the absorption efficiency of 3D GMAs by enriching their

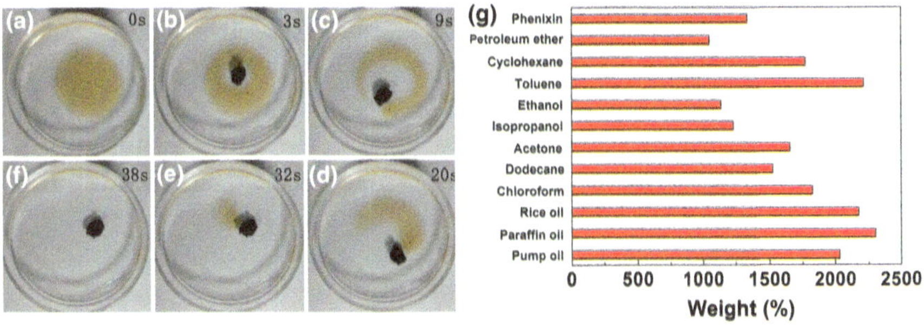

Fig. 9 **a–f** Rapid absorption of dodecane from water by BN-GA. The dodecane was dyed with Sudan red for clear observation. **g** Absorption capacity of BN-GA for various oils and organic solvents. Reproduced from Liu et al. (2017), Copyright 2017, with permission of the Royal Society of Chemistry

surface hydrophobicity. Notably, Cong et al. (2012) uniformly anchored magnetic Fe_3O_4 nanoparticles or α-FeOOH nanorods on the surface of GA through an in situ growth and self-assembly scheme. Owing to the excellent synergy between superhydrophobic π–π stacking and capillary effects, the functionalized graphene-based monoliths (i.e., Fe_3O_4@GA and FeOOH@GA) exhibited decent absorption abilities and high degrees of recyclability for several oils and organic solvents (including cyclohexane, toluene, gasoline, paraffin oil, vegetable oil and phenoxin), and could therefore be readily adopted for the cleanup of oil spills and industrial chemical leaks. Attractively, Zhou et al. (2015) fabricated an ultralight (5 mg cm^{-3}) and highly porous polystyrene/Fe_3O_4/graphene hybrid aerogels (CPFAs) by an environmentally friendly and inexpensive solvothermal approach (Fig. 10a). The effective deposition of porous Fe_3O_4 nanoparticles and polystyrene microspheres synergistically enhanced the hydrophobicity of the aerogel, which could be further upgraded by raising the magnetite concentration (Fig. 10b). Consequentially, the absorption capacity of the as-developed CPFAs for lubricating oil and crude oil were about 37–40 times their own weights after 10 cycles of oil/water separation (Fig. 10c), which is amongst the highest ever recorded for oil absorbents. Furthermore, the oil-soaked composite could be easily recovered from the aqueous media using an external magnetic field to squeeze out the absorbate before the next cycle (Fig. 10d). Meanwhile, Ye and Feng (2014) and Qiu et al. (2015) independently incorporated Fe_2O_3 nanoparticles into GAs to obtain hybrid aerogels possessing high hydrophobicity, good compressibility and strong magnetic properties. Similar to the aforementioned composite aerogel, the Fe_2O_3/GAs presented outstanding absorbance capabilities for a broad spectrum of organic solvents (50–130 times their own weight) (Ye and Feng 2014), and could be easily recycled by burning off the absorbed fluid (Qiu et al. 2015). In a more recent study, Hong et al. (2015) introduced fluorine-containing substituents into GAs using a simple solution–immersion process. By virtue of its superior physicochemical characteristics, including low density (bulk density of 14 mg cm^{-3}), high porosity

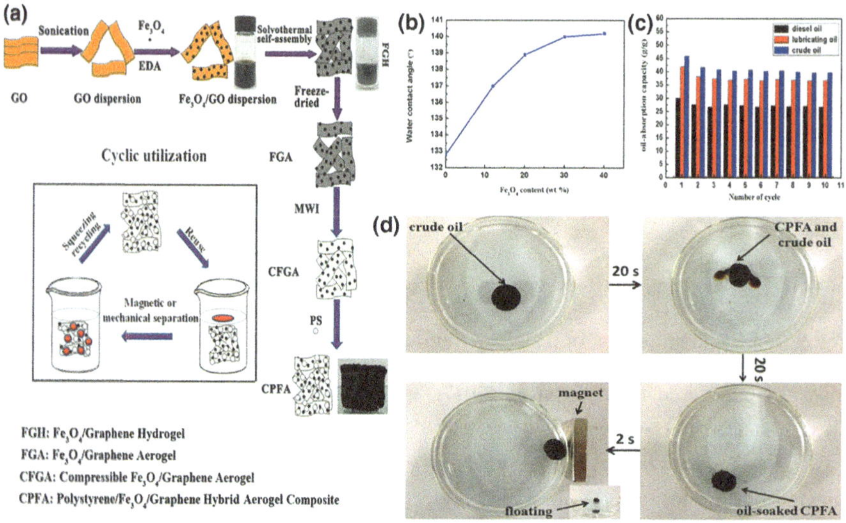

Fig. 10 **a** Illustration of the fabrication process of the CPFA and its cyclic utilization for removal of oil from water. **b** Wettability of CPFA as a function of Fe_3O_4 loading. **c** Cyclic oil absorption capacity of CPFA. **d** Removal of crude oil from water with CPFA under magnetic field. After absorption, the oil-soaked CPFA could be readily separated from the aqueous phase using a magnetic bar. Reproduced from Zhou et al. (2015), Copyright 2015, with permission of the American Chemical Society

(>87%), mechanical stability (supports at least 2600 times its own weight) and hydrophobicity (WCA of 144°), the fluorinated GA could not only eliminate spilled oils and other toxic organic solvents from aqueous environment (up to 11,200% of its own weight), but also displayed a remarkable regeneration capability with no obvious change in absorption efficiency after 10 cycles.

Additionally, functionalization of 3D GMAs with organic molecules or polymers can also significantly reform their absorption capabilities, as have been extensively reported in recent years (Li et al. 2012, 2013, 2014b; Cheng et al. 2013; Tao et al. 2014; Luo et al. 2016; Zhu et al. 2015a; Periasamy et al. 2017). The presence of aromatic functional moieties can substantially strengthen the intermolecular interactions between the randomly-oriented graphene constituents *via* the formation of chemical cross-links (Campbell et al. 2015). As a result, 3D graphene monoliths are obtained that feature mechanical stiffness and temperature resistance orders of magnitude higher than those that are typically held together only through physical interactions (e.g., van der Waals forces) (Worsley et al. 2010, 2012); making these materials viable candidates for continuous operation under rigorous reprocessing, including combustion, thermal treatment and mechanical extrusion (Wan et al. 2016). A representative illustration is the work of Tao et al. (2014), who covalently cross-linked GO sheets into a monolith by conjugating the oxygen functionalities on GO with the hydroxyl groups of polyvinyl alcohol through a dehydration reaction. The obtained GO hydrogel was then freeze-dried

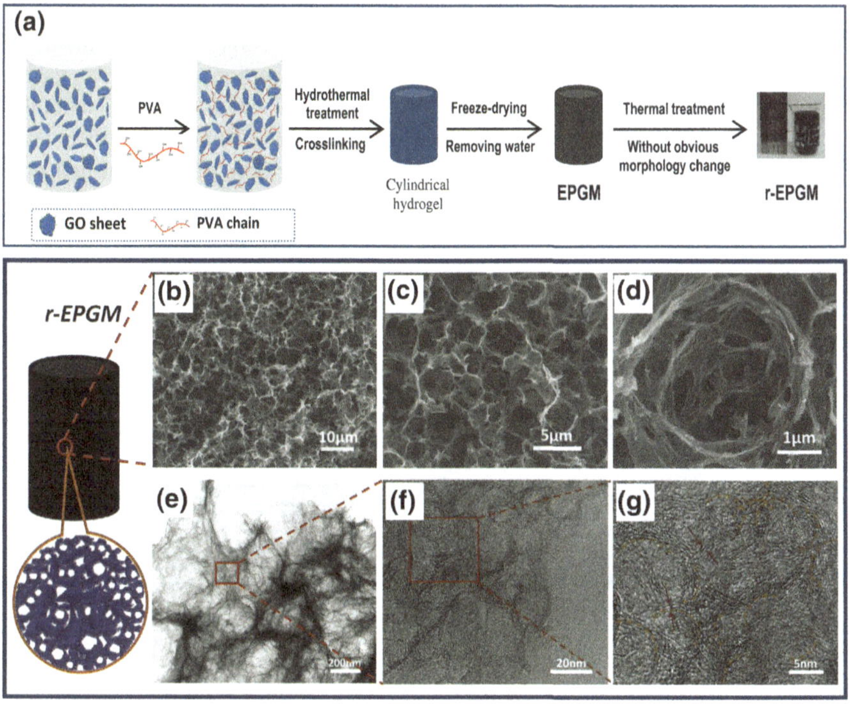

Fig. 11 **a** Protocol for the fabrication of r-EPGM. SEM **b–d** and transmission electron microscopy **e–g** images of r-EPGM. A schematic of the spheroidal and hierarchical pore structure of r-EPGM is shown on the left. Reproduced from Tao et al. (2014), Copyright 2013, with permission of Elsevier

and converted into reduced GO (rGO) aerogel by thermal treatment (Fig. 11a). This robust graphene-based hydrophobic porous macroassembly (r-EPGM) could rapidly remove significant amounts of both oil and petroleum hydrocarbons from water because of its unique spheroidal and trimodal (micro-meso-macro) pore structure (Fig. 11b). In another exciting study, Zhu et al. (2015a) grafted an amphiphilic copolymer comprising a block of poly(2-vinylpyridine) and polyhexadecyl acrylate onto GF, resulting in an advanced macroscale material with switchable superoleophilicity (alkaline) and superoleophobicity (acidic) in aqueous media, which they referred to as 'smart GF' (ss-GF). The as-developed ss-GF effectively absorbed oils and organic solvents from aqueous media because of its superoleophilic surface at a neutral pH of 7. Upon decreasing the pH to 3, the ss-GF underwent protonation and completely released the absorbate molecules for successful reusage over multiple cycles. In a similar way, a host of different organic compounds, such as triisocyanate (Li et al. 2012), polydopamine (Cheng et al. 2013), polypyrrole (Li et al. 2013), polyvinylidene fluoride (Li et al. 2014b), polyethylenimine (Periasamy et al. 2017) and tannic acid (Luo et al. 2016), have been employed as cross-linkers to maximize the application potential of 3D graphene architectures for practical oil/water separation in complex environments.

Apart from the aforementioned techniques, the insertion of different carbon nanoarchitectures, including porous carbons (Zhu et al. 2016), carbon nanotubes (CNTs) (Dong et al. 2012; Sun et al. 2013; Hu et al. 2014), carbon nanofibers (Wan et al. 2015) and carbon cryogels (Wei et al. 2013), have been lately identified as an equally promising strategy to expand the oil absorption profile of 3D GMAs. The intercalation of carbon materials with different geometries is believed to enlarge the basal spacing between graphene sheets, leading to a more accessible surface area

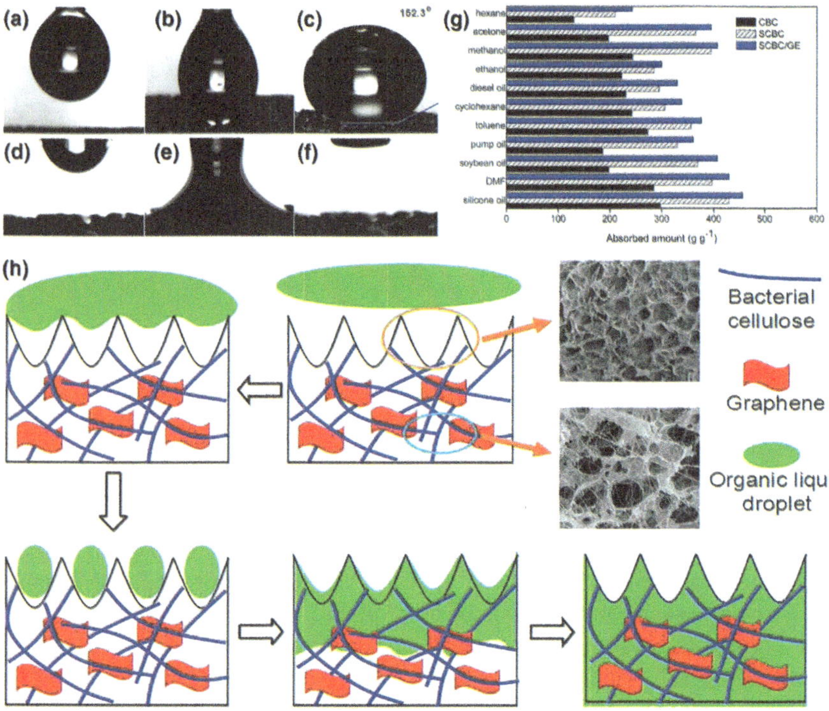

Fig. 12 Video snapshots of the wetting behavior of **a–c** water and **d–f** compressor oil on the as-prepared CNT@GF. Reproduced from Dong et al. (2012), Copyright 2012, with permission of the Royal Society of Chemistry. **g** Absorption capacities of carbonized bacterial cellulose nanofibers (CBC), SCBC sphere-like carbonized bacterial cellulose (SCBC) and SCBC/GE aerogel for various organic liquids. **h** Plausible mechanism of organic liquid absorption onto the SCBC/GE monolith. The 3D interconnected honeycomb-like open cavities and ridges of SCBC/GE improved the absorption capacity by increasing the specific surface area and providing additional space for effective diffusion of organic liquids. In addition, the ridges reduced the molecular size of the hydrocarbons by trimming the chain length, thus facilitating their interaction with the hydrophobic absorbent. Meanwhile, the cavities trapped the absorbates and redirected them to the underlying porous channels. Reproduced from Wan et al. (2015), Copyright 2015, with permission of the Royal Society of Chemistry

and additional pore channels (Chowdhury and Balasubramanian 2017). This in turn allows a greater penetration of organic liquids, eventually resulting in their increased uptake. For example, Dong et al. (2012) constructed a CNT@GF through the direct growth of CNT forests on GF using the CVD method. Due to its intense macroporosity, surface nanoroughness, nanoscopic voids and superhydrophobicity (WCA = 152°), the superoleophilic hybrid foam could selectively remove oils and organic solvents from the surface of water (Fig. 12a–f). These findings were later corroborated by coating the porous interconnected networks of GA with vertically aligned multiwall CNTs (Hu et al. 2014). More interestingly, Wan et al. (2015) manufactured a novel, all-carbon, spherical, hybrid aerogel through *in situ* assembly of sphere-shaped carbonized bacterial cellulose (SCBC) nanofibers and graphene (GE) nanosheets *via* a facile and scalable one-pot biosynthetic pathway, followed by carbonization under an inert atmosphere. The free-standing SCBC/GE aerogel had a high porosity (>98%), large specific surface area (514 m^2 g^{-1}), strong hydrophobicity and good elasticity, thereby demonstrating superhigh water-repelling and oil-absorbing abilities, with a maximum potential of up to 243–457 times its own dry weight (Fig. 12g). This extraordinary performance was attributed to the unique, 3D hierarchically interconnected, honeycomb-like open structure of the SCBC/GE monolith, as explained in Fig. 12h.

3 Summary and Outlook

The development of robust materials with high absorption capacity and excellent oil/water selectivity are of utmost importance to tackle the multifaceted environmental challenges arising from frequent oil spill events. As a versatile new class of porous carbon materials, self-assembled 3D graphene macrostructures, such as hydrogels, aerogels, foams, sponges, etc., possess immense potential as advanced absorbents in oil spill response and cleanup operations. Their intrinsic hydrophobicity, seamlessly interconnected porous channels and superior graphene building blocks endow a plethora of exciting features that make them extremely suitable for absorbing oils and petroleum-based liquid hydrocarbons, with maximum capacities several orders of magnitude higher than that of the peers. Moreover, attributing to their intense porosity, they can serve as ideal scaffolds for functionalization with heteroatoms, functional polymers, inorganic nanostructures, as well as a whole range of topologically different carbon architectures, leading to the conceptualization of new material systems with superlative properties and novel functionalities, thereby outspreading their application potential. However, in spite of this, a number of fundamental issues and technological bottlenecks persist that require immediate attention to make even more significant progress in this exciting new research field of topical interest. First, a comprehensive understanding of the interactions of graphene sheets at nanoscales to assemble into hierarchical 3D macroscopic structures is still at a nascent stage. Consequently, the precise architecture of these graphene macroassemblies remains largely unknown, precluding the ability to tailor their internal path length for greater molecular diffusion and mobility. Supramolecular engineering (i.e., the design, synthesis and self-assembly of molecular modules with pre-programmed conformations and custom-made properties) of such 3D graphene macroarchitectures may pose to be an exciting and promising new research avenue to the development of functional graphene-based macroscale materials with optimized structures and tailored morphologies *via* a controllable and scalable self-assembly approach (Ciesielski et al. 2010). This advancement should significantly improve the mass transfer rates and mechanical properties of 3D GMAs and assist their integration into practical oil spill control technologies. Second, even though multiple strategies exist for synthesis of 3D GMAs, including hydrothermal/solvothermal reduction, chemical reduction, electrochemical reduction, CVD, etc., it is still a challenge to produce them on a commercial scale. Hence, a more concerted research effort should be invested to explore cost-effective, sustainable and environmentally benign techniques for the mass production of 3D GMAs, which would in turn aid and accelerate their widespread adoption and deployment. Last but not the least, since liquid infiltration and defiltration is predominantly affected by the pore size, with larger pores facilitating hydrocarbon migration and fluid distribution throughout the bulk, the synthesis of macroporous graphene monoliths with a uniform pore size distribution

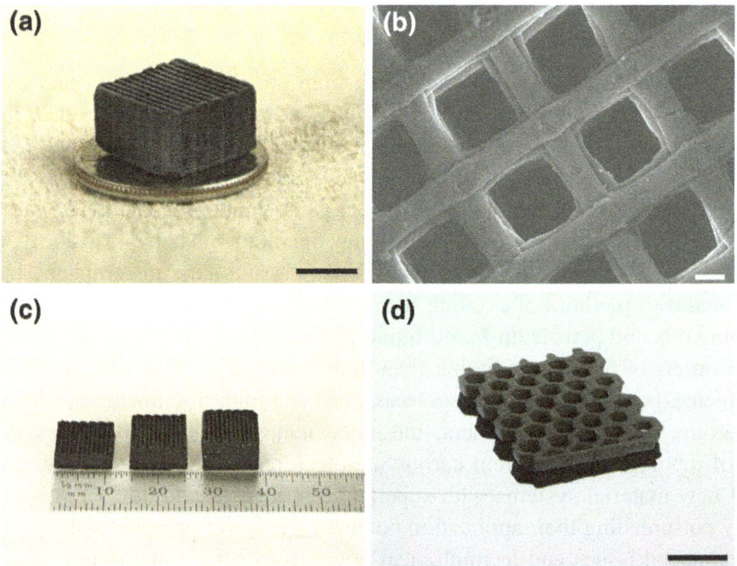

Fig. 13 **a** Digital picture and **b** SEM image of a 3D printed GA. Optical image of **c** 3D printed GAs with varying thickness and **d** a 3D printed GA honeycomb. Adapting the 3D printing technique to GAs realizes the possibility of fabricating a broad class of 3D macroscopic GAs of predetermined geometries and engineered porosity for predictable fluid flow characteristics. Reproduced from Zhu et al. (2015b), Copyright 2015, with permission of Nature Publishing Group

is highly desirable. In this context, a promising avenue could be direct 3D printing of graphene macroassemblies with predesigned macroporosity and permeability (Fig. 13). In comparison to the contemporary 3D graphene scaffolds with randomly distributed pores, we believe the well-ordered macroporous structure would significantly improve the mass transfer rate and reduce the diffusion resistance, thereby enhancing the liquid infiltration speed as well as the absorption efficiency. We anticipate that the ongoing interdisciplinary and integrated research initiatives by the global scientific community will resolve these issues, paving the way for a technological breakthrough to effectively manage the environmental and economic consequences of a major oil spill in the foreseeable future.

Acknowledgements SC and RB gratefully acknowledge the financial support provided by the National Research Foundation (NRF), Prime Minister's Office, Singapore through the Centre for Advanced 2D Materials (CA2DM) at the National University of Singapore to carry out this work.

References

Aguilera F, Méndez J, Pásaro E, Laffon B (2010) Review on the effects of exposure to spilled oils on human health. J Appl Toxicol 30:291–301. https://doi.org/10.1002/jat.1521

Al-Majed AA, Adebayo AR, Hossain ME (2012) A sustainable approach to controlling oil spills. J Environ Manage 113:213–227. https://doi.org/10.1016/j.jenvman.2012.07.034

Balasubramanian R, Chowdhury S (2015) Recent advances and progress in the development of graphene-based adsorbents for CO_2 capture. J Mater Chem A 3:21968–21989. https://doi.org/10.1039/c5ta04822b

Beyer J, Trannum HC, Bakke T, Hodson PV, Collier TK (2016) Environmental effects of the Deepwater Horizon oil spill: a review. Marine Poll Bull 110:28–51. https://doi.org/10.1016/j.marpolbul.2016.06.027

Bi H, Yin K, Xie X, Zhou Y, Wan N, Xu F, Banhart F, Sun L, Ruoff RS (2012a) Low temperature casting of graphene with high compressive strength. Adv Mater 24:5124–5129. https://doi.org/10.1002/adma.201201519

Bi H, Xie X, Yin K, Zhou Y, Wan S, He L, Xu F, Banhart F, Sun L, Ruoff RS (2012b) Spongy graphene as a highly efficient and recyclable sorbent for oils and organic solvents. Adv Funct Mater 22:4421–4425. https://doi.org/10.1002/adfm.201200888

Bi H, Xie X, Yin K, Zhou Y, Wan S, Ruoff RS, Sun L (2014) Highly enhanced performance of spongy graphene as an oil sorbent. J Mater Chem A 2:1652–1656. https://doi.org/10.1039/c3ta14112h

Broje V, Keller AA (2006) Improved mechanical oil spill recovery using an optimized geometry for the skimmer surface. Environ Sci Technol 40:7914–7918. https://doi.org/10.1021/es061842m

Buist I, McCourt J, Potter S, Ross S, Trudel K (1999) In situ burning. Pure Appl Chem 71:43–65. https://doi.org/10.1351/pac199971010043

Campbell PG, Worsley MA, Hiszpanski AM, Baumann TF, Biener J (2015) Synthesis and functionalization of 3D nano-graphene materials: graphene aerogels and graphene macro assemblies. J Vis Exp 105:e53235. https://doi.org/10.3791/53235

Chen Z, Ren W, Gao L, Liu B, Pei S, Chen H-M (2011) Three-dimensional flexible and conductive interconnected graphene networks grown by chemical vapor deposition. Nat Mater 10:424–428. https://doi.org/10.1038/nmat3001

Cheng C, Li S, Zhao J, Li X, Liu Z, Ma L, Zhang X, Sun S, Zhao C (2013) Biomimetic assembly of polydopamine-layer on graphene: mechanisms, versatile 2D and 3D architectures and pollutant disposal. Chem Eng J 228:468–481. https://doi.org/10.1016/j.cej.2013.05.019

Cheng Y, Zhou S, Hu P, Zhao G, Li Y, Zhang X, Han W (2017) Enhanced mechanical, thermal, and electric properties of graphene aerogels via supercritical ethanol drying and high-temperature thermal reduction. Sci Rep 7:1439. https://doi.org/10.1038/s41598-017-01601-x

Chowdhury S, Balasubramanian R (2014a) Recent advances in the use of graphene-family nanoadsorbents for removal of toxic pollutants from wastewater. Adv Colloid Interface Sci 204:35–56. https://doi.org/10.1016/j.cis.2013.12.005

Chowdhury S, Balasubramanian R (2014b) Graphene/semiconductor nanocomposites (GSNs) for heterogeneous photocatalytic decolorization of wastewaters contaminated with synthetic dyes: a review. Appl Catal B 160–161:307–324. https://doi.org/10.1016/j.apcatb.2014.05.035

Chowdhury S, Balasubramanian R (2016) Highly efficient, rapid and selective CO_2 capture by thermally treated graphene nanosheets. J CO_2 Util 13:50–60. https://doi.org/10.1016/j.jcou.2015.12.001

Chowdhury S, Balasubramanian R (2017) Three-dimensional graphene-based macrostructures for sustainable energy applications and climate change mitigation. Prog Mater Sci 90:224–275. https://doi.org/10.1016/j.pmatsci.2017.07.001

Ciesielski A, Palma C-A, Bonini M, Samori P (2010) Towards supramolecular engineering of functional nanomaterials: preprogramming multi-component 2D self-assembly at solid-liquid interfaces. Adv Mater 22:3506–3520. https://doi.org/10.1002/adma.201001582

Cong H-P, Ren X-C, Wang P, Yu S-H (2012) Macroscopic multifunctional graphene-based hydrogels and aerogels by a metal ion induced self-assembly process. ACS Nano 6:2693–2703. https://doi.org/10.1021/nn300082k

Cong H-P, Chen J-F, Yu S-H (2014) Graphene-based macroscopic assemblies and architectures: an emerging material system. Chem Soc Rev 43:7295–7325. https://doi.org/10.1039/c4cs00181h

Dong X, Chen J, Ma Y, Wang J, Chan-Park MB, Liu X, Wang L, Huang W, Chen P (2012) Superhydrophobic and superoleophilic hybrid foam of graphene and carbon nanotube for selective removal of oils or organic solvents from the surface of water. Chem Commun 48:10660–10662. https://doi.org/10.1039/c2cc35844a

Du X, Liu H-Y, Mai Y-W (2016) Ultrafast synthesis of multifunctional N-doped graphene foam in an ethanol flame. ACS Nano 10:453–462. https://doi.org/10.1021/acsnano.5b05373

El-Kady MF, Shao Y, Kaner RB (2016) Graphene for batteries, supercapacitors and beyond. Nat Rev Mater 1:16033. https://doi.org/10.1038/natrevmats.2016.33

Fan X, Chen X, Dai L (2015) 3D graphene based materials for energy storage. Curr Opin Colloid Interface Sci 20:429–438. https://doi.org/10.1016/j.cocis.2015.11.005

Ge J, Shi L-A, Wang Y-C, Zhao H-Y, Yao H-B, Zhu Y-B, Zhang Y, Zhu H-W, Wu H-A, Yu S-H (2017) Joule-heated graphene-wrapped sponge enables fast clean-up of viscous crude-oil spill. Nat Nanotechnol 12:434–440. https://doi.org/10.1038/nnano.2017.33

Gorgolis G, Galiotis C (2017) Graphene aerogels: a review. 2D Mater 4:032001. https://doi.org/10.1088/2053-1583/aa7883

Guo X, Bi H, Zafar A, Liang Z, Shi Z, Sun L, Ni Z (2016) Investigation of dodecane in three-dimensional porous graphene sponge by Raman mapping. 27:055702. https://doi.org/10.1088/0957-4484/27/5/055702

Gupta S, Tai N-H (2016) Carbon materials as oil sorbents: a review on the synthesis and performance. J Mater Chem A 4:1550–1565. https://doi.org/10.1039/c5ta08321d

He Y, Liu Y, Wu T, Ma J, Wang X, Gong X, Kong W, Xing F, Liu Y, Gao J (2013) An environmentally friendly method for the fabrication of reduced graphene oxide foam with a super oil absorption capacity. J Hazard Mater 260:796–805. https://doi.org/10.1016/j.jhazmat.2013.06.042

Hong J-Y, Sohn E-H, Park S, Park HS (2015) Highly-efficient and recyclable oil absorbing performance of functionalized graphene aerogel. Chem Eng J 269:229–235. https://doi.org/10.1016/j.cej.2015.01.066

Hu H, Zhao Z, Gogotsi Y, Qiu J (2014) Compressible carbon nanotube-graphene hybrid aerogels with superhydrophobicity and superoleophilicity for oil sorption. Environ Sci Technol Lett 1:214–220. https://doi.org/10.1021/ez500021w

Kujawinski EB, Kido Soule MC, Valentine DL, Boysen AK, Longnecker K, Redmond MC (2011) Fate of dispersants associated with the Deepwater Horizon oil spill. Environ Sci Technol 45:1298–1306. https://doi.org/10.1021/es103838p

Kumari B, Singh DP (2016) A review on multifaceted application of nanoparticles in the field of bioremediation of petroleum hydrocarbons. Ecol Eng 97:98–105. https://doi.org/10.1016/j.ecoleng.2016.08.006

Li J, Wang F, C-y Liu (2012) Tri-isocyanate reinforced graphene aerogel and its use for crude oil adsorption. J Colloid Interface Sci 382:13–16. https://doi.org/10.1016/j.jcis.2012.05.040

Li H, Liu L, Yang F (2013) Covalent assembly of 3D graphene/polypyrrole foams for oil spill cleanup. J Mater Chem A 1:3446–3453. https://doi.org/10.1039/c3ta00166k

Li J, Li J, Meng H, Xie S, Zhang B, Li L, Ma H, Zhang J, Yu M (2014a) Ultra-light, compressible and fire-resistant graphene aerogel as a highly efficient and recyclable absorbent for organic liquids. J Mater Chem A 2:2934–2941. https://doi.org/10.1039/c3ta14725h

Li R, Chen C, Li J, Xu L, Xiao G, Yan D (2014b) A facile approach to superhydrophobic and superoleophilic graphene/polymer aerogels. J Mater Chem A 2:3057–3064. https://doi.org/10.1039/c3ta14262k

Liu T, Huang M, Li X, Wang C, Gui C-X, Yu Z-Z (2016) Highly compressible anisotropic graphene aerogels fabricated by directional freezing for efficient absorption of organic liquids. Carbon 100:456–464. https://doi.org/10.1016/j.carbon.2016.01.038

Liu Y, Xiang M, Hong L (2017) Three-dimensional nitrogen and boron codoped graphene for carbon dioxide and oils adsorption. RSC Adv 7:6467–6473. https://doi.org/10.1039/c6ra22297h

Luo J, Lai J, Zhang N, Liu Y, Liu R, Liu X (2016) Tannic acid induced self-assembly of three-dimensional graphene with good adsorption and antibacterial properties. ACS Sustain Chem Eng 4:1404–1413. https://doi.org/10.1021/acssuschemeng.5b01407

Ma Y, Chen Y (2015) Three-dimensional graphene networks: synthesis, properties and applications. Nat Sci Rev 2:40–53. https://doi.org/10.1093/nsr/nwu072

Muttin F (2008) Structural analysis of oil spill containment booms in coastal and estuary waters. Appl Ocean Res 30:107–112. https://doi.org/10.1016/j.apor.2008.07.001

Niu Z, Chen J, Hng HH, Ma J, Chen X (2012) A leavening strategy to prepare reduced graphene oxide foams. Adv Mat 24:4144–4150. https://doi.org/10.1002/adma.201200197

Periasamy AP, Wu W-P, Ravindranath R, Roy P, Lin G-L, Chang H-T (2017) Polymer/reduced graphene oxide functionalized sponges as superabsorbents for oil removal and recovery. Marine Pol Bull 114:888–895. https://doi.org/10.1016/j.marpolbul.2016.11.005

Qian Y, Ismail IM, Stein A (2013) Ultralight, high-surface-area, multifunctional graphene-based aerogels from self-assembly of graphene oxide and resol. Carbon 68:221–231. https://doi.org/10.1016/j.carbon.2013.10.082

Qiu B, Xing M, Zhang J (2015) Stöber-like method to synthesize ultralight, porous, stretchable Fe_2O_3/graphene aerogels for excellent performance in photo-Fenton reaction and electrochemical capacitors. J Mater Chem A 3:12820–12827. https://doi.org/10.1039/c5ta02675j

Ren H, Shi X, Zhu J, Zhang Y, Bi Y, Zhang L (2016) Facile synthesis of N-doped graphene aerogel and its application for organic solvent adsorption. J Mater Sci 51:6419–6427. https://doi.org/10.1007/s10853-016-9939-y

Ren R-P, Li W, Lv Y-K (2017) A robust, superhydrophobic graphene aerogel as a recyclable sorbent for oils and organic solvents at various temperatures. J Colloid Interface Sci 500:63–68. https://doi.org/10.1016/j.jcis.2017.01.071

Ron EZ, Rosenberg E (2014) Enhanced bioremediation of oil spills in the sea. Curr Opin Biotechnol 27:191–194. https://doi.org/10.1016/j.copbio.2014.02.004

Shen Y, Fang Q, Chen B (2015) Environmental applications of three-dimensional graphene-based macrostructures: adsorption, transformation, and detection. Environ Sci Technol 49:67–84. https://doi.org/10.1021/es504421y

Shi L, Chen K, Du R, Bachmatiuk A, Rümmeli MH, Xie K, Huang Y, Zhang Y, Liu Z (2016) Scalable seashell-based chemical vapor deposition growth of three-dimensional graphene foams for oil-water separation. J Am Chem Soc 138:6360–6363. https://doi.org/10.1021/jacs.6b02262

Subrati A, Mondal S, Ali M, Alhindi A, Ghazi R, Abdala A, Reinalda D, Alhasaan S (2017) Developing hydrophobic graphene foam for oil spill cleanup. Ind Eng Chem Res 56:6945–6951. https://doi.org/10.1021/acs.iecr.7b00716

Sun H, Xu Z, Gao C (2013) Multifunctional, ultra-flyweight, synergistically assembled carbon aerogels. Adv Mater 25:2554–2560. https://doi.org/10.1002/adma.201204576

Tao Y, Kong D, Zhang C, Lv W, Wang M, Li B, Huang Z-H, Kang F, Yang Q-H (2014) Monolithic carbons with spheroidal and hierarchical pores produced by the linkage of functionalized graphene sheets. Carbon 69:169–177. https://doi.org/10.1016/j.carbon.2013.12.003

Vidyasagar A, Handore K, Suresham KM (2011) Soft optical devices from self-healing gels formed by oil and sugar-based organogelators. Angew Chem Int Ed 50:8021–8024. https://doi.org/10.1002/anie.201103584

Wahi R, Chuah LA, Choong TSY, Ngaini Z, Nourouzi MM (2013) Oil removal from aqueous state by natural fibrous sorbent: an overview. Sep Purif Technol 113:51–63. https://doi.org/10.1016/j.seppur.2013.04.015

Wan Y, Zhang F, Li Cm Xiong G, Zhu Y, Luo H (2015) Facile and scalable production of three-dimensional spherical carbonized bacterial cellulose/graphene nanocomposites with a honeycomb-like surface pattern as potential superior absorbents. J Mater Chem A 3:24389–24396. https://doi.org/10.1039/c5ta07464a

Wan W, Lin Y, Prakash A, Zhou Y (2016) Three-dimensional carbon-based architectures for oil remediation: from synthesis and modification to functionalization. J Mater Chem A 4:18687–18705. https://doi.org/10.1039/c6ta07211a

Wang J, Shi Z, Fan J, Ge Y, Yin J, Hu G (2012) Self-assembly of graphene into three-dimensional structures promoted by natural phenolic acids. J Mater Chem 22:22459–22466. https://doi.org/10.1039/c2jm35024f

Wei G, Miao Y-E, Zhang C, Yang Z, Liu Z, Tiju WW, Liu T (2013) Ni-doped graphene/carbon cryogels and their applications as versatile sorbents for water purification. ACS Appl Mater Interfaces 5:7584–7591. https://doi.org/10.1021/am401887g

Worsley MA, Pauzauskie PJ, Olson TY, Biener J, Satcher JH Jr, Baumann TF (2010) Synthesis of graphene aerogel with high electrical conductivity. J Am Chem Soc 132:14067–14069. https://doi.org/10.1021/ja1072299

Worsley MA, Kucheyev SO, Mason HE, Merrill MD, Mayer BP, Lewicki J, Valdez CA, Suss ME, Stadermann M, Pauzauskie PJ, Satcher JH, Biener J, Baumann TF (2012) Mechanically robust 3D graphene macroassembly with high surface area. Chem Commun 48:8428–8430. https://doi.org/10.1039/c2cc33979j

Wu Z-S, Winter A, Chen L, Sun Y, Turchanin A, Feng X, Müllen K (2012) Three-dimensional nitrogen and boron co-doped graphene for high-performance all-solid-state supercapacitors. Adv Mater 24:5130–5135. https://doi.org/10.1002/adma.201201948

Wu T, Chen M, Zhang L, Xu X, Liu Y, Yan J, Wang W, Gao J (2013) Three-dimensional graphene-based aerogels prepared by a self-assembly process and its excellent catalytic and absorbing performance. J Mater Chem A 1:7612–7621. https://doi.org/10.1039/c3ta10989e

Wu Y, Yi N, Huang L, Zhang T, Fang S, Chang H, Li N, Oh J, Lee JA, Kozlov M, Chipara AC, Terrones H, Xiao P, Long G, Huang Y, Zhang F, Zhang L, Lepró X, Haines C, Lima MD, Lopez NP, Rajukumar LP, Elias AL, Feng S, Kim SJ, Narayanan NT, Ajayan PM, Terrones M, Aliev A, Chu P, Zhang Z, Baughman RH, Chen Y (2015) Three-dimensionally bonded spongy graphene material with super compressive elasticity and near-zero Poisson's ratio. Nat Commun 6:6141. https://doi.org/10.1038/ncomms7141

Wu R, Yu B, Liu X, Li H, Wang W, Chen L, Bai Y, Ming Z, Yang S-T (2016) One-pot hydrothermal preparation of graphene sponge for the removal of oils and organic solvents. Appl Surf Sci 32:56–62. https://doi.org/10.1016/j.apsusc.2015.11.215

Xie X, Zhou Y, Bi H, Yin K, Wan S, Sun L (2013) Large-range control of the microstructures and properties of three-dimensional porous graphene. Sci Rep 3:2117. https://doi.org/10.1038/srep02117

Xu Y, Sheng K, Li C, Shi G (2010) Self-assembled graphene hydrogel via a one-step hydrothermal process. ACS Nano 4:4324–4330. https://doi.org/10.1021/nn101187z

Xu Y, Shi G, Duan X (2015a) Self-assembled three-dimensional graphene macrostructures: synthesis and applications in supercapacitors. Acc Chem Res 48:1666–1675. https://doi.org/10.1021/acs.accounts.5b00117

Xu L, Xiao G, Chen C, Li R, Mai Y, Sun G, Yan D (2015b) Superhydrophobic and superoleophilic graphene aerogel prepared by facile chemical reduction. J Mater Chem A 3:7498–7504. https://doi.org/10.1039/c5ta00383k

Yan Z, Yao W, Hu L, Liu D, Wang C, Lee C-S (2015) Progress in the preparation and application of three-dimensional graphene-based porous nanocomposites. Nanoscale 7:5563–5577. https://doi.org/10.1039/c5nr00030k

Yang S, Chen L, Mu L, Hao B, Chen J, Ma P-C (2016a) Graphene foam with hierarchical structures for the removal of organic pollutants from water. RSC Adv 6:4889–4898. https://doi.org/10.1039/c5ra23820j

Yang W, Gao H, Zhao Y, Bi K, Li X (2016b) Facile preparation of nitrogen-doped graphene sponge as a highly efficient oil absorption material. Mater Lett 178:95–99

Yao X, Zhao Y (2017) Three-dimensional porous graphene networks and hybrids for lithium-ion batteries and supercapacitors. Chem 2:171–200. https://doi.org/10.1016/j.chempr.2017.01.010

Ye S, Feng J (2014) Towards three-dimensional, multi-functional graphene-based nanocomposite aerogels by hydrophobicity-driven absorption. J Mater Chem A 2:10365–10369. https://doi.org/10.1039/c4ta01392a

Yin S, Niu Z, Chen X (2012) Assembly of graphene sheets into 3D macroscopic structures. Small 8:2458–2463. https://doi.org/10.1002/smll.201102614

Zhang R, Cao Y, Li P, Zang X, Sun P, Wang K, Zhong M, Wei J, Wu D, Kang F, Zhu H (2014) Three-dimensional porous graphene sponges assembled with the combination of surfactant and freeze-drying. Nano Res 7:1477–1487. https://doi.org/10.1007/s12274-014-0508-x

Zhang B, Zhang J, Sang X, Liu C, Luo T, Peng L, Han B, Tan X, Ma Wang D, Zhao N (2016a) Cellular graphene aerogel combines ultralow weight and high mechanical strength: a highly efficient reactor for catalytic hydrogenation. Sci Rep 6:25830. https://doi.org/10.1038/srep25830

Zhang W, Sun Y, Liu T, Li D, Hou C, Gao L, Liu Y (2016b) Preparation of graphene foam with high performance by modified self-assembly method. Appl Phys A 122:259. https://doi.org/10.1007/s00339-016-9684-8

Zhang L, Li H, Lai X, Su X, Liang T, Zeng X (2017a) Thiolated graphene-based superhydrophobic sponges for oil-water separation. Chem Eng J 316:736–743. https://doi.org/10.1016/j.cej.2017.02.030

Zhang X, Liu D, Ma Y, Nie J, Sui G (2017b) Super-hydrophobic graphene coated polyurethane (GN@PU) sponge with great oil-water separation performance. Appl Surf Sci 422:116–124. https://doi.org/10.1016/j.apsusc.2017.06.009

Zhao J, Ren W, Chen H-M (2012a) Graphene sponge for efficient and repeatable adsorption and desorption of water contaminants. J Mater Chem 22:20197–20202. https://doi.org/10.1039/c2jm34128j

Zhao Y, Hu C, Hu Y, Cheng H, Shi G, Qu L (2012b) A versatile, ultralight, nitrogen-doped graphene framework. Angew Chem Int Ed 51:11371–11375. https://doi.org/10.1002/anie.201206554

Zhou Z, Jiang W, Wang T, Lu Y (2015) Highly hydrophobic, compressible, and magnetic polystyrene/Fe_3O_4/graphene aerogel composite for oil-water separation. Ind Eng Chem Res 54:5460–5467. https://doi.org/10.1021/acs.iecr.5b00296

Zhu H, Chen D, Li N, Xu Q, Li H, He J, Lu J (2015a) Graphene foam with switchable oil wettability for oil and organic solvents recovery. Adv Funct Mater 25:597–605. https://doi.org/10.1002/adfm.201403864

Zhu C, Han TY, Duoss EB, Golobic AM, Kuntz JD, Spadaccini CM, Worsley MA (2015b) Highly compressible 3D periodic graphene aerogel microlattices. Nat Commun 6:6962. https://doi.org/10.1038/ncomms7962

Zhu J, Yang X, Fu Z, He J, Wang C, Wu W, Zhang L (2016) Three-dimensional macroassembly of sandwich-like, hierarchical, porous carbon/graphene nanosheets towards ultralight, superhigh surface area, multifunctional aerogels. Chem Eur J 22:2515–2524. https://doi.org/10.1002/chem.201504309

Chapter 4
Green and Ecofriendly Materials for the Remediation of Inorganic and Organic Pollutants in Water

Tetiana Tatarchuk, Mohamed Bououdina, Basma Al-Najar and Rajesh Babu Bitra

Abstract The widespread of organic and inorganic pollutants in wastewater from various industries, are responsible for serious environmental problems meanwhile represent a danger for human being. Therefore, the search of cost-effective methods of wastewater treatment containing in particular heavy metals and dyes, become of great importance. Noteworthy, adsorption has proven to be most effective technology for purification of wastewater from organic and inorganic pollutants. In this review, different types of green and ecofriendly materials (biosorbents, graphene-based composites, metal oxides, etc.) for dyes and heavy metals adsorption will be discussed. The biosorbents such as agricultural waste materials (waste seeds, orange peel, exhausted coffee ground powder, wood apple shell, sweet potato peels, wheat straws, etc.), activated carbon prepared from different types of agricultural waste (coconut husk, forest and wood-processing residues, papaya seeds, magnetic biochar etc.), graphene-based adsorbents and their derivatives, obtained by eco-friendly green synthesis, have been discussed and their adsorption activity has been described in details.

T. Tatarchuk (✉)
Educational and Scientific Center of Materials Science and Nanotechnology,
Vasyl Stefanyk Precarpathian National University, 57, Shevchenko Str.,
Ivano-Frankivsk 76018, Ukraine
e-mail: tatarchuk.tetyana@gmail.com

M. Bououdina · B. Al-Najar
Department of Physics, College of Science, University of Bahrain,
Sakheer PO Box 32038 Bahrain
e-mail: mboudina@gmail.com

B. Al-Najar
e-mail: balnajar85@gmail.com

R. B. Bitra
Department of Physics, G.V.P. College of Engineering for Women,
Visakhapatnam 530041, Andhra Pradesh, India
e-mail: rajeshbabu.bitra@gmail.com

M. Naushad (ed.), *A New Generation Material Graphene: Applications in Water Technology*, https://doi.org/10.1007/978-3-319-75484-0_4

Keywords Adsorption · Biosorbent · Heavy metal · Dye · Green synthesis
Agricultural waste

1 Introduction

Green and eco-friendly synthesis of nanoparticles (NPs) have emerged as an efficient and environmental friendly route which makes use of naturally formed materials such as plants (Li et al. 2017a, b; Al-Shehri et al. 2017) and agricultural waste (Adebisi et al. 2017; Vaibhav et al. 2015). This approach provides an easy, cheap and highly sustainable method for the preparation of various NPs. Substances formed in nature were proved to be a good alternative to other hazardous chemical compounds. Studies showed that many plants can be a perfect alternative of chemical reducing agents. Surface reduction process is very essential in the synthesis of oxides and graphene based NPs. Hence, substituting these chemical surface moderators by plants is much safer resulting degradable outcomes that keeps environment clean. It is very interesting that the amount of green substance could directly affect the morphology, properties and hence applications of the synthesized NPs (Amiri et al. 2017). Green synthesized NPs have shown potential in various applications including electrocatalytic (Al-Shehri et al. 2017), biomedical applications (Choudhary et al. 2017; Anand et al. 2017), electronics (Baca and Cheong 2015), antibacterial applications (Ghidan et al. 2016; Karthik et al. 2017a, b), energy production (Archana et al. 2017) and waste water treatment (Upadhyay et al. 2015).

Wastewater represents a major problem that threatens living species because it contains harmful compounds that cannot be degraded or destroyed, thereby causing considerable pollution issues (Awual et al. 2015). This includes, mainly, bacteria (Ocsoy et al. 2017), heavy metals (Papa et al. 2017) and other organic compounds such as benzene groups (Wang et al. 2017a, b) and dies (Sreekanth et al. 2016). Many studies showed effective ways of the removal of heavy metals and dies by applying various NPs such as oxides (Xia et al. 2017) and carbon based NPs (Peng et al. 2017). However, the preparation method of these NPs includes some chemical substances that could be hazardous which in turn affect environment too (Kuang et al. 2013). In this chapter, the type and preparation of green and eco-friendly oxides and graphene-based NPs are discussed as well as the application of these NPs in heavy metals removal and other organic pollutants from wastewater.

2 Types and Preparation of Green and Eco-friendly Materials

The development of the industry creates serious environmental impact on water resources, due to the use of toxic inorganic and organic pollutants. Various dyes and toxic metals are widely used in textile, paper, pharmaceutical, plastic and leather industries, which produced a big volume of wastewater mixed with these pollutants. This is why the adsorption onto sorbents among various techniques such as precipitation, osmosis, chemical oxidation, and ion exchange is more effective method due to its simple operation, least waste generation and economic reliability. But the founding of the effective sorbents for the remediation of inorganic and organic pollutants in water still attract great attention of researchers (Naushad et al. 2014; Hafshejani et al. 2017; Carolin et al. 2017; Awual et al. 2016a, b; Khan et al. 2015; Kumar et al. 2015, 2017; Shahat et al. 2015). On the other hand, in the adsorption process, the high cost of adsorbents represents the main challenge for its industrial usage. Nevertheless, the adsorbent cost can be reduced if prepared from some waste materials, for example, from agricultural waste. So the development of low-cost adsorbent for the removal of pollutants from contaminated wastewater is currently relevant.

2.1 Biosorbents

Biosorption consists on the uptake of heavy metals, radionuclides and dyes from aqueous environments by biological materials, such as agricultural waste, algae, bacteria, fungi, plant leaves and root tissues, which can be used as biosorbents for the remediation of inorganic and organic pollutants in water (Ali and Hassan 2017; Adegoke and Bello 2015). There are many different investigations of natural biosorbents as adsorbents for pollutants in water purification: *agricultural waste* (Wang and Zhang 2017; 2017a, b), *algal biomass* (Barquilha et al. 2017; Tran et al. 2016; Verma et al. 2016; Anastopoulos and Kyzas 2015), *chitosan* (Gokila et al. 2017), *fungal biomass* (Smily and Sumithra 2017) etc.

During the last years, a huge number of ***agricultural waste materials*** are being studied for the removal of different dyes from aqueous solutions under different operating conditions. Agricultural waste materials are low-cost and ecofriendly materials, that are used in the removal of pollutants either in their natural form or after some physical or chemical modification (Adegoke and Bello 2015). Agricultural waste such as *orange peel* (Ali et al. 2016), *banana peel* (Ali et al. 2016, 2017), *rice husk* (Hubadillah et al. 2017), *bamboo* (Jiao et al. 2016), *peat* (Bartczak et al. 2015), *coconut* (Johari et al. 2016; Etim et al. 2016), *tea* (Shen et al. 2017), *coffee* (Alhogbi 2017; Anastopoulos et al. 2017), *eggshell waste* (Lam et al. 2017; Angelis et al. 2017), *sawdust* (Banerjee and Chattopadhyaya 2017), *sesame husk* (Feng et al. 2017), *cashew NUT shell* (Subramaniam and Ponnusamy 2015) have previously been studied as sorbents for the removal of an anionic and cationic dyes and heavy metals. The use of low-cost, easily obtained and eco-friendly adsorbents could be as an alternative to the current expensive methods of removing toxic pollutants from wastewater.

Removal of cationic dyes such as rhodamine B (RhB) and methylene blue (MB) by ***waste seeds Aleurites moluccana*** (WAM) was studied by (Postai et al. 2016). The plant has the advantage of large-scale production of seeds all year round, low cost, and easy access. The maximum adsorption capacities of the dyes were 178 mg/g for MB and 117 mg/g for RhB. Shehzad et al. (2018) reported about waste of ***orange peel*** modified with magnetic nanoparticles as a novel adsorbent of As(III) from aqueous solutions. It exhibited superior As(III) adsorption capacity of 10.3 mg/g (experimental conditions: adsorbent dose = 1 g/L, temperature = 25 °C, pH = 7.0). The use of low-cost, abundantly available, highly efficient and eco-friendly adsorbent ***wood apple shell*** (WAS) has been reported by Sartape et al. (2017) as an alternative to the current expensive methods of removing malachite green (MG) dye from aqueous solution: the maximum adsorption capacity at 299 K was 80.645 mg/g. The ***exhausted coffee ground powder*** (CGP) was proved to be an efficient adsorbent for the removal of Rhodamine dyes (i.e. Rhodamine B and Rhodamine 6G) from aqueous solutions (Shen and Goudal 2017). It was shown that the maximum adsorption capacities of RhB and Rh6G were 5.255 and 17.369 μmol g^{-1} calculated by Langmuir model fitting.

Nayak and Pal (2017) investigated the adsorptive removal of methylene blue (MB) dye from aqueous solution using different parts of abundantly available agricultural product ***Abelmoschus esculentus seed*** (LFSP) (Fig. 1a). It was observed that the maximum removal (~96%) of MB dye was achieved at the biosorbent dose of 5 g/L (Fig. 1b). A dye removal efficiency of 95.60% was obtained at 50 mg/L, and then decreased to 76.26% at 1000 mg/L (Fig. 1c). So the cost-effective nature of Abelmoschus esculentus seed represents an efficient and cost-effective biosorbent for the MB biosorption process.

The low-cost adsorbent ***sweet potato peels*** reported by Asuquo and Martin (2016), was used for the removal of Cd(II) from aqueous solutions. The maximum loading capacity (q_{max}) was 18 mg g^{-1} at 25 °C. The removal of tartrazine from aqueous solutions using ***masau stone*** (MS) as a novel low-cost biosorbent was investigated by Albadarin et al. (2017). The maximum biosorption capacities of tartrazine were between 0.096 mmol/g (51.3 mg/g) at 20 °C and 0.126 mmol/g (65.1 mg/g) at 60 °C. The maximum efficiency is approximately 87% at C_o = 100 mg/L was found at pH 2.

A novel bio-based polymeric adsorbent ***curcumin formaldehyde resin*** (CFR) synthesized via polycondensation of curcumin (a natural bis-phenol) and formaldehyde was investigated by Alshehri et al. (2014). The CFR was effectively used for the removal of phenol; a removal percentage of 92.8% after 60 min at concentration of phenol 100 mg L^{-1}, pH 6.5 and temperature 55 °C, was achieved. This resin showed a high adsorption capacity for phenol 357.14 mg/g. The regeneration studies showed that CFR could be efficiently utilized for the removal

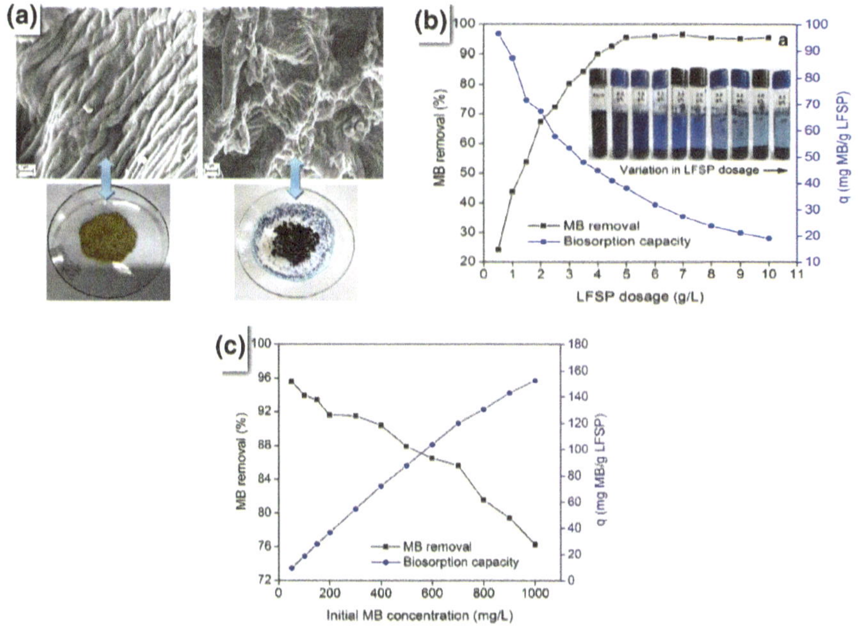

Fig. 1 Abelmoschus esculentus seed as adsorbent for MB: **a** SEM micrographs and photographs of LFSP before and after biosorption of MB dye; **b** the effect of biosorbent dosage on MB remediation onto LFSP (conditions: initial MB concentration = 200 mg/L, pH = ~6.74, agitation speed = 150 rpm, contact time = 120 min, temperature = 303 K); **c** effect of initial concentration on removal of MB by LFSP (conditions: biosorbent dosage = 5 g/L, pH = ~6.74, agitation speed = 150 rpm, contact time = 120 min, T = 303 K) (Reprinted from Nayak and Pal (2017), Copyright (2017), with permission from Elsevier)

and recovery of phenol because the adsorption was reduced from 92 to 80% after four consecutive cycles and the recovery was reduced from 85 to 74%.

A novel carbonization-functionalization of nitrogen-doped porous carbon sheets (NPCS) derived from wheat straws via a facile molten salt synthesis method was described in Yang et al. (2017). ***Wheat straws*** were used as a precursor for the synthesis of powders (Fig. 2). Powders of wheat straws and metal chloride salts (LiCl/KCl = 45/55 by weight, the melting point is 352 °C) were mixed and homogenized with a ball mill in the weight ratio of reactants (LiCl + KCl) = 1:10 with addition of $LiNO_3$. The calcination carried out at various temperatures (650, 750 and 850 °C) under N_2 flow (200 mL min^{-1}) for 2 h. The black solid product was subsequently immersed in a solution containing HCl. The as-prepared samples exhibited excellent adsorption performance with the adsorption capacity of 82.8 mg/g for atrazine removal and quick adsorption rate of 1.43 L/(g h) due to their unique microporous structure and the introduction of nitrogen atoms.

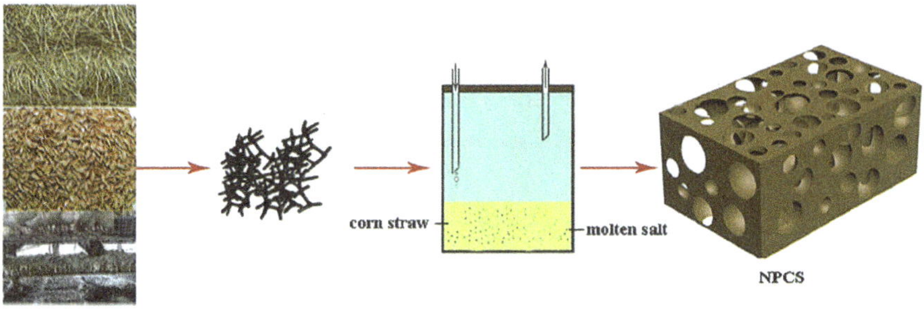

Fig. 2 Schematic illustration of the formation of nitrogen-doped porous carbon sheets (NPCS) (Reprinted from Yang et al. (2017), Copyright (2017), with permission from Elsevier)

2.2 Activated Carbon Prepared from an Agricultural Waste

Activated carbon (AC) is the most popular material for the removal of pollutants from wastewater. AC is an amorphous carbon with the high degree of porosity and very large surface area, thereby allowing toxic molecules or heavy metals to be adsorbed on its surface. The preparation of AC consists from two main steps: (i) the carbonization of the carbonaceous raw material below 1000 °C, in an inert, usually nitrogen, atmosphere; (ii) the activation of the carbonized product (char), which is either physical or chemical (Adegoke and Bello 2015). Physical activation involves contacting the char with an oxidizing gas, such as carbon dioxide, steam, air or their mixtures, in the temperature range between 600 and 900 °C, which results in the removal of the more disorganized carbon and the formation of a well-developed micropore structure. Chemical activation allow to obtain AC at lower temperature and involves the use of chemical reagents such as KOH, NaOH, K_2CO_3, Na_2CO_3, H_3PO_4, H_2SO_4, $ZnCl_2$ etc. during the carbonization (Adegoke and Bello 2015; Okman et al. 2014; Kumar and Jena 2015, Kumar and Jena 2016; Hayashi et al. 2000).

The preparation of ***AC from coconut husk*** with H_2SO_4 activation and its ability to remove textile dyes (maxilon blue GRL, and direct yellow DY 12), from aqueous solutions were described by Aljeboree et al. (2017). H_2SO_4/AC sample was prepared by carbonization of coconut shells at 500 °C for 2 h in the absence of air and then was soaked with certain weight of H_2SO_4 and finally activated at 600 °C for 4 h.

Activated carbon prepared from forest and wood-processing residues was developed by Aghababaei et al. (2017) and used for the adsorptive removal of antibiotic drug oxytetracycline (OTC) and toxic heavy metal cadmium (Cd) from aqueous solution. H_3PO_4 and NaOH-treated biochars showed the highest removal efficiencies of oxytetracycline (263.8 mg/g) and cadmium (79.30 mg/g). Kumar and Jena 2017a reported about ***AC prepared from Fox nutshell*** (FNAC) by $ZnCl_2$ activation. BET surface area and total pore volume of FNAC were found to be 2869 m^2/g and 1.96 cm^3/g, respectively (Fig. 3). FNAC was investigated as eco-friendly adsorbent for Cr(VI); with a maximum adsorption capacity of

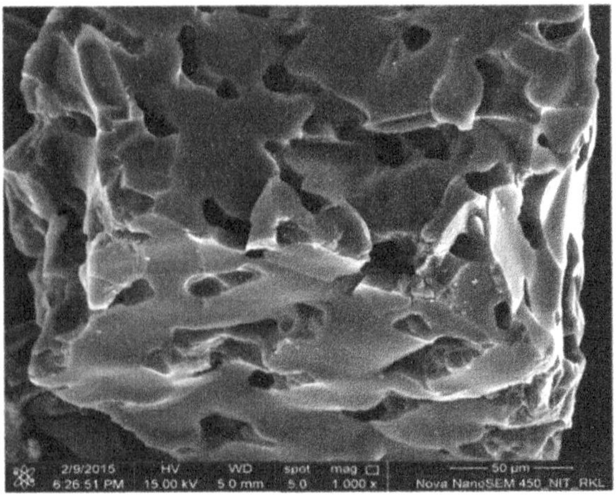

Fig. 3 FESEM of activated carbon prepared from Fox nutshell (Reprinted from Kumar et al. (2017), Copyright (2017), with permission from Elsevier)

43.45 mg/g. A very high removal efficiency of 99.08% for of 10 mg/L concentration was achieved at pH = 2 and temperature of 30 °C for 3 h. Therefore, FNAC can be used as an efficient adsorbent in the treatment of water and wastewater with no need for further filtering and centrifugation.

The same research group (Kumar and Jena 2017b 2017) reported about AC prepared from Fox nutshell (FNAC) by chemical activation with H_3PO_4 at an impregnation ratio of 1.5 and activation temperature of 700 °C under N_2 atmosphere. Adsorption studies of Cr(VI) onto FNAC-700-1.5 showed a maximum adsorption of 74.95 mg/g at the operating condition of 35 mg/L of initial concentration of Cr(VI) with a pH of 2.0, temperature 45 °C and contact time of 3 h.

Various different sources were also used for the preparation of low-cost AC, such as ***coconut leaf*** (Jawad et al. 2017), ***papaya seeds*** (Krishnaiah et al. 2017), ***babassu coconut and bone*** (Reck et al. 2018), ***aamla seed*** (Hameed et al. 2017), ***jambul seed*** (Hameed et al. 2017), ***tamarind seed*** (Hameed et al. 2017), ***soapnut*** (Hameed et al. 2017) etc. for the removal of dyes and heavy metals from wastewater.

Biochar is a porous carbonaceous material derived from pyrolyzing organic biomass under oxygen-limited conditions. In the recent years, biochar also has attracted attention as ecofriendly and green adsorbent due to its mechanical and thermal stability, and the presence of a highly porous structure and functional groups. Various types of biochars derived from marine macro-algae waste (Anastopoulos and Kyzas 2015), corn straw (Zhou et al. 2018; Zhang et al. 2018) etc. were investigated.

Son et al. (2018) reported about ***magnetic biochar*** prepared by pyrolyzing of ***marine macro-algae waste*** for heavy metal adsorption. The development of magnetic adsorbents for water and wastewater treatment is very widely investigated because magnetic adsorbents can be easily manipulated by external magnetic field and easily recovered from contaminated water by magnetic separation. Magnetic macro-algae biochar showed high selectivity for Cu (69.37 mg/g for kelp magnetic biochar and 63.52 mg/g for hijikia magnetic biochar) with plentiful O-containing groups. The high heavy metal removal performance was attributed to the presence of functional groups –COOH and –OH on the magnetic biochar, which serve as potential adsorption sites for heavy metals.

Magnetic oak wood biochar (MOWBC) and ***magnetic oak bark biochar*** (MOBBC) were obtained from oak wood and oak bark biochars made by fast pyrolysis during bio-oil production (Mohan et al. 2014) (Fig. 4). Biochars were obtained from fast pyrolysis at 400 and 450 °C in an augerfed reactor and then magnetized by mixing aqueous biochar suspensions with aqueous Fe^{3+}/Fe^{2+} solutions, followed by treatment with NaOH. They were investigated as potential green adsorbents for cadmium and lead remediation from water. Maximum lead and cadmium removal occurred at pH 4–5. The adsorption capacity was found to be 30 mg/g Pb^{2+} and 7 mg/g Cd^{2+} by magnetic oak bark biochar at pH 5.0 and adsorbent concentration of 3 g/L at 25 °C.

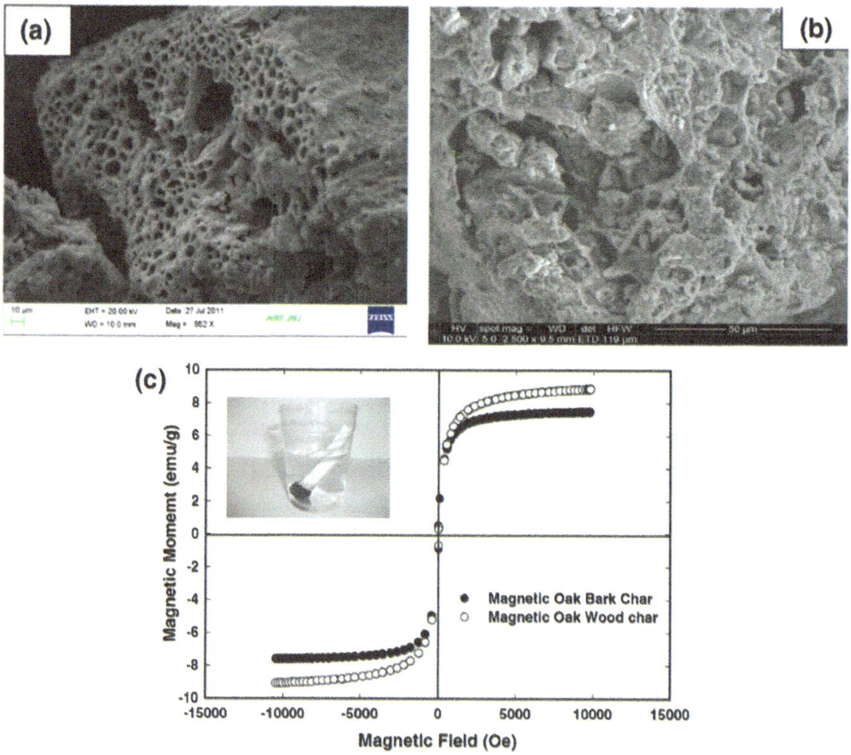

Fig. 4 Magnetic oak wood biochar and magnetic oak bark biochar: **a** SEM micrograph of magnetic oak wood char at 952× magnification; **b** SEM micrograph of magnetic oak bark char at 2500× magnification; **c** magnetic moment of magnetic oak bark char and magnetic oak wood char at 23.7 °C (Reprinted from Mohan et al. (2014), Copyright (2014), with permission from Elsevier)

Naushad et al. (2017) reported about ***nickel ferrite bearing nitrogen-doped mesoporous carbon*** ($NiFe_2O_4$-NC) which was prepared using polymer bimetal complexes and used for the removal of Hg^{2+} from aqueous medium. Urea-formaldehyde pre-polymer was used as a source of carbon and nitrogen as well as structural templates for magnetic ferrite nanoparticles. The adsorbent showed high surface area of 147.4 $m^2\ g^{-1}$, with particle size in the range of 8–10 nm and a saturation magnetization of 53.40 emu g^{-1} (strong enough to be efficiently separated from a solution using external magnetic field). The results showed a maximum adsorption capacity of 476.2 mg g^{-1} at 25 °C, indicating that $NiFe_2O_4$-NC can be used as potential sorbent for the removal of Hg^{2+} from aqueous environment.

Another magnetic nanocomposite Fe_3O_4@TAS prepared by Alqadami et al. (2017) (Fig. 5) revealed excellent magnetization and adsorption capacity for the removal of Cd(II), Cr(III) and Co(II) metal ions. The nanocomposite show high surface area (210.34 $m^2\ g^{-1}$) and high saturation magnetization (41.4 emu g^{-1}). The maximum adsorption capacities were 286, 370 and 270 mg g^{-1} for Cd(II), Cr (III) and Co(II), respectively. It was found that the adsorption onto Fe_3O_4@TAS was monolayer and after saturation of this layer no further adsorption took place.

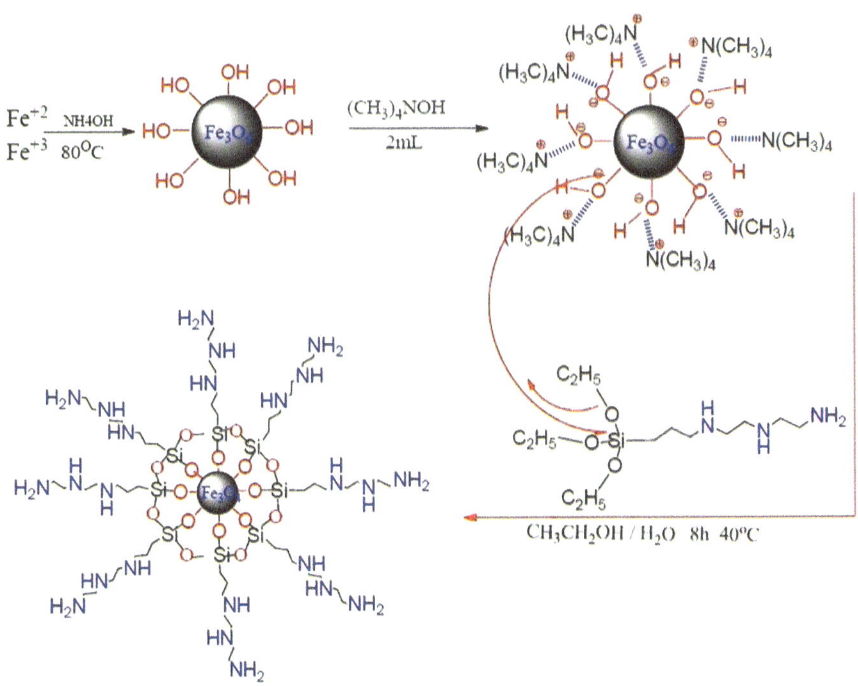

Fig. 5 Synthesis of Fe_3O_4@TAS nanocomposite as sorbent for Cd(II), Cr(III) and Co(II) from liquid environment (Reprinted from Alqadami et al. (2017), Copyright (2017), with permission from Elsevier)

2.3 Graphene-Based Adsorbents and Their Derivatives

Graphene is a two dimensional material with a unique structure and large specific surface area. The modification of graphene and the combination with other materials leads to new important composites with enhanced properties. For example, graphene based composites and its derivatives (graphene oxide (GO), reduced graphene oxide (rGO)) have been studied in the past years as adsorbents for the sorption of organic dyes, heavy metals, and others pollutants. ***Graphene oxide (GO)*** is one of the derivative of carbon-based materials, which can be obtained from graphite using Hummer's method. The surface of GO contains different functional groups (i.e., carboxyl –COOH, carbonyl C=O, hydroxyl –OH groups), which allow to use GO as potential sorbent for organic and inorganic pollutants, for example for dye molecules and heavy metals adsorption. However, GO sheets exhibit a high dispersibility in water, which prevents the efficient separation of dye-adsorbed GO sheets from an aqueous environment (Guo et al. 2015). Thus, many efforts have been carried out in recent years for developing of GO-based adsorbent composites, which can be easily separated from aqueous solutions. This is why magnetic-graphene-based nanocomposites have been widely applied for wastewater treatment (Roy et al. 2016; Jia et al. 2017).

A new stretchable and graphene-oxide-based hydrogel Fe_3O_4/RGO/PAM was developed by Dong et al. (2017) for removing organic pollutants and heavy metal ions. The hydrogels exhibit excellent photo-Fenton in the degradation of 20 mg/L RhB for 90% within 60 min under visible light irradiation, and even after 10 times cycle test, the degradation rate for RhB still at 90%. This adsorbent can also remove 34.8–66.3% heavy metal ions after continual two days adsorption (Fig. 6).

The spinel ferrite systems (Tatarchuk et al. 2017a, b, c, d; Reddy and Yun 2016) very often used as magnetic carriers for graphene-based adsorbents and their derivatives, such as GO/Fe_3O_4-*g*-PMAAM nanohybrid (Ma et al. 2017), Fe_3O_4/porous graphene nanocomposites (Bharath et al. 2017), magnetic 3D rGO-loaded hydrogels (Halouane et al. 2017), poly(m-phenylenediamine)/reduced graphene oxide/nickel ferrite (Wang et al. 2017a, b) etc.

For example, Guo et al. (2015) successfully reported the preparation of GO/polyethylenimine (PEI) hydrogels by chemical route as efficient dye adsorbents for wastewater treatment (Fig. 7a). Figure 7b shows the calculated dye removal rate versus time plots for both MB and RhB. The dye removal rates can reach nearly 100% for both MB and RhB within approximately 4 h in accordance with the pseudo-second-order model, suggesting the as-prepared GO/PEI hydrogels as efficient dye adsorbents. The PEI was incorporated to facilitate the gelation process of GO sheets and it allow conveniently to separate dye-adsorbed hydrogels from an aqueous environment.

Yang et al. (2018) fabricated a Fe_3O_4-PSS@ZIF-67 composite as novel type of metal-organic frameworks (MOF) magnetic porous composite materials (Fig. 8). The ZIF-67 (Zeolitic imidazolate framework-67) nanocrystals were selected as MOF. The results show that the equilibrium adsorption capacity for MO is up to

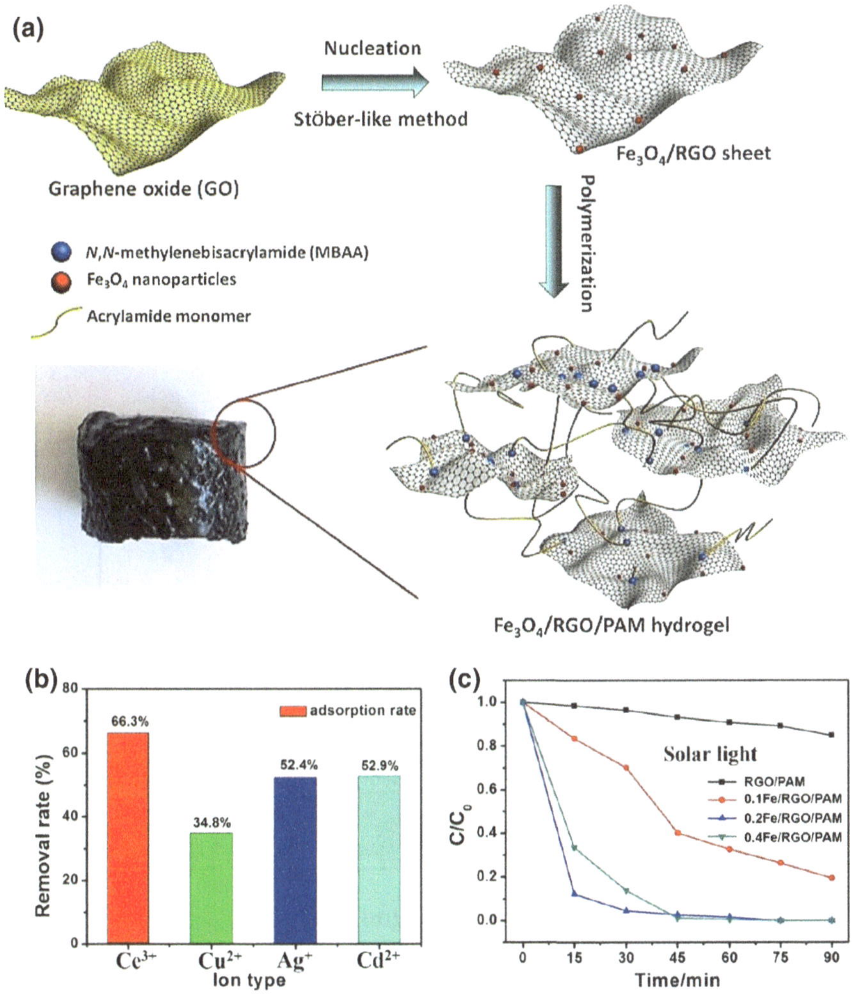

Fig. 6 Graphene-oxide-based Fe_3O_4/RGO/PAM hydrogel for the removal of organic pollutants and heavy metal ions: **a** schematic steps of preparation of Fe_3O_4/RGO/PAM hydrogels; **b** removal rate of heavy metal ions after the degradation of RhB over 0.2Fe/RGO/PAM, hydrogel (adsorption in the dark for two days); **c** photo-Fenton-reaction for the degradation of RhB over different samples (20 mg/L RhB, pH = 4.5, simulated solar light: 300 W Xe with AM 1.5 filter) (Reprinted from Dong et al. (2017), Copyright (2017), with permission from Elsevier)

738 mg g^{-1} (pH = 8.0, contact time 7 h, adsorbent dose 5 mg and initial MO concentration 400 mg L^{-1}). The authors stated that MZIF-67 composites could selectively separate MO from the mixture solution of MO and MB. The removal rate of MO is up to 92%. The results suggest that MZIF-67 composites could be a good candidate for treatment of dye-bearing wastewater (Yang et al. 2018).

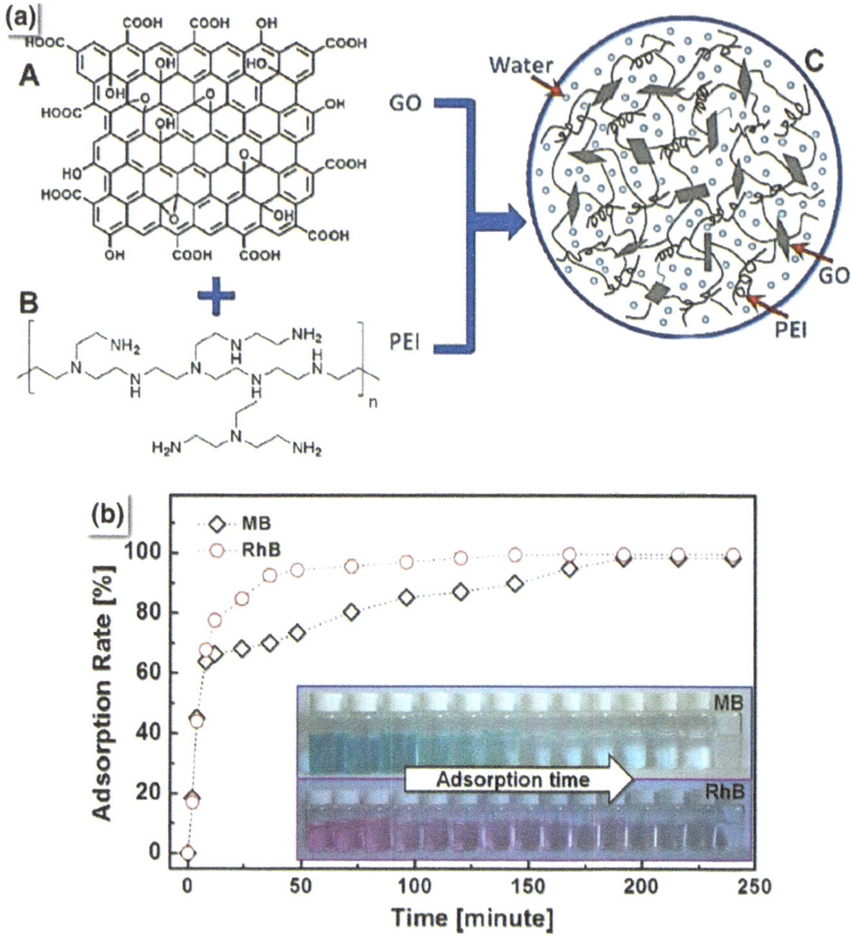

Fig. 7 GO/polyethylenimine (PEI) hydrogels as efficient dye adsorbents: **a** preparation (**A** GO and **B** amine-rich PEI were combined to give **C** GO/PEI hydrogels); **b** the calculated dye removal rates versus time plots for both Methylene Blue and Rhodamine B (Guo et al. 2015)

The new magnetic nanocomposite amino-functionalized $SiO_2@CoFe_2O_4$ nanoparticles decorated on graphene oxide were synthesized using solvothermal and sol-gel processes by Santhosh et al. (2017). The obtained $SiO_2@CoFe_2O_4$ nanocomposite was functionalized with amino groups by adding 3-aminop ropyltriethoxysilane, then was dispersed in 100 mL DI water with 100 mg of GO and stirred for 2 h at room temperature. After, the obtained product was washed with water and ethanol and dried at vacuum oven at 45 °C overnight. The prepared nanocomposite was used as magnetic adsorbents for the removal of organic and inorganic pollutants (acid black 1 dye and Cr(VI) ions as model pollutants) from aqueous solution. The monolayer adsorption capacity of AB 1 and Cr(VI) ions onto

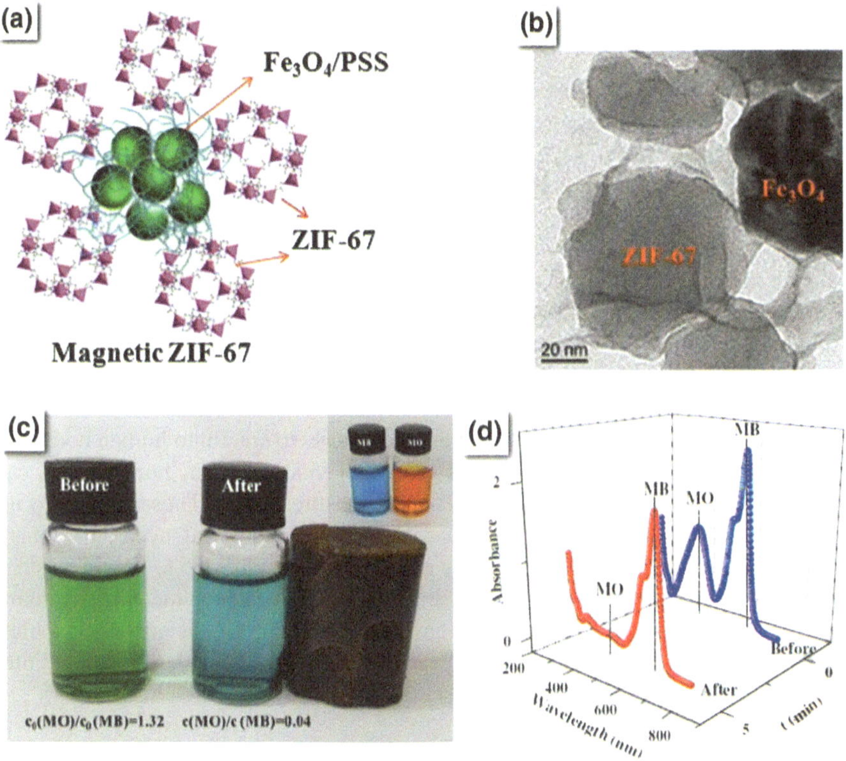

Fig. 8 Porous magnetic MOF composites Fe_3O_4-PSS@ZIF-67: **a** structure; **b** TEM images of MZIF-67; **c** separation of MB/MO mixture (photographs of MB/MO mixture solution before and after magnetic separation; inset—photograph of MO and MB solution before mixed); **d** UV-vis spectra of MB/MO mixture solution before and after magnetic separation (Reprinted from Yang et al. (2018), Copyright (2018), with permission from Elsevier)

aminofunctionalized SiO_2@$CoFe_2O_4$-GO was found to be 130.74 and 136.40 mg g^{-1} at pH = 2 and pH = 1 respectively.

3 Removal of Heavy Metals

Water is the source of existence of life on earth. The continuous increase in population and industrialization, cause adverse affect on human health and ecosystems, due to severe contamination of water and surrounding soils by hazardous and toxic pollutants. Nowadays, water pollution represents one of the most serious problem, as 70–80% of all illness in developing countries are related to water contamination (WHO 2000).

3.1 Sources and Their Adverse Effects

In developing countries, various industries like, chemical, textiles, oil refinery, etc., directly or indirectly discharge the effluents containing toxic pollutants (organic and inorganic) into surface water bodies and nearby land areas. Among the water pollutants, metallic elements having high density like Zn(II), Pb(II), Cr(III), Cd(II), Ni(II), Co(II), Mn(II)… etc. are well-known hazardous pollutants due to their tendency to accumulate in living organisms, especially at higher concentrations causing health problems in human beings and raising serious environmental issues (Chowdhury et al. 2016). In general, these metals exist naturally either alone or in combination with other elements, but anthropogenic activities like mining, landfill leaches, urban runoff and wastewater released by industries enhances their concentration (Ebtesam et al. 2013).

Water and food are the gateway for these elements to enter into human body, and damage liver, nerves, block functional groups of vital enzymes, bones, renal, cardiovascular, neurological disorders and even life threatening. These metals do not only affect the humans but also, causes damage to biological cycles of aquatic species, cell morphology and inhibit the cytoplasmic enzyme due to oxidative stress (Bulgariu and Bulgariu 2014). Table 1 shows the limit of heavy metals on drinking water and their adverse affects on human health, if the concentration is high. To-date, only 10% of the produced wastewater is remediated and the residual

Table 1 Heavy metal and their limit in drinking water and their adverse affects on human health

Pollutant	Permissible limits for drinking water (in mg/L) (WHO 2000)	Disorders
Arsenic	0.01	Hypo-pigmentation, cancer, produce liver tumors and skin gastrointesinal effects
Mercury	0.006	Corrosive to skin, eyes, and muscle membrane, dermatitis, anorexia, kidney damage, respiratory disorder
Cadmium	0.003	Carcinogenic, cause lung fibrosis, weight loss, kidney damage
Lead	0.01	High blood pressure, loss of apetite, anemia, muscle and joint pains, kidney problem
Chromium	0.05	Producing lung tumors, and allergic dermatitis
Nickel	0.07	Chronic bronchitis, reduced lung function, disorder of bones
Zinc	3	Cause short-term illness called "metal fume fever"
Copper	2	Long term exposure causes irritation of nose, mouth, and eyes
Uranium	No relaxation	Mental retardation, estrogenic effect
Pesticides	No relaxation	Cancer
Nitrate	No relaxation	Methemoglobinemia

wastewater is released into water resources and diffused into groundwater (Al-Zoubi et al. 2015). Therefore, remediation of wastewater is important before its discharge into water resources such as, rivers and seas. In general, heavy metals may easily dissolve in solutions due to their water soluble property. This causes difficulties in their elimination via physical and chemical separation methods.

3.2 *Current Remediation Methods*

There are several techniques employed for the removal of heavy metal ions from wastewater and shown in Scheme 1. The conventional methods are more or less successful in control of removal of heavy metals. However, they suffer several drawbacks such as high-energy requirement, incomplete pollutant removal and generation of toxic sludge. On the other hand, adsorption method has been proven to be most effective and regenerative metal recovery technology for purification of wastewater from heavy metal ions.

3.2.1 Chemical Precipitation Method

The most commonly used conventional method for the treatment of industrial wastewater containing heavy metals is the chemical precipitation method. In this process, positively charged molecules (cations) are combined with negatively charged molecules (anions). Precipitation of dissolved metals (cations) is

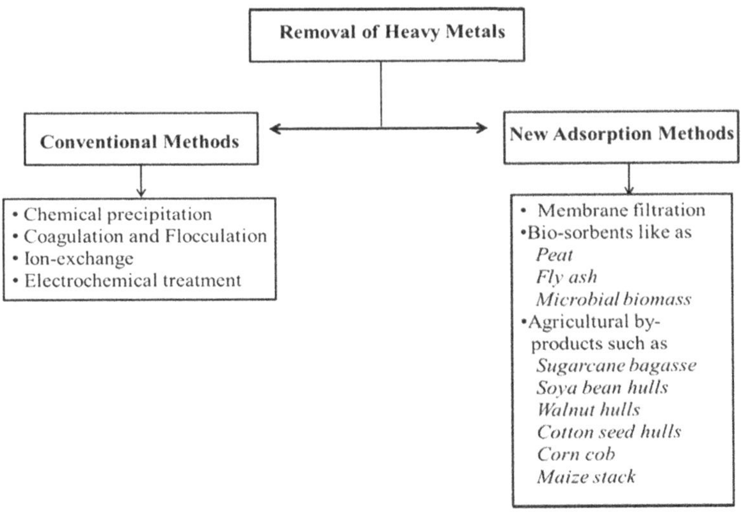

Scheme 1 Different techniques are used for the removal of heavy metals from wastewater

accomplished by increasing the concentration of soluble anions. For most metals, this can be achieved by the addition of hydroxides (Harper and Kingham 1992). However, this process requires a large amount of treatment chemicals to decrease the heavy metals to levels imposed by the regulations. In addition, the sludge produced from the precipitation process has to be subjected to dewatering and disposal into landfills, which adds an additional cost to the treatment process.

3.2.2 Coagulation and Flocculation

Coagulation-flocculation process is well-known and simple chemical method used to treat industrial wastewaters. The main purpose of this method is to enhance the ability of a treatment process prior to sedimentation and filtration (Parson and Jefferson 2006). Coagulation is a process used to neutralize the charges and form a gelatinous mass to trap the particles, thus forming a large mass enough to settle or be trapped in the filter. Flocculation is gentle stirring or agitation to encourage the particles thus formed to agglomerate into masses large enough to settle or be filtered from solution. Due to its easy operation, relatively simple design and low energy consumption, coagulation-flocculation has been successfully employed in different type of industries (Almubaddal et al. 2009). Moreover, coagulation-flocculation can be used as a pre-treatment, post-treatment, or even as the main treatment of wastewater due to the versatility of the treatment process.

3.2.3 Ion-Exchange

Ion exchange is used for the removal of heavy metals from effluent. An ion exchange is a solid (synthetic organic ion exchange resins) capable of exchanging either cations or anions from the surrounding materials (Barakat 2011). Even though this method is successful in removing toxic elements, it cannot handle concentrated metal solution as the matrix gets easily fouled by organics and other solids in the wastewater. Moreover, ion exchange is non-selective and is highly sensitive to the pH of the solution.

3.2.4 Membrane Filtration

Membrane filtration is a promising technique for the treatment of industrial wastewater as it requires small space, low pressure and high separation selectivity. It is capable of removing particulate material, pathogens, nutrients, organic and inorganic contaminants and dissolved substances. Different types of filtration techniques are employed depending upon the size of the particle/pore size of the membrane and type of the effluent to be treated. Microfiltration (MF) is based on the sieving mechanism, where as particles with a higher size than the membrane pore dimension are retained and smaller particles pass through the membrane. In

Ultrafiltration (UF) technique, adsorption-sieving mechanism is applied for the separation of macromolecules or colloidal particles from liquid. Reverse Osmosis (RO) demonstrates a high heavy metal rejection, compared to ion exchange, but it operates in a continuous flow mode and has considerably lower reagent consumption. RO technique of heavy metal cationic wastewater treatment can provide potable-grade water or high purity water for technological operations. Nanofiltration (NF) is the separation process, using specially designed membranes with electrically charged surface (Kagramanov et al. 2009). Membrane technologies provide an important solution in reuse and recovery of water. The separation in MF and UF is based on the mechanical sieving while in NF and RO membranes is based on the capillary flow or diffusion of the solute.

3.2.5 Adsorption

Adsorption is the process of accumulating substances that are in solution on a suitable interface. The factors affecting the adsorption process are mainly the amount of sorbent, initial concentration of metal ion, pH of the medium and operating temperature. During adsorption, molecules of a substance (adsorbate) collect on the surface of another substance (adsorbent). Therefore, adsorption is considered to be a mass transfer operation, usually from a fluid phase to solid phase. In adsorption method, activated carbon (AC) is used as adsorbent and has been synthesized from coconut husk, sawdust and rice husk, jackfruit peel, sugarcane bagasse, mango nuts, pumpkin seed shell. However, the use of AC is inadequate owing to its higher cost (Wang et al. 2003; Dai et al. 2012). Therefore, several low-cost sorbents like fly ash, silica gel, zeolite, lignin, seaweed, wool wastes, agricultural wastes (maize, coconut shell, rice husk), chitin, chitosan (CTS), and clay materials (kaolinite, montmorillonite, diatomaceous, and bentonite) are investigated as they are efficient when compared to activated carbon (Jafari et al. 2014; Debbaudt et al. 2004; Zhu et al. 2012).

Biosorption method is another competent and economic method for the remediation of heavy metals from industrial effluents (Hossain et al. 2012). Bio-wastes generally contain carboxylic, alcohols, aldehydes, ketones, and phenolic functional groups which have the capability to interact and bind with heavy metals by forming complexes in solution and thus can be utilized as potential sorbents for the removal of heavy metals (Demirbas 2008).

3.3 Removal of Heavy Metals

3.3.1 Arsenic (As)

The occurrence of arsenic in ground water is naturally due to the weathering of rocks and sediments. In addition to the above anthropogenic activities like agriculture, industrial and domestic endeavors has increased the concentration level in the water. The arsenic problem is found mainly in India, Bangladesh, Taiwan, Magnolia, China, Chile, New Zealand, Hungary, Japan, Mexico, and Poland (Mondal and Garg 2017). The permissible limit of arsenic in water is about 0.01 mg/L (Islam and Patel 2008; Yadav et al. 2006). It exist in As(III) and As(V) oxidation states whereas As(III) is easily mobile and is about 25–60 times toxic than As(V). The removal of these elements depends upon pH of the solution and redox potential. As(V) usually adsorbs and reacts more strongly than As(III) (Sharma et al. 2007).

Agents that can bind to arsenic through adsorption or through formation of insoluble compounds with arsenic by precipitation or co-precipitation are referred to sorbents (Table 2). The AC is most commonly used as an adsorbent for water purification. However, the removal of As depends on the pH of the solution. The adsorbents fabricated from char's and AC were synthesized from sawdust, fly ash and agriculture by-products. Chuang et al. (2005) reported oat hulls based AC and observed that adsorption capacity was largely affected by the initial pH with decreasing adsorption capacity from 3.09 to 1.57 mg/g when initial pH value was increased from 5 to 8. Numerous reports are available in the literature for the development of various low-cost adsorbents and concluded that their use may contribute to the sustainability (Babel and Kurniawan 2003). Several authors employed bio-sorption process (biomaterials including plant biomass, agricultural waste, bacteria, fungi and algae) to eliminate As from water (Martinez-Juarez et al.

Table 2 Different types of sorbent materials and their efficiency for the removal of Arsenic

Material/ Sorbent used	Amount of sorbent/conditions	Efficiency	References
Lanthanum compounds	Lanthanum hydroxide (LH), lanthanum (LC) and basic lanthanum carbonate (BLC). The optimum pH range was 3–8, 4–7 and 2–4 for LH, LC and BLC, respectively	0.001 mM	Tokunaga et al. (1997)
Aqua bind	–	100%	Senapati and Alam (2001)
Ferric chloride	100 mg/L of ferric chloride and 1.4 mg/L of $KMnO_4$	>90%, often 95% or more	Ahmed (2001)
Aluminum alum	20 mg/L	96%	Ahmed (2001)
Iron filings type	1 g	99% (4 days, 2 mg/L)	Su and Puls (2001)
Alumina manganese oxide	1 g	94%	Kepner et al. (1998)
Calcium oxide	0.1% lime/water ratio by weight	99.90%	Hussain et al. (2001)
Wood charcoal	600–757	97–99%	Hussain et al. (2001)

2013; Bhargavi and Savitha 2014; Zhang et al. 2015; Aryal et al. 2010; Vinh et al. 2015; Raj et al. 2013).

Coagulation and flocculation are the most simple and less expensive methods that have been used for As removal. It was reported that chitosan as natural coagulant supports the removal of As by coagulation process using $FeCl_3$. It was found that the removal percentage of As(V) and As(III) was 90 and 60%, respectively, when $FeCl_3$ was used alone and it was enhanced almost to 100 and 80% when used together with chitosan (Hesami et al. 2013). Recently, nanoscale based material are used in the removal of As from wastewaters. Jungcharoen et al. (2017) reported lignin-modified nanoscale zerovalent iron (nZVI) for the removal of As from wastewaters. A lignin-modified (L-nZVI) and pulp-modified nZVI (P-nZVI) nanoparticles were synthesized using the borohydride method. Both L-nZVI and P-nZVI could remove 1 mg/L of As with the removal efficiency of 88.66 and 90.79%, respectively at an initial pH of 7. Adio et al. (2017) also reported nanoscale zero-valent iron (Fe) using plant extract as a reducing agent and about 95% of As removal was recorded using 100 mg adsorbent at solution pH 3.

3.3.2 Lead (Pb)

Lead is a very toxic heavy metal, and in human body it causes damage to the bones, brain, blood, kidneys, and the thyroid glands. The major source of lead in water is from the discharge of effluents of industries, i.e., electroplating, paint, pigment, basic steel work, textile industries, metal finishing and electric accumulators' batteries. Pb(II) is used in batteries due to its specific characteristics such as conductivity, resistance to corrosion and a reversible reaction between lead oxide and sulfuric acid (Bahadir et al. 2007). Its presence also shows adverse effect on the living creatures received through waters. Even a very low concentration of heavy metals in water can be very toxic to aquatic life.

Wen et al. (2016) reported reusable electrospun zein nanoribbons for treatment of lead contained wastewater. The modified coaxial electrospinning process was conducted permitting the smooth and continuous generation of zein nanoribbons. Adsorption results indicated that equilibrium was obtained in 60 min for Pb(II) solutions with initial concentrations of 100, 150 and 200 mg/L. Desorption results showed that the adsorption capacity can remain up to 82.3% even after 5 cycles of re-use. Badawi et al. (2017) developed three modified biopolymers by reaction of chitosan and tannic acid in molar ration 1:1, 1:3 and 1:6, according to their equivalent molecular weights of (17.01 g/1.73 g of 1:1), (17.01 g/5.19 g of 1:3) and (17.01 g/10.38 g of 1:6) to adsorb Pb(II) metal ions from industrial wastewater. The absorption efficiency as a function of bisorbent amount is shown in Fig. 9. The maximum adsorption efficiencies were obtained at 99.6, 99.8 and 99.2% for Pb(II) metal ion and at 98.9, 99.8 and 99.5% by using 0.5 g of TC1, TC3, and TC6 biosorbents respectively.

A novel series of polyamine/CNT composites were synthesized via a single step polycondensation reaction of melamine, paraformaldehyde, various alkyldiamines

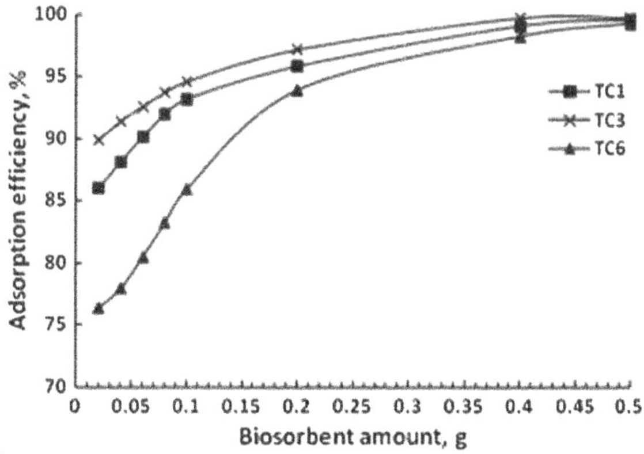

Fig. 9 Effect of Bisorbent amounts on the adsorption efficiency of Pb(II) removal from waste water (Reprinted from Badawi et al. (2017), Copyright (2017), with permission from Elsevier)

and chlorinated carbon nanotubes (CNT) at optimized reaction conditions in the presence of N, N-Dimethyl formamide as a solvent. Wastewater treatment revealed the high efficacy of the synthesized polyamine/CNT composite in the removal of ~99% of Lead ions in wastewater samples. Awual (2016a, b) designed a new re-usable conjugate adsorbent that can simultaneously detect and remove of toxic Pb(II) even at lower concentrations without using any sophisticated instruments. The detection of Pb(II) ions was 1.2 1 g/L and was found to be relatively fast. The maximum adsorption capacity was determined to be 188.67 mg/g. Moreover, this technique achieved residual Pb(II) concentration less than 10 mg/L, which is acceptable by water quality regulations. In addition, the conjugate adsorbent was regenerated by 0.10 M HCl treatment. In addition, the Pb(II) ion adsorption efficiency was retained after nine adsorption-elution-regeneration cycles. Vasudevan et al. (2012) used electrocoagulation method for the removal of Pb from water. The results showed that the maximum removal efficiency was achieved with magnesium alloy as anode and galvanized iron as cathode at a current density of 0.15 $A/dm^2$and pH of 7.0. Recently, Stevens and Batlokwa (2017) used fish scales waste remains as an environmentally friendly to remove 81.97% Pb(II) from wastewater. This is an alternative method for agricultural and industrial waste management, particularly for fisheries and food industries. Tinas et al. (2016) prepared and used Fe_3O_4@montmorillonite composite for the separation of Pb and Cd from aqueous solutions. Under optimized conditions, Pb and Cd were quantitatively removed from wastewater (between 95 and 98%) for a contact time of less than 5 min. Recently, Yap et al. (2017) synthesized magnetic biochar (MB) with discarded materials such as coconut shell (CS) using the microwave technique and achieved maximum removal of Pb at pH 4.5 of the solution. The different adsorbent for the removal of Lead and their capacities are shown in Table 3.

Table 3 Adsorbent for the removal of lead and their capacity

Adsorbent	Capacity	References
Coconut activated carbon	3.91	Gueu et al. (2007)
Activated carbon	13.05	Imamoglu and Tekir (2008)
Sugarcane dust	13.9/50.4	Ho et al. (2005)
Olive stone	23.47	Alslaibi et al. (2014)
Activated carbon	32.17	Machida et al. (2005)
EFB based activated carbon	48.96	Wahi and Ngaini (2009)
Sugarcane dust	50.4	Ho et al. (2005)
Coconut husk	69.06	Sulak et al. (2007)
Activated carbon	99.5	Li et al. (2009)
Carbon nanotubes	112.36	Xu et al. (2008)
Peanut husks	113.96	Ricordel et al. (2001)
CNT's	120.6	Tofighy and Mohammadi (2012)
Activated carbon	134.22	Singh et al. (2008)
Flax shive	147.1	El-Shafey et al. (2002)
Coconut shell	162.75	Yap et al. (2017)
Sugarcane bagasse pith	200	Yap et al. (2017)
Sawdust	200	Krishnan and Anirudhan (2002)
Bagasse	227.3	Ayyappan et al. (2005)
Coir pith	263	Kadirvelu and Namasivayam (2000)
E. rigida	279.72	Gercel and Gercel (2007)
Wild almond shells	823.1	Thitame and Shukla (2017)

3.3.3 Mercury (Hg)

Mercury Hg(II) is also considered as one of the most harmful metal ions to the human and environment. It has been proved to have toxic effects on reproduction, liver, kidneys and central nervous system, and it can cause psycho-logical and sensory impairments (Harris et al. 2003). The maximum acceptable level of Hg(II) in drinking waters is recommended 1 $\mu g\ L^{-1}$ according to World Health Organization (WHO). Attari et al. (2017) developed a cost-effective Hg removal approach based on a synthesized zeolite Linde Type A(LTA) from coal fly ash (CFA-ZA). Adsorption experiments at room temperature achieved an average removal efficiency of 94% with 10 mg/L initial concentration; that is comparable with activated carbon. Devani et al. (2017) repoted biosorption of Hg(II) from solutions by normal and chemically modified husk of *Cajanus cajan* (Pigeon pea). The maximum metal uptake was obtained as 68 and 82 mg/g for an initial Hg(II) concentration of 150 mg/L for normal and chemically activated biosorbents, respectively, at a most favorable solution pH of 5.5. Amidoamine functionalized multi-walled carbon nanotubes (MWCNT-AA) has been investigated for the separation of Hg(II) ions by solid-liquid extraction (Deb et al. 2016). The adsorbent showed superior selectivity with high adsorption capacity towards Hg as compared

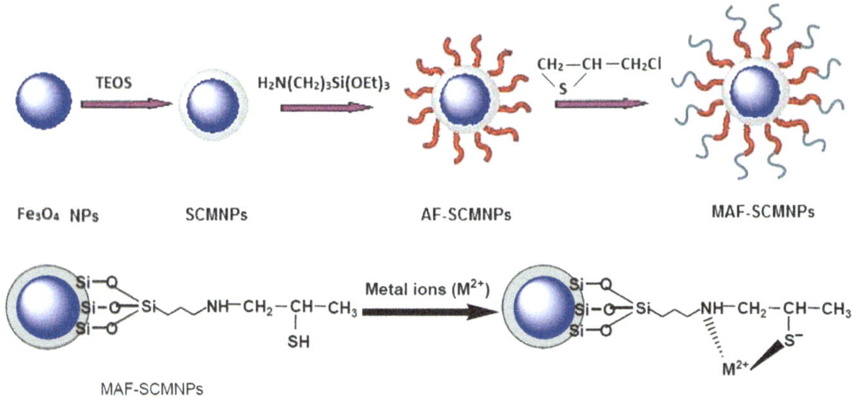

Fig. 10 Preparation of mercaptoamine-functionalised silica-coated magnetic nano-adsorbents and a possible mechanism for adsorption of Hg(II) or Pb(II) on MAF-SCMNPs (Reprinted from Bao et al. (2016), Copyright (2016), with permission from Elsevier)

to other metal ions present in industrial wastewater containing Hg. A novel magnetic hybrid material was fabricated using mercaptoamine-functionalized silica-coated magnetic nanoparticles (MAF-SCMNPs) for the extraction and recovery of mercury was initiated by Bao et al. (2016) (Fig. 10). The maximum adsorption capacities was found to be 355 mg/g at pH 5–6.

Also, adsorbents with thiol or poly(1-vinylimidazole) functionalized superparamagnetic nanoparticles were found highly selective for adsorption of Hg(II) as reported by Hakami et al. (2015). Awual et al. (2016a, b) explored the ligands *N,N*-disalicylidene-4,5-dimethyl-phenylenedene impregnated highly ordered mesoporous silica based optical nanocomposite material for sensitive, selective detection and removal of mercury (Hg(II)) from polluted water solutions. The Hg(II) ions adsorption reached an equilibrium within short time and the maximum monolayer adsorption capacity was 179.74 mg/g. The comparison of adsorption capacities of nanocomposite materials with other adsorbents reported in the literature for Hg(II) ion in aqueous solutions are shown in Table 4.

Table 4 Comparison of adsorption capacities of nanocomposite materials with other adsorbents reported in the literature for Hg(II) ion in aqueous solutions

Adsorbent	Adsorbent capacity (mg/g)	References
Magnetic iron oxide nanoparticle	0.125	Cho et al. (2012)
Alkynyl monolayers silica gel	174.3	Choi et al. (2016)
Chitosan coated magnetic nanoparticles	10	Awual et al. (2016a, b)
Functionalized SBA-15	122.36	Mureseanu et al. (2010)
AC-based pistachio-nut shells	147.1	Asasian et al. (2012)
Pyrite	180.53	Duan et al. (2016)
Functionalized MCM-41	140	Quintanilla et al. (2006)

3.3.4 Cadmium (Cd)

Al Hamouz et al. (2017a, b) developed new series of pyrrole based dithiocarbamate functionalized polymers via one-pot polycondensation reaction. The developed polymer (PYEDCS2) is comparatively good absorbent for the removal of Cd ions from wastewater compared to the reported adsorbent (Q_m 14.18 mg g^{-1}) materials like Na-montmorillonite 5.2 mg g^{-1} (Abollino et al. 2003), Biosorption 0.71 mg g^{-1} (Fawzy 2016), magnetic GO 5.34 mg g^{-1} (Zare-Dorabei et al. 2016), Red mud 10.6 mg g^{-1} (Sahu et al. 2011) and activated carbon 8 mg g^{-1} (Leyva-Ramos et al. 1997). *Rhodobacter sphaeroides* was used for bioremediation of wastewater polluted with Cd. The results showed that Rhodobacter sphaeroides had relatively higher remediation effect about 97.92%, when compared with other bioremediation methods. Recently, Verma and Sarkar (2017) investigated simultaneous removal of Cd^{2+} and p-cresol by micellar-enhanced ultrafiltration (MEUF) with rhamnolipid (RHL) as a biosurfactant. The maximum rejection coefficient of 98.8 and 25% for Cd^{2+} and p-cresol, respectively were obtained under the following optimal conditions: RHL: 370 mg/L, Cd^{2+}: 60 mg/L, p-cresol: 75 mg/L and pH 7.8 (Verma and Sarkar 2017).

Several authors reported that natural and agricultural waste materials that are biodegradable, readily available at little or no cost, in replacement of conventional

Table 5 Sorbent material and their sorption capacity

Sorbent material (natural/ agriculture waste)	Sorption capacity (mg g^{-1})	References
Wheat bran	0.7	Singh et al. (2006)
Modified walnut sawdust	4.51	Bulut and Tez (2003)
Hazelnut shells	5.42	Bulut and Tez (2007)
Abies sachalinensis bark/*Hardwickia binata* bark	6.72/33.6	Seki et al. (1997)
Coffee husks	6.9	Oliveira et al. (2008)
Pinus pinaster bark	7.84	Kumar (2006)
Olive stone waste	8.99	Fiol et al. (2006)
Modified sawdust	9.29	Taty-Costodes et al. (2003)
Olive cake	10.56	Doyurum and Celik (2006)
Ceratonia siliqua bark	14.27	Farhan et al. (2012)
Cassava waste	14.3	Abia et al. (2003)
Bark of Eucalyptus	14.56	Ghodbane et al. (2008)
Romanina *Pinus sylvestris* L. Bark	27.32	Tofan et al. (2012)
Eriobotrya japonica Loquat bark	28.8	Salem et al. (2012)
Waste tea leaves	31.48	Ahluwalia and Goyal (2005)
Cassia siamea bark	37.7	John et al. (2011)

sorbents, is in conformity with green chemistry and green environment. Fraxinus tree leaves were successfully used to remove ternary mixture of Cd(II) from an aqueous solution in a batch system under the optimal conditions of pH = 5, sorbent mass of 0.05 g, and initial Cd(II) concentration of 129.1 mg/L (Zolgharnein et al. 2017). A comparison of different natural biodegradable sorbents/agricultural waste and their adsorption capacity is listed in Table 5.

4 Removal of Organic Pollutants in Water

Organic compounds such as dies, alcohol and benzene groups are found in wastewater causing serious problems due to their toxicity. Dyes are one of the main organic compounds that found excessively in wastewater, usually as an output from industrial effluents (Vidya et al. 2016). Organic dyes as Methylene Blue (MB) (Nava et al. 2017a, b) Rhodamine B (RhB) (Sharma et al. 2017), Remazol Brilliant Orange (RBO) (Bibi et al. 2017) and Malachite Green (MG) (Upadhyay et al. 2015) are reported to have harmful effect on the environment. One of the best approach for the elimination of such dyes consists on their degradation into less harmful components. This can be achieved using catalysis reaction, which is a series of chemical reactions that excite the catalyst atoms creating electrons and holes forming reactive oxygen species that cause degradation of organic compounds. The creation of electrons and holes depends on the power of the irradiation as well as the catalyst. Other factors such as the initial concentration of the dye and pH of the solution also affect the efficiency of the catalysis procedure (Bibi et al. 2017). To evaluate the efficiency of the degradation, the following equation is usually applied:

$$R(\%) = \frac{C_0 - C_t}{C_0} \times 100 \quad (1)$$

where R (%) is the compound/dye removal efficiency, C_0 (mg/L) is the initial concentration of compound/dye in the solution, and C_t (mg/L) is the concentration of compound/dye at a certain time in min (Kuang et al. 2013).

There exists several approaches to enhance the efficiency of the catalysis process, this includes creating a heterogenous catalyst by using mixed compounds (Amiri et al. 2017; Kuang et al. 2013) as well as doping (Wang et al. 2017a, b) resulting in an increase in the generation of electron- hole pairs. Moreover, adding a nobel metal such as Pd (Wang et al. 2017a, b) and Ag (Vendamani et al. 2017) are found to be very effective in improving the catalyst reaction by surface plasmon resonance (SPR) which occurs when nobel metal atoms accumulate on the surface of the catalyst creating a larger electron yield. Most of the catalysis reactions are performed under irradiation (photocatalysis) to enhance the excitation process of atoms.

Recently, researchers have applied a successful green route to synthesise different iron oxide NPs (α-FeOOH, β-FeOOH and Fe_2O_3) using "*Sapindus mukorossi*" which is a plant that biodegrade in nature with no harmful effect (Jassal et al. 2016). These nature friendly iron oxides have shown a good binding ability with

3-Aminopyridine which is an organic compound that is used in pharmacology and may have harmful effect on environment. Experiments showed that the binding percentage of surface of 3-Aminopyridine is 94.27, 92.36 and 88.31% using the synthesized α-FeOOH, β-FeOOH and Fe_2O_3 NPs (Jassal et al. 2016).

Zinc Oxide (ZnO) NPs is an important semiconductor oxide that has been used in catalytic reactions. Recently, pure nanocrystalline ZnO NPs have been prepared using Jackfruit leaf extract (Vidya et al. 2016). SEM images showed aggregation of particles while HRTEM images showed a hexagonal-shaped morphology with particle size in the range 15–25 nm (Fig. 11a). Degradation mechanism was summarized in the following equations:

$$RB_1 \xrightarrow{hv} RB_2 \quad (2)$$

$$ZnO \xrightarrow{hv} e^-(CB) + h^+(VB) \quad (3)$$

$$h^+ + OH^- \rightarrow OH^* \quad (4)$$

$$RB_2 + OH^* \rightarrow Products \quad (5)$$

The obtained results revealed that under irradiation, RB dye start to excite into different chemical states while the green synthesized ZnO NPs create electron–hole pairs, as electrons are in the conduction band while the holes are in the valance band. A hole causes electron removed from the medium causing OH– which eventually tends to the radical OH*. This radical causes the RB dye to decompose into different products that are considered to be harmless to the surroundings. It was also shown that the pH of the medium affected the degradation efficiency; a pH 8 was the optimum value with a maximum degradation efficiency of 85%. This is

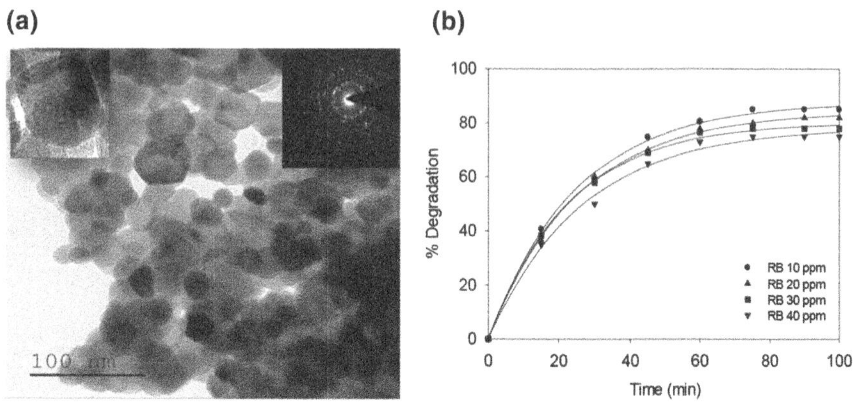

Fig. 11 ZnO nanoparticles synthesised by a simple method using leaf extract of jackfruit (*Artocarpus heterophyllus*): **a** TEM image of ZnO NPs; **b** effect of initial dye concentration on 250 mL of dye solution with 60 mg ZnO NPs at pH 8 (Reprinted from Vidya et al. (2016), Copyright (2016), with permission from Elsevier)

because at pH 8, the anionic medium reacts with the dye causing the creation of more holes, however, at higher pH (9 and 10) the medium becomes more negatively charged which allows recombination and hence resulted in less holes that minimized degradation process. The initial concentration of the RB dye has also a direct influence on the degradation rate too (Fig. 11b). The maximum degradation (85%) was related to the minimum concentration (10 ppm), and was attributed to the effective catalyst surface that relatively increases with decreasing the dye concentration favouring more catalysis reactions (Vidya et al. 2016).

Moreover, ZnO NPs synthesized using *Camellia sinensis* extract (CSE) showed a considerable efficiency in the degradation of MB from water (Nava et al. 2017a). XRD results confirmed the formation of pure ZnO phase. It was shown that using different concentrations of CSE had a direct effect on the crystallite size as it decreased from 17.47 to 9.04 nm when CSE amount was increased from 10 to 40 mL per gram of Zinc nitrate. This even affected the shape and homogeneity of the nanocrystals as shown in HRTEM image (Fig. 12a). The samples with smaller crystallite size (40 mL) chosen for photocatalysis experiment for the degradation of MB dye showed an efficiency of 84.37% at 120 min under UV irradiation (Fig. 12b) (Nava et al. 2017a).

Furthermore, photocatalytic activity was tested for ZnO NPs synthesized using peel extract from different fruits including tomatoes (*Lycopersicon esculentum*), orange (*Citrus sinensis*), grapefruit (*Citrus paradisi*) and lemon (*Citrus aurantifolia*) making use of the aromatic hydroxyle groups that cause nucleation process ending up with forming ZnO NPs named as M1, M2, M3 and M4 respectively (Nava et al. 2017b). The crystallite size measured from XRD was found to be 9.01, 12.55, 19.66 and 11.39 nm which was confirmed by HRTEM too (Fig. 13a). Photocatalytic experiment under UV light showed the degradation of MB dye using the "fruit peel" prepared ZnO (Fig. 13b). It was obvious that ZnO prepared from

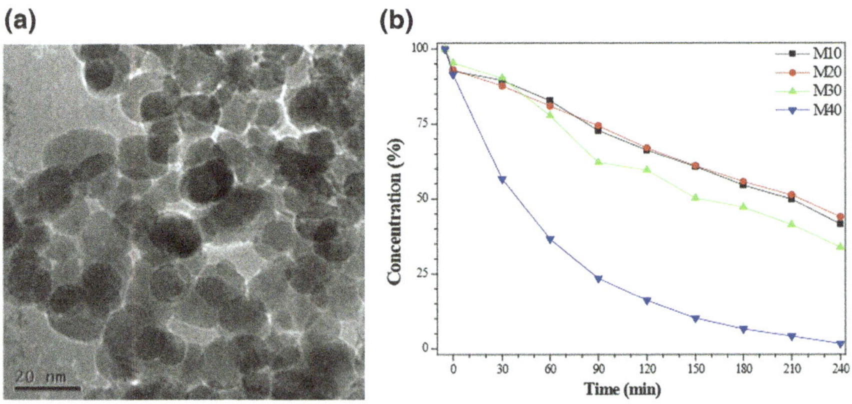

Fig. 12 ZnO nanoparticles prepared using *Camellia sinensis* extract: **a** morphology of the ZnO NPs (40 mL); **b** MB degradation through photocatalytic activity of the ZnO NPs (Reprinted from Nava et al. (2017a), Copyright (2017), with permission from Elsevier)

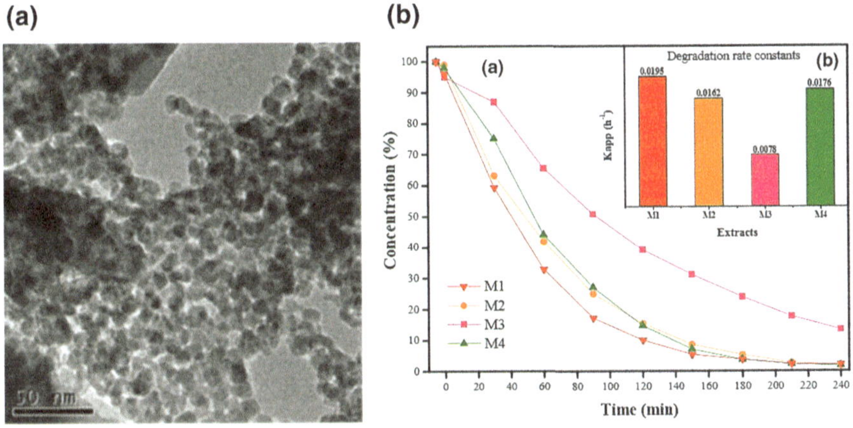

Fig. 13 ZnO nanoparticles prepared via green synthesis using different fruit peel extracts (tomato, orange, grapefruit, lemon): **a** TEM micrographs of ZnO NPs (M1); **b** photocatalytic activity and degradation rate constants of the ZnO NPs (Reprinted from Nava et al. (2017b), Copyright (2017), with permission from Elsevier)

tomato peel has the best degradation rate (97% in 180 min) as well as the highest k value (0.0195). This is mainly due to the relatively small crystallite size (Nava et al. 2017b).

Cerium Oxide (CeO_2) NPs were synthesized through a cost effective green method using "*Azadirachta indica*" plant leaf extract (Sharma et al. 2017). This resulted in agglomerated spherical particles with a uniform size distribution in the range 10–15 nm as shown in FESEM (Fig. 14a). The as-prepared green CeO_2 NPs have shown great effect in the degradation of RhB dye into less harmful radicals (Fig. 14b), under visible light. Photocatalysis experiment showed that 96% of the dye was degraded from water after 120 min of treatment (Sharma et al. 2017).

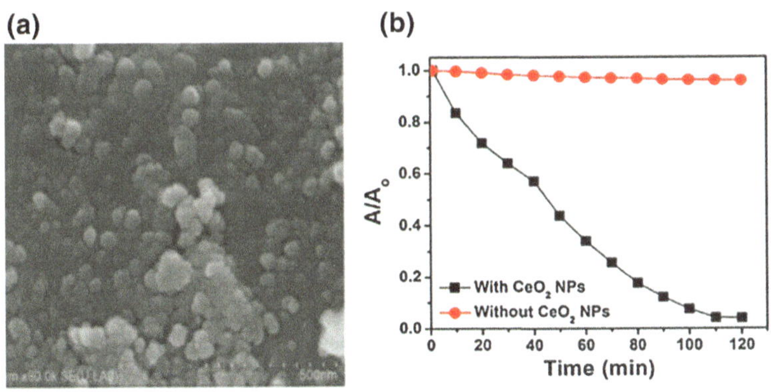

Fig. 14 CeO_2 nanoparticles synthesized by eco-friendly green synthesis using *Azadirachta indica* leaf extract: **a** FESEM of the green synthesized CeO_2 NPs; **b** curve of A/A_o versus time interval (Reprinted from Sharma et al. (2017), Copyright (2017), with permission from Elsevier)

Moreover, a successful mechanism of biosynthesis of cobalt oxide (CoO) was achieved using "*P. granatum*" extracts which can be found easily in local markets in Pakistan (Bibi et al. 2017). The reaction process monitored by UV-visible spectra confirmed that "*P. granatum*" played a key role in surface reduction and hence, the formation of nanoscale CoO NPs (Fig. 15a). Pure face centered cubic crystalline phase was confirmed by XRD with an estimated average size in the range 43.78–73.10 nm. SEM image showed spherical agglomerated particles with an average size of 80 nm. These biosynthesized CoO NPs was used for the photocatalytic

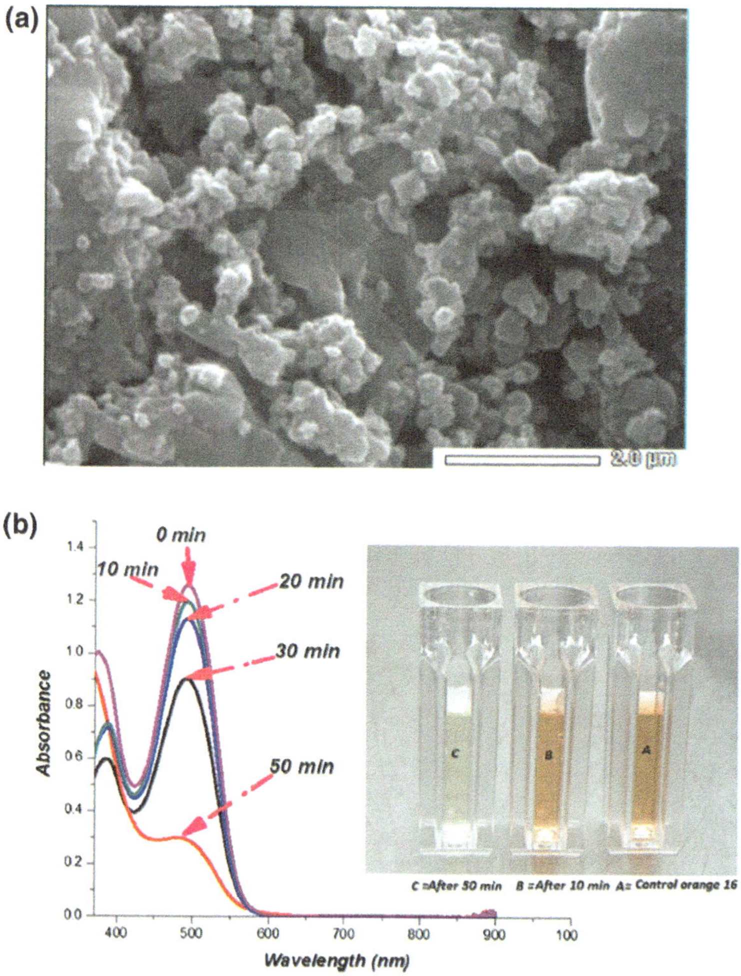

Fig. 15 CoO NPs fabricated using *Punica granatum* peel extract: **a** SEM image of CoO NPs **b** UV absorption spectra of dye before and after treatment using CoO NPs as a catalyst under solar light irradiation and (A–C) visual observation of dye before and after treatment (Reprinted from Bibi et al. (2017), Copyright (2017), with permission from Elsevier)

degradation of RBO in water under solar light (Bibi et al. 2017). A 78.45% of RBO dye was degraded after only 50 min, indicating a considerable efficiency of these biosynthesized CoO NPs in the elimination of such organic compound (Fig. 15b).

Among green NPs, ferrite system has shown efficiency in organic compounds removal (Amiri et al. 2017). Green sol-gel was carried out to synthesize Cobalt Ferrite Silica ($CoFe_2O_4$-SiO_2) composite using various amount of "*Salix alba* bark (SA)" extract (3, 5, 7, 10, 15, 20 mg). The average particle size was found to decrease from 51.02 to 15.41 nm with the amount of SA extract increasing from 3 to 20 mg. Consequently, room temperature saturation magnetization was found to raise from 0.14 to 2.89 emu/g. Hence, the sample with the highest amount of SA extract (20 mg) was chosen as photocatalyst for the degradation of malachite green (MG) and the adsorption capacity reached 75.5 mg/g indicating good degradation process (Amiri et al. 2017).

Carbon based NPs are known to have high surface area which can be easily modified, as well as their good thermal stability which makes them good candidates for catalysis (Choudhary et al. 2017; Wang et al. 2017a, b). Studies showed that green synthesized graphene oxide can be used in water treatment. Grape extract was applied to obtain an environmental friendly GO NPs that can be used efficiently in the removal of dyes from water (Upadhyay et al. 2015).

Nitrogen doped CNTs with Palladium (Pd/N-CNT) were prepared through green routes using different ratios of Polyvinylpyrrolidone (PVP, a water soluble polymer) (Wang et al. 2017a, b). High magnification TEM image (Fig. 16a) showed Pd nanoparticles clearly distributed onto CNT. XRD and EDS techniques confirmed the purity and chemical composition of the samples with average diameter ranging from 30 to 60 nm. Theses samples were shown to be an excellent catalyst for the reaction of iodobenzene (an organic iodine compound that might found in wastewater). The effect of temperature was tested too as experiments showed a that at 50 °C, the efficiency of the reaction reached 92.8% in prolonged time where at higher temperature the efficiency reached almost 100% in only 2 h. Also, Pd/N-CNT have shown a great effect on the catalysis of other organic compounds such as Benzyl alcohol and Benzaldehyde as it reached 85 and 100%, respectively, at Pd:PVP ratio of 8.6 wt% (Fig. 16b). It is noteworthy to mention the effect of Pd as a nobel metal in the catalysis process as it enhances the efficiency by surface plasmonic resonance (SPR) effect. However, exceeding a certain amount of Pd could actually disturb the flow of electrons allowing recombination with holes which lower the efficiency of the catalysis. This can be shown clearly evidenced from Fig. 4 as the efficiency of the reaction was found to decrease for the Benzaldehyde when the Pd:PVP ratio is increased to 9 wt% (Wang et al. 2017a, b).

Recently, green synthesis of graphene oxide (GO) has been reported for catalytic applications (Sreekanth et al. 2016). Nanosheets of GO decorated with Ag NPs (GO-Ag) was synthesized using "*Picrasma quassioides*" plant extract. Morphology and chemical composition were confirmed by SEM, TEM and EDAX, revealing monodispersed spheres with no noticeable agglomeration. It was shown that these nanosheets caused the degradation of more than 80% of MB dye from water within 30 min with a rate constant (k) of 0.038 min^{-1} (Sreekanth et al. 2016).

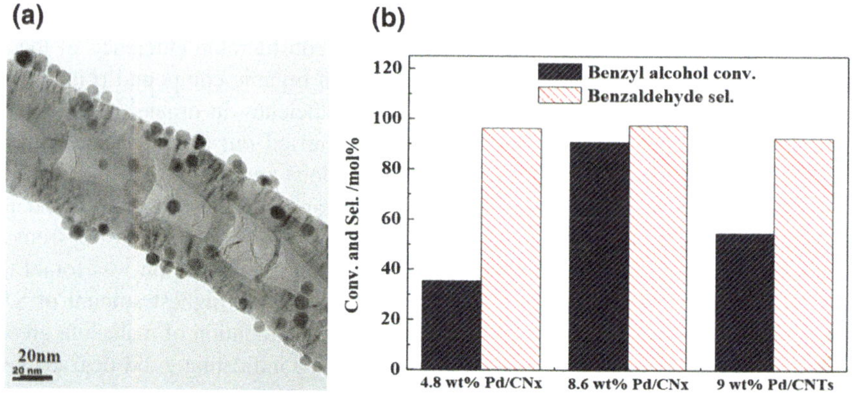

Fig. 16 N-doped CNTs with Pd (Pd/N-CNT) synthesized through green route: **a** high-magnification TEM image; **b** catalytic results for the oxidation of benzyl alcohol with O_2 over different catalyst. Reaction conditions: 0.02 g catalyst, 120 °C, 3 h, 5 mL benzyl alcohol, 20 mL/min O_2 (Reprinted from Wang et al. (2017a, b), Copyright (2017), with permission from Elsevier)

Moreover, another research work has successfully reported the synthesis of Ag reduced graphene (Ag-rGO) using the extract of another plant called "*Psidium guajava*" with different concentrations of $AgNO_3$ (1, 2, 5 and 10 mM in 1 mg) and hence obtained different amount of Ag NPs. Such green synthesis method is claimed to have less toxic outcomes in comparison with other conventional methods (Vendamani et al. 2017). The observed sharp and high peaks of XRD pattern confirm the crystalline nature, while the intensity of Ag-peaks increases with the variation in Ag amount, as seen in Ag 10-rGO. Chemical composition was also confirmed by FTIR analysis whereas TEM revealed layers of graphene oxide nanosheets with homogenous distribution of Ag NPs. Degradation studies showed that these synthesized samples are efficient for the degradation of MB dye from water (Vendamani et al. 2017). The quenching constant k represents a measure of photoluminescence (PL) quenching efficiency of a given system and was calculated using the following equation:

$$I_o/I = e^k \quad (6)$$

where (I_o/I) is the MB intensity ratio. The k value is found to be 32.5 and 13.4 $(mg/ml)^{-1}$ for rGO and Ag10-rGO respectively (Vendamani et al. 2017).

Sapindus-mukorossi is a plant used as a natural surfactant in the synthesis of iron hexacyanoferrate (Shanker et al. 2017). The formation of pure crystals of cubic phase was confirmed through the sharp high peaks in XRD pattern. An average particle size of 50 nm was observed by FESEM and the chemical composition was confirmed too by energy dispersive spectroscopy (EDS). TEM revealed rod- and spherical-shaped NPs with particle size in the range 10–60 nm (Fig. 17a). These NPs was tested for the adsorption of several toxic organic compounds including

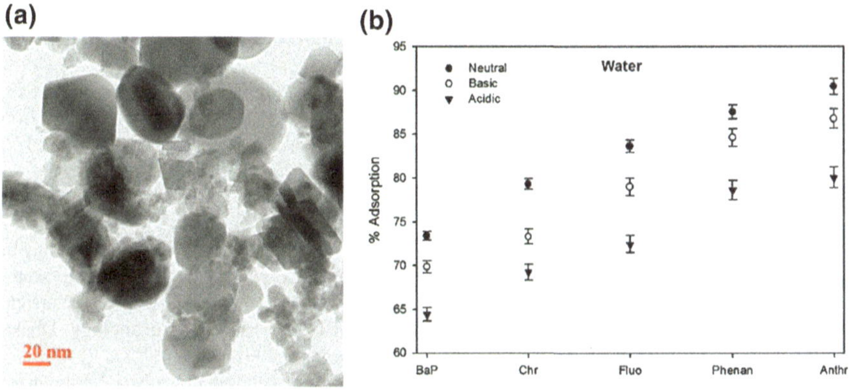

Fig. 17 Iron hexacyanoferrate nanoparticles synthesized green route using *Sapindus-mukorossi* as natural surfactant: **a** FE-SEM image of iron hexacyanoferrate NPs; **b** adsorption of different organic compounds at different pH (Reprinted from Shanker et al. (2017), Copyright (2017), with permission from Elsevier)

anthracene, phenanthrene chrysene and benzopyrene in water. A good adsorption rate was found above 70% for all compounds. This efficiency was found to be directly proportional to the initial concentration of the compound. This was explained with the enhancement of the surface area with the concentration, which creates more active sites of catalysts and hence results in higher adsorption rate (Fig. 17b). The effect of pH was also examined too, showing that the maximum adsorption rate was found in a neutral pH medium, where proceeding the reaction in an acidic medium will decrease the adsorption for the tested organic compounds (Fig. 17b). This was explained as follow: the created holes during the catalysis reaction may interact with the compounds at acidic medium limiting the formation of ROS and lowering the adsorption efficiency (Shanker et al. 2017). It is very important to mention that the effect of pH depends on the chemical behavior of the introduced compounds as well as the catalyst used within the medium.

References

Abia AA, Harsfall M, Didi O (2003) The use of chemically modified and unmodified cassava waste for the removal of Cd, Cu and Zn ions from aqueous solution. Bioresour Technol 90:345–348

Abollino O, Aceto M, Malandrino M et al (2003) Adsorption of heavy metals on Na-montmorillonite. Effect of pH and organic substances. Water Res 37:1619–1627

Adebisi JA, Agunsoye JO, Bello SA et al (2017) Potential of producing solar grade silicon nanoparticles from selected agro-wastes. Sol Energy 142:68–86

Adegoke KA, Bello OS (2015) Dye sequestration using agricultural wastes as adsorbents. Water Res Ind 12:8–24. https://doi.org/10.1016/j.wri.2015.09.002

Adio SO, Omar MH, Asif M et al (2017) Arsenic and selenium removal from water using biosynthesized nanoscale zero-valent iron: a factorial design analysis. Process Saf Environ Prot 107:518–527

Aghababaei A, Ncibi MC, Sillanpää M (2017) Optimized removal of oxytetracycline and cadmium from contaminated waters using chemically-activated and pyrolyzed biochars from forest and wood-processing residues. Bioresour Technol 239:28–36. https://doi.org/10.1016/j.biortech.2017.04.119

Ahluwalia SS, Goyal D (2005) Removal of heavy metals by waste tea leaves from aqueous solution. Eng Life Sci 5:158–162

Ahmed FM (2001) An overview of arsenic removal technologies in Bangladesh and India. In: Feroze Ahmed M et al (eds) Technologies for arsenic removal from drinking water. A compilation of papers presented at the international workshop on technologies for arsenic removal from drinking water. Bangladesh University of Engineering and Technology, Dhaka, Bangladesh and the United Nations University, Tokyo, May 2001

Al Hamouz OCS, Adelabu IO, Saleh TA (2017a) Novel cross-linked melamine based polyamine/CNT composites for lead ions removal. J Environ Manage 192:163–170

Al Hamouz OCS, Estatie M, Tawfik A (2017b) Saleh removal of cadmium ions from wastewater by dithiocarbamate functionalized pyrrole based terpolymers. Sep Purif Technol 177:101–109

Albadarin AB, Charara M, Tarboush BJA et al (2017) Mechanism analysis of tartrazine biosorption onto masau stones; a low cost by-product from semi-arid regions. J Mol Liq 242:478–483. https://doi.org/10.1016/j.molliq.2017.07.045

Alhogbi BG (2017) Potential of coffee husk biomass waste for the adsorption of Pb(II) ion from aqueous solutions. Sustain Chem Pharm 6:21–25. https://doi.org/10.1016/j.scp.2017.06.004

Ali A (2017) Removal of Mn(II) from water using chemically modified banana peels as efficient adsorbent. Environ Nanotech Monit Manage 7:57–63. https://doi.org/10.1016/j.enmm.2016.12.004

Ali HR, Hassaan MA (2017) Applications of bio-waste materials as green synthesis of nanoparticles and water purification. Adv Mater Chem 1(1):6–22. https://doi.org/10.11648/j.amc.20170101.12

Ali A, Saeed K, Mabood F (2016) Removal of chromium (VI) from aqueous medium using chemically modified banana peels as efficient low-cost adsorbent. Alex Eng J 55(3):2933–2942. https://doi.org/10.1016/j.aej.2016.05.011

Aljeboree AM, Alshirifi AN, Alkaim AF (2017) Kinetics and equilibrium study for the adsorption of textile dyes on coconut shell activated carbon. Arab J Chem 10(2):S3381–S3393. https://doi.org/10.1016/j.arabjc.2014.01.020

Almubaddal F, Alrumaihi K, Ajbar A (2009) Performance optimization of coagulation/flocculation in the treatment of wastewater from a polyvinyl chloride plant. J Hazard Mater 161:431–438

Alqadami AA, Naushad M, Abdalla MA et al (2017) Efficient removal of toxic metal ions from wastewater using a recyclable nanocomposite: a study of adsorption parameters and interaction mechanism. J Clean Prod 156:426–436. https://doi.org/10.1016/j.jclepro.2017.04.085

Al-Shehri S, Al-Senany N, Altuwirqi R et al (2017) Green synthesis of Cu_xO nanoscale MOS capacitors processed at low temperatures. Surf Coat Technol 320:246–251

Alshehri SM, Naushad M, Ahamad T et al (2014) Synthesis, characterization of curcumin based ecofriendly antimicrobial bio-adsorbent for the removal of phenol from aqueous medium. Chem Eng J 254:181–189. https://doi.org/10.1016/j.cej.2014.05.100

Alslaibi TM, Abustan I, Ahmad MA et al (2014) Comparison of activated carbon prepared from olive stones by microwave and conventional heating for iron (II), lead (II), and copper (II) removal from synthetic wastewater. Environ Prog Sustain Energy 33:1074–1085

Al-Zoubi H, Ibrahim KA, Abu-Sbeih KA (2015) Removal of heavy metals from wastewater by economical polymeric collectors using dissolved air flotation process. J Water Process Eng 8:19–27

Amiri M, Salavati-Niasari M, Akbari A et al (2017) Removal of malachite green (a toxic dye) from water by cobalt ferrite silica magnetic nanocomposite: herbal and green sol-gel autocombustion synthesis. Int J Hydrogen Energy 4:24846–24860

Anand K, Tilokea C, Naidoo P et al (2017) Phytonanotherapy for management of diabetes using green synthesis nanoparticles. J Photochem Photobiol, B: Biol 173:626–639
Anastopoulos I, Kyzas GZ (2015) Progress in batch biosorption of heavy metals onto algae. J Mol Liq 209:77–86. https://doi.org/10.1016/j.molliq.2015.05.023
Anastopoulos I, Karamesouti M, Mitropoulos AC et al (2017) A review for coffee adsorbents. J Mol Liq 229:555–565. https://doi.org/10.1016/j.molliq.2016.12.096
Angelis GD, Medeghini L, Conte AM et al (2017) Recycling of eggshell waste into low-cost adsorbent for Ni removal from wastewater. J Clean Prod 164:1497–1506. https://doi.org/10.1016/j.jclepro.2017.07.085
Archana B, Manjunath K, Nagaraju G et al (2017) Enhanced photocatalytic hydrogen generation and photostability of ZnO nanoparticles obtained via green synthesis. Int J Hydrogen Energy 42:5125–5131
Aryal M, Ziagova M, Liakopoulou-Kyriakides M (2010) Study on arsenic biosorption using Fe (III)-treated biomass of *Staphylococcus xylosus*. Chem Eng J 162:178–185
Asasian N, Kaghazchi T, Soleimani M (2012) Elimination of mercury by adsorption onto activated carbon prepared from the biomass material. J Ind Eng Chem 18:283–289
Asuquo ED, Martin AD (2016) Sorption of cadmium (II) ion from aqueous solution onto sweet potato (*Ipomoea batatas* L.) peel adsorbent: characterisation, kinetic and isotherm studies. J Environ Chem Eng 4(4):4207–4228. https://doi.org/10.1016/j.jece.2016.09.024
Attari M, Bukhari SS, Kazemian H et al (2017) A low-cost adsorbent from coal fly ash for mercury removal from industrial wastewater. J Environ Chem Eng 5:391–399
Awual MR (2016a) Assessing of lead(III) capturing from contaminated wastewater using ligand doped conjugate adsorbent. Chem Eng J 289:65–73
Awual MR (2016b) Novel nanocomposite materials for efficient and selective mercury ions capturing from wastewater. Chem Eng J 307:456–465
Awual MR, Hasan MM, Shahat A et al (2015) Investigation of ligand immobilized nano-composite adsorbent for efficient cerium(III) detection and recovery. Chem Eng J 265:210–218. https://doi.org/10.1016/j.cej.2014.12.052
Awual MR, Hasan MM, Eldesoky GE et al (2016a) Facile mercury detection and removal from aqueous media involving ligand impregnated conjugate nanomaterials. Chem Eng J 290:243–251
Awual MR, Hasan MM, Khaleque MA et al (2016b) Treatment of copper(II) containing wastewater by a newly developed ligand based facial conjugate materials. Chem Eng J 288:368–376
Ayyappan R, Sophia CA, Swaminathan K et al (2005) Removal of Pb (II) from aqueous solution using carbon derived from agricultural wastes. Process Biochem 4:1293–1299
Babel S, Kurniawan TA (2003) Low-cost adsorbents for heavy metal uptake from contaminated water: a review. J Hazard Mater 97:219–243
Baca R, Cheong KY (2015) Green synthesis of iron oxide thin-films grown from recycled iron foils. Mater Sci Semicond Process 29:294–299
Badawi MA, Negm NA, Abou Kana MTH et al (2017) Adsorption of aluminum and lead from wastewater by chitosan-tannic acid modified biopolymers: isotherms, kinetics, thermodynamics and process mechanism. Int J Biomacromol. https://doi.org/10.1016/j.ijbiomac.2017.03.003
Bahadir T, Bakan G, Altas L et al (2007) The investigation of lead removal by biosorption: an application at storage battery industry wastewaters. Enzyme Microb Technol 41:98–102
Banerjee S, Chattopadhyaya MC (2017) Adsorption characteristics for the removal of a toxic dye, tartrazine from aqueous solutions by a low cost agricultural by-product. Arab J Chem 10: S1629–S1638. https://doi.org/10.1016/j.arabjc.2013.06.005
Bao S, Li K, Ning P et al (2016) Highly effective removal of mercury and lead ions from wastewater by mercaptoamine-functionalised silica-coated magnetic nano-adsorbents: behaviours and mechanisms. Appl Surf Sci 393:457–466
Barakat MA (2011) New trends in removing heavy metals from industrial wastewater. Arab J Chem 4:361–377

Barquilha CER, Cossich ES, Tavares CRG et al (2017) Biosorption of nickel(II) and copper(II) ions in batch and fixed-bed columns by free and immobilized marine algae Sargassum sp. J Clean Prod 150:58–64. https://doi.org/10.1016/j.jclepro.2017.02.199

Bartczak P, Norman M, Klapiszewski Ł et al (2015) Removal of nickel(II) and lead(II) ions from aqueous solution using peat as a low-cost adsorbent: a kinetic and equilibrium study. Arab J Chem. https://doi.org/10.1016/j.arabjc.2015.07.018

Bharath G, Alhseinat E, Ponpandian N et al (2017) Development of adsorption and electrosorption techniques for removal of organic and inorganic pollutants from wastewater using novel magnetite/porous graphene-based nanocomposites. Sep Purif Technol 188:206–218. https://doi.org/10.1016/j.seppur.2017.07.024

Bhargavi SD, Savitha J (2014) Arsenate resistant *Penicillium coffeae*: a potential fungus for soil bioremediation. Bull Environ Contam Toxicol 92:369–373

Bibi I, Nazar N, Iqbal M, Kamal S, Nawaz H, Nouren S, Safa Y, Jilani K, Sultan M, Ata S, Rehman F, Abbas M (2017) Green and eco-friendly synthesis of cobalt-oxide nanoparticle: characterization and photo-catalytic activity. Adv Powder Technol 28:2035–2043

Bulgariu L, Bulgariu D (2014) Enhancing biosorption characteristics of marine green algae (*Ulva lactuca*) for heavy metals removal by alkaline treatment. J Bioprocess Biotechniques 4:146

Bulut Y, Tez Z (2003) Removal of heavy metal ions by modified sawdust of walnut. Fresen Environ Bull 12:1499–1504

Bulut Y, Tez Z (2007) Adsorption studies on ground shells of hazelnut and almond. J Hazard Mater 149:35–41

Carolin CF, Kumar PS, Saravanan A et al (2017) Efficient techniques for the removal of toxic heavy metals from aquatic environment: a review. J Environ Chem Eng 5:2782–2799

Cho ES, Kim J, Tejerina B et al (2012) Ultrasensitive detection of toxic cations through changes in the tunnelling current across films of striped nanoparticles. Nat Mater 11(2012):978–985

Choi JM, Jeong D, Cho E et al (2016) Chemically functionalized silica gel with alkynyl terminated monolayers as an efficient new material for removal of mercury ions from water. J Ind Eng Chem 35:376–382

Choudhary R, Patra S, Madhuria R et al (2017) Designing of carbon based fluorescent nanosea-urchin via green-synthesis approach for live cell detection of zinc oxide nanoparticle. Biosens Bioelectron 91:472–481

Chowdhury S, Mazumder MAJ, Al-Attas O et al (2016) Heavy metals in drinking water: occurrences, implications, and future needs in developing countries. Sci Total Environ 569–570:476–488

Chuang CL, Fan M, Xu M et al (2005) Adsorption of arsenic(V) by activated carbon prepared from oat hulls. Chemosphere 61:478–483

Dai B, Cao M, Fang G et al (2012) Schiff base-chitosan grafted multiwalled carbon nanotubes as a novel solid-phase extraction adsorbent for determination of heavy metal by ICP-MS. J Hazard Mater 219–220:103–110

Deb KS, Dwivedi V, Dasgupta K et al (2016) Novel Amidoamine functionalized multi-walled carbon nanotubes for removal of mercury(II) ions from wastewater: combined experimental and density functional theoretical approach. Chem Eng J 313:899

Debbaudt AL, Ferreira ML, Gschaider ME (2004) Theoretical and experimental study of M2 adsorption on biopolymers III: comparative kinetic pattern of Pb, Hg and Cd. Carbohydr Polym 56:321–332

Demirbas A (2008) Heavy metal adsorption onto agro-based waste materials: a review. J Hazard Mater 157:220–229

Devani MA, Munshi B, Oubagaranadin JUK et al (2017) Remediation of Hg(II) from solutions using *Cajanus cajan* husk as a new sorbent. Environ Technol 38:1878–1886

Dong C, Lu J, Qiu B et al (2017) Developing stretchable and graphene-oxide-based hydrogel for the removal of organic pollutants and heavy metal ions. Appl Catal B: Environ. https://doi.org/10.1016/j.apcatb.2017.10.011

Doyurum S, Celik A (2006) Pb(II) and Cd(II) removal from aqueous solutions by olive cake. J Hazard Mater 138:22–28

Duan Y, Han DS, Batchelor B et al (2016) Synthesis, characterization, and application of pyrite for removal of mercury. Colloid Surf Aspects 490:326–335
Ebtesam EB, Helmy S, Hussien H et al (2013) Bioremediation of heavy metal-contaminated effluent using optimized activated sludge bacteria. Appl Water Sci 3:181–192
El-Shafey EI, Cox M, Pichugin AA et al (2002) Application of a carbon sorbent for the removal of cadmium and other heavy metal ions from aqueous solution. J Chem Tech Biotechnol 77:429–436
Etim UJ, Umoren SA, Eduok UM (2016) Coconut coir dust as a low cost adsorbent for the removal of cationic dye from aqueous solution. J Saud Chem Soc 20:S67–S76. https://doi.org/10.1016/j.jscs.2012.09.014
Farhan AM, Salem NM, Ahmad AL et al (2012) Kinetic, equilibrium and thermodynamic studies on the biosorption of heavy metals by *Ceratonia siliqua* bark. Am J Chem 2:335–342
Fawzy MA (2016) Phycoremediation and adsorption isotherms of cadmium and copper ions by *Merismopedia tenuissima* and their effect on growth and metabolism. Environ Toxicol Pharmacol 46:116–121
Feng Y, Liu Y, Xue L et al (2017) Carboxylic acid functionalized sesame straw: a sustainable cost-effective bioadsorbent with superior dye adsorption capacity. Bioresour Technol 238:675–683. https://doi.org/10.1016/j.biortech.2017.04.066
Fiol N, Villaescusa I, Martinez M et al (2006) Sorption of Pb(II), Ni(II), Cu(II) and Cd(II) from aqueous solution by olive stone waste. Sep Purif Technol 50:132–140
Gercel O, Gercel HF (2007) Adsorption of lead(II) ions from aqueous solutions by activated carbon prepared from biomass plant material of *Euphorbia rigida*. Chem Eng J 132:289–297
Ghidan AY, Al-Antary TM, Awwad AM (2016) Green synthesis of copper oxide nanoparticles using *Punica granatum* peels extract: effect on green peach. Environ Nanotech, Monit & Manage 6:95–98
Ghodbane I, Nouri K, Hamdaoui O et al (2008) Kinetic and equilibrium study for the sorption of cadmium (II) ions from aqueous phase by eucalyptus bark. J Hazard Mater 152:148–158
Gokila S, Gomathi T, Sudha PN et al (2017) Removal of the heavy metal ion chromium (VI) using chitosan and alginate nanocomposites. Int J Biol Macromol 104(B):1459–1468. https://doi.org/10.1016/j.ijbiomac.2017.05.117
Gueu S, Yao B, Adouby K et al (2007) Kinetics and thermodynamics study of lead adsorption on to activated carbons from coconut and seed hull of the palm tree. Int J Environ Sci Technol 4:11–17
Guo H, Jiao T, Zhang Q et al (2015) Preparation of graphene oxide-based hydrogels as efficient dye adsorbents for wastewater treatment. Nanoscale Res Lett 10:272. https://doi.org/10.1186/s11671-015-0931-2
Hafshejani LD, Tangsir S, Daneshvaret E et al (2017) Optimization of fluoride removal from aqueous solution by Al_2O_3 nanoparticles. J Mol Liq 238:254–262
Hakami O, Zhang Y, Banks CJ (2015) Thiol-functionalised mesoporous silica-coated magnetite nanoparticles for high efficiency removal and recovery of Hg from water. Water Res 46:3913–3922
Halouane F, Oz Y, Meziane D et al (2017) Magnetic reduced graphene oxide loaded hydrogels: highly versatile and efficient adsorbents for dyes and selective Cr(VI) ions removal. J Colloid Interface Sci 507:360–369. https://doi.org/10.1016/j.jcis.2017.07.075
Hameed KS, Muthirulan P, Sundaram MM (2017) Adsorption of chromotrope dye onto activated carbons obtained from the seeds of various plants: equilibrium and kinetics studies. Arab J Chem 10(2):S2225–S2233. https://doi.org/10.1016/j.arabjc.2013.07.058
Harper TR, Kingham NW (1992) Removal of arsenic from wastewater using chemical precipitation methods. Water Environ Res 64:200–203
Harris HH, Pickering IJ, George GN (2003) The chemical form of mercury in fish. Science 301 (5637):1203
Hayashi J, Kazehaya A, Muroyama K et al (2000) Preparation of activated carbon from lignin by chemical activation. Carbon 38(13):1873–1878. https://doi.org/10.1016/S0008-6223(00)00027-0

Hesami F, Bina B, Ebrahimi A, Amin MM (2013) Arsenic removal by coagulation using ferric chloride and chitosan from water. Int J Environ Health Eng 2(17). https://doi.org/10.4103/2277-9183.110170

Ho YS, Chiu WT, Wang CC (2005) Equilibrium isotherms and kinetic studies of removal of methylene blue dye by adsorption onto Miswak leaves as a natural adsorbent. Bioresour Technol 96:1285–1291

Hossain M, Ngo HH, Guo WS (2012) Adsorption and desorption of copper(II) ions onto garden grass. Bioresour Technol 121:386–395

Hubadillah SK, Othman MHD, Harun Z et al (2017) A novel green ceramic hollow fiber membrane (CHFM) derived from rice husk ash as combined adsorbent-separator for efficient heavy metals removal. Ceram Int 43(5):4716–4720. https://doi.org/10.1016/j.ceramint.2016.12.122

Hussain MD, Haque MA, Islam MM et al (2001) Approaches for removal of arsenic from tubewell water for drinking purpose. In: Feroze Ahmed M et al (eds) Technologies for arsenic removal from drinking water. A compilation of papers presented at the international workshop on technologies for arsenic removal from drinking water. Bangladesh University of Engineering and Technology, Dhaka, Bangladesh and the United Nations University, Tokyo, May 2001

Imamoglu M, Tekir O (2008) Removal of copper (II) and lead (II) ions from aqueous solutions by adsorption on activated carbon from a new precursor hazelnut husks. Desalination 228:108–113

Islam M, Patel RK (2008) Polyacrylamide thorium (IV) phosphate as an important lead selective fibrous ion exchanger: synthesis, characterization and removal study. J Hazard Mater 156:509–520

Jafari MH, Mahvi A, Jonidi JA (2014) Removal of lead and zinc from battery industry wastewater using electrocoagulation process: influence of direct and alternating current by using iron and stainless steel rod electrodes. Sep Purif Technol 135:165–175

Jassal V, Shanker U, Gahlota S (2016) Green synthesis of some iron oxide nanoparticles and their interaction with 2-Amino, 3-Amino and 4-aminopyridines. Mater Today: Proc 3:1874–1882

Jawad AH, Sabar S, Ishak MAM et al (2017) Microwave-assisted preparation of mesoporous-activated carbon from coconut (*Cocos nucifera*) leaf by H_3PO_4 activation for methylene blue adsorption. Chem Eng Commun 204(10):1143–1156

Jia Y, Wu C, Lee BW et al (2017) Magnetically separable sulfur-doped $SnFe_2O_4$/graphene nanohybrids for effective photocatalytic purification of wastewater under visible light. J Hazard Mater 338:447–457. https://doi.org/10.1016/j.jhazmat.2017.05.057

Jiao Y, Wan C, Li J (2016) Synthesis of carbon fiber aerogel from natural bamboo fiber and its application as a green high-efficiency and recyclable adsorbent. Mater Design 107:26–32. https://doi.org/10.1016/j.matdes.2016.06.015

Johari K, Saman N, Song ST (2016) Adsorption enhancement of elemental mercury by various surface modified coconut husk as eco-friendly low-cost adsorbents. Int Biodeterior Biodegradation 109:45–52. https://doi.org/10.1016/j.ibiod.2016.01.004

John AC, Ibironke LO, Adedeji V et al (2011) Equilibrium and kinetic studies of the biosorption of heavy metals (cadmium) on *Cassia siamea* bark. Am-Eur J Sci Res 6:123–130

Jungcharoen P, O'Carroll D, Anotai J et al (2017) Synthesis of lignin-modified nanoscale zerovalent iron applied to arsenic removal. Full Paper Proc ECBA 3(9):1–6

Kadirvelu K, Namasivayam C (2000) Agricultural byproduct as metal adsorbent: sorption of lead (II) from aqueous solution onto coir pith carbon. Environ Tech 21:1091–1097

Kagramanov GG, Farnosova EN, Kandelaki GI (2009) Heavy metal cationic wastewater treatment with membrane methods. In: Václavíková M, Vitale K, Gallios GP, Ivaničová L (eds) Water treatment technologies for the removal of high-toxicity pollutants. NATO science for peace and security series C: environmental security. Springer, Dordrecht, pp 61–84

Karthik K, Dhanuskodi S, Gobinath C et al (2017a) Andrographis paniculata extract mediated green synthesis of CdO nanoparticles and its electrochemical and antibacterial studies. J Mater Sci: Mater Electron. https://doi.org/10.1007/s10854-017-6503-8

Karthik K, Dhanuskodi S, Kumar SP (2017b) Microwave assisted green synthesis of MgO nanorods and their antibacterial and anti-breast cancer activities. Mater Lett 206:217–220. https://doi.org/10.1016/j.matlet.2017.07.004

Kepner B, Spotts J, Mintz E et al (1998) Removal of arsenic from drinking water with enhanced hybrid aluminas and composite metal oxide particles. In: International conference on arsenic pollution of groundwater: causes, effects, remedies, Dhaka Community Hospital, Dhaka, Bangladesh

Khan MA, Alam MM, Naushad Mu et al (2015) Sol–gel assisted synthesis of porous nano-crystalline $CoFe_2O_4$ composite and its application in the removal of brilliant blue-R from aqueous phase: an ecofriendly and economical approach. Chem Eng J 279:416–424

Krishnaiah D, Joseph CG, Anisuzzaman SM et al (2017) Removal of chlorinated phenol from aqueous solution utilizing activated carbon derived from papaya (*Carica papaya*) seeds. Korean J Chem Eng 34:1377. https://doi.org/10.1007/s11814-016-0337-6

Krishnan KA, Anirudhan TS (2002) Uptake of heavy metals in batch systems by sulfurized steam activated carbon prepared from sugarcane bagasse pith. Ind Eng Chem Res 41:5085–5093

Kuang Y, Wang Q, Chen Z et al (2013) Heterogeneous Fenton-like oxidation of monochlorobenzene using green synthesis of iron nanoparticles. J Colloid Interface Sci 410:67–73

Kumar A, Jena HM (2015) High surface area microporous activated carbons prepared from Fox nut (*Euryale ferox*) shell by zinc chloride activation. Appl Surf Sci 356:753–761. https://doi.org/10.1016/j.apsusc.2015.08.074

Kumar A, Jena HM (2016) Preparation and characterization of high surface area activated carbon from Fox nut (*Euryale ferox*) shell by chemical activation with H_3PO_4. Results Phys 6:651–658. https://doi.org/10.1016/j.rinp.2016.09.012

Kumar A, Jena HM (2017a) Adsorption of Cr(VI) from aqueous phase by high surface area activated carbon prepared by chemical activation with $ZnCl_2$. Process Saf Environ Prot 109:63–71. https://doi.org/10.1016/j.psep.2017.03.032

Kumar A, Jena HM (2017b) Adsorption of Cr(VI) from aqueous solution by prepared high surface area activated carbon from Fox nutshell by chemical activation with H_3PO_4. J Environ Chem Eng 5(2):2032–2041. https://doi.org/10.1016/j.jece.2017.03.035

Kumar U (2006) Agricultural products and by-products as a low cost adsorbent for heavy metal removal from water and waste-water: a review. Sci Res Essays 1:33–37

Kumar A, Sharma G, Naushad Mu et al (2015) SPION/β-cyclodextrin core–shell nanostructures for oil spill remediation and organic pollutant removal from waste water. Chem Eng J 280:175–187

Kumar A, Kumar A, Sharma G et al (2017) Sustainable nano-hybrids of magnetic biochar supported $gC_3N_4/FeVO_4$ for solar powered degradation of noxious pollutants-synergism of adsorption, photocatalysis & photo-ozonation. J Clean Prod 165:431–451

Lam SS, Liew RK, Wong YM et al (2017) Microwave-assisted pyrolysis with chemical activation, an innovative method to convert orange peel into activated carbon with improved properties as dye adsorbent. J Clean Prod 162:1376–1387. https://doi.org/10.1016/j.jclepro.2017.06.131

Leyva-Ramos R, Rangel-Mendez JR, Mendoza-Barron J et al (1997) Adsorption of cadmium (II) from aqueous solution onto activated carbon. Water Sci Technol 35:205–211

Li K, Wang X (2008) Adsorptive removal of Pb(II) by activated carbon prepared from *Spartina alterniflora*: equilibrium, kinetics and thermodynamics. Bioresour Technol 100 (2009):2810–2815

Li C, Zhuang Z, Jin X, Chena Z (2017a) A facile and green preparation of reduced graphene oxide using *Eucalyptus* leaf extract. Appl Surf Sci 422:469–474

Li X, Peng W, Jia Y et al (2017b) Removal of cadmium and zinc from contaminated wastewater using *Rhodobacter sphaeroides*. Water Sci Techno 75:2489–2498

Ma Y-X, Kou Y-L, Xing D et al (2017) Synthesis of magnetic graphene oxide grafted polymaleicamide dendrimer nanohybrids for adsorption of Pb(II) in aqueous solution. J Hazard Mater 340:407–416. https://doi.org/10.1016/j.jhazmat.2017.07.026

Machida M, Yamazaki R, Aikawa M et al (2005) Role of minerals in carbonaceous adsorbents for removal of Pb(II) ions from aqueous solution. Sep Purif Technol 46:88–94

Martinez-Juarez RVM, Cardenas-Gonzalez JF, Moctezuma-Zarate MG (2013) Biosorption of arsenic(III) from aqueous solutions by modified fungal biomass of Paecilomyces sp. Bioinorg Chem. https://doi.org/10.1155/2013/376780

Mohan D, Kumar H, Sarswat A et al (2014) Cadmium and lead remediation using magnetic oak wood and oak bark fast pyrolysis bio-chars. Chem Eng J 236:513–528. https://doi.org/10.1016/j.cej.2013.09.057

Mondal MK, Garg R (2017) A comprehensive review on removal of arsenic using activated carbon prepared from easily available waste materials. Environ Sci Pollut Res 24:13295–13306

Mureseanu M, Reiss A, Cioatera N et al (2010) Mesoporous silica functionalized with 1-furoyl thiourea urea for Hg(II) adsorption from aqueous media. J Hazard Mater 182:197–203

Naushad M, ALOthman ZA, Khan MR et al (2014) Equilibrium, kinetics and thermodynamic studies for the removal of organophosphorus pesticide using Amberlyst-15 resin: quantitative analysis by liquid chromatography-mass spectrometry. J Ind Eng Chem 20:4393–4400. https://doi.org/10.1016/j.jiec.2014.02.006

Naushad M, Ahamad T, Al-Maswari BM et al (2017) Nickel ferrite bearing nitrogen-doped mesoporous carbon as efficient adsorbent for the removal of highly toxic metal ion from aqueous medium. Chem Eng J 330:1351–1360. https://doi.org/10.1016/j.cej.2017.08.079

Nava OJ, Luque PA, Gomez-Gutierrez CM et al (2017a) Influence of *Camellia sinensis* extract on Zinc Oxide nanoparticle green synthesis. J Mol Struct 1134:121–125

Nava OJ, Soto-Robles CA, Gomez-Gutierrez CM et al (2017b) Fruit peel extract mediated green synthesis of zinc oxide nanoparticles. J Mol Struct 1147:1–6

Nayak AK, Pal A (2017) Green and efficient biosorptive removal of methylene blue by *Abelmoschus esculentus* seed: process optimization and multi-variate modeling. J Environ Manage 200:145–159. https://doi.org/10.1016/j.jenvman.2017.05.045

Ocsoy I, Temiz M, Celik C et al (2017) Green approach for formation of silver nanoparticles on magnetic graphene oxide and highly effective antimicrobial activity and reusability. J Mol Liq 227:147–152

Okman I, Karagöz S, Tay T et al (2014) Activated carbons from grape seeds by chemical activation with potassium carbonate and potassium hydroxide. Appl Surf Sci 293:138–142. https://doi.org/10.1016/j.apsusc.2013.12.117

Oliveira WE, Franca AS, Oliveira LS, Rocha SD (2008) Untreated coffee husks as biosorbents for the removal of heavy metals from aqueous solutions. J Hazard Mater 152:1073–1081

Papa S, Šolević T, Jelena K et al (2017) Utilization of fruit processing industry waste as green activated carbon for the treatment of heavy metals and chlorophenols contaminated water. J Clean Prod 162:958–972

Parson S, Jefferson B (2006) Introduction to potable water treatment processes. Blackwell Publishing Ltd, UK

Peng W, Li H, Liu Y, Song S (2017) A review on heavy metal ions adsorption from water by graphene oxide and its composites. J Mol Liq 230:496–504

Postai DL, Demarchi CA, Zanatta F (2016) Adsorption of Rhodamine B and methylene blue dyes using waste of seeds of *Aleurites moluccana*, a low cost adsorbent. Alex Eng J 55(2):1713–1723. https://doi.org/10.1016/j.aej.2016.03.017

Quintanilla DP, Hierro I, Fajardo M et al (2006) 2-Mercaptothiazoline modified mesoporous silica for mercury removal from aqueous media. J Hazard Mater 134:245–256

Raj KR, Kardam A, Srivastava S (2013) PEI modified *Leucaena leucocephala* seed powder, a potential biosorbent for the decontamination of arsenic species from water bodies: bioremediation. Appl Water Sci 3:327–333

Reck IM, Paixão RM, Bergamasco R et al (2018) Removal of tartrazine from aqueous solutions using adsorbents based on activated carbon and *Moringa oleifera* seeds. J Clean Prod 171:85–97. https://doi.org/10.1016/j.jclepro.2017.09.237

Reddy DHK, Yun Y-S (2016) Spinel ferrite magnetic adsorbents: alternative future materials for water purification? Coord Chem Rev 315:90–111. https://doi.org/10.1016/j.ccr.2016.01.012

Ricordel S, Taha S, Cisse I et al (2001) Heavy metals removal by adsorption onto peanut husks carbon: characterization, kinetic study and modeling. Sep Purif Technol 24:389–401

Roy E, Patra S, Madhuri R et al (2016) Europium doped magnetic graphene oxide-MWCNT nanohybrid for estimation and removal of arsenate and arsenite from real water samples. Chem Eng J 299:244–254. https://doi.org/10.1016/j.cej.2016.04.051

Sahu RC, Patel R, Ray BC (2011) Adsorption of Zn(II) on activated red mud: neutralized by CO_2. Desalination 266:93–97

Salem NM, Awwad AM, Al-Dujahi AH (2012) Biosorption of Pb(II), Zn(II) and Cd(II) by *Eriobotrya japonica* Loquat bark. Int J Environ Prot 2:1–7

Santhosh C, Daneshvar E, Kollu P et al (2017) Magnetic SiO_2@$CoFe_2O_4$ nanoparticles decorated on graphene oxide as efficient adsorbents for the removal of anionic pollutants from water. Chem Eng J 322:472–487. https://doi.org/10.1016/j.cej.2017.03.144

Sartape AS, Mandhare AM, Jadhav VV (2017) Removal of malachite green dye from aqueous solution with adsorption technique using *Limonia acidissima* (wood apple) shell as low cost adsorbent. Arab J Chem 10:S3229–S3238. https://doi.org/10.1016/j.arabjc.2013.12.019

Seki K, Saito N, Aoyama M (1997) Removal of heavy metal ions from solutions by coniferous barks. Wood Sci Technol 31:441–447

Senapati K, Alam I (2001) Apyron arsenic treatment unit—reliable technology for arsenic safe water. In: Feroze Ahmed M et al (eds) Technologies for arsenic removal from drinking water. A compilation of papers presented at the international workshop on technologies for arsenic removal from drinking water. Bangladesh University of Engineering and Technology, Dhaka, Bangladesh and the United Nations University, Tokyo, May 2001

Shahat A, Awual MR, Khaleque MA et al (2015) Large-pore diameter nano-adsorbent and its application for rapid lead(II) detection and removal from aqueous media. Chem Eng J 273:286–295

Shanker U, Jassal V, Rani M (2017) Green synthesis of iron hexacyanoferrate nanoparticles: potential candidate for the degradation of toxic PAHs. J Environ Chem Eng 5:4108–4120

Sharma VK, Dutta PK, Ray AK (2007) Review of kinetics of chemical and arsenic(III) as influenced by pH. Environ Sci Health 42:997

Sharma JK, Srivastavaa P, Ameen S et al (2017) Phytoconstituents assisted green synthesis of cerium oxide nanoparticles for thermal decomposition and dye remediation. Mat Res Bull 91:98–107

Shehzad K, Xie C, He J et al (2018) Facile synthesis of novel calcined magnetic orange peel composites for efficient removal of arsenite through simultaneous oxidation and adsorption. J Colloid Int Sci 511:155–164. https://doi.org/10.1016/j.jcis.2017.09.110

Shen K, Gondal MA (2017) Removal of hazardous Rhodamine dye from water by adsorption onto exhausted coffee ground. J Saudi Chem Soc 21(1):S120–S127. https://doi.org/10.1016/j.jscs.2013.11.005

Shen B, Tian L, Li F et al (2017) Elemental mercury removal by the modified bio-char from waste tea. Fuel 187:189–196. https://doi.org/10.1016/j.fuel.2016.09.059

Singh KK, Singh AK, Hasan SH (2006) Low cost bio-sorbent 'wheat bran' for the removal of cadmium from wastewater: kinetic and equilibrium studies. Bioresour Technol 97:994–1001

Singh C, Sahu J, Mahalik K et al (2008) Studies on the removal of Pb(II) from wastewater by activated carbon developed from *Tamarind wood* activated with sulphuric acid. J Hazard Mater 153:221–228

Smily JRMB, Sumithra PA (2017) Optimization of chromium biosorption by fungal adsorbent, *Trichoderma* sp. BSCR02 and its desorption studies. HAYATI J Biosci. https://doi.org/10.1016/j.hjb.2017.08.005

Son E-B, Poo K-M, Chang J-S et al (2018) Heavy metal removal from aqueous solutions using engineered magnetic biochars derived from waste marine macro-algal biomass. Sci Total Environ 615:161–168. https://doi.org/10.1016/j.scitotenv.2017.09.171

Sreekanth TVM, Jung M, Eom I (2016) Green synthesis of silver nanoparticles, decorated on graphene oxide nanosheets and their catalytic activity. Appl Surf Sci 361:102–106

Stevens MGF, Batlokwa BS (2017) Environmentally friendly and cheap removal of lead (II) and zinc (II) from wastewater with fish scales waste remains. Int J Chem 9(4). https://doi.org/10.5539/ijc.v9n4p22

Su C, Puls RW (2001) Arsenate and arsenite removal by zerovalent iron: kinetics, redox transformation, and implications for in situ groundwater remediation. Environ Sci Technol 35:1487–1492

Subramaniam R, Ponnusamy SK (2015) Novel adsorbent from agricultural waste (cashew NUT shell) for methylene blue dye removal: optimization by response surface methodology. Water Res Ind 11:64–70. https://doi.org/10.1016/j.wri.2015.07.002

Sulak MT, Demirbas E, Kobya M (2007) Removal of Astrazon Yellow 7GL from aqueous solutions by adsorption onto wheat bran. Bioresour Technol 98:2590–2598

Tatarchuk T, Bououdina M, Macyk W et al (2017a) Structural, optical, and magnetic properties of Zn-doped $CoFe_2O_4$ nanoparticles. Nanoscale Res Lett 12(1):141–151. https://doi.org/10.1186/s11671-017-1899-x

Tatarchuk T, Bououdina M, Paliychuk N et al (2017b) Structural characterization and antistructure modeling of cobalt-substituted zinc ferrites. J Alloys Compd 694:777–791. https://doi.org/10.1016/j.jallcom.2016.10.067

Tatarchuk T, Bououdina M, Vijaya JJ et al (2017c) Spinel ferrite nanoparticles: synthesis, crystal structure, properties, and perspective applications. In: Fesenko O, Yatsenko L (eds) Nanophysics, nanomaterials, interface studies, and applications. NANO 2016. In: Springer proceedings in physics, vol 195. Springer, pp 305–325. https://doi.org/10.1007/978-3-319-56422-7_22

Tatarchuk TR, Paliychuk ND, Bououdina M et al (2017d) Effect of cobalt substitution on structural, elastic, magnetic and optical properties of zinc ferrite nanoparticles. J Alloys Compd. https://doi.org/10.1016/j.jallcom.2017.10.103

Taty-Costodes VC, Fauduet H, Porte C et al (2003) Removal of Cd(II) and Pb(II) ions from aqueous solutions by adsorption onto sawdust of *Pinus sylvestris*. J Hazard Mater B 105:121–142

Thitame PV, Shukla SR (2017) Removal of lead (II) from synthetic solution and industry wastewater using almond shell activated carbon. Environ Prog Sustain Energy. https://doi.org/10.1002/ep.12616

Tinas H, Caliskan E, Ozbek N et al (2016) Preparation of Fe_3O_4@montmorillonite composite as an effective sorbent for the removal of lead and cadmium from wastewater samples. Turk J Chem 40:974–978

Tofan L, Paduraru C, Robu B et al (2012) Removal of Cd(II) ions from aqueous solution by retention on pine bark. Environ Eng Manage J 11:199–205

Tofighy MA, Mohammadi T (2012) Chem Eng Res Des 90:1815–1822

Tokunaga S, Wasay SA, Park SW (1997) Removal of arsenic (V) ion from aqueous solutions by Lanthanum compounds. Water Sci Technol 35(7):71–78

Tran HT, Vu ND, Matsukawa M et al (2016) Heavy metal biosorption from aqueous solutions by algae inhabiting rice paddies in Vietnam. J Environ Chem Eng 4(2):2529–2535. https://doi.org/10.1016/j.jece.2016.04.038

Upadhyay RK, Soin N, Bhattacharya G, Saha S, Barman A, Roy SS (2015) Grape extract assisted green synthesis of reduced graphene oxide for water treatment application. Mater Lett 160:355–358

Vaibhav V, Vijayalakshmi U, Mohana S, Amiri M, Salavati-Niasari M, Akbari A, Gholami T (2015) Agricultural waste as a source for the production of silica nanoparticles. Spectrochim Acta Part A Mol Biomol Spectrosc 139:515–520

Vasudevan S, Lakshmi J, Sozhan G (2012) Optimization of electrocoagulation process for the simultaneous removal of mercury, lead, and nickel from contaminated water. Environ Sci Pollut Res 19(7):2734–2744

Vendamanib VS, Tripathia A, Singh MK, Pathak AP, Tiwaria A, Chettri P (2017) Green synthesis of silver nanoparticle-reduced graphene oxide using *Psidium guajava* and its application in SERS for the detection of methylene blue. Appl Surf Sci 406:312–318

Verma SP, Sarkar B (2017) Rhamnolipid based micellar-enhanced ultrafiltration for simultaneous removal of Cd(II) and phenolic compound from wastewater. Chem Eng J 319:131–142

Verma A, Kumar S, Kumar S (2016) Biosorption of lead ions from the aqueous solution by *Sargassum filipendula*: equilibrium and kinetic studies. J Environ Chem Eng 4(4):4587–4599. https://doi.org/10.1016/j.jece.2016.10.026

Vidya C, Chandra Prabhab MN, Antony Raj MAL (2016) Green mediated synthesis of zinc oxide nanoparticles for the photocatalytic degradation of Rose Bengal dye. Environ Nanotechnol, Monit Manage 6:134–138

Vinh NV, Zafar M, Behera SK et al (2015) Arsenic(III) removal from aqueous solution by raw and zinc loaded pine cone biochar: equilibrium, kinetics, and thermodynamics studies. Int J Environ Sci Technol 12:1283–1294

Wahi R, Ngaini Z (2009) Removal of mercury, lead and copper from aqueous solution by activated carbon of palm oil empty fruit bunch. World Appl Sci J 5:84–91

Wang L, Zhang F-S (2017) Characterization of a novel sound absorption material derived from waste agricultural film. Constr Build Mater 157:237–243. https://doi.org/10.1016/j.conbuildmat.2017.07.192

Wang H, Lin SH, Juang RS (2003) Removal of heavy metals from aqueous solutions using various low-cost adsorbents. J Hazard Mater 102:291–302

Wang L, Zhu L, Bing N, Wang L (2017a) Facile green synthesis of Pd/N-doped carbon nanotubes catalysts and their application in Heck reaction and oxidation of benzyl alcohol. J Phys Chem Solids 107:125–130

Wang W, Cai K, Wu X et al (2017b) A novel poly(m-phenylenediamine)/reduced graphene oxide/nickel ferrite magnetic adsorbent with excellent removal ability of dyes and Cr(VI). J Alloys Compd 722:532–543. https://doi.org/10.1016/j.jallcom.2017.06.069

Wen H-F, Yang C, Yu D-G et al (2016) Electrospun zein nanoribbons for treatment of lead-contained wastewater. Chem Eng J 290:263–272

WHO/UNICEF (2000) Global water supply and sanitation assessment report 2000, WHO, Geneva, Switzerland

Xia Z, Baird L, Zimmerman N, Yeager M (2017) Heavy metal ion removal by thiol functionalized aluminum oxide hydroxide nanowhiskers. Appl Surf Sci 416:565–573

Xu D, Tan X, Chen C et al (2008) Removal of Pb(II) from aqueous solution by oxidized multiwalled carbon nanotube. J Hazard Mater 154:407–416

Yadav AK, Kaushik CP, Haritash AK et al (2006) Defluoridation of groundwater using brick powder as an adsorben. J Hazard Mater 128:289–293

Yang F, Sun L, Xie W et al (2017) Nitrogen-functionalization biochars derived from wheat straws via molten salt synthesis: an efficient adsorbent for atrazine removal. Sci Total Environ 607–608:1391–1399. https://doi.org/10.1016/j.scitotenv.2017.07.020

Yang Q, Ren SS, Zhao Q et al (2018) Selective separation of methyl orange from water using magnetic ZIF-67 composites. Chem Eng J 333:49–57. https://doi.org/10.1016/j.cej.2017.09.099

Yap MW, Mubarak NM, Sahu JN et al (2017) Microwave induced synthesis of magnetic biochar from agricultural biomass for removal of lead and cadmium from wastewater. J Ind Eng Chem 45:287–295

Zare-Dorabei R, Ferdowsi SM, Barzin A et al (2016) Highly efficient simultaneous ultrasonic-assisted adsorption of Pb(II), Cd(II), Ni(II) and Cu (II) ions from aqueous solutions by graphene oxide modified with 2,20-dipyridylamine: central composite design optimization. Ultrason Sonochem 32:265–276

Zhang J, Ding T, Zhang Z et al (2015) Enhanced adsorption of trivalent arsenic from water by functionalized diatom silica shells. PLoS ONE 10(4):e0123395. https://doi.org/10.1371/journal.pone.0123395A

Zhang Y, Cao B, Zhao L et al (2018) Biochar-supported reduced graphene oxide composite for adsorption and coadsorption of atrazine and lead ions. Appl Surf Sci 427(A):147–155. https://doi.org/10.1016/j.apsusc.2017.07.237

Zhou Q, Liao B, Lin L et al (2018) Adsorption of Cu(II) and Cd(II) from aqueous solutions by ferromanganese binary oxide–biochar composites. Sci Total Environ 615:115–122. https://doi.org/10.1016/j.scitotenv.2017.09.220

Zhu Y, Hu J, Wang J (2012) Competitive adsorption of Pb(II), Cu(II) and Zn(II) onto xanthate modified magnetic chitosan. J Hazard Mater 221:155–161

Zolgharnein J, Shahmoradi A, Bagtash M et al (2017) Chemometrics optimization for simultaneous adsorptive removal of ternary mixture of Cu(II), Cd(II), and Pb(II) by Fraxinus tree leaves. J Chemom. https://doi.org/10.1002/cem.2935

Chapter 5
Graphene Characterization and Its Use to Reduce Trihalomethanes (THMs) in Drinking Water in Puerto Rico

Jorge L. Hernández Bourdón

Abstract This chapter mainly focuses on the reduction of trihalomethanes (THMs) in drinking water in Puerto Rico. Three different nanostructured materials (Graphene, mordenite and multiwalled carbon nanotubes) were used to reduce the THMs formation by adsorption in specific contact time. The results showed that graphene is the best nanomaterial to reduce THMs in drinking water. Graphene can reduce THMs 80 parts per billion (ppb) in about 2 h.

Keywords THMs (trihalomethanes) · ppb (parts per billion) · Graphene Mordenite · Multiwalled carbon nanotubes

1 Introduction

The processes that occur in the surfaces of crystals depend on many factors such as crystal structure and composition, conditions of a medium where the crystal surface exists, and others physical and chemical factors. The appearance of a crystal surface is the result of a complexity of interactions between the crystal surface and the environment. There are a lot of computational techniques that allow us to predict the changes in surface morphology of minerals that are influenced by chemical and physical conditions (temperature, pressure, pH, etc.), of the solution when it's reacting with the surface (Barron et al. 2017).

The changes in the surface structure can be studied through the observation of the crystal surface topography using a micro-scope. Microscopic observation allows us to study small changes and estimates the rate of the processes by observing changes in the crystal surface topography.

Methods of structure determination can be classified in two categories: direct space methods and reciprocal space based methods. Direct space methods allow the visualization of the atomic arrangement in nanometer-sized regions. Examples of

J. L. H. Bourdón (✉)
School of Science and Technology, Universidad Del Turabo, Gurabo, PR, USA
e-mail: jhernandezbourdon@gmail.com

M. Naushad (ed.), *A New Generation Material Graphene: Applications in Water Technology*, https://doi.org/10.1007/978-3-319-75484-0_5

these methods are: Transmission Electron Microscopy (TEM), Scanning Electron Microscopy (SEM), and Atomic Force Microscopy (AFM). Reciprocal space based methods exploit interference and diffraction effects of photons or electrons to provide sample averaged information about structure (Wang 2000). This chapter has been divided in three sections: the first one corresponds to the characterization techniques (section A), the second one has been focused on the characterization results (section B) and finally the section C, is focused on Trihalomethanes absorption by graphene.

The following section presents the conventional techniques used to characterized graphene to reduce thrihalomethanes (THMs) in drinking water of Puerto Rico. The techniques are: (1) Field Emission Transmission Electron Microscopy (FE-TEM) (2) Atomic Force Microscopy (AFM) (3) X-Ray diffraction (XRD), (4) Brunauer, Emmett and Teller (BET Method), (5) Raman Spectroscopy and Thermogravimetric Analysis (TGA).

2 Field Emission Transmission Electron Microscopy (FE-TEM)

It is possible to employ the fine-probe/scanning technique with a thin sample and record, instead of secondary electrons, the electrons that emerge (in a particular direction) from the opposite side of the specimen. The resulting is a scanning-transmission electron microscope (STEM). The first STEM was constructed by von Ardenne in 1938 by adding scanning coils to a TEM, and nowadays many TEMs are equipped with scanning attachments, making them dual-mode (TEM/STEM) instruments. For this purpose, the hot-filament electron source that is often used in the SEM (and TEM) must be replaced by a field-emission source, in which electrons are released from a very sharp tungsten tip under the application of an intense electric field. The field-emission gun requires ultra-high vacuum conditions (UHV), involving pressures of around 10^{-8} Pa. After many years of development, this type of instrument has produced the first-ever images of single atoms, visible as bright dots on a dark background.

TEM is a technique where an electron beam interacts and passes through a specimen. The electron beam is emitted by a source and is focused and magnified by a system of magnetic lenses (Voutou and Stefanaki 2008). The electron beam is confined by two condenser lenses which also control the brightness of the beam, passes the condenser aperture and “hits” the sample surface. The electrons that are elastically scattered consist of the transmitted beams which pass through the objective lens. In addition, the instrument can be used to produce electron-diffraction patterns, useful for analyzing the properties of a crystalline specimen. This overall flexibility is achieved with an electron-optical system containing an electron gun (which produces the beam of electrons) and several magnetic lenses, stacked vertically to form a lens column. The electron gun has a Wehnelt cylinder, which is a metal electrode that can be easily removed (to allow changing the filament or LaB_6 source) but which normally surrounds the filament completely except for a small hole (<1 mm diameter) through which the electron beam emerges. The function of the Wehnelt electrode is to control the emission current of the electron gun. For this purpose, its potential is made more negative than that of the cathode. This negative potential prevents electrons from leaving the cathode unless they are emitted from a region near to its tip, which is located immediately above the hole in the Wehnelt electrode, where the electrostatic potential is less negative (Egerton 2005).

TEM is composed of various systems: an illumination source, a specimen stage, an objective lens, the magnification system, the data recording, and the chemical analysis system. The illumination system includes the condenser lenses that are important for forming a fine electron probe. The specimen, for carrying out structure analysis induced by annealing electric field or mechanical stress, to characterize the physical properties of individual nanostructure. The objective lens is the heart of TEM because they determine the limit of image resolution (Wang 2000). The magnification system consists of two lenses, (intermediate and projection) lenses, and it gives a magnification up to 1.5 million. The data recording

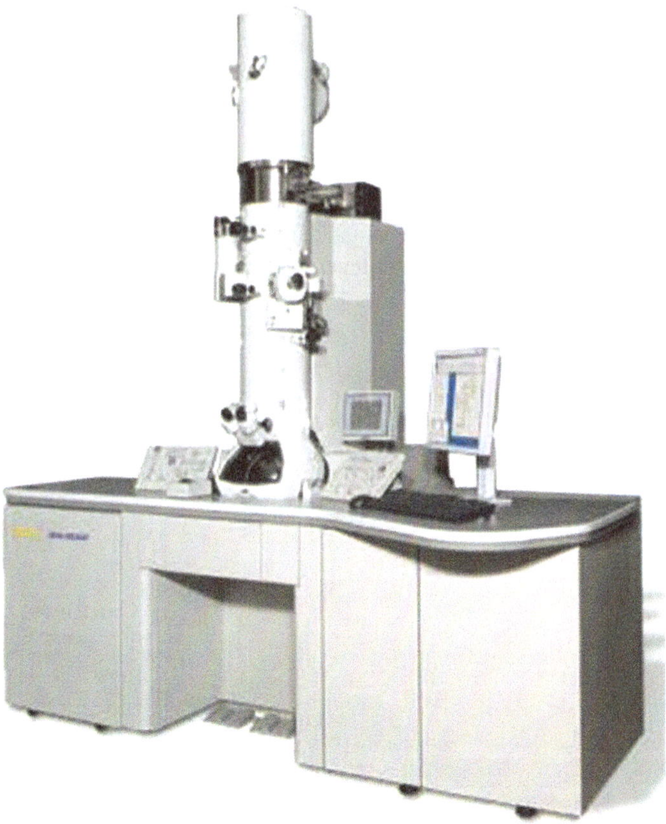

Fig. 1 JEOL 3000 field emission transmission electron microscope (Hernández 2016)

system tends to be digital with the use of a charge couple device (CCD). Finally, the chemical analysis system is provided by using the energy dispersive X-ray spectroscopy (EDS) and electron energy-loss spectroscopy (EELS) (Wang 2000). TEM provides a direct image of the specimen and is unique because it can provide a real space image on the atom distribution in nanocrystals. In addition TEM provides atomic resolution lattice images and also chemical information at a spatial resolution of 1 nanometer (nm) or better, allowing direct identification of the chemistry of a single nanocrystal (Fig. 1).

3 Atomic Force Microscopy (AFM)

AFM is a very high-resolution scanning probe microscopy of a demonstrated resolution of fractions of a nanometer, more than 1000 times better than the optical diffraction limit (Ahn et al. 2004). The precursor of the AFM is the scanning tunneling microscope.

The AFM consists of a cantilever with a sharp tip at its end that is used to scan the sample surface, is typically made of silicon nitride or silicon with a tip radius of curvature of nanometers (Fig. 2). When the tip is brought into to a sample surface, forces (mechanical contact forces, Van der Waals forces (VdW), capillary forces and electrostatic forces) between the tip and the sample lead to a deflection of the cantilever. The deflection is measured using a laser spot reflected from the top surface of the cantilever into an array of photodiodes. The AFM can be operated in a number of modes (contact modes or static and non-contact or dynamic) depending on its application and where the cantilever is oscillated externally at or close to its

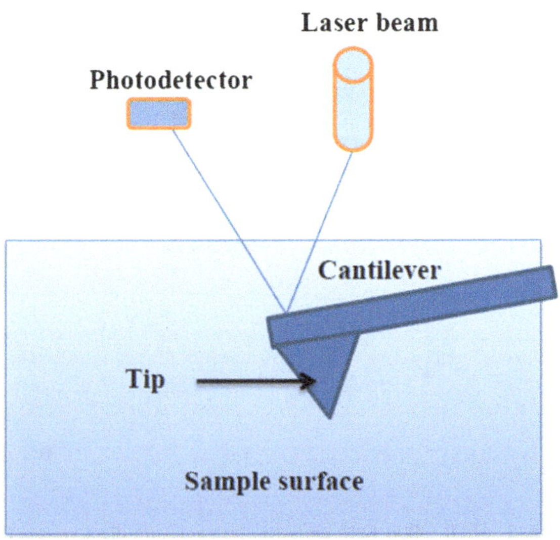

Fig. 2 Schematic drawing of an AFM apparatus. Adapted from Barron et al. (2017)

fundamental resonance frequency or harmonic. The oscillation amplitude and the phase and resonance frequency are modified by the tip-sample interaction forces; these changes in oscillation, in comparison with external reference oscillation provide information about the sample's characteristics (Ahn et al. 2004).

There are 3 primary imaging modes in AFM: (1) Contact AFM < 0.5 nm probe-surface separation, (2) Intermittent contact (tapping mode AFM) 0.5–2 nm probe-surface separation, and (3) Non-contact AFM 0.1–10 nm probe-surface separation (Wilson et al. 2006).

3.1 Contact Mode AFM

This contact occurs when the spring constant of the cantilever is less than the surface and the cantilever bends. The force on the tip is repulsive. By maintaining a constant cantilever deflection the force between the probe and the sample remains constant and an image of the surface is showed. Some advantages include the fast scanning, the potential use with rough samples and the use in friction analysis. As disadvantages is that forces can damage or deform soft samples.

3.2 Intermittent Mode (Tapping)

The probe "taps" lightly on the sample surface during scanning, contacting the surface at the bottom of its swing. A constant tip-sample interaction is maintained and an image of the surface is obtained by maintaining constant oscillation amplitude, the image is similar to contact mode. Advantages: allows high resolution of samples that are easily damaged and/or loosely held to a surface. Disadvantages: more challenging to image in liquids, slower scan speeds needed.

3.3 Non-contact Mode

The probe does not contact the sample surface but oscillates above the adsorbed fluid layer on the surface during scanning. Using a feedback loop to monitor changes in the amplitude due to attractive VdW forces, the surface topography can be measured. Advantages: very low force exerted on the sample (10^{-12} N) and extended probe lifetime. Disadvantages: generally lower resolution, contaminant layer on surface can interfere with oscillation, usually need ultra-high vacuum to have the best images. In a force curve analysis, the probe is brought repeatedly toward the surface and then retracted. Force curve analyses can be used to determine mechanical and chemical properties such as adhesion, elasticity, hardness, and rupture bond lengths (Wilson et al. 2006).

Fig. 3 AFM instrument (Hernández 2016)

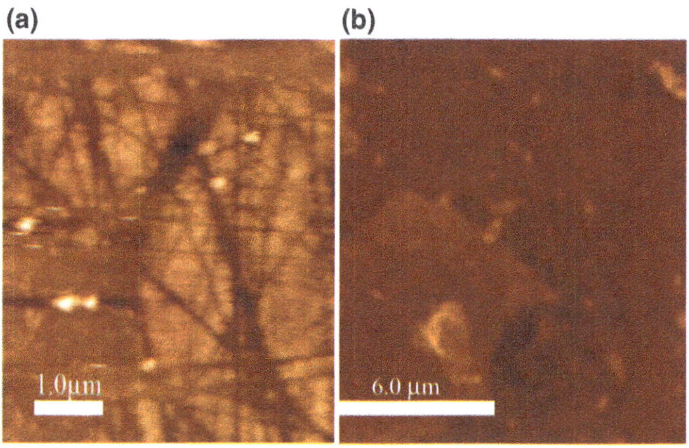

Fig. 4 **a** AFM corresponding to the surface of a graphene flakes **b** AFM of an isolated graphene flake deposited on a Si (100) wafer (Hernández 2016)

The samples morphology was determine using this technique. The image was taken using a Nanotec Electronica equipment of the Applied Physics Department at UAM, in tapping mode at room temperature in atmospheric conditions (Fig. 3). Cantilevers Olympus used in this research are made (RC800PSA, ORC8) of Si_3N_4, and are characterized by a typical V-shaped. The tip has a pyramidal form with a radio less than of 20 nm. Finally the images were processed with the WSxM program (software available on www.nanotec.es) (Hernández 2016).

Figures 4 show different graphene images obtained by atomic force microscopy (AFM). Figure 4a corresponds to the surface of a graphene flake of several micrometers of diameter. Figure 4b shows a sample of graphene previously dispersed in ethanol (in a soft ultrasound bath) and deposited on a Si (100) surface. As can be seen there, after dispersion the graphene is characterized by having micrometric dimensions. The measured surface profile was about ± 3 nm (not shown), which may be caused by folding of the GO flake and also to the presence of solvent molecules below the graphene surface.

4 X-Ray Diffraction (XRD)

Max von Laue discovered that crystalline substances act as three-dimensional diffraction gratings for X-ray wavelengths equal to the spacing of planes in a crystal lattice. X-ray diffraction is a common technique for the study of crystal structures and chemical compositions of materials, identification of multiple phases in microcrystalline mixtures and structural analysis of clay minerals, etc. The XRD is a well-developed, versatile, and powerful technique for non-destructive ex situ characterization of semiconductor structures. Structure information such as crystalline quality, strain relaxation, and lattice constant can be obtained accurately with high resolution XRD. The distribution of X-ray intensity, scattered by a finite-sized atomic aggregate, takes on a simple form, provided that aggregates within the sample are uniformly and randomly oriented with respect to the incident beam. Under these conditions, a radially symmetric powder diffraction pattern can be observed. Powder diffraction patterns, for individual particle structures, can be calculated using the Debye equation of kinematic diffraction. This equation is expressed as follows:

$$I_N(S) = I_0 N f^2(S)\left(1 + \frac{D}{N}\right)\left(\sum_{n \neq m} \frac{(\sin 2\pi s r_{mn})}{2\pi s r_{mn}}\right) \quad (1)$$

where I_o is the intensity of the incident beam and $I_N(s)$ is the power scattered per unit soil angle in the direction defined by $s = 2 \sin(\Theta)/\lambda_p$, with Θ equal to half the scattering angle and λ_p the wavelength of the radiation. The scattering factor f(s) determines the single-atom contribution of scattering and is available in tabulated form. N is the number of atoms in the cluster and r_{mn} is the distance between atom m and atom n.

The Debye-Waller factor, D, damps the interference terms and so expresses a degree of disorder in the sample (Wang 2000). This disorder may be dynamic, due to thermal vibrations, or static, as defects in the structure. A simple model assumes that the displacement of atoms is random and isotropic about their equilibrium positions. The peak positions in the X-ray diffraction pattern can be used to determine the chemical composition and the crystal phase structure of different materials, including catalysts as zeolites, nanowires, etc. (Ahn et al. 2004). X-ray powder diffraction patterns (XRD) were collected using an X′Pert PRO X-ray

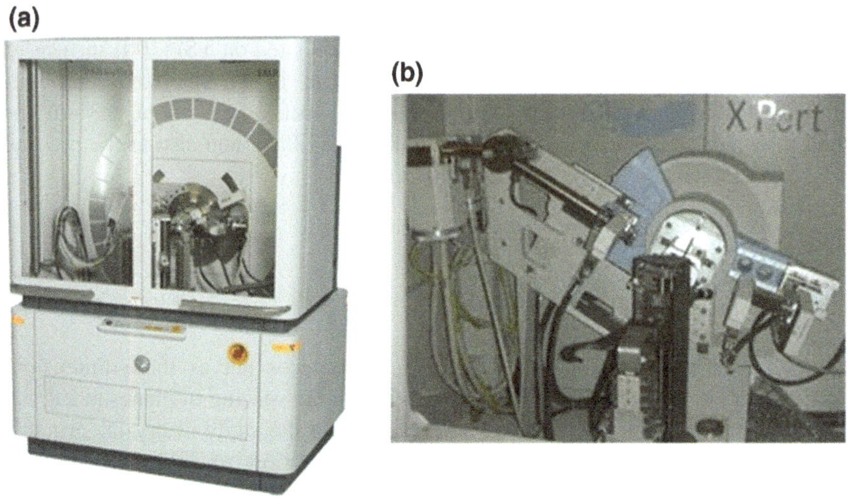

Fig. 5 Image of the analytical XRD system (a), and detail of the goniometer (b) (Hernández 2016)

diffractometer (PANalytical, The Netherlands) in Bragg-Brentano goniometer configuration. X-ray radiation source was a ceramic X-ray diffraction Cu anode tube type Empyrean of 2.2 kW Angular measurements ($\theta - 2\theta$) were made with reproducibility of: ±0.0001°, applying steps of 0.05° from 5° to 60°. Figure 5a, shows an image of the XRD diffractometer. A detailed view of the goniometer is shown in Fig. 5b.

5 Brunauer, Emmett and Teller (BET Method)

The best known and most commonly used method for studying specific surface areas of solid materials is the BET method based on the theory of Stephen Brunauer, Paul Hugh Emmett, and Edward Teller published in 1938 (Mclaughlin et al. 2012). The BET method is used to determine the amount of the adsorptive gas required to cover the external and the internal pore surfaces of a solid with a complete monolayer of adsorbate. This monolayer capacity can be calculated from the adsorption isotherm by means of the BET equation. The gases used as adsorptives have to be physically adsorbed on the surface of the solid by weak bonds (VdW forces) and can be desorbed by a decrease of pressure at the same temperature. The most common gas is nitrogen at its boiling temperature (77.3 K) and krypton can be used at 77.3 K in the case of small surface area (below 1 m^2/g) (Klobes 2011).

To determine the adsorption isotherm, known amounts of adsorptive are admitted stepwise into the sample cell containing the sample previously outgassed by heating with vacuum and dried. The amount of gas adsorbed is the difference of gas admitted and the amount of gas filling the dead volume (free space in the sample cell). The adsorption isotherm is the plot of the amount gas adsorbed in the process (in mmol/g) as a function of the relative pressure p/p_0; a BET diagram is plotted with as ordinate against $x = p/p_0$ as abscissa (Klobes 2011).

$$y_{\mathrm{BET}} = \frac{p/p_0}{n_a(1 - p/p_0)} = \frac{C_{\mathrm{BET}} - 1}{n_{\mathrm{mono}} \cdot C_{\mathrm{BET}}} \cdot (p/p_0) + \frac{1}{n_{\mathrm{mono}} \cdot C_{\mathrm{BET}}} = f(x) = i + kx \quad (2)$$

The plot should give a straight line $y_{\mathrm{BET}} = i + kx$ within the so-called BET relative pressure range. The intercept a, and the slope b, which are determined by means of linear regression, must be positive. From these values, the monolayer capacity n_{mono} (in mmol/g) should be calculated according to $n_{\mathrm{mono}} = 1/(i + k)$. The specific surface area a_{BET} is calculated from the monolayer capacity by assessing a value σ for the average area occupied by each molecule in the complete monolayer (Klobes 2011).

$$a_{\mathrm{BET}} = n_{\mathrm{mono}} \cdot \sigma \cdot N_A$$

N_A is the Avogadro constant. Molecular cross sectional area values σ for krypton (Kr) and nitrogen (N) are recommended by the standard ISO 9277:2010 and the IUPAC recommendations of 1984 as $\sigma = 0.162\ \mathrm{nm}^2$ for N_2 and $\sigma = 0.202\ \mathrm{nm}^2$ for Kr, respectively (Klobes 2011). The specific surface areas of the nanomaterials used in the present research were determined by the BET method using a Micromeritics ASAP 2020 (Fig. 6).

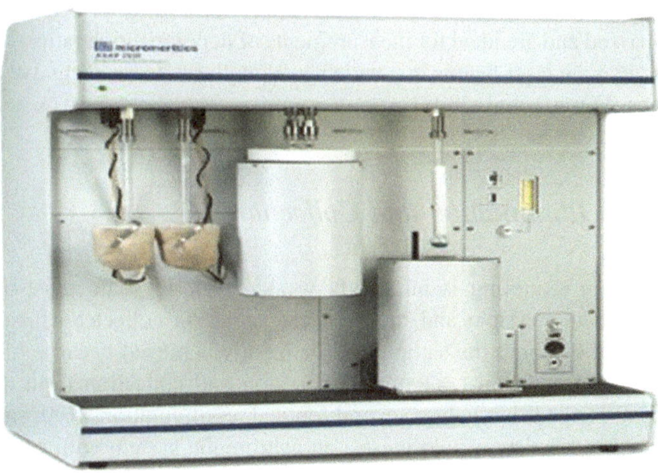

Fig. 6 Micromeritics ASAP 2020 Accelerated Surface area and porosimetry (Hernández 2016)

6 Raman Spectroscopy

In 1928, Chandrasekhra Venkata Raman discovered the phenomenon that bears his name. Raman used sunlight as the source and a telescope as the collector and the detectors were his eyes. Gradually, improvements in the various components of Raman instrumentation took place (Ferraro et al. 2003). The Raman spectrometer consists of four major components, (1) exciting sources (2) sample illumination and collection system (3) wavelength selector and detection and (4) computer control/processing system (Ferraro et al. 2003).

6.1 Exciting Sources

Elaborate systems were developed to supply a single wavelength with enough power to produce Raman scattering. The mercury arc lamp eventually became the standard source for Raman spectroscopy; this strange-looking lamp consisted of a 2-in. helical tube, ~4 ft. (length) coiled approximately of a 6-in. radius connecting two large pools of mercury covering with high-voltage electrodes. Continuous-wave (CW) lasers such as (Argon) Ar^+ (351.1–514.5 nm), Kr^+ (337.4–676.4 nm), and (Helium–Neon) He–Ne (632.8 nm) are now commonly used for Raman spectroscopy. More recently, pulsed lasers such as Nd:YAG, diode, and excimer lasers have been used. Lasers are ideal excitation sources for Raman spectroscopy due to (1) single lines from large CW lasers can provide 1–2 W of power, and pulsed lasers produce huge peak powers on the order of 10–100 MW, (2) laser beams are highly monochromatic (band width $\sim 0.1\ cm^{-1}$ for Ar^+ laser) (Ferraro et al. 2003), (3) most laser beams have small diameters (1–2 mm), which can be reduced to 0.1 mm, (4) laser beams are almost linearly polarized and are ideal for measurements of depolarization ratios, and (5) it is possible to generate laser beams in a wide wavelength range by using dye lasers and other devices (Ferraro et al. 2003).

6.2 Sample Illumination and Collection

Since the Raman scattering is inherently weak, the laser beam must be focused properly toward the sample and the scattered radiation collected efficiently. The focusing of the laser onto the sample can be readily achieved because of the small diameter of the laser beam (~1 mm). Excitation and collection from the sample can be accomplished by using several optical configurations (90° and 180°). Collection optics consists of two lenses, an achromatic system with collecting and focusing lens. An oblique illumination angle is a common configuration for illuminating the front surface of a sample with the incident laser beam at ~85° to normal, an off-axis ellipsoid is used to collect the scattered light and focus it on the entrance slits (Ferraro et al. 2003).

6.3 Wavelength Selector and Detection

Wavelength selectors can be classified into several categories. The simplest device is an interference filter, which depends on its two optically flat surfaces to create a constructive interference and transmit an integer number of wavelengths and is constructed for single wavelengths. Variable, wedge-shaped interference filters are available for selecting desired wavelengths; the spectral resolution of these devices is generally too large for Raman spectra. Recently, acoustic and liquid-crystal tunable filters have been successfully used for measuring both Raman spectra and spectral images. Both prism and grating monochromators and spectrographs have been used extensively for measuring Raman spectra. While monochromators are still the mainstay of Raman instrumentation, FT-Raman has made considerable advancement in recent years and is now considered to be competitive with monochromators (Ferraro et al. 2003).

6.4 Computer Control/Processing System

For spectrometers equipped with monochromators and single detectors, the Raman scattered light in the focal plane and exiting through the slits of the monochromator is collected and focused on a photomultiplier (PM) tube, to converts photons into an electrical signal. The PM tube consists of a photocathode that emits electrons when photons strike it; a series of dynodes, each of which emits a number of secondary electrons when struck by an electron; and an anode that collects these electrons as an output signal. The photons strike the cathode, which is at the largest negative voltage and each dynodes represented by a curved surface is at a less negative voltage. For example, the cathode might be at −1100 V dc, the first dynode at −1000 V dc, the second dynode at −900 V dc, etc. A photon striking the cathode releases an electron, which is attracted to the less negative first dynode and causes the release of several electrons, these electrons are attracted to the second dynode, where they each release several electrons. The electrons released by the single photon are multiplied many-fold by passing from one dynode to the next until they all reach the anode (Ferraro et al. 2003).

The Raman method is the complementary method of IR spectroscopy. As in the case of the IR spectrum, the features of a Raman spectrum (number of Raman bands, their intensities, and their shapes) are related directly to the molecular structure of a compound (Gauglitz and Vo-Dinh 2003). Raman spectrum provides information on the samples composition, structure and vibrational modes that is complementary to the infrared spectrum.

This instrument provides information about the sample, allowing for spectral interpretation, library searching, data manipulation, and the application of chemometric methods. Raman spectroscopy has been shown to be one of the best techniques for non- destructive and quick characterization of samples (Ni et al. 2008).

The Raman spectroscopy can be used for identification of compounds due principally to some advantages that include: (i) two molecules do not give exactly the same Raman spectrum, and (ii) the intensity of the scattered light is proportional to the amount of material present.

The sample preparation is easy, there is no need to dissolve solids, press pellets, compress the sample, or alter its physical or chemical structure; the sample can be measured without manipulation, this makes Raman spectroscopy ideal for the investigation of physical properties such as crystallinity, phase transitions, and polymorphs.

The lack of sample preparation minimizes cleanup and the possibility of cross-contamination. Raman spectroscopy has several additional advantages because its operational wavelength range is usually independent of the vibrational modes being studied. Since Raman spectroscopy measures the shift in frequency of the exciting laser, it can be performed using any operating range from UV to NIR. It permits access to vibrational mode information associated normally with wavelengths ranging from 2 to 200 μm. Nowadays, Raman spectroscopy is one of the most powerful tools for the study of carbonaceous materials, especially sp^2 or close to sp^2 carbon materials such as graphite, carbon nanotubes, and graphene. Through resonance effects the Raman spectrum also gives information about the electronic structure of the material. The Raman spectra reflect simply the phonon density of states of the material (Zolyomi et al. 2011).

In this research, Raman spectra were collected using a micro-Raman Renishaw RM2000 single grating spectrograph (Fig. 7) equipped with 532 and 785 nm

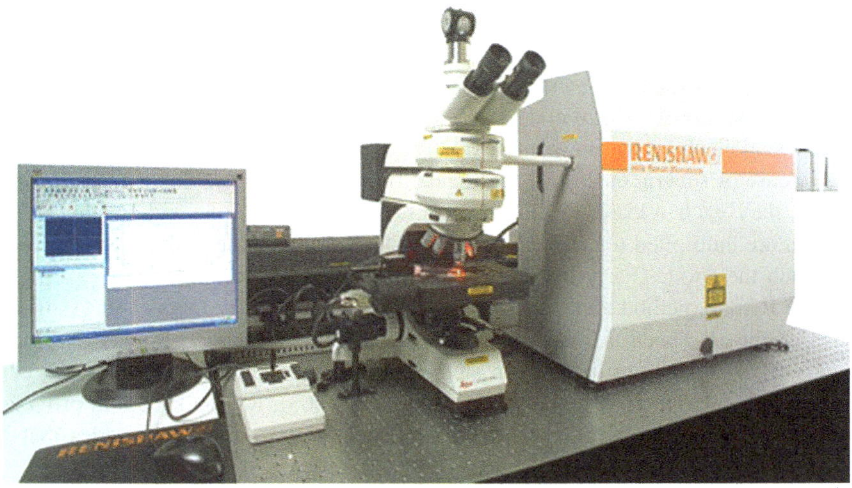

Fig. 7 Image of the micro Raman scattering spectrometer (Hernández 2016)

excitation sources. Raman spectra were acquired in the spectral range of 3200–100 cm^{-1}. The acquisition time for each measurement was 20 s and a defocused laser power level in the range of 10–60 mW was used to prevent the possible thermal effects on the samples.

7 Thermogravimetric Analysis (TGA)

TGA is a method to determine changes in mass with change in temperature. TGA is commonly employed in the research of the dehydration and decomposition point and then to determine the thermal stability of samples. The principal function of TGA is the monitoring of the thermal stability of a sample by recording the change in mass of the sample with respect to the temperature. There are two pans inside the TGA: a reference pan and a sample pan. The pan material can be either aluminum or platinum. The type of pan used depends specifically on the maximum temperature of a given run. As platinum melts at 1760 °C and aluminum melts at 660 °C, platinum pans are chosen when the maximum temperature exceeds approximately 600 °C (Frein and Barron 2009).

Each pan is balanced on a balance arm before the start of each run. The balance arms should be calibrated to compensate the deferential thermal expansion between each arm. To calibrate the system, the empty pans are placed on the balance arms and the pans are weighed and zeroed. As well as recording the change in mass, the heat flow into the sample pan (deferential scanning calorimetry, DSC) can also be measured and the difference in temperature between the sample and reference pan (differential thermal analysis, DTA). The DTA means how the sample phase changed. If the DTA is constant, this means that there was not a phase change (Frein and Barron 2009).

When the sample melts, the DTA dips which signifies an endotherm. When the sample is melting, it requires energy from the system and the temperature of the sample pan decreases compared with the temperature of the reference pan. When the sample has melted, the temperature of the sample pan increases as the sample is releasing energy. Finally, constant DTA occur when the temperatures of the reference and sample pans equilibrate. When the sample evaporates, there is a peak in the DTA. Typically, the sample mass range should be between 0.1 and 10 mg, and the heating rate should be between 3 and 5 °C/min (Frein and Barron 2009).

In this research, the thermogravimetric analyses were done with a TGA Q-500 instrument (TA Instruments) under an inert atmosphere of nitrogen. The heating rampage was of 5 °C/min from 100 to 700 °C. Figure 8 shows the instrument used for these measurements.

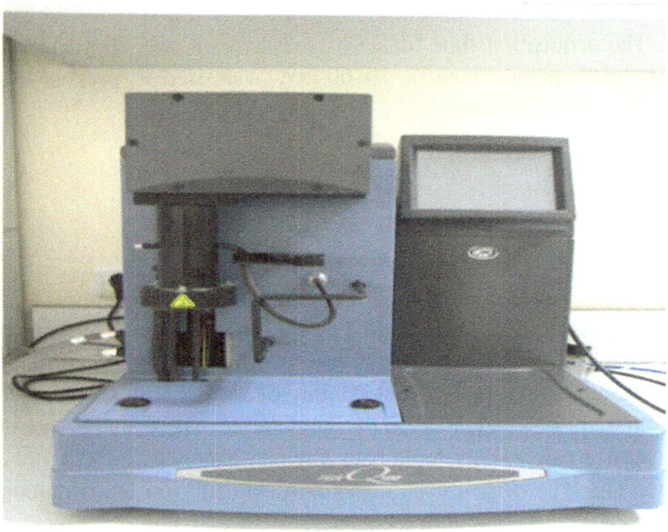

Fig. 8 Thermal gravimetric analysis (TGA), TA instrument, Q500 (Hernández 2016)

8 Characterization Results

Different techniques have been used to characterize graphene. All of these techniques were discuss on section A. This section has been focused on characterization results graphene.

8.1 Graphene

Graphene flakes used in this research showed the highest specific surface area BET with a value of 1850 m^2/g. The theoretical specific surface area for graphene is (2630 m^2/g) (Rao et al. 2011) although our experimental value is lower than the theoretical value, due possibly to the presence of defects in the structure and the presence of parallel layers of graphene. This experimental value is higher in comparison with other BET values described for graphene; for example 925 and 520 m^2/g were found by Subrahmanyam et al. (2008). This BET result means that graphene has potentially more adsorption capacity.

TEM and HRTEM micrographs of graphene flakes are illustrated in Fig. 9a, b. In Fig. 9a, graphene flakes are clearly observed in different orientations and Fig. 9a shows different layers of graphene and, as can be seen there, disaggregation apparently was not fully accomplished during the ultrasound treatment in ethanol before depositing of the solution onto copper grids. In this image, each horizontal line represents a graphene sheet of one atom of thickness. The flexibility of this

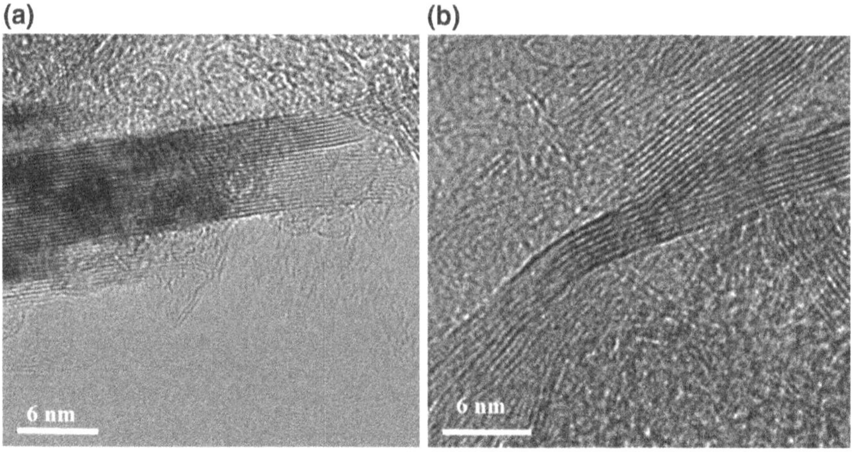

Fig. 9 **a** HRTEM image of graphene flakes bar = 6 nm, **b** different HRTEM imqges of graphene flakes showing different orientation of their layers (Hernández 2016)

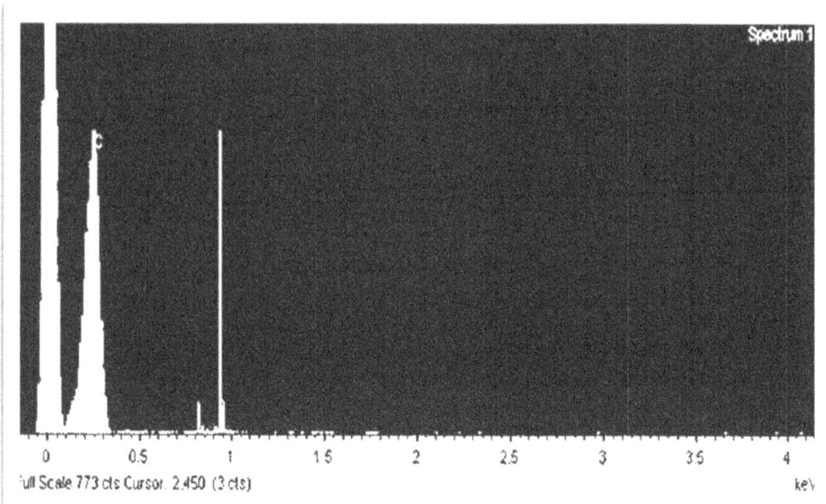

Fig. 10 EDX analysis obtained in a JEOL 3000 FE transmission electron microscope FE-HRTEM, indicating that carbon is the only element in the sample (Hernández 2016)

nanomaterial is also observed in Fig. 9b. In this figure, graphene appears in different orientations.

Graphene was also analyzed by Energy Dispersive X ray analysis (EDX) to determine the chemical composition of the sample. Figure 10, shows the EDX analysis of graphene previously observed by TEM and HRTEM. As can be seen

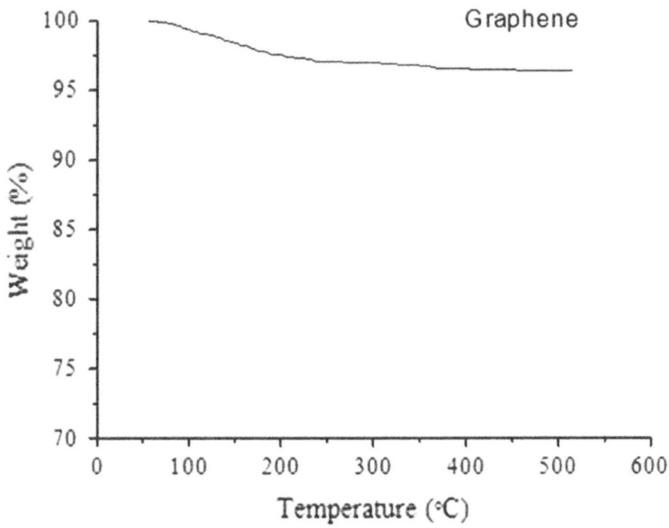

Fig. 11 TGA results for graphene (Hernández 2016)

there graphene practically does not show oxygen, indicating that graphene flakes used in this research after the experiment treatment, is reduced.

Graphene is a very stable nanomaterial. Figure 11 shows the thermogravimetric analysis (TGA) of graphene. As can be seen there graphene experiences a low decrease of weight ranging from room temperature to ca. 450 °C. This decrease of weight is mainly produced by the loss of water and possibly other carbonaceous materials.

Figure 12 shows the Raman spectrum measured for graphene. The spectrum is dominated by two bands (G and D bands); G-band appears at ca. 1580 cm^{-1} and can be correlated with the symmetry of the hexagonal lattice and, the D-band, at ca. 2700 cm^{-1}, due to the breathing modes of sp2 rings. D-band is not related to the number of graphene layers but to the amount of disorder and defects in the structure. This result is comparable with the Raman spectrum measured previously by Zolyomi et al. (2011).

Figure 13 shows the X ray powder diffraction (XRD) of graphene. The sharp peak at approximately 27.0°, (002) reflection, indicates a high organized crystal structure. In addition this peak may be attributed to a mixture powder of graphene sheets with different number of graphene layers. The broad peak at 21.1°, is indicative of material with lower crystallinity and some defects. The peak at ca. 11.3° may be due to the diffraction by the inter-few-layered graphene. This result is comparable with previously XRD results obtained for graphene (Wu et al. 2012; Wang et al. 2011).

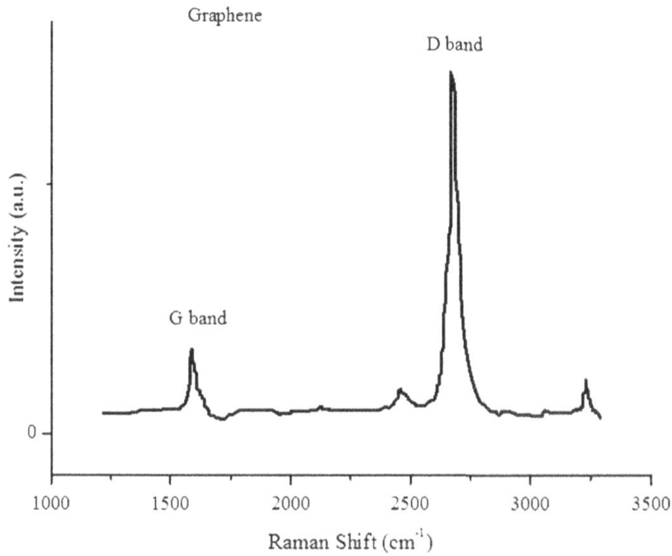

Fig. 12 Measured Raman spectrum of a sample of graphene previously dispersed in ethanol and deposited on a Si (100) wafer (Hernández 2016)

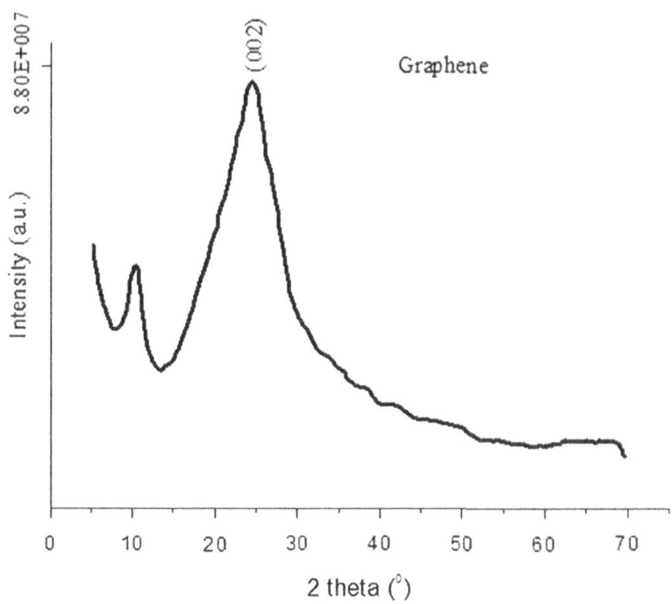

Fig. 13 XRD of graphene (Hernández 2016)

9 Experimental Procedure

With the aim to test graphene adsorbers of THMs, four solutions of each THM were prepared. Solutions of chloroform, bromoform, diclorobromomethane and dibrochloromethane with a concentration of 90 ppb were prepared using CO_2-free water and subsequently mixed with the different THMs in an Erlenmeyer flask of 1000 mL. Next, 25 mg of graphene were added to each solution. In this step different contact times were measured (30, 60, 120, 180 and 240 min, respectively). After completed each contact time, approximately 90 mL of each sample was transferred to a centrifuge tub and the mixture was centrifuged for 15 min at 3000 revolution per minute (rpm). With the aim to remove all possible residues of graphene, the reaction mixture was also filtered by using disk filters of 0.45 μm.

Samples were analyzed with a TOC analyzer Phoenix 8000 Kekmar Dohrmann and GS-MS Shimadzu GC-MS-QP5050A purge and trap system. Each sample was run in duplicates and the mean value of each sample was reported.

10 THMs Adsorption by Graphene

The adsorption of graphene was measure in different contact times of each THMs. Graphene was found to be the nanomaterial with highest adsorption capacity for THMs. In the first 30 min graphene adsorbed 61.2 ppb of $CHCl_3$, 66.5 ppb of $CHBr_3$, 59.1 ppb of Br_2CHCl and 51.2 ppb of $BrCHCl_2$ (Table 1). After 1 h, graphene adsorbed 78.2, 80.2, 75.7, and 66.6 ppb of $CHCl_3$, $CHBr_3$ Br_2CHCl, and $BrCHCl_2$ respectively (Table 1).

The adsorption process in graphene is very fast during the first 120 min, where graphene adsorbed the majority of THMs. At contact time of 120 min graphene was able to adsorb the following concentrations of THMs: 84.2 ppb of $CHCl_3$, 85.8 ppb of $CHBr_3$, 88.3 ppb of Br_2CHCl and 80 ppb of $BrCHCl_2$ (Fig. 14) (Table 1). At this contact time is when graphene adsorbed 80 ppb or higher, being this concentration the MCL of THMs in drinking water as per The Environmental Protection Agency (USEPA). The completed adsorption of each THMs occurs in the range of 180–240 min. The reason for the extremely high adsorption capacity of

Table 1 THMs adsorption by graphene at different contact times (in ppb) (Hernández 2016)

Contac times (min)	$CHCl_3$	$CHBr_3$	Br_2CHCl	$BrCHCl_2$
0	89.8	89.3	90.2	88.5
30	28.6	22.8	31.1	37.3
60	11.6	9.1	14.5	21.9
120	5.6	3.5	1.9	8.5
180	3.6	0.7	1.9	1.5
240	2.4	0.3	0.7	0.3

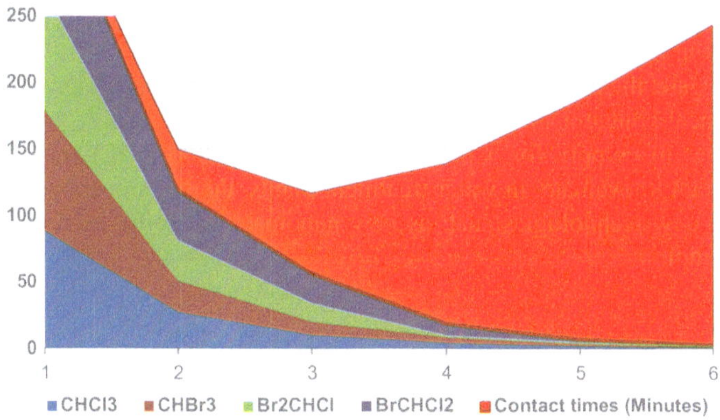

Fig. 14 THMs absorption by graphene at different contact times (in ppb) (Hernández 2016)

graphene is possibly due to the high specific surface area. Graphene is formed by flakes of single layers or associations of no more than 6–7 sheets (as was observed by HR-TEM), being this fact the reason for the high surface area and the high adsorption capacity.

11 Conclusion

THMs concentrations exceeding MCL established by USEPA are present in drinking water of Puerto Rico. The evaluation of historical data of drinking water from PRASA shows a serious problem of water quality in terms of VOC in drinking water. The principal municipalities with this problem in Puerto Rico are: Cidra, Aibonito, Aguas Buenas, Bayamon, Caguas, Canovanas, Gurabo, Isabela, San Juan, Toa Baja, and Quebradillas. In these municipalities THMs were detectable more than two times in a year.

Graphene has an extremely high surface area. This enormous surface area, together with the special structure of this material, increases the exposure of graphene to contaminants. Graphene adsorbs 80 ppb of each THMs after 120 min of contact time. Therefore the MCL of THMs in water (0.080 mg/L) established by USEPA could be accomplished by adsorption after two hours if graphene is used as adsorbent in water treatment plants. Graphene is the best nanomaterial to reduce THMs in drinking water treatment plants.

Recommendations

To evaluate the formation of THMs in drinking water and to study the potential adsorption of THMs by graphene, some recommendations have been done:

(1) Determine the exposure levels of THMs in humans.

(2) Design a nano filter using graphene to experiment on site the reduction of THMs in drinking water.
(3) Investigate, the possibility of graphene to reduce another common chemical and biological contaminants present in water.
(4) In order to comply for a possible new MCL (0.060 mg/L) for THMs, it is important to evaluate, in water treatment plants, the process and the possibility to use new technologies such as new nanomaterials to be used as filters (i.e. graphene).

References

Ahn CH, Anczykowski B, Atashbar MZ, Bacsa W, Bainbridge WS, Baldi A, Barnes PD, Batteas JD, Bennewitz R, Berman AD et al (2004) [CD-ROM]. Germany: Springer handbook of nanotechnology (1st edn), pp 331–332. Springer, Berlin. ISBN 3-540-01218-4

Barron AR, Bratt A, Dominguez M, Fransworth M, Fernández M, Fisher E, Gangoli VS, Hamilton CE, Harrison L, Holden J et al (2017) Nanotechnology for the oil and gas industry. Rice University, Houston, TX; 2010. [Internet]. [Cited 2017 May 31]. Available from: http://www.cnx.org/content/col10700/1.11/

Egerton RF (2005) Physical principles of electron microscopy: an introduction to TEM, SEM and AEM (1st edn). Springer Science Business Madia Inc., New York [cited 2017 May 6]. Available from: http://chem.wzu.edu.cn/20082613122640.pdf

Ferraro JR, Nakamoto K, Brown CW (2003) Introductory Raman spectroscopy (2nd edn). [Internet]. Elsevier Science (USA): [cited 2017 May 5]. Available from: http://www.fulviofrisone.com/attachments/article/406/introductory ramanspectroscopy.pdf

Frein C, Barron AR (2009) Thermogravimetric analysis of single walled carbon anotubes. [Cited 2017 May 10]. Available from: http://cnx.org/content/m22972/latest/

Gauglitz G, Vo-Dinh T (2003) Handbook of spectroscopy. [CD-ROM]. Wiley_Vch Verlag Gmbh & Co. KGaA, Weinheim. ISBN 3-527-29782-0

Hernández J (2016) Trihalomethanes in drinking water of Puerto Rico and their reduction using nanomaterials. Editorial Académica Española. ISBN: 978-3-8417-5946-7

Klobes P (2011) Precision measurement of the specific surface area of solids by gas adsorption, Chapter 4: microprobing and microstructures analysis [Internet]. BAM Federal Institute for Materials Research and Testing, Berlin, Germany. [Cited 2017 May 20]. Available from http://www.bam.de/en/basiskontakte/index.htm

Mclaughlin H, Shields F, Jagiello J, Thiele G (2012) Analytical options for biochar adsorption and surface area. Presented at the 2012 US biochar conference session on characterization

Ni Z, Wang Y, Yu T, Shen Z (2008) Raman spectroscopy and imaging of graphene. Nano Res 1:273–291

Rao CNR, Subrahmanyam KS, Ramakrishna HSS, Govindaraj A (2011) Graphene: synthesis, functionalization and properties. Modern Phys Lett 25(7):427–451

Subrahmanyam KS, Vivekchand SRC, Govindaraj A, Rao CNR (2008) A study of graphenes prepared by different methods: characterization, properties, and solubilization. J Mater Chem 18:1517–1523

United States Environmental Protection Agency (2012) 2012 Edition of drinking water standards and health advisories. EPA 822-S-12-001. Office of Water U.S. Environmental Protection Agency Washington, DC. [Cited 2017 Apr 12]. Available from: http://water.epa.gov/drink/contaminants/index.cfm.Secondary

Voutou B, Stefanaki EC (2008) Electron microscopy: the basics. 2008. Aristotle University of Thessaloniki, Greece. [Internet]. [Cited 2017 Apr 14]. Available from: http://www.mansic.eu/documents/PAM1/Giannakopoulos1.pdf

Wang ZL (2000) Characterization of nanophase materials. Part I technical approaches. [CD-ROM]. Wiley-VCH Verlag GmbH. ISBNs: 3-527-29837-1

Wang Y, Fan W, Wang G, Ji M (2011) New insight into the graphene based films prepared from carbon fibers. Mater Sci Appl 2:834–838

Wilson RA, Bullen HA (2006) Introduction to scanning probe microscope. Basic theory atomic force microscopy (AFM). Department of Chemistry, Northern Kentucky University, Highland Heights, KY

Wu W, Chen W, Lin D, Yang K (2012) Influence of surface oxidation of multiwalled carbon nanotubes on the adsorption affinity and capacity of polar and nonpolar organic compounds in aqueous phase. Environ Sci Technol 46:5446–5454

Zolyomi V, Koltai J, Kurti J (2011) Resonance Raman spectroscopy of graphite and graphene. Phys Status Solidi 11:2435–2444

Chapter 6
Functionalized Nanosize Graphene and Its Derivatives for Removal of Contaminations and Water Treatment

Rajesh Kumar, Rajesh K. Singh, Vinod Kumar and Stanislav A. Moshkalev

Abstract The physical, chemical, biological, or radiological substance or matter in water makes it contaminant leaving no longer suitable as a drinking water as it may produce severe health risk. Two-dimensional (2D) carbon based material "Graphene" have attracted attention in a variety of research fields such as physics, material science, chemistry and engineering. Graphene with exceptional physical, chemical, mechanical and thermal properties, now has been widely investigated as next-generation adsorbents in both water and waste water treatment. The chemically modified graphene containing various functional groups, offer potential applications for water treatment due to adsorption of various inorganic pollutants and organic dyes. Also, the derivatives of graphene or graphene based materials such as graphene oxide (GO)/reduced graphene oxide (rGO)/functionalized GO are extensively used for water treatment as an adsorbent for heavy metal ions and organic contaminants. GO membranes are brilliant candidate for water treatment, such as sewage purification and desalination owing to very high mechanical strength, superior flexibility and hydrophilic property. Several past studies have been specially focused on the efficient exploitation of chemically modified GO, GO-polymer

R. Kumar (✉) · S. A. Moshkalev
Centre for Semiconductor Components and Nanotechnology (CCS Nano),
University of Campinas (UNICAMP), Campinas, Sao Paulo 13083-870, Brazil
e-mail: rajeshbhu1@gmail.com

S. A. Moshkalev
e-mail: stanisla@unicamp.br

R. K. Singh (✉)
School of Physical & Material Sciences, Central University of Himachal Pradesh (CUHP),
Kangra, Dharamshala 176215, India
e-mail: rksbhu@gmail.com

V. Kumar
Department of Materials Engineering, Ben Gurion University of the Negev,
P.O.B. 653, 84105 Beer-Sheva, Israel
e-mail: mail2vinod2@gmail.com

M. Naushad (ed.), *A New Generation Material Graphene: Applications in Water Technology*, https://doi.org/10.1007/978-3-319-75484-0_6

composite, GO-semiconductor composite and metal oxide nanoparticle attached with GO for contaminant removal from water. These GO based composites have shown outstanding performance for the removal of water contaminants as well as good adsorbent of various types of inorganic pollutants like cadmium, chromium, arsenic, mercury, antimony, lead, fluoride, zinc, copper etc. and organic dyes as methylene blue (MB), methyl violet (MV), methyl orange (MO), rhodamine B (RB) etc. This chapter is focused on the recent advances in water treatment using chemically modified graphenene/GO, and aims to provide insights into the developments in water contamination removal technologies based on these novel nanomaterials.

Keywords Graphene oxide · Water treatment · Heavy metal ions
Organic dyes · Pollutants · Contaminations

1 Introduction

Drinking water is one of the primary requirements for survival of human beings and its contamination has caused a huge scarcity all around the world (Chen et al. 2017; Lee et al. 2014; Perreault et al. 2015; Poornima Parvathi et al. 2015; Unuabonah et al. 2013; Ye et al. 2017). Water contamination is a global concern that requires progressive and rationalized evaluation at all levels as it is the main causes of infection and mortality of living organisms (2017; Luh et al. 2017; Radojčić Redovniković et al. 2017; Singh et al. 2017b). The rapid expansion in transportation and increasing industrialization accelerated the discharge of waste water and toxic chemicals, which aggravates the environmental problems leading to many harmful effects and severe health issues (LeClair 1997; Whelton et al. 2015). The toxic chemicals released from industrial waste water generally contains high amount of organic and inorganic matters in varying concentrations (Anirudhan and Ramachandran 2008; Busetti et al. 2005; Castillo et al. 1998; Köhler et al. 2006). Mixing of these industrial toxic waste water into ground drinking water, or drainage into sea through various sources, contaminate ground drinking water with heavy metals ions such as As, Cr, Cd, Hg, Cu, Pb, Ni Hg and Zn etc. (Alothman et al. 2013; Berg et al. 2001; Cohen 1986). The other reason for contamination of drinking water is due to the hazardous materials released from the plastic, leather, synthetic dyes and textile industries (Arshad et al. 2017; Mukherjee et al. 2017; Pathania et al. 2015; Van Hoa et al. 2016). The textile industry normally uses a large amount of water for dying/washing and resultant effluents from the dying process often have a certain amount of color compounds, which cause the pollution of the water (Javadian et al. 2014; Sharma et al. 2015; Van Hoa et al. 2016).

The treatment of water pollutants including industrial effluents, domestic waste agricultural discharge, etc., is being researched world-wide (Aboubaraka et al. 2017; McCoy 1998; Mudassir et al. 2017; Peng et al. 2016b; Wei et al. 2013; Zhang et al. 2016). Highly adsorbent materials play a very important role in the treatment of industrial waste water discharge and oil-spill accidents (Cao et al. 2017). For the removal of heavy metal ions by adsorption/treatment of toxicity in water, several main issues like low efficiency and expensive bulk production are the key concerns in the performance on nanotechnology for waste water treatment (Naushad et al. 2015). To overcome these issues, engineering of nanomaterials in terms of cost effectiveness; activity, non-toxicity and stability over continuous use are the most important necessities for water treatment. The conventional adsorbent materials such as carbon nanotubes, active carbon, silica gel, clay materials, polypropylene sponge, zeolites, wool fibers and other microporous materials exhibit many drawbacks such as low separation efficiency and selectivity, difficult oil recovery and poor material durability etc., which seriously restrict their long term practical application (Cao et al. 2017; Wang et al. 2015; Zhu et al. 2016).

Since it's first isolation in 2004, graphene has become one of the hottest topics in the research field of materials sciences, contains one atom thick sheet of sp^2 hybridized hexagonal carbon lattice with delocalized π electrons (Geim and Novoselov 2007; Singh et al. 2011; Zhao et al. 2017; Zhong et al. 2017). Diverse studies have shown that functionalization of graphene could greatly improve the photocatalytic and adsorption properties owing to its high carrier mobility (20,000 $cm^2\ v^{-1}\ s^{-1}$), large surface area (2600 $m^2\ g^{-1}$), good optical transparency, high transmittance, strong absorption and mechanical flexibility (Balandin 2011; Balandin et al. 2008; Bolotin et al. 2008; Kumar et al. 2015a; Nair et al. 2008; Stoller et al. 2008). High elasticity and unusual magnetic properties of graphene also find its use in water treatment (Peng et al. 2017; Singh et al. 2017a; Suárez-Iglesias et al. 2017; Upadhyay et al. 2015). The derivative of graphene i.e. graphene oxide (GO) can be obtained easily from the oxidation of graphite and its exfoliation. GO is conventionally prepared by chemical oxidation of graphite using Brodie, Staudenmaier, or Hummers methods, or some variations of these methods and subsequent exfoliation of graphite oxide using various reduction techniques like thermal, microwave, laser etc. (Kumar et al. 2015b, 2017a, b, c; Singh et al. 2016; Zaaba et al. 2017). Unlike the special synthesis procedures for carbon based nanomaterials, the synthesis of GO from oxidation of graphite is basically easier and offers the potential of large-scale production, cost effectiveness, and biocompatibility.

2 Graphene and Its Derivatives for Water Treatment

Recently, GO is rapidly used as an alternative option for waste water treatment and adsorption of heavy metal ions attributed to their large surface areas and more functionalized active sites (An et al. 2017). Also, the studies for water treatment have concentrated on chemically modified GO and GO-metal/semiconductor oxide composites providing novel properties of interest. The oxygen containing functional groups like hydroxyl, carboxyl, carbonyl, and epoxy on GO surfaces, help to absorb the heavy metal ions and improve the capabilities to remove the pollutants from

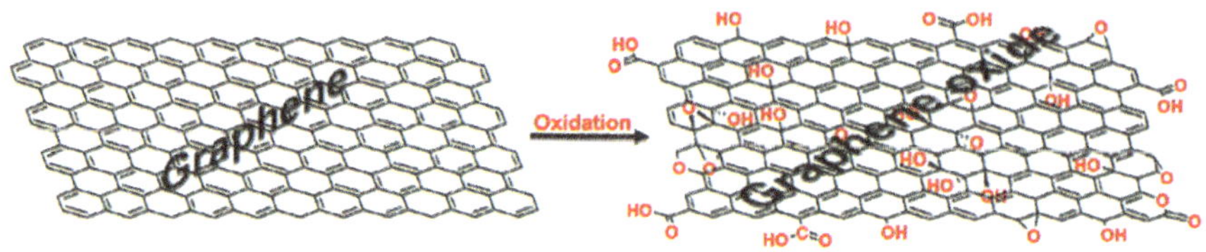

Fig. 1 Oxidation of graphene sheet into graphene oxide (Georgakilas et al. 2016). Reprinted (adapted) with permission from Georgakilas et al. (2016). Copyright (2016) American Chemical Society

water (Gadipelli and Guo 2015; Liu et al. 2016a; Vilela et al. 2016; Wu et al. 2016b). Moreover, GO based composites exhibit not only high adsorption capacity for metal ions and organic solvents, but also efficient release of adsorbents owing excellent recyclability. GO have been used in the waste water treatment containing heavy metal ions, toxic organic solvents, dyes, virus, bacteria, and radionuclides following its enhanced active adsorptive sites (Chandra et al. 2010; Kemp et al. 2013). The bare graphene/GO have been also used for phosphate removal from water (Vasudevan and Lakshmi 2012). The GO functionalized with β-FeOOH, ZrO_2, La_2O_3, and LaOH have shown high aqueous phosphate removal capacity (Chen et al. 2016a; Harijan and Chandra 2017; Luo et al. 2016; Tran et al. 2015; Zhang et al. 2014a). The magnetic nanoparticles attached GO have attracted attention for application in the removal of dyes because of the ease and simplicity in recovering adsorbent from the liquid phase. The chemically bonded Fe_3O_4 nanoparticles with GO surface has been utilized for the removal of dye from aqueous solution by the adsorption process with high removal efficiency (Raghu et al. 2017). Hierarchical mesoporous graphene@Fe_3O_4@chitosan composites have shown the rapid adsorption of MB with high capacity and easy separation in a less contact time (Van Hoa et al. 2016). Various metal oxides attached with GO has been utilized for the waste water treatment and removal of toxic elements (Mehta et al. 2015). Figure 1 shows the schematic illustration of graphene and its oxidation towards the formation of GO, which contains many oxygen functional groups useful for the adsorption of heavy metal ions and dyes (Georgakilas et al. 2016).

3 Heavy Metal Ions Removal from Waste Water

The challenging industrialization around the world discharges a vast amount of effluents containing numerous heavy metals which is one of the main sources of water pollution and contaminations (Bhaumik et al. 2016; Singh et al. 2017a). The excessive use of heavy metals in industries such as leather tanning, metallurgy, petrochemicals, battery, paper manufacturing and paint industries increases concern and severe threat to human health and the surroundings (Fu and Wang 2011; Taseidifar et al. 2017). These heavy metals like As(III), Cd(II), Hg(II), Cr(VI) and

Pb(II) are resistant to conventional treatment methods due to strong bonding with organic ligands, which increases their solubility and mobility (El-Sherif et al. 2013). This creates trouble for public health, since these toxic compounds are carcinogenic and highly toxic for cellular organisms (Zou et al. 2016). The heavy metals contamination in water has critically emphasized from past few years as a raising environmental hazard.

3.1 *Removal of Heavy Metal Ions by Chemically Modified Graphene/Graphene Oxide*

Heavy metal ions in underground water pose a severe threat to public health and ecological systems. The water pollution containing metal ions such as Cr, As, Cd, Cu, Pb, Ni, Hg and Zn and others heavy metal ions from industrial and domestic wastes is becoming one of the most severe environmental troubles worldwide (Samadi et al. 2017). These heavy metal ions contribute as pollutant in geochemical systems, and have non-repairable dangerous effects. They create rigorous threats to human health and ecological balance. Therefore, it is vital to eliminate these ions from contaminated water/aqueous media. Immense studies have been conceded to develop cost-effective heavy metal removal techniques. Several groups of researchers conducted experiments with graphene and its derivatives GO/rGO/ functionalized GO to study removal of heavy metals from water. Kireeti et al. (2016) offered a facile route to synthesize a hydrophobic sodium modified rGO-magnetic iron oxide (SMGI) nanocomposite which was used as an adsorbent material for the removal of Pb(II) from aqueous solutions. Figure 2a shows the schematic of the adsorption mechanism of Pb(II) on SMGI nanocomposite. The nanocomposite possesses superparamagnetic property, which was exploited for the easy recovery of the SMGI nanocomposite from water after Pb(II) adsorption. The maximum adsorption capacity of 1666.66 ± 59 mg g^{-1} was obtained. The adsorbent showed more than 75% removal efficiency of Pb(II) after 3 cycles of regeneration (Fig. 2b).

Recently in 2017, Raghubanshi et al. (Raghubanshi et al. 2017) reported the synthesis of GO and used it for the removal of Pb^{2+} from aqueous solution. The adsorption experiments were carried out in batches to study the effect of initial concentration, contact time, and temperature on the adsorption of Pb^{2+}. The schematic diagrams for the adsorption of Pb^{2+} on GO is presented in Fig. 3a. Further the wrinkles nature of GO was accounted for the enhancement of the adsorption of Pb^{2+} over the surface of GO. Figure 3b shows that the Pb^{2+} adsorption on GO increases with the increase in the initial concentration of Pb^{2+} and the maximum adsorption of 120 mg g^{-1} took place at initial concentration of 60 mg L^{-1} Pb^{2+}.

The Cu are found in many waste water sources including, printed circuit board manufacturing, electronics plating, paint manufacturing etc. Cu can be removed

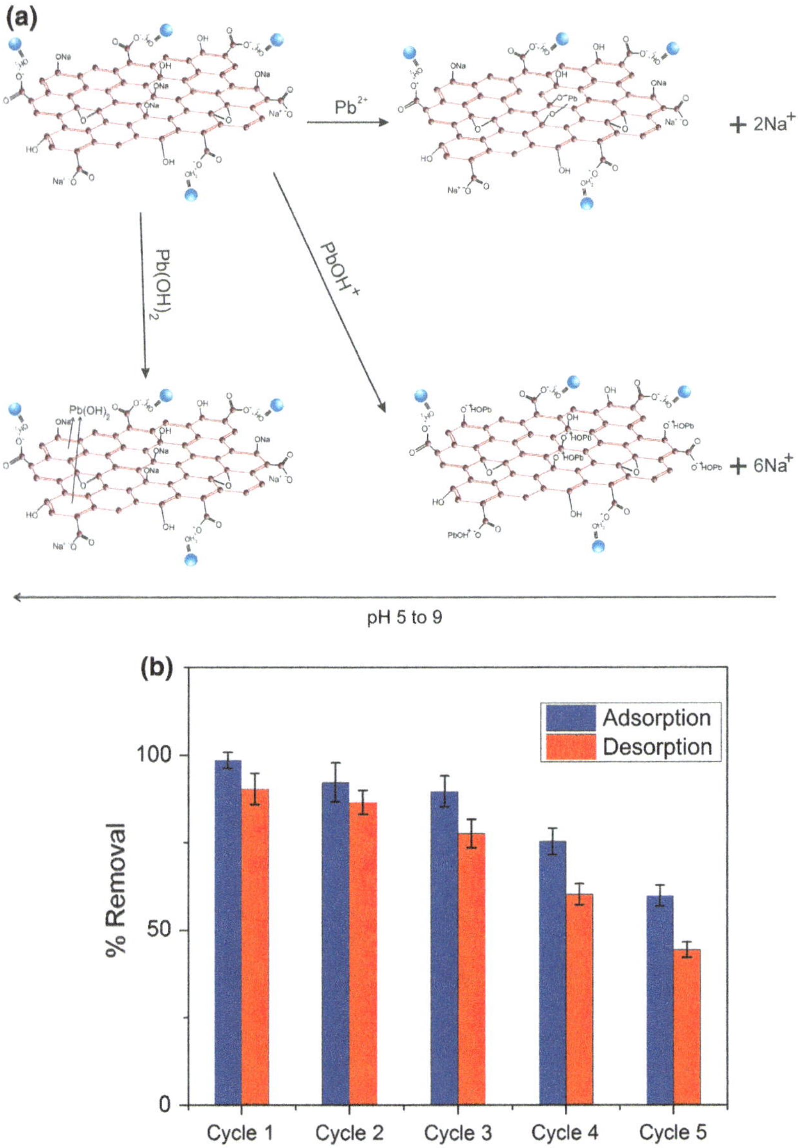

Fig. 2 **a** Adsorption mechanism of Pb(II) on SMGI nanocomposite and **b** reusability graph (Co: 100 ppm, adsorbent: 4 mg, contact time: 300 min, temperature: 303 K) (Kireeti et al. 2016). Reprinted (adapted) with permission from Kireeti et al. (2016). Copyright (2016) The Royal Society of Chemistry

from waste water by GO using various techniques. Yang et al. (2010) demonstrated the synthesis of folded/aggregated GO and its application in Cu^{2+} ion removal. The folding/aggregation triggered the coordination between GO and Cu^{2+}. The

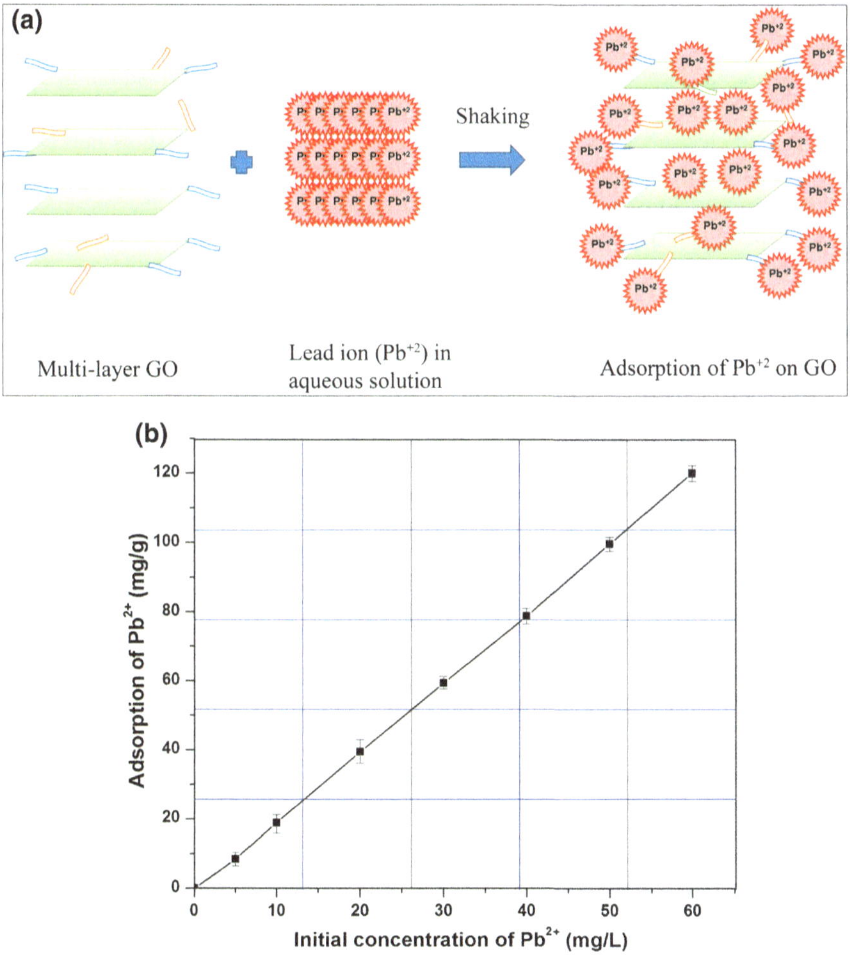

Fig. 3 **a** Schematic diagram for the adsorption of Pb^{+2} on GO and **b** adsorption isotherm of GO toward Pb^{2+} (GO: 0.01 g, Pb^{2+} solution: 0.02 L, 25 °C, 200 rpm, 30 min) (Raghubanshi et al. 2017). Reprinted (adapted) with permission from Raghubanshi et al. (2017). Copyright © 2017 Elsevier B. V. All rights reserved

equilibrium Cu^{2+} concentrations and equilibrium absorption capacity of GO shows the maximum absorption capacity of GO for Cu^{2+} which is around 10 times of that of active carbon. The obtained results shows excellent Cu^{2+} absorption capacity of GO, ranking GO among the most effective absorbents for Cu^{2+} removal. In 2013, Sulfonated magnetic GO composite (SMGO) was synthesized from GO and used as adsorbent for removing Cu(II) ions from aqueous solution (Hu et al. 2013). Optimum Cu(II) uptake of 62.73 mg g^{-1} was achieved at pH 4.68, Cu(II) concentrations 73.71 mg L^{-1}, and temperature 50 °C. The thermodynamic parameters calculated from the temperature-dependent isotherms indicated that the adsorption reaction was an endothermic and spontaneous process.

Also the adsorption of Cd(II) onto magnetic GO-supported sulfanilic acid (MGO–SA) in aqueous solutions containing different types and concentrations of background electrolytes were studied (Hu et al. 2014). The results show that Cd(II) adsorption is strongly dependent on pH and adversly affected by background electrolytes and ionic strength. The Cd(II) removal decreases in the presence of background electrolyte cations (Na^+, K^+, Ca^{2+}, Mg^{2+}, Mn^{2+}, Zn^{2+}, and Ni^{2+}), and the divalent cations exerted more influences on the Cd(II) uptake than the monovalent cations at pH 6. The Cl^- and NO_3^- had negative effects on Cd(II) adsorption because they can form water-soluble metal–anion complexes with Cd(II) ions. The presence of 0.01 mol L^{-1} Na_3PO_4 reduced the removal percentage of Cd(II) at pH < 5 but extremely enhanced the Cd(II) removal at pH > 5. The Cd(II) adsorption was sensitive to changes in the concentration of NaCl, $NaNO_3$, $NaClO_4$, and Na_3PO_4.

Wang et al. (2013b) used magnetic mesoporous silica and GO (MMSP–GO) composites as adsorbents for the removal of heavy metal (Pb(II) and Cd (II)) ions and humic acid (HA). Figure 4a represents the schematized pathway for the synthesis of polyethylenimine-modified MMSP-GO composites. The maximum adsorption capacities of MMSP-GO for Pb(II) and Cd (II) were 333 and 167 mg g^{-1}, calculated by Langmuir model, respectively as shown in Fig. 4b. The HA enhances adsorption of heavy metals by MMSP-GO composites and show potential for its use as adsorbents for the simultaneous removal of heavy metals and HA in waste water treatment processes.

In 2014, A simple method for the synthesis of hierarchical polyaniline modified GO (PANI/GO) composites using 1D (uniform aligned PANI nanorods) and 2D (GO nanosheets) nanocomponents by dilute polymerization was developed and applied for the adsorption of heavy metals from aqueous solutions under ambient conditions (Liu et al. 2014). The maximum adsorption capacities of Co(II), Ni(II), Pb(II) and U(VI) ions on PANI/GO composites calculated from Langmuir models are 22.28, 25.67, 65.40 and 1552.31 mg g^{-1}, respectively. The excellent adsorption capacity suggests that PANI/GO composites can be applied as a promising adsorbent in heavy metal removal. Hur et al. (2015) demonstrated competitive adsorption isotherms of Pb(II), Cd(II) and Cu(II) examined on a magnetic GO, MWCNTs, and powered activated carbon (PAC). Magnetic GO synthesized by a simple ultrasonification method gives the maximum adsorption capacities of the adsorbents in the order of Pb(II) > Cu(II) > Cd(II) (Fig. 5), which corroborates with the degree of the electronegativity and the hydrated radius of the metals, suggesting that the metal adsorption may be governed by an ion exchange between positively charged metals and negatively charged surfaces, as well as diffusion of metals into the surface layer. Comparison of the q_{max} values between the single and multi-metal systems of Cu(II) and Pb(II) revealed that the competitive adsorption was greater in the order of the magnetic GO > MWCNT > PAC. The antagonistic effects were influenced by pH as well as the metals type, and were in the order, magnetic GO > MWCNT > PAC.

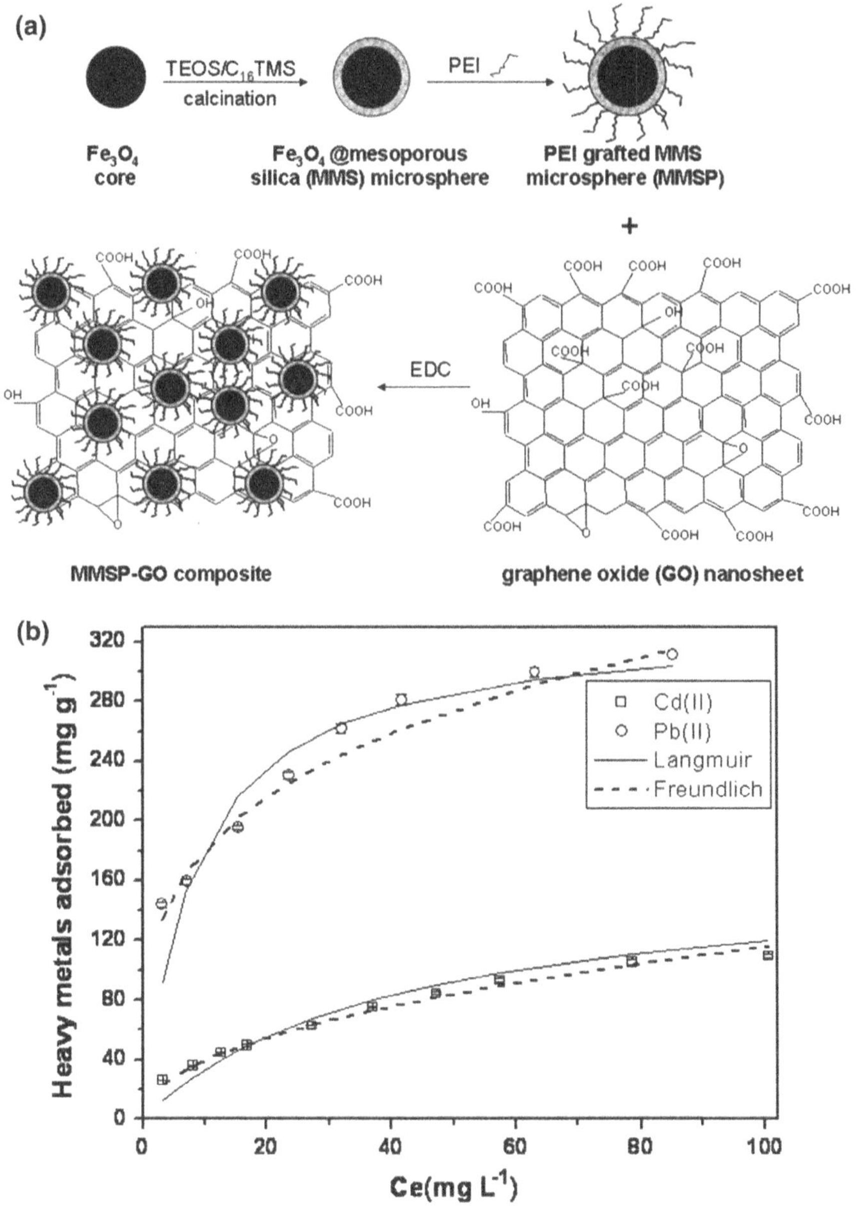

Fig. 4 **a** Schematic illustration of pathway for preparation of polyethylenimine-modified MMSP–GO composites and **b** Pb(II) and Cd(II) adsorption isotherms on MMSP–GO (adsorbent loading: 100 mg L21; pH: 7.1; contact time: 24 h) (Wang et al. 2013b). Reprinted (adapted) with permission from Wang et al. (2013b).

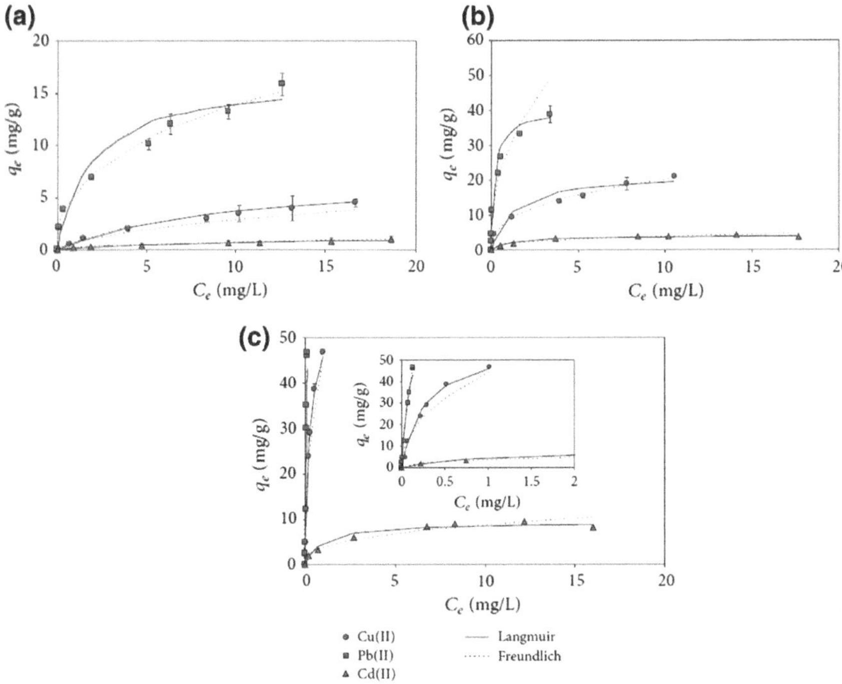

Fig. 5 Adsorption isotherms of Cu(II), Pb(II), and Cd(II) on magnetic GO at pH 4.0 (**a**), 6.0 (**b**), and 7.0 (**c**) under the single-metal systems (Hur et al. 2015). Reprinted (adapted) with permission from Hur et al. (2015). Copyright 2015; Hindawi Publishing Corporation

Zhao et al. (2011b) reported a few-layered GO nanosheets as superior sorbents of heavy metal and used for removal of Cd(II) and Co(II) ions from large volumes of aqueous solutions. The results indicated that Cd(II) and Co(II) adsorption on GO nanosheets depends strongly on pH and weakly on ionic strength. The presence of humic acid reduces the adsorption of Cd(II) and Co(II) on GO nanosheets at pH < 8. The maximum adsorption capacities (C_{smax}) of Cd(II) and Co(II) on GO nanosheets at pH 6.0 ± 0.1 and T = 303 K were about 106.3 and 68.2 mg g^{-1}, respectively as shown in Fig. 6. The thermodynamic parameters calculated from temperature-dependent adsorption isotherms suggested that Cd(II) and Co(II) adsorptions on GO nanosheets were endothermic and spontaneous processes.

The adsorption of divalent metal ions like Cu(II), Zn(II), Cd(II) and Pb(II) from aqueous solutions using GO was studied (Sitko et al. 2013). The obtained results indicate that maximum adsorption is achieved in broad pH ranges: 3–7 for Cu(II), 5–8 for Zn(II), 4–8 for Cd(II), 3–7 for Pb(II). The maximum adsorption capacities of Cu(II), Zn(II), Cd(II) and Pb(II) on GO at pH = 5 are 294, 345, 530, 1119 mg g^{-1}, respectively. The single and competitive adsorption of Cu(II), Zn(II), Cd(II) and Pb(II) shows that the affinities of GO for these metal ions follow the order Pb(II) > Cu(II) > > Cd(II) > Zn(II). Adsorption isotherms and kinetic studies

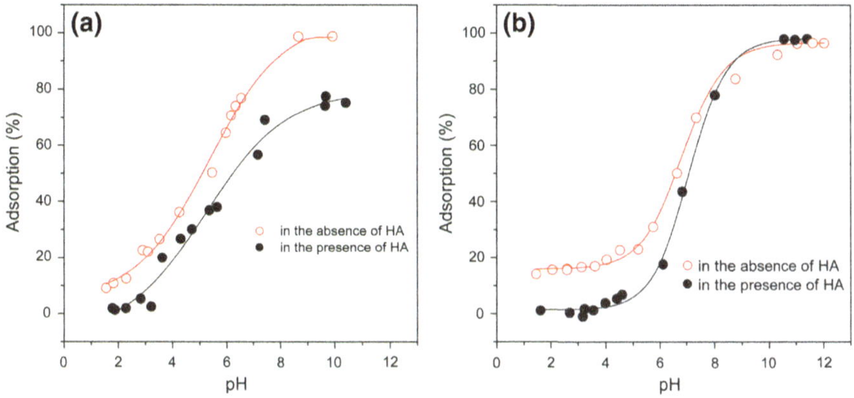

Fig. 6 Effect of humic acid on the adsorption of Cd(II) (A) and Co(II) (B) on GO nanosheets. C $[HA]_{initial}$ = 10 mg L^{-1}, $C[Co(II)]_{initial}$ = 30 mg L^{-1}, $C[Cd(II)]_{initial}$ = 20 mg L^{-1}, I = 0.01 M $NaClO_4$, T = 303 K, m/V = 0.1 g L^{-1} (Zhao et al. 2011b). Reprinted (adapted) with permission from Zhao et al. (2011b). Copyright (2011) American Chemical Society

suggested that adsorption of metal ions on GO nanosheets is monolayer coverage and is controlled by chemical adsorption involving the strong surface complexation of metal ions with the oxygen-containing groups on the surface of GO.

Chen et al. (2016b) reported adsorption of heavy metals by GO/cellulose hydrogel prepared from NaOH/urea aqueous solution. A simple and novel method was used to prepare GO/cellulose hydrogel using cellulose and GO to adsorb metal ions. Adsorption capacity of Cu^{2+} significantly increases with an increase in the GO/cellulose ratio. The adsorption capacities at equilibrium (q_e^{cale}) for GO/cellulose hydrogel (GO: cellulose = 20:100 in weight) was up to 94.34 mg g^{-1} as shown in Fig. 7. The GO/cellulose hydrogel exhibited high efficient regeneration and metal ion recovery, and high adsorption capacity for Zn^{2+}, Fe^{3+}, and Pb^{2+}.

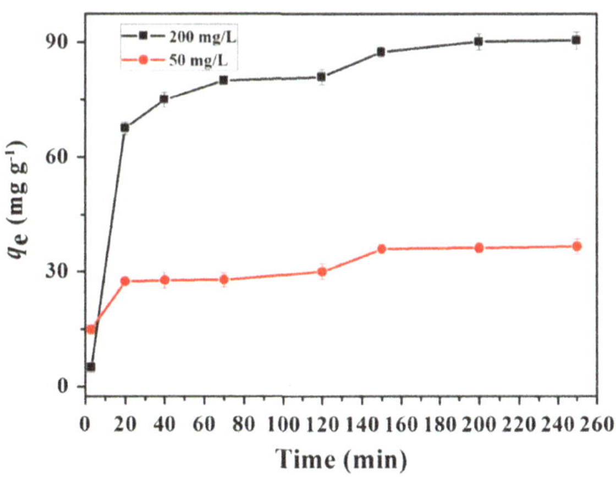

Fig. 7 Adsorption of Cu^{2+} on the GO(20)/cellulose(100) hydrogel as a function of contact time (the error bars represent standard deviations based on three measurements) (Chen et al. 2016b). Reprinted (adapted) with permission from Chen et al. (2016b). Copyright © 2016 MDPI. All rights reserved

Wu et al. (2016a) synthezed three dimensional rGO (3-D rGO) hydrogel by hydrothermal method from GO, de-ionized water, and oxalic acid dehydrate. The synthesized 3-D rGO aerogel was used as adsorbent for the removal of the inorganic ions, Hg^{2+}, and F^{-}, from aqueous solutions and showed excellent removal capabilities. The maximum adsorption capability of Hg^{2+} and F^{-} approached 185 and 31.3 mg g^{-1}, respectively, indicating that the 3-D rGO aerogel is a very suitable material for environmental pollution management.

3.2 Removal of Heavy Metal Ions by Graphene Oxide-Metal/Oxide Nanoparticles

GO/Fe_3O_4 composites were synthesized via a Cu catalyzed azide-alkyne cyclo addition reaction and further modified with polyacrylic acid (PAA) (Zhang et al. 2013a). Figure 8a shows the SEM image of PAA/GO/Fe_3O_4 nanocomposites having high surface area, excellent complex ability and superparamagnetism, were used as nano-adsorbents for recyclable removal of Cu^{2+}, Cd^{2+} and Pb^{2+} ions from aqueous solution. The PAA/GO/Fe_3O_4 nanocomposites show extraordinary removal capacity for Cu^{2+}, Cd^{2+} and Pb^{2+} ions as shown in Fig. 8b. The PAA/GO/Fe_3O_4 nanocomposites can be easily separated and recycled due to the superparamagnetism of Fe_3O_4. After five cycles, the removal efficacy of the nano adsorbents for Cu^{2+}, Cd^{2+} and Pb^{2+} ions is over 85%.

Sun et al. (2011) reported an one step solvothermal strategy using non-toxic and cost-effective precursors to prepare magnetite/rGO (MrGO) nanocomposites for removal of MrGO nanocomposites using GO, $FeCl_3 \cdot 6H_2O$, sodium acrylate (CH_2=CHCOONa) and sodium acetate (NaOAc) in a mixture of EG and diethylene glycol (DEG) as shown in Fig. 9a. Graphene with magnetic nanoparticles exhibits

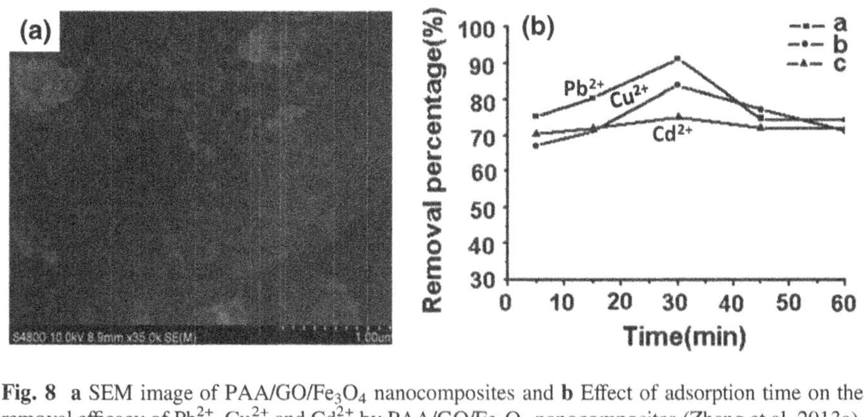

Fig. 8 **a** SEM image of PAA/GO/Fe_3O_4 nanocomposites and **b** Effect of adsorption time on the removal efficacy of Pb^{2+}, Cu^{2+} and Cd^{2+} by PAA/GO/Fe_3O_4 nanocomposites (Zhang et al. 2013a). Reprinted (adapted) with permission from Zhang et al. (2013a). Copyright (2013) The Royal Society of Chemistry

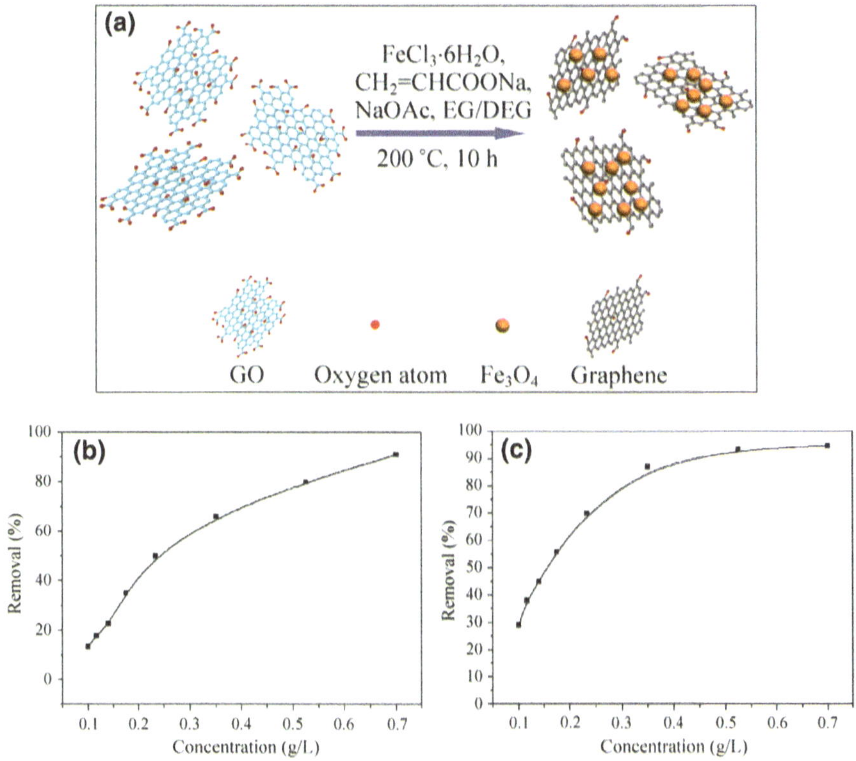

Fig. 9 **a** Schamatice representation of MrGO by the solvothermal reduction method and the removal of **b** RB and **c** malachite green at different concentrations of MrGO-4 (Sun et al. 2011). Reprinted (adapted) with permission from Sun et al. (2011). Copyright © 2011, Springer-Verlag Gmbh and Co. KG

excellent removal efficiency and rapid separation from aqueous solution by an external magnetic field. When the amount of MRGO-4 (4.4 emu/g) was increased to 0.7 g L^{-1}, over 91% of RB and over 94% of malachite green can be removed within 2 h at room temperature (Fig. 9b and c, respectively). The performances of the MrGO composites strongly depend on loading of Fe_3O_4 and pH value. Further real samples including industrial waste water and lake water was also treated using the MrGO composites.

Liu et al. (2011) demonstrated the synthesis of magnetite/GO (M/GO) composite and used it for the removal of heavy metal ions from aqueous solutions. The sorption of Co(II) on the M/GO composite was carried out by varying conditions like contact time, sorbent content, pH, ionic strength, foreign ions, and temperature. The adsorption isotherms of Co(II) on the M/GO composite. The adsorption of Co (II) on M/GO described by the Langmuir model shows the influence of foreign ions and the thermo dynamical parameters calculated indicate that the adsorption reaction of Co(II) on the M/GO composite is endothermic and spontaneous process.

M/GO-bound Co(II) can be quickly separated and recovered from a solution by magnetic separation.

Vuong Hoan et al. (2016) reported synthesis of rGO modified by magnetic iron oxide (Fe_3O_4/rGO) and its application for heavy metals removal. The formal kinetics and adsorption isotherms of As(V), Ni(II), and Pb(II) over obtained Fe_3O_4/rGO was investigated. The adsorption of As(V), Ni(II), and Pb(II) on the Fe_3O_4/rGO was well fitted to the *pseudo*-second-order kinetic model and obeyed both Langmuir and Freundlich models, which indicates surface heterogeneity and monolayer adsorption of the adsorbents. The Fe_3O_4/rGO exhibits excellent heavy metals adsorption. Figure 10 shows the adsorption data of different concentrations for As(V), Ni(II), and Pb(V) with time and represent that the adsorption capacity of Fe_3O_4/rGO for As(V) increases with the initial As(V) concentration from 13.1 to 75.8 ppm. Also, the maximum monolayer adsorption capacities calculated by Langmuir equation are 58.48 mg g^{-1} for As(V), 65.79 mg g^{-1} for Pb(II), and 76.34 mg g^{-1} for Ni(II).

The adsorption behaviors of M(II) (M=Cu, Cd) by Mn-doped Fe(III) oxide magnetic nanoparticle implanted graphene (GMIO) at different pH, contact times, and concentrations were investigated (Nandi et al. 2013). Monolayer adsorption

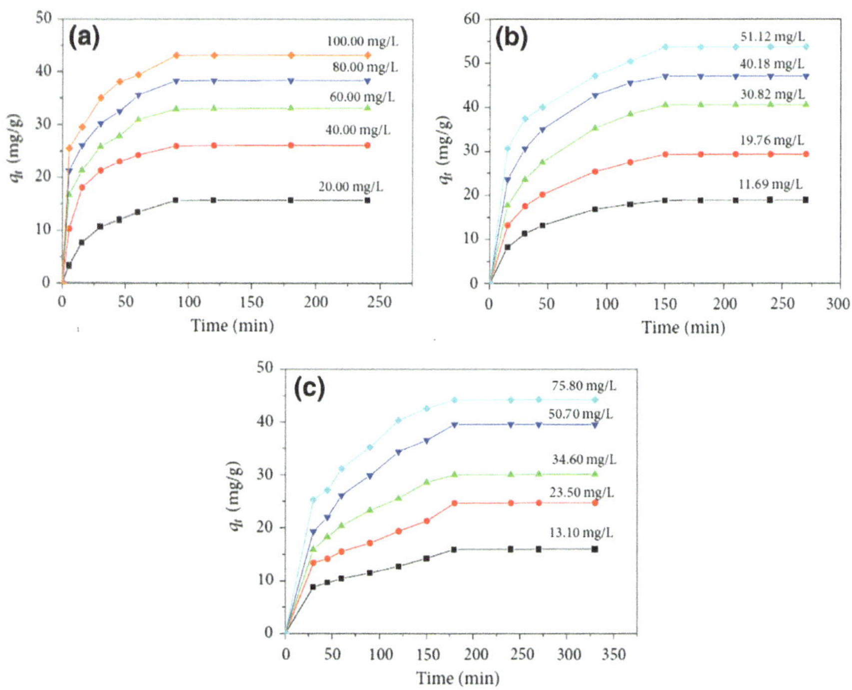

Fig. 10 Adsorption kinetics of As(V) (**a**), Ni(II) (**b**), and Pb(II) (**c**) over Fe_3O_4/rGO with different concentrations (Vuong Hoan et al. 2016). Reprinted (adapted) with permission from Vuong Hoan et al. (2016). Copyright 2016; Hindawi Publishing Corporation

capacity values of GMIO for M(II) increases (88–127 mg g^{-1} for Cd; 130–144 mg g^{-1} for Cu) with increasing temperature from 288 to 333 K. Negative ΔG^0 values indicate spontaneous adsorption nature despite positive ΔH^0 values owing to an entropy increase (positive ΔS^0) at the solid-liquid interface. Zhang et al. (Zhang et al. 2014b) reported the synthesis of magnetic cobalt ferrite ($CoFe_2O_4$)–rGO nanocomposites ($CoFe_2O_4$–rGO) as adsorbent to remove Pb(II) and Hg(II) from aqueous solution. The adsorption kinetics and isotherm model fitting studies, demonstrated that the data fit well to pseudosecond order kinetics and Langmuir isotherm models. The highest adsorption equilibrium for Pb(II) were 299.4 mg g^{-1} at pH 5.3 and 25 °C; while for Hg(II) were 157.9 mg g^{-1} at pH of 4.6 and 25 °C. Due to magnetic property, metal adsorbed $CoFe_2O_4$–rGO could be simply recovered from water by magnetic separations at low magnetic field within 2 min.

Zong et al. (2013) reported the synthesis of magnetic graphene/iron oxides composite (Fe_3O_4/GO) and used it as adsorbent and solidification of U(VI) ions from aqueous solutions. The sorption behavior of U(VI) on the surface of Fe_3O_4/GO was carried out for different contact time, pH and ionic strength according to concentration of U(VI)initial = 1.12×10^{-4} mol L^{-1}. Figure 11 shows the sorption of U(VI) on Fe_3O_4/GO as a function of contact time (at pH 5.5). The removal percentage of U(VI) increases sharply for first 2 h, and then attains equilibrium after 4 h. The maximum adsorption capacity of U(VI) on Fe_3O_4/GO at T = 293 K and pH = 5.5 ± 0.1 was about 69.49 mg g^{-1} higher than majority of materials. This

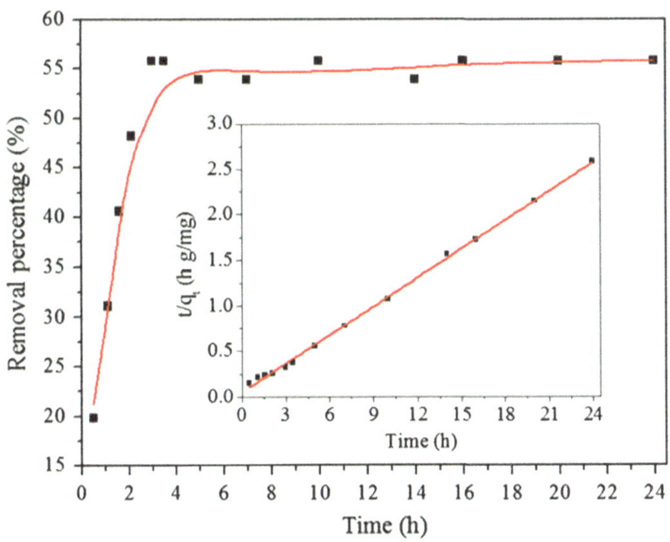

Fig. 11 Influence of contact time on U(VI) adsorption onto Fe_3O_4/GO and the pseudo second-order rate equation fit (inset). T = 293 K, pH = 5.5 ± 0.1, $C_{U(VI)initial} = 1.12 \times 10^{-4}$ mol/L, m/V = 0.3 g/L, I = 0.01 mol/L KNO_3 (Zong et al. 2013). Reprinted (adapted) with permission from Zong et al. (2013). Copyright (2013) The Royal Society of Chemistry

shows that the Fe_3O_4/GO is a promising adsorbent for the removal of radio nuclides and heavy metal ions from large volumes of aqueous solution.

One-step macroscopic multifunctional graphene-based hydrogels utilizing the synergistic effects of the reduction of GO by ferrous ions and in situ simultaneous deposition of nanoparticles on graphene sheets was fabricated (Cong et al. 2012). The pure free-standing α-Fe_2O_3 monolith with elegant microstructures composed of interconnected nanorods was prepared by burning the oil-saturated graphene/ α-FeOOH aerogel directly. The functional graphene-based hydrogels exhibit good capability for removal of pollutants. The multifunctional graphene-based hydrogels and aerogels are ideal candidates as adsorbents with high adsorption capacity for removal of heavy ions and oils from industrial water. Ren et al. (2011) synthesized graphene nanosheet/δ-MnO_2 (GNS/MnO_2) composite by a microwave-assisted method and used it for Ni (II) removal. The saturated adsorption capacity of Ni (II) onto GNS/MnO_2 was 46.6 mg g^{-1} at room temperature, which is 1.5 and 15 times higher than pure δ-MnO_2 and GNS, respectively. The positive values of ΔH and ΔS shows an endothermic reaction and increase in randomness at the solid-liquid interface during Ni (II) adsorption process. The negative ΔG values indicate a spontaneous adsorption process and GNS/MnO_2 reused for 5 times with recovery rate of 91%.

Sreeprasad et al. (2011) described a versatile, and simple synthetic route for the synthesis of a range of rGO-metal/metal oxide composites and their application in water purification. The rGO and its composites (rGO–MnO_2 and rGO–Ag) illustrated excellent Hg(II) uptake capacity. The rGO and its composites give a high distribution coefficient (K_d), greater than 10 L g^{-1} for Hg(II) uptake. The K_d values for the composites were about an order of magnitude higher compared to parent rGO and GO for this application.

Lee and Yang (2012) have described a hydrothermal synthesis of flower-like TiO_2–GO (GO–TiO_2) hybrid and used it for the removal of heavy metal ions (Cd^{2+}, Pb^{2+}, Zn^{2+}) from water. The flower-like TiO_2 on GO structure significantly improved the removal efficiency of heavy metals. The GO–TiO_2 hybrid adsorption capacities of heavy metal ions, after 6 and 12 h of hydrothermal treatment at 100 ° C, were respectively 44.8 ± 3.4 and 88.9 ± 3.3 mg g^{-1} for Zn^{2+} removal, 65.1 ± 4.4 and 72.8 ± 1.6 mg g^{-1} for Cd^{2+} removal, and 45.0 ± 3.8 and 65.6 ± 2.7 mg g^{-1} for Pb^{2+} removal at pH 5.6. In contrast, colloidal GO under identical condition showed removal capacities of 30.1 ± 2.5 (Zn^{2+}), 14.9 ± 1.5 (Cd^{2+}), and 35.6 ± 1.3 mg g^{-1} (Pb^{2+}). TiO_2 blossoms markedly formed upon GO as the hydrothermal treatment time at 100 °C increased from 6 to 12 h.

Hao et al. (2012) presented a simple two-step reaction for the synthesis of SiO_2/ graphene composite. This composite shows high efficiency and high selectivity toward Pb(II) ion. The adsorption capacity of SiO_2/graphene composite for Pb(II) ion represent 113.6 mg g^{-1}, which was higher than that of bare SiO_2 nanoparticles.

4 Removal of Arsenic

Arsenic is a obviously occurring metalloid and in terms of oxidation state, can exist in four forms, which are arsenite (As(III)), arsenate (As(V)), arsenic (As(0)), and arsine (As(III)). Among these four arsenic species, the most prevalent forms, which are commonly found in water, are the inorganic arsenite and arsenate (Nicomel et al. 2016; Pous et al. 2015). The groundwater contamination by arsenic poses imperative health risks and has prompted research efforts for its removal from water. Arsenic is a toxic carcinogen widely present in drinking water predominantly as As (III) (H_3AsO_3, $H_2AsO_3^-$, and $HAsO_3^{2-}$) and As (V) (H_3AsO_4, $H_2AsO_4^-$, and $HAsO_3^{2-}$) (Mohan and Pittman 2007). The magnetite-rGO (M–rGO) hybrids formed via chemical reaction shows superparamagnetic behavior at room temperature and was used for arsenic removal (Chandra et al. 2010). The hybrids demonstrated high binding capacity for As(III) and As(V) as compared to bare magnetite particles. Their high binding capacity is due to the increased adsorption sites in the M–rGO hybrids which occur by reducing the aggregation of bare magnetite. Since the composites show complete (over 99.9%) As removal within 1 ppb, and hence can be practically usable for arsenic separation from water.

Zhang et al. (2010) employed a series of composites (GO–Fe-1, GO–Fe-2, GO–Fe-3, GO–Fe-4, GO–Fe-5) based on GO cross-linked with ferric hydroxide for the removal of arsenic from contaminated drinking water. The ferrous compound cross-linked with GO was in situ oxidized to ferric compound by hydrogen peroxide, and treated with ammonium hydroxide as shown in Fig. 12a. For the water with arsenate concentration at 51.14 ppm, more than 95% of arsenate was absorbed by GO–Fe-5 composite with an absorption capacity of 23.78 mg arsenate/g of composite. Effective arsenate removal occurred in a wide range of pH from 4 to 9 and the efficiency of arsenate removal decreases when pH was increased higher than 8 as shown in Fig. 12b.

Zhu et al. (2012a) synthesized magnetic graphene nanoplatelet composites (MGNCs) decorated with core-shell Fe–Fe_2O_3 nanoparticles (NPs) using a facile one-pot thermal decomposition method. The graphene nanoplatelets (GNPs) decorated with uniformly dispersed NPs are observed to exhibit a strong magnetization and can be separated from the liquid mixture by a permanent magnet. These MGNCs shows adsorption of As(III) from the polluted water due to the increased adsorption sites. The adsorption behavior fitted well with Langmuir and Freundlich models and shows a significantly higher adsorption capacity (11.34 mg g^{-1}) than the other values reported on the conventional iron oxide based adsorbents ($\sim$1 mg g^{-1}). The obtained results show a nearly complete removal of As(III) within 1 ppb. A highly versatile and one-pot microwave route for the mass production of 3D graphene-carbon nanotube-iron oxide nanostructures was used and employed for the removal of arsenic from contaminated water (Vadahanambi et al. 2013). The three-dimensional nanostructures show excellent absorption of arsenic from contaminated water, due to its high surface-to-volume ratio and open pore network.

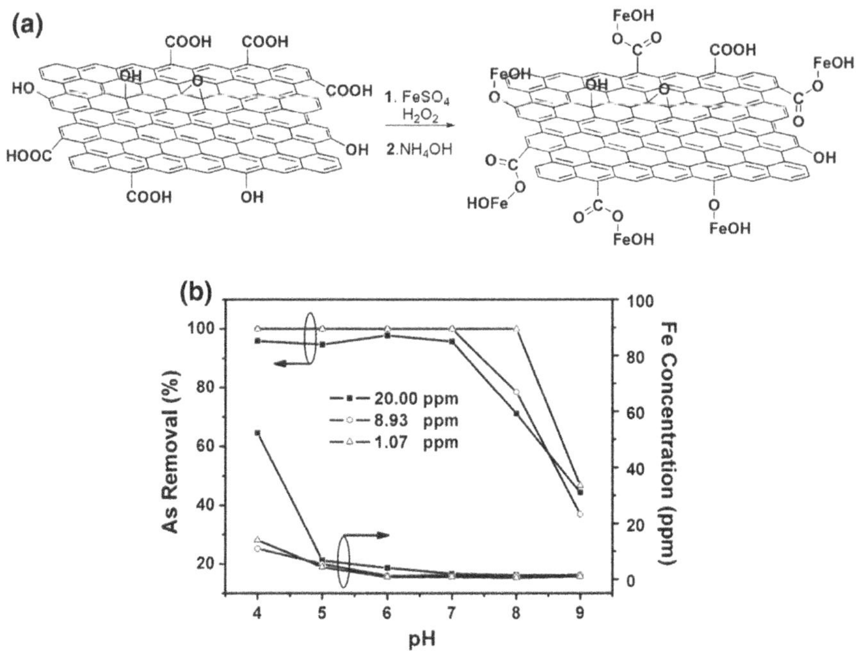

Fig. 12 **a** Illustration of the process for preparation of GO–Fe composites and **b** Effects of pH on arsenate removal at different initial arsenate concentration (square: 20.00 ppm, circle: 8.93 ppm, triangle: 1.07 ppm) and the ferric concentration in the filtrate (Zhang et al. 2010). Reprinted (adapted) with permission from Zhang et al. (2010).

The as-synthesized 3D Fe_3O_4/graphene aerogels (GAs) was used for the removal of arsenic ions from water (Ye et al. 2015). Figure 13a shows the SEM image of Fe_3O_4/GAs having interconnected pores constituting 3D graphene and Fe_3O_4 NPs anchored homogeneously on the graphene surface. Figure 13b shows that Fe_3O_4/GAs have a capacity of up to 40.048 mg g^{-1} As(V) ions adsorption due to their remarkable 3D structure and existence of magnetic Fe_3O_4 nanoparticles for magnetic separation. In addition to the excellent adsorption capability, Fe_3O_4/GAs can be easily and effectively separated from water, indicating potential applications in water treatment.

Mishra and Ramaprabhu (2011) reported the synthesis of graphene sheets by hydrogen induced exfoliation of graphitic oxide followed by functionalization and applied it successfully for simultaneous removal of high concentration of inorganic species of arsenic (both trivalent and pentavalent) and sodium from aqueous solution using supercapacitor based water filter. These functionalized graphene sheets based water filter was used for desalination of sea water. Maximum adsorption capacities, using Langmuir isotherm, for arsenate, arsenite and sodium were found to be nearly 142, 139 and 122 mg g^{-1}, respectively. High adsorption

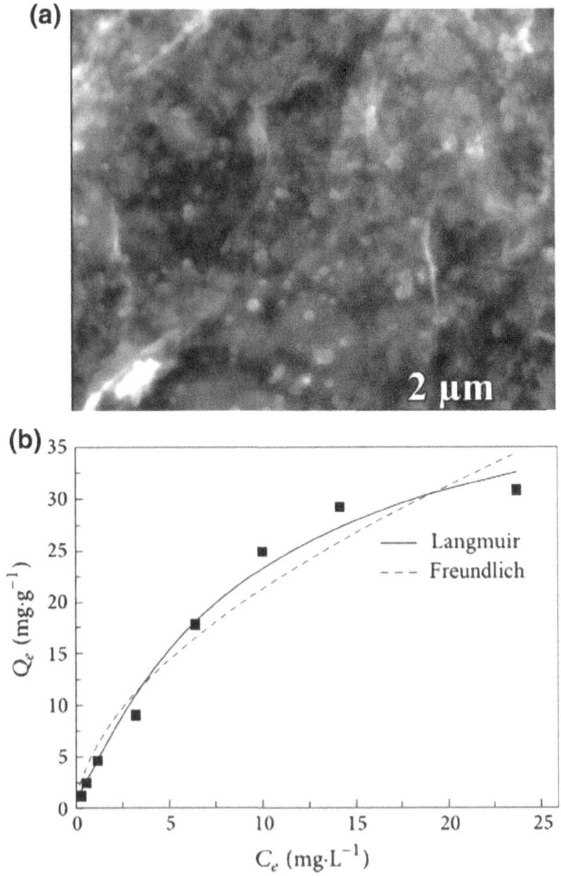

Fig. 13 **a** SEM image of Fe_3O_4/GAs and **b** Adsorption isotherm of As on Fe_3O_4/GAs at room temperature (Ye et al. 2015). Reprinted (adapted) with permission from Ye et al. (2015). Copyright 2015; Hindawi Publishing Corporation

capacity for both inorganic species of arsenic and sodium along with desalination ability of graphene based supercapacitor provides a solution for commercially feasible water filter.

5 Removal of Chromium(Cr)

The presence of Cr in drinking water sources in many countries and water pollution by chromium concerns the societies and regulation authorities around the world (Atieh et al. 2010; Sharma et al. 2008). Different industries such as metal plating, textile dyeing, leather tanning, printing inks, paints and pigments, and wood preservation, discharges huge quantities of waste water containing trivalent and hexavalent chromium into the environment (Dubey et al. 2015; El-Taweel et al. 2015; Singh et al. 2017a). Bhowmik et al. (2014) described a simple synthesis route

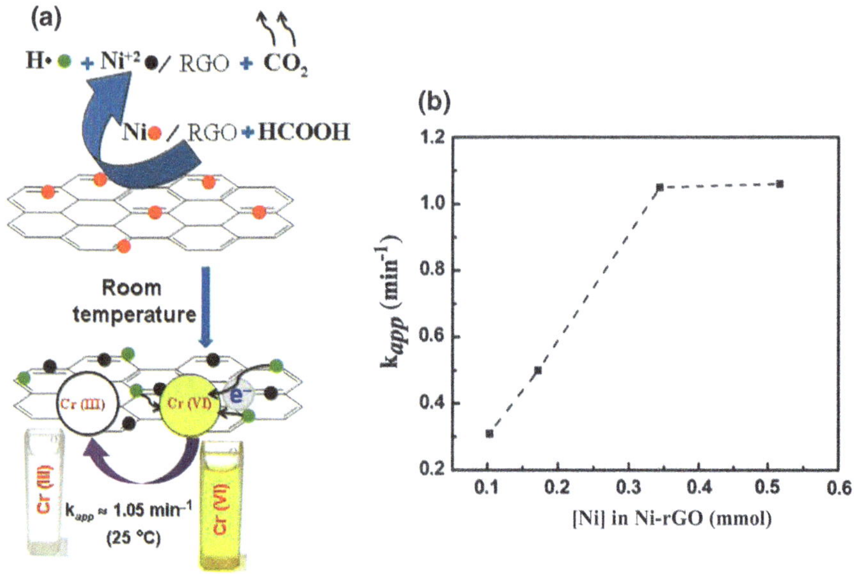

Fig. 14 **a** Schematic representation of the reaction mechanism and **b** k_{app} (min^{-1}) versus amount of Ni–rGO used in a fixed amount of reaction mixture (Bhowmik et al. 2014). Reprinted (adapted) with permission from Bhowmik et al. (2014). Copyright (2014) American Chemical Society

for the formation of stable Ni–rGO composite and applied in the formic acid-induced reduction of highly toxic aqueous Cr(VI) at room temperature. The rate of the reaction was slow at room temperature and enhances significantly in the presence of Ni–rGO by the introduction of an intermediate redox step with formic acid. The HCOOH gets adsorbed on the surface of Ni–rGO and the redox reaction occurs between Ni NPs and HCOOH on the graphene surface (as shown in Fig. 14a) and the HCOOH-induced Cr(VI) reduction indicated that varying amount of Ni–rGO with respect to Ni concentration resulted in a significant increase in the apparent rate constant (k_{app}) value up to 1.05 min^{-1} for 0.34 mmol of Ni–rGO (Fig. 14b).

Chemically reduced and functionalized GO by refluxing of GO with ethylenediamine (ED) using dimethyl formamide (DMF) as solvent was prepared (Zhang et al. 2013b). The ED and DMF contributed to the reduction and functionalization of GO and the resulting adsorbent (ED–DMF–rGO) with attached amine groups was highly efficient in removing Cr(VI) from its aqueous solution and could be easily separated by filtration. The Cr(VI) removal capacity of ED–DMF–rGO at optimum pH 2.0 was 92.15 mg g^{-1}, which was about 27 times higher than that of activated carbon, even nearly 4–8 times higher than that of various modified activated carbons. The Cr(VI) reduces to low-toxic Cr(III) during the adsorption process, by following an indirect reduction mechanism. Both the Cr(VI) adsorption and subsequent reduction of adsorbed Cr(VI) to Cr(III) contributed to the Cr(VI)

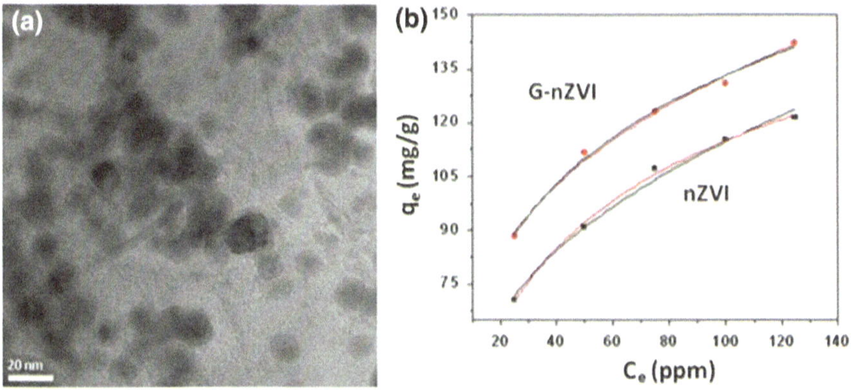

Fig. 15 **a** TEM image of G-nZVI and **b** effect of initial chromium concentration on the chromium adsorption for nZVI and G-nZVI (Jabeen et al. 2011). Reprinted (adapted) with permission from Jabeen et al. (2011). Copyright (2011) The Royal Society of Chemistry

removal. The iron particles decorated graphene sheets synthesized via sodium borohydride reduction of GO was reported for the removal of Cr(VI) (Jabeen et al. 2011). The TEM image (Fig. 15a) of zero valent iron nanoparticles decorated on graphene (G-nZVI) nanoparticles shows chain like aggregates due to the interaction among the Fe particles with particle size 17 nm. The maximum adsorption capacity for nZVI and G-nZVI was 148 and 162 mg g^{-1}, respectively as shown in Fig. 15b. The reduction of hexavalent chromium to trivalent chromium is higher for G-nZVI than for nZVI.

Zhu et al. (2012b) presented a facile thermo-decomposition process to synthesize magnetic graphene nanocomposites (MGNCs) for removal of Cr(VI) from waste water with a high removal efficiency within 5 min. The large saturation magnetization (96.3 emu g^{-1}) of the synthesized nanoparticles allowed fast separation of the MGNCs from liquid suspension. Chemical reduction and removal of Cr(VI) from acidic aqueous solution by ethylenediamine (ED)–rGO was also reported (Ma et al. 2012). The ED–rGO sheets were synthesized by simple refluxing of GO solution with ED, which effectively reduces the toxic Cr(VI) to less toxic Cr(III) by an indirect reduction mechanism with the assistance of π electrons on the carbocyclic six-membered ring of ED–rGO. The ED–rGO can be applied as a novel adsorbent for Cr(VI) removal from waste water.

In another report, polypyrrole(PPy)/GO composite nanosheets were fabricated and applied for Cr(VI) removal in aqueous solution (Li et al. 2012a). The synthesis of PPy/GO composite was achieved via sacrificial-template polymerization method. The MnO_2 nano slices were chosen as a sacrificial-template to deposit PPy, which served as the oxidant. The as synthesized PPy/GO composite exhibited an enhanced properties for Cr(VI) ions removal in aqueous solution based on the synergetic effect. The adsorption capacity of the PPy/GO composite was about two times as large as that of conventional PPy nanoparticles. Zhang et al. (2012a) analysed that

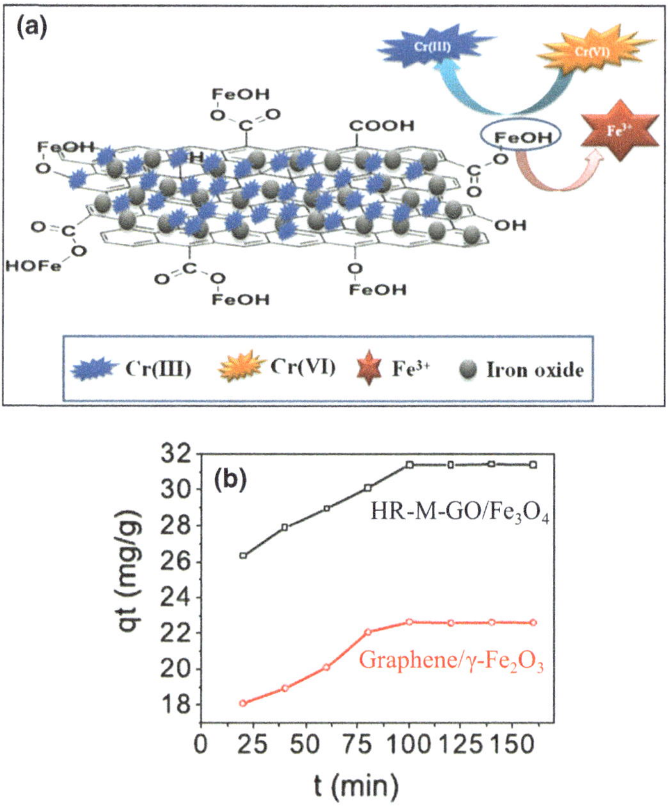

Fig. 16 **a** Mechanism of Cr(VI) adsorption on HR–M–GO/Fe_3O_4 composite and **b** Time-dependent sorption capacities (q_t) for Cr(VI) removed with HR–M–GO/Fe_3O_4 and graphene/γ-Fe_2O_3 (temperature: 25 °C; adsorbent dose: 0.5 g L^{-1}; Cr(VI) concentration: 50 ppm, and pH = 7) (Ting et al. 2016). Reprinted (adapted) with permission from Ting et al. (2016). Copyright 2016, IOP Publishing Ltd.

homogeneous anchoring of TiO_2 nanoparticles on graphene sheets (TiO_2–AGS) can help in waste water treatment by Cr(VI) reduction. The covalent anchoring of TiO_2 and rGO in the TiO_2–AGS composite leads to a reduction in electron-hole recombination which in turn leads to an enhanced Cr(VI) reduction ability. Due to the shift in the absorption band of the TiO_2–AGS composite Cr(VI) reduction occurs under natural sunlight which makes it a potential candidate for Cr(VI) remediation in waste water.

In 2016, Ting et al. (2016) reported the synthesis of a highly reductive and magnetic graphene/Fe_3O_4 composite (HR–M–GO/Fe_3O_4) via GO oxidizing $FeCl_2$ and this synthesized superparamagnetic composite was used efficiently for the removal of Cr(VI) from waste water conveniently by applying an external magnet. A possible mechanism for Cr(VI) removal by the HR–M–GO/Fe_3O_4 is presented in Fig. 16a and it was depicted that Fe^{2+} ions in HR–M–GO/Fe_3O_4 gets partly oxidized to form Fe^{3+} ions and Cr(VI) ions is reduced into Cr(III). Figure 16b shows the time-dependent sorption capacities (q_t) for Cr(VI) by HR–M–GO/Fe_3O_4 and graphene/γ-Fe_2O_3 composites. The equilibrium adsorption of Cr(VI) was reached after 100 min and demonstrates the higher sorption capacities for HR–M–GO/Fe_3O_4 at same condition. The maximum adsorption capacity of the HR–M–GO/Fe_3O_4 for Cr(VI) reaches 31.8 mg g^{-1}, which was greater than the graphene/γ-Fe_2O_3 composite.

6 Removal of Other Toxic Ions from Water (NO_3^-, SO_4^{2-}, ClO_4^-, PO_4^{3-} and F^-)

Electrodeposition method was presented as a promising method for large scale water treatment for the removal of perchlorate (ClO_4^-) ions using functionalized graphene-polypyrrole (Ppy) composites(Zhang et al. 2011). The functionalized graphene sheets were employed as the scaffold to synthesize a novel graphene-Ppy nanocomposite and it serves as an excellent electrically switched ion exchanger for ClO_4^- removal. The 3D nanostructured graphene-Ppy nanocomposite exhibits improved uptake capacity for ClO_4^- compared with Ppy film alone. Graphene itself shows the adsorption behavior for ClO_4^- ion and this property was applied for the removal of ClO_4^- from water (Lakshmi and Vasudevan 2013). It can effectively remove ClO_4^- in water in a wide range of pH = 6.0–8.0. The percentage of ClO_4^- removal by graphene can reach 99.2% (at pH = 7.0) as shown in Fig. 17a, proving it to be a good adsorbent with an adsorbent capacity of up to 0.024 mg g^{-1} at initial ClO_4^- concentration of 2 mg L^{-1} and at 298 K. The effect of initial ClO_4^- concentrations at varying initial concentration (0.5 to 2.5 mg L^{-1}) can be seen in Fig. 17b. The adsorption amount increases with increase in initial concentration from 0.5 to 2.5 mg L^{-1} and nearly shows equilibrium after 20 min for all of the concentrations.

Graphene prepared by a facile liquid phase exfoliation was investigated for phosphate adsorption studies at varying pH, ionic strength and temperature (Vasudevan and Lakshmi 2012). The experimental findings showed that the adsorption capacity increases with pH in the acidic range and reaches the maximum removal efficiency of 99.1% at pH 7.0 as shown in Fig. 18a. In addition, the adsorption capacity increases from 89.37 to 92.36 mg g^{-1} at a phosphate equilibrium concentration of 100 mg L^{-1}, as the temperature rises from 303 to 343 K (Fig. 18b). The phosphate adsorption performance of the as-prepared zirconium-loaded rGO (rGO–Zr) nanocomposites formed by one-step green hydrothermal method was investigated in aqueous environment under different conditions (Luo et al. 2016). The data from equilibrium phosphate adsorption on rGO–Zr revealed that the adsorption kinetics followed a pseudo-second-order kinetic model, where the adsorption isotherm followed Langmuir isotherm model with a maximum adsorption capacity of 27.71 mg P/g at pH = 5 and 298 K. The phosphate adsorption remains insensitive to the increase in pH, while the adsorption

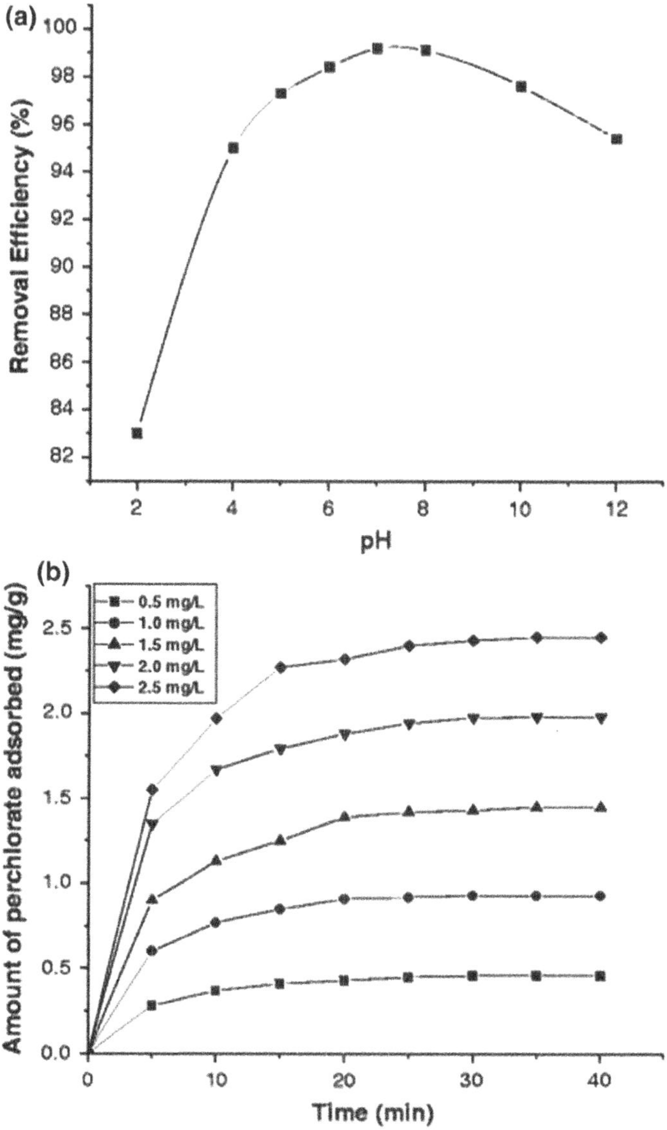

Fig. 17 **a** Plot between pH with removal efficiency of ClO_4^- (conditions: concentration, 2 mg L^{-1}) and **b** Effect of agitation time and initial concentration of ClO_4^- for the adsorption of ClO_4^- on graphene (conditions: concentration, 2 mg L^{-1}; pH, 7.0) (Lakshmi and Vasudevan 2013). Reprinted (adapted) with permission from Lakshmi and Vasudevan (2013). Copyright © 2013, Springer-Verlag Gmbh and Co. KG

capacity enhances with increase in temperature. The coexisting interfering anions SO_4^{2-}, F^-, Cl^-, NO^{3-} and CO_3^{2-} also affects the phosphate adsorption capacities of rGO–Zr.

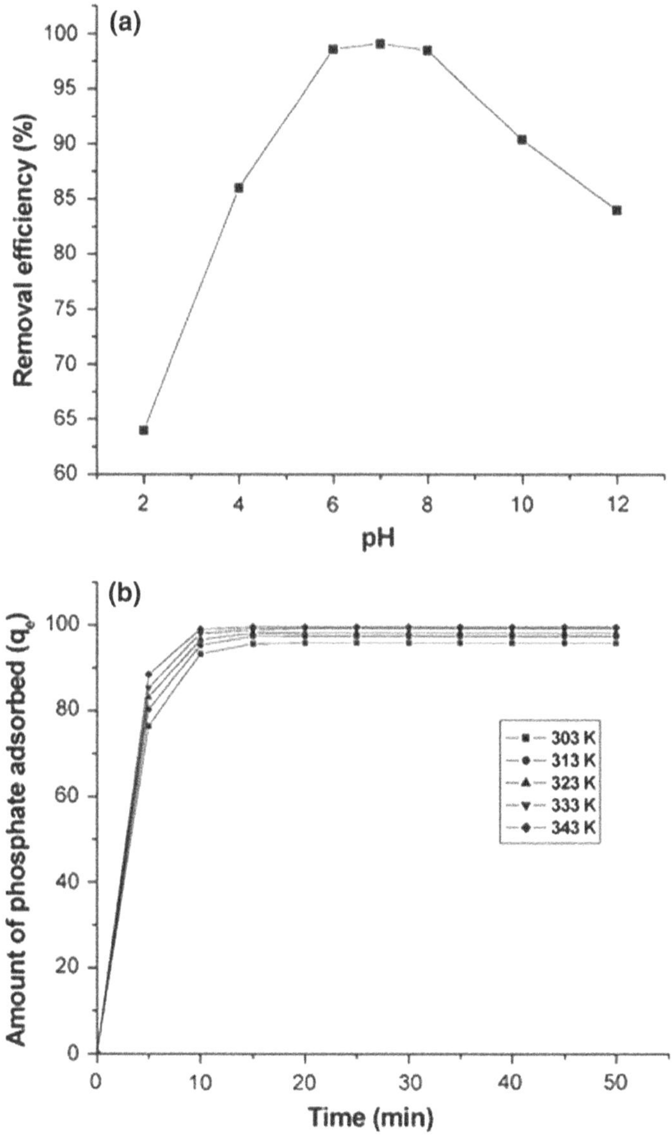

Fig. 18 **a** The effect of initial pH on phosphate adsorption by graphene (conditions: concentration = 100 mg L^{-1}) and **b** The effect of time on the amount of phosphate adsorbed at various temperatures (conditions: concentration = 100 mg L^{-1}; pH = 7.0) (Vasudevan and Lakshmi 2012). Reprinted (adapted) with permission from Vasudevan and Lakshmi (2012). Copyright (2012) The Royal Society of Chemistry

Graphene prepared by liquid phase exfoliation shows an excellent nitrate (NO_3^-) adsorption with an adsorption capacity of up to 89.97 mg g^{-1} at an initial NO_3^- concentration of 500 mg L^{-1} and temp 303 K (Ganesan et al. 2013). The Langmuir and Freundlich models were used to describe the equilibrium isotherms and the isotherm constants. Thermodynamic studies revealed that the adsorption reaction was spontaneous and was an endothermic process. Motamedi et al. (Motamedi et al. 2014) used GO coated Ni and Co nanoparticles (NPs) for removal of nitrate from water. Specifically, $NaBH_4$ reduction of Fe^{3+}, Ni^{2+} and Co^{2+}, in the presence of GO affords quantitative yields of their corresponding magnetically functionalized nanocomposites: Ni@r–GO, Co@rGO as well as Fe@rGO. Efficiencies of Ni@r–GO and Co@rGO were proved to be comparable to that of Fe@rGO for the removal of nitrate from water. The latter demonstrated a superior efficiency over the previously reported unanchored FeNPs.

Li et al. (2011b) reported a batch adsorption system to investigate the adsorption of fluoride from aqueous solution by graphene. The adsorption capacities and rates of fluoride onto graphene at different initial pH, contact time, and temperature were evaluated. Fluoride adsorption by graphene is sensitive to pH variations and Fig. 19a shows the variation of adsorption capacity of fluoride adsorbed by graphene at various initial and final pH values. The results also showed that graphene

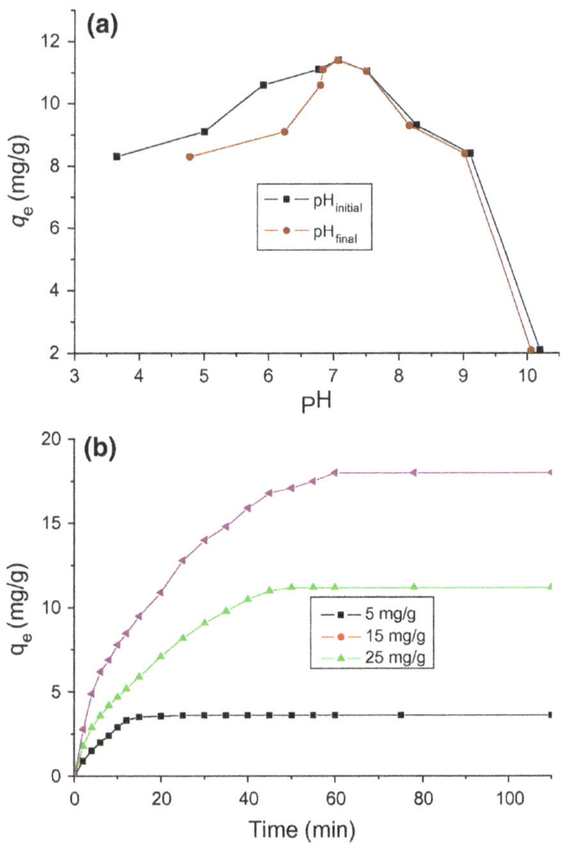

Fig. 19 **a** Effect of initial and final pH on fluoride adsorption by graphene and **b** Time effect on fluoride adsorption by graphene (Li et al. 2011b). Reprinted (adapted) with permission from Li et al. (2011b). Copyright © 2011 Elsevier B. V. All rights reserved

is a good fluoride adsorbent with an adsorption capacity of up to 17.65 mg g^{-1} at initial fluoride concentration of 25 mg L^{-1} and temperature of 298 K (Fig. 19b). The monolayer adsorption capacity of fluoride by graphene was found to be 35.59 mg g^{-1} at pH = 7.0 and at 298 K.

7 Removal of Organic Pollutants

7.1 Dye Removal from Graphene/Graphene Oxide Based Materials

Graphite oxide was used as an adsorbent for the removal of dyes in aqueous solution (Bradder et al. 2011). The amount of the dyes, MB and malachite green, adsorbed on the graphite oxide was higher than that on graphite, and the adsorption capacity based on the Langmuir isotherm was (351 and 248) mg g^{-1}, respectively, much higher than activated carbon. Also in another study, GO was used as an highly effective absorbent for removal of MB from aqueous solution (Yang et al. 2011) with an absorption capacity of was 714 mg g^{-1}. At initial MB concentrations lower than 250 mg L^{-1}, the removal efficiency is higher than 99% and the solution decolorized to nearly colorless. The removal process was fast and more efficient at lower temperatures and higher pH values. The increase of ionic strength and the presence of dissolved organic matters further enhance the removal process at higher MB concentrations.

Li et al. (2011a) synthesized $Mg(OH)_2$@rGO composite (MGC) via a chemical deposition method and used it as an adsorbent for the removal of MB from water. Figure 20a shows the schematic illustration of MGC synthesis from GO. MGC exhibited outstanding adsorption behavior for MB. The $Mg(OH)_2$ nano plates assist rGO to maintain a high specific surface area and form the mesopore structure. Compared to GO and rGO, MGC can be easily separated from solution after adsorption, which is useful for industrial application. Figure 20b shows the MB adsorption by $Mg(OH)_2$,GO, rGO and Fig. 20c shows the adsorption for MGC which is much higher as compared to $Mg(OH)_2$,GO, rGO.

3D rGO-based hydrogels formed by the reduction of GO using sodium ascorbate were used for the removal of the organic dyes, MB and RB (Tiwari et al. 2013). The synthesized 3D RGO-based hydrogels with large surface area, and a uniform pore size distribution helped in removal of MB and RB, from aqueous solutions. The 3D rGO-based hydrogels showed outstanding removal capabilities for MB ($\sim$100%) and RB ($\sim$97%) due to adsorption through strong π–π stacking and anion-cation interactions. Jiao et al. (2015) prepared GO-based magnetic hybrids composed of GO and Fe_3O_4 nanoparticles with various geometrical nanostructures for fast and highly efficient removal of organic dyes. The as-prepared and regenerated nanohybrids demonstrate a nearly 100% removal rate for MB and an impressively high removal rate for RB.

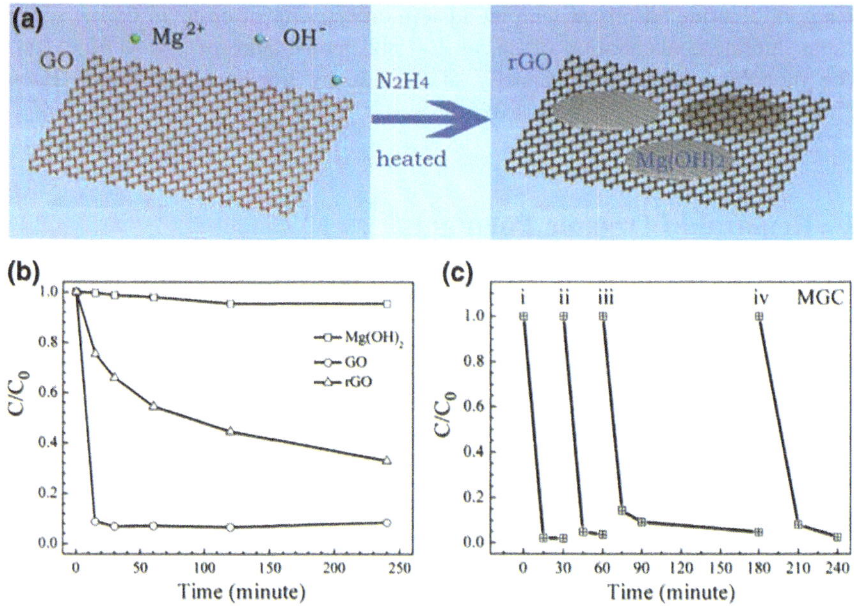

Fig. 20 **a** Illustration of synthesis of MGC from GO. The C/C_0 versus time plots for adsorption of dye solution (MB, 2.0×10^{-5} M, 50 ml) with **b** $Mg(OH)_2$, GO, and rGO for various times, and **c** MGC in various cycles: (i) 1st, (ii) 2nd, (ii) 3rd, and (iv) 4th cycle (Li et al. 2011a). Reprinted (adapted) with permission from Li et al. (2011a). Copyright (2011) The Royal Society of Chemistry

Liu et al. reported synthesis of for fast and efficient water-soluble dye removal. The authors demonstrated the potential of 3D GO nanostructure as environmental pollutant adsorbents by utilizing the characteristics of ultra large surface area and strong π–π interaction on the surface was demonstrated (Liu et al. 2012). The synthetic scheme of the synthesis of 3D GO sponge is illustrated in Fig. 21a and b reflects the flexibility of the GO sponge. The extruded 3D GO structure recovered its original shape in a short time. The 3D GO sponge was used to remove the MB and MV dyes. The efficiency and speed of dye (MB and MV) adsorption on a GO sponge was investigated under various parameters such as stirring speed, temperature, and pH as shown in Fig. 21c–h. The adsorption process shows that 99.1% of MB and 98.8% of MV gets removed and the equilibrium status reaches in 2 min. The 3D GO sponge displays adsorption capacity as high as 397 and 467 mg g^{-1} for MB and MV dye, respectively, and the kinetic data reveal that the adsorption process of MB and MV dyes is well-matched with the pseudo second-order model.

Fan et al. (2012) demonstrated the fabrication of novel magnetic chitosan grafted GO with enhanced adsorption properties for MB. The adsorption of MB onto magnetic composite bio adsorbent composed of magnetic chitosan and GO (MCGO) was investigated with respect to pH, adsorption time, initial MB concentration and temperature. Kinetics data and adsorption isotherm at the optimum

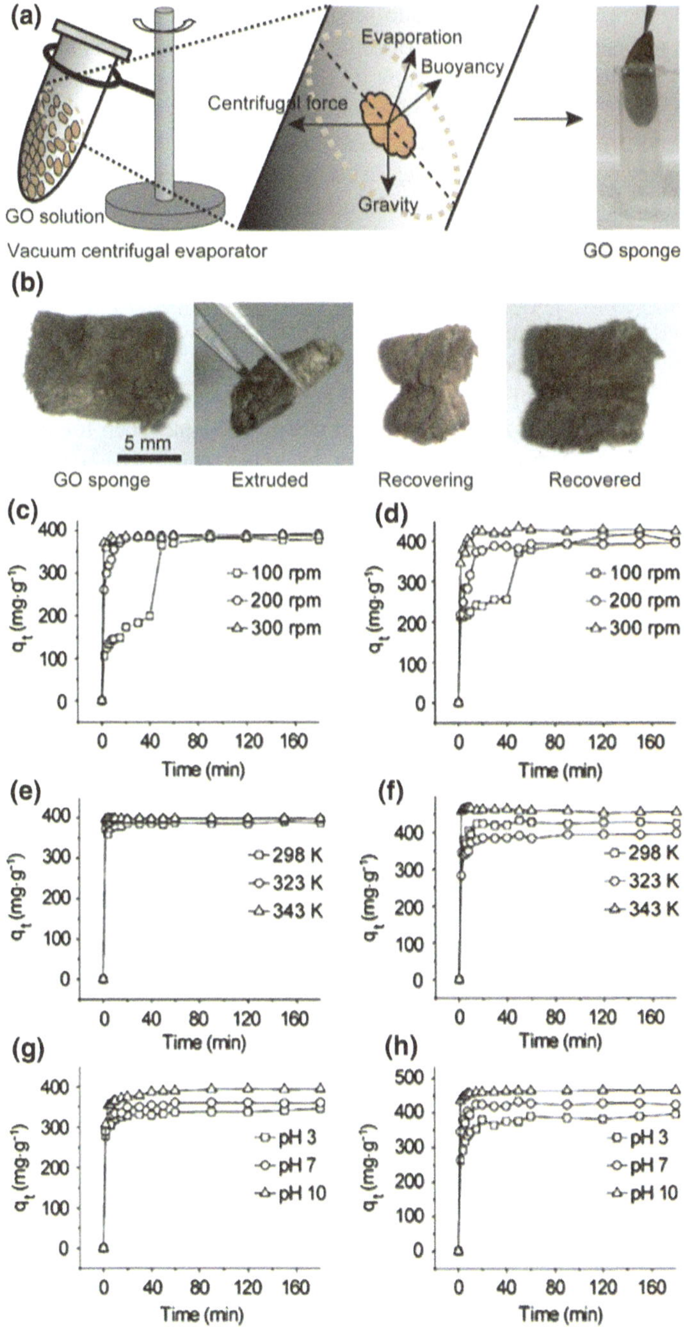

Fig. 21 **a** Synthetic scheme of a 3D GO sponge. **b** Flexibility test of a 3D GO sponge. Effect of stirring rate on the adsorption rate of **c** MB and **d** MV dyes. The effect of temperature on the adsorption rate of **e** MB and **f** MV dyes. The effect of pH on the adsorption rate of **g** MB and **h** MV dyes (Liu et al. 2012). Reprinted (adapted) with permission from Liu et al. (2012). Copyright (2012) American Chemical Society

pH = 5.3, were well fitted by pseudo-second-order kinetic model and by Langmuir isotherm, respectively. The values of activation parameters such as free energy (ΔG, −0.74 to −1.46 kJ mol^{-1}), enthalpy (ΔH, −10.28 kJ mol^{-1}) and entropy (ΔS, −36.35 J mol^{-1} K^{-1}) were determined. The MCGO was stable and easily recovered, the adsorption capacity was about 90% of the initial saturation adsorption capacity after being used four times. Ai et al. (2011) demonstrated the removal of MB from aqueous solution by a solvothermal-synthesized graphene nanosheet/magnetite (GNS/Fe_3O_4) composite. The synthesized GNS/Fe_3O_4 composite shows extraordinary adsorption capacity and fast adsorption rates for removal of MB in water.

Magnetic Fe_3O_4@graphene composite (FGC) was synthesized and used in the dye removal from aqueous media (Yao et al. 2012). Adsorption isotherm and kinetics of MB and Congo red (CR) onto FGC were studied in a batch system. The maximum adsorption capacity of MB and CR on FGC has been found to be 45.27 and 33.66 mg g^{-1}, respectively. Guo et al. (Guo et al. 2012) synthesized Fe nanoparticles@graphene composites (FGC) for environmental applications. The as synthesized FGC was used for decolorization of MB solution. Compared with bare Fe particles, the high removal capacities of FGC are due to the increased adsorption sites in the hybrids, which are achieved by inhibiting the particle aggregation and reducing the size of Fe nanoparticles. Superparamagnetic GO–Fe_3O_4 (GO–Fe_3O_4) composite synthesized via a facile chemical method was also applied in the removal of dyes from aqueous solution (Xie et al. 2012). Figure 22a–d shows the TEM images of as synthesized GO–Fe_3O_4 composite. The magnetic GO–Fe_3O_4 composite was introduced as adsorbent for the magnetic separation of dye contaminants from water. The adsorption test of dyes MB and neutral red (NR)) demonstrated that it only takes 30 min for MB and 90 min for NR to attain equilibrium. The adsorption isotherms are shown in Fig. 22e and it shows that the maximum adsorption capacities for MB and NR in the concentration range are 167.2 and 171.3 mg g^{-1}, respectively.

Graphene-melamine-sponge (GMS) which can readily submerge into water solution with the help of a superhydrophilic melamine skeleton shows good adsorption capabilities for MB and orange G (OG) (Du et al. 2016). The maximum adsorption capacities for MB and OG obtained from the Langmuir isotherm equation are 286.5 and 80.51 mg g^{-1}, respectively. The as-synthesized GMS can be easily regenerated and it maintains 95% adsorption capacity after 10 cycles. Pan and Liu (2012) reported a new approach for the synthesis of CdS-graphene (CdS-G) nanocomposite via a hydrothermal method. The CdS-G reveals a high photodegradation rate under visible light irradiation for MO. The noteworthy enhancement in photoactivity can be ascribed to the smaller morphology of the CdS nanoparticles in the nanocomposite and the introduction of graphene which reinforced the adsorptivity of the photocatalysts, suppressed the electron-hole pair recombination, and extended the light absorption range in the as-synthesized CdS-G nanocomposite photocatalyst.

Gong et al. (2017) synthesized GO nanosheets by electrochemical exfoliation and examined it for adsorption of MO, in aqueous solution at different pHs and

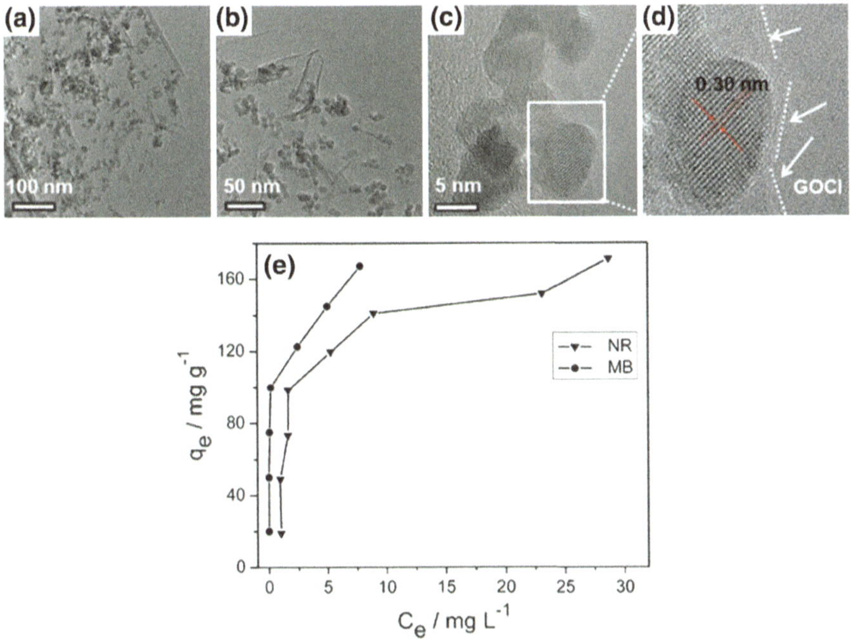

Fig. 22 **a–d** TEM images of GO–Fe_3O_4 composite and **e** Adsorption isotherms of dyes onto the GO–Fe_3O_4 composite (Xie et al. 2012). Reprinted (adapted) with permission from Xie et al. (2012). Copyright (2012) The Royal Society of Chemistry

temperatures. The maximum adsorption capacity of MO on GO nanosheets obtained from the Langmuir isotherm was 138.69 mg g^{-1} at pH 2. The large adsorption affinity of GO nanosheets to MO might be due to the presence of hydrogen bonding and π–π interaction between MO and GO nanosheets and thermodynamic studies indicated that the adsorption reaction was a spontaneous physisorption process. The adsorption behavior GO-poly(2-hydroxyethyl methacrylate) (GO-PHEMA) composite synthesized by dispersion polymerization in supercritical CO_2 was utilized for the removal of organic dye (Kharismadewi et al. 2016). The adsorption parameters were found to fit well into the Freundlich adsorption isotherm with a correlation coefficient of 0.975 and a maximum predicted adsorption capacity of 39.41 mg g^{-1} at 25 °C. The thermodynamics studies showed that the adsorption of MB on GO-PHEMA composite followed spontaneous and endothermic adsorption process with an efficient adsorption temperature at 45 °C. The GO-PHEMA composite removes 99.8% of the dye in 45 min. Figure 23a–d shows the relation amongst dye removal with concentration, contact time, pH, and dye concentration effects.

Nguyen-Phan et al. (2012) synthesized rGO–titanate (rGO–Ti) hybrids by alkali-solvothermal treatment and used it for water purification. The hybrids in sheet and tubular titanate structures possess larger surface areas (> 350 m^2 g^{-1}) and

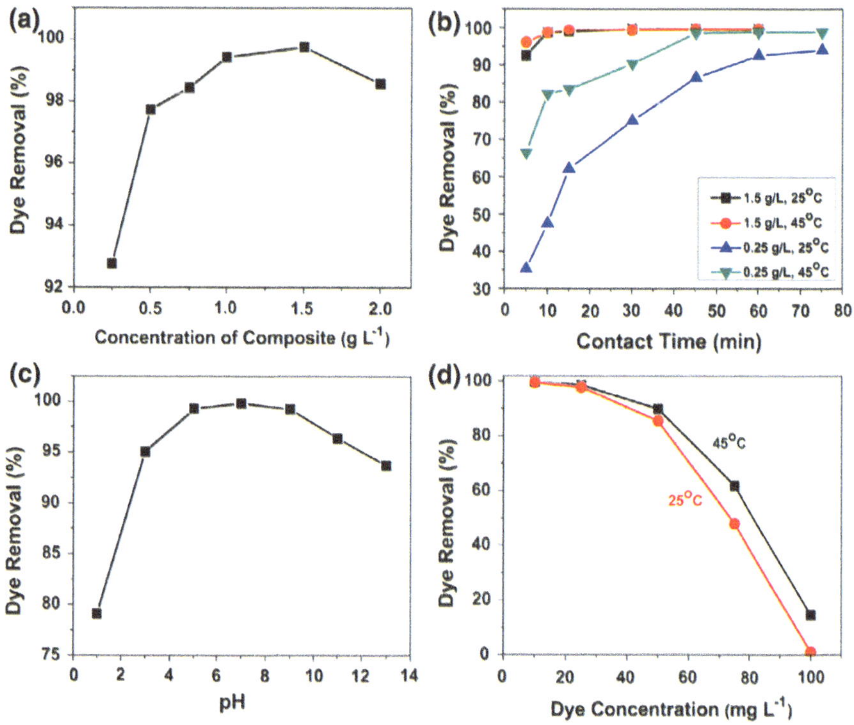

Fig. 23 Effects of **a** adsorbent concentration (t = 60 min, [dye] = 10 mg L^{-1}, pH 7, T = 25 °C), **b** contact time ([composite] = 0.25, 1.5 g L^{-1}, [dye] = 10 mg L^{-1}, pH 7), **c** pH ([composite] = 1.5 g L^{-1}, [dye] = 10 mg L^{-1}, t = 45 min, T = 25 °C), and **d** dye concentration ([composite] = 1.5 g L^{-1}, t = 45 min, pH 7) (Kharismadewi et al. 2016). Reprinted (adapted) with permission from Kharismadewi et al. (2016). Copyright (2016) Taylor & Francis

higher pore volumes (>1 cm^3 g^{-1}). The presence of rGO sheets as a 2D platform for the deposition of titanate significantly promoted the adsorptivity of dye contaminants compared to pure materials. In another work, magnetite/rGO (MrGO) nanocomposites were synthesized using solvothermal strategy for the removal of dye pollutants (Sun et al. 2011). The synthesized MrGO nanocomposites exhibit outstanding removal efficiency (over 91% for RB and over 94% for malachite green) and can be rapidly separated from aqueous solution by an external magnetic field. The performance of the MrGO composites strongly depend on both the loading of Fe_3O_4 and the pH value. Wang et al. (2011) also prepared graphene-based magnetic nanocomposite for the removal of an organic dye fuchsine from aqueous solution. The graphene magnetic nanocomposite proved to be highly efficient adsorbent and can be readily separated. The adsorption kinetics, adsorption capacity of the adsorbent, and the effect of the adsorbent dosage and solution pH on the removal efficiency of fuchsine were investigated. GO is a highly effective adsorbent, and its absorbing capability can be further enhanced through its

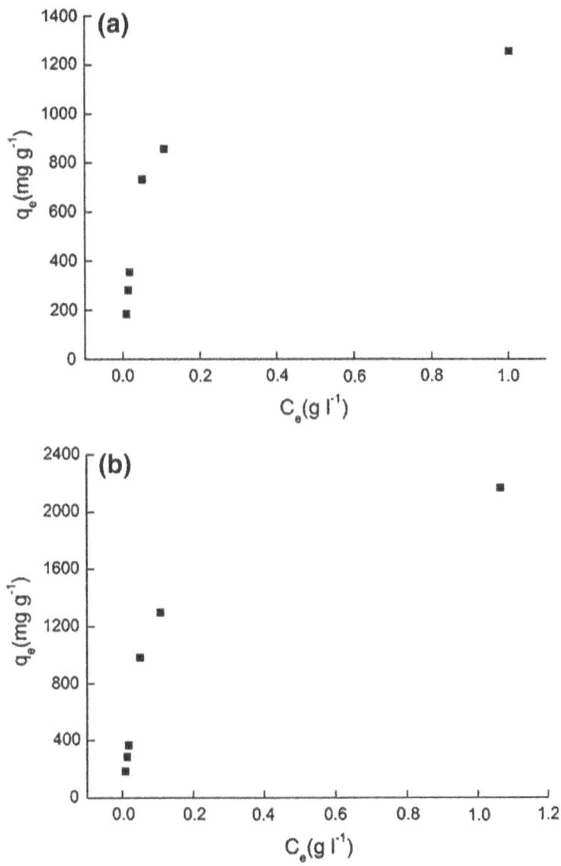

Fig. 24 Adsorption isotherms of **a** GO and **b** SRGO (Sun et al. 2012). Reprinted (adapted) with permission from Sun et al. (2012). Copyright © 2012 Elsevier B. V. All rights reserved

in situ reduction with sodium hydrosulfite as the reductant (Sun et al. 2012). Acridine orange (AO) is the selected adsorbent target to eliminate with GO. Under similar conditions, GO (without the in situ reduction) showed a maximum adsorption capacity of 1.4 g g^{-1}, and SRGO (with the in situ reduction) provided a maximum adsorption capacity of 3.3 g g^{-1}as shown in Fig. 24a, b.

GO-iron oxide and rGO-iron oxide hybrid were synthesized and used for the removal of organic and inorganic pollutants (Yang et al. 2012). The adsorption of Pb(II), 1-naphthol, and 1-naphthylamine, as representatives of inorganic and organic pollutants, on GO-iron oxides and rGO-iron oxides was investigated. The GO-iron oxide material act as high-quality adsorbent for Pb(II) but not for 1-naphthol and 1-naphthylamine due to oxygen-containing groups on the surface, whereas the rGO-iron oxide material was a good adsorbent for 1-naphthol and 1-naphthylamine but not for Pb(II). Also, the adsorption of 1-naphthol and 1-naphthylamine on rGO-iron oxides was an endothermic and spontaneous process. Konicki et al. (Konicki et al. 2017) studied that GO can also be used for the adsorption of anionic azo-dyes such as acid orange 8 (AO8) and direct red 23

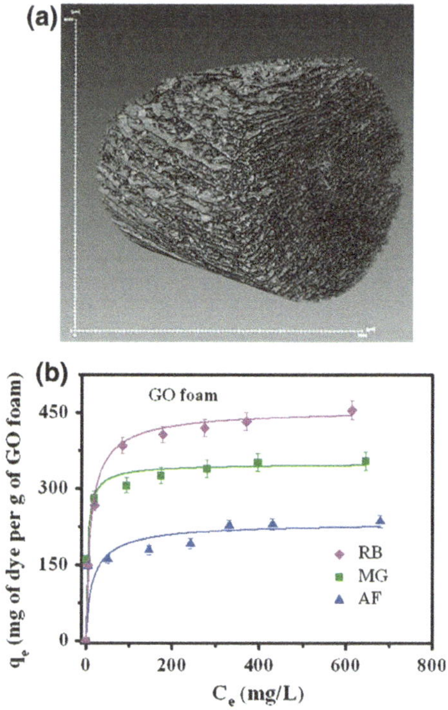

Fig. 25 **a** Three dimensional XT micrographs of GO foam (scale bar is 7000 mm) and **b** Adsorption isotherms and the corresponding fitting plots of RB, MG and AF dye molecules on GO foam (Jayanthi et al. 2016). Reprinted (adapted) with permission from Jayanthi et al. (2016). Copyright (2016) The Royal Society of Chemistry

(DR23) from aqueous solutions. The dye initial concentration, temperature and pH influences the AO8 and DR23 adsorption onto GO. Thermodynamics parameters, ΔG^0, ΔH^0 and ΔS^0, were calculated, indicating that the adsorption of AO8 and DR23 onto GO was spontaneous process. Cheng et al. (2012) demonstrated a facile method for 3D chitosan-graphene mesostructures synthesis and removal of reactive black 5 (RB5) from aqueous solution applying these structures. A hydrophilic and biocompatible 3D chitosan–graphene mesostructures with large specific surface area (603.2 m^2 g^{-1}) and unique mesoporosity were prepared by simply thermal treatment of GO nanosheets and chitosan. The investigations showed that the removal efficiency was 97.5% at initial RB5 concentrations of 1.0 mg mL^{-1}.

Magnetic rGO is also an efficient adsorbent for organic pollutants like triphenylmethane (TPM) dye (Sun et al. 2014). The magnetic rGO showed fast adsorption rate and high adsorption capacity towards different TPM dyes (the Langmuir monolayer adsorption capacity is 64.93 mg g^{-1} for adsorption of crystal violet). The magnetic rGO shows excellent recycling and regeneration capabilities and. Highly porous, ultralight GO foams were prepared by cost effective lyophilization technique and applied in environmental applications (Jayanthi et al. 2016). The 3D architecture of GO foams allowed their direct use in the removal of aqueous pollutants without pretreatment. Figure 25a shows that macropores of 3D GO possess excellent adsorption abilities towards carcinogenic dyes. The 3D GO foams display respective adsorption capacities of 446, 321 and 228 mg g^{-1} for RB, malachite green (MG) and acriflavine (AF) as shown in Fig. 25b. These foams were also investigated for antibacterial activities against *E. coli* bacteria in aqueous and nutrient growth media.

In another approach, green synthesis of carbon nanotubes-graphene hybrid aerogels were performed by supercritical CO_2 drying and used it as versatile agents for water purification (Sui et al. 2012). The synthesized carbon nanotubes-graphene hybrid aerogels show promising performance in water purification including capacitive deionization of light metal salts, removal of organic dyes and enrichment of heavy metal ions. The resulting graphene-CNT hybrid aerogels show high desalination capacity (633.3 mg g^{-1}) with the NaCl concentration up to 35 g L^{-1}. The resulting graphene–CNT hybrid aerogels also display high binding capacities to some heavy metal ions including Pb^{2+}, Ag^{+} etc., and high adsorption capacities for dye stuffs, especially for basic dyes with well-controlled desorption behavior.

7.2 Removal of Aromatic Compounds and Gasoline

Removal of various aromatic compound and gasoline contaminants including inorganic and organic compounds from the environment is a big challenge. GO and graphene are the advanced nanomaterials for adsorptive treatment of environmental pollution. GO possesses several functional groups and strong acidity, exhibiting high adsorption for basic compounds and cations and presents high adsorption to chemicals due to strong π–π interaction. Modification of GO/graphene with different kind of metal oxides/organics can produce various nanocomposites, enhancing adsorption capacity and separation efficiency.

Wang et al. (2013a) reported an adsorptive remediation of environmental pollutants using novel graphene-based nanomaterials. Cong et al. (2012) reported the synthesis of macroscopic multifunctional graphene-based hydrogels and aerogels by a metal ion induced self-assembly process. The synthesized material has interconnected networks due to the simultaneous reduction of GO sheets by ferrous ions and in situ deposition of nanoparticles on graphene sheets. The functional group attached on graphene-based hydrogels exhibit excellent capability for removal of pollutants and used as promising adsorbents for water purification. Hartono et al. (Hartono et al. 2009) synthesized the layered structured graphite oxide as a novel adsorbent for humic acid from aqueous solution. Figure 26a shows the TEM image of graphite oxide and containing thin slice of carbon with random spatial arrangement and crumpled conformation. The synthesized graphite oxide was tested for humic acid (HA) adsorption in aqueous solution and it was found that the graphite oxide exhibits strong and higher adsorption capacity of HA than graphite. Based on the Langmuir isotherm, the maximum adsorption of HA on graphite and graphite oxide was 16.5 and 190 mg g^{-1}, respectively (Fig. 26b).

Graphene-based sponges fabricated using a facile dip coating method show the superhydrophobic and superoleophilic properties and are efficient absorbents for a

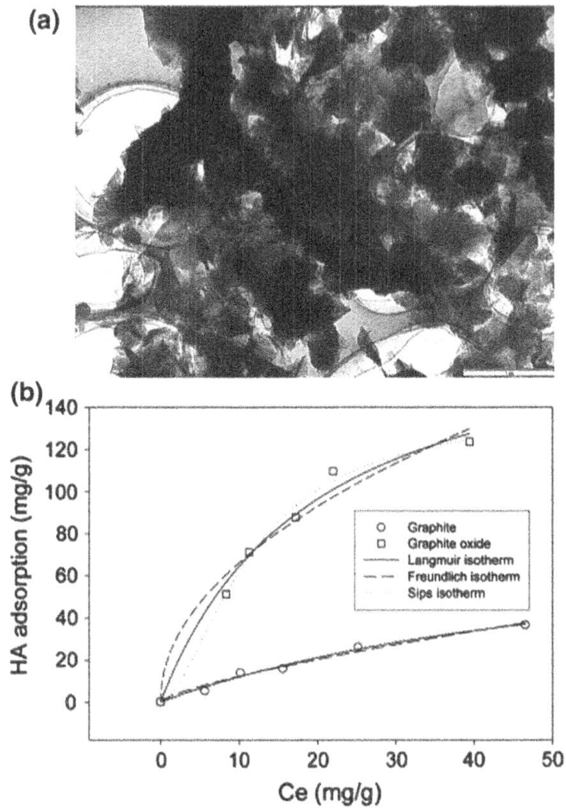

Fig. 26 **a** TEM image of graphite oxide and **b** adsorption isotherm of HA adsorption on graphite and graphite oxide (Hartono et al. 2009). Reprinted (adapted) with permission from Hartono et al. (2009).

broad range of oils and organic solvents with high selectivity, good recyclability, and excellent absorption capacities up to 165 times their own weight (Nguyen et al. 2012).

Traces of pharmaceuticals have been classified as emerging pollutants due to their persistence in the aquatic ecosystem. The adsorption is a promising method for the removal of micro-pollutants because of its low-cost, high efficiency, etc. Researchers have shown that graphene materials can be used as photo-catalysts, adsorbents, and as disinfectants in water treatment. Carmalin Sophia et al. (2016) reported the application of graphene based materials for the adsorption of pharmaceutical traces from water and waste water and explains the possible mechanism of adsorption.

The GO is a potential effective absorbent for tetracycline antibiotics and can be used to remove them from aqueous solution (Gao et al. 2012). Tetracycline strongly deposits on the GO surface via π–π interaction and cation–π bonding. The adsorption isotherm fits well with Langmuir and Temkin models, and the theoretical maximum adsorption capacity calculated by Langmuir model is 313 mg g^{-1}. The kinetics of adsorption fits pseudo-second-order model perfectly, and gives a better

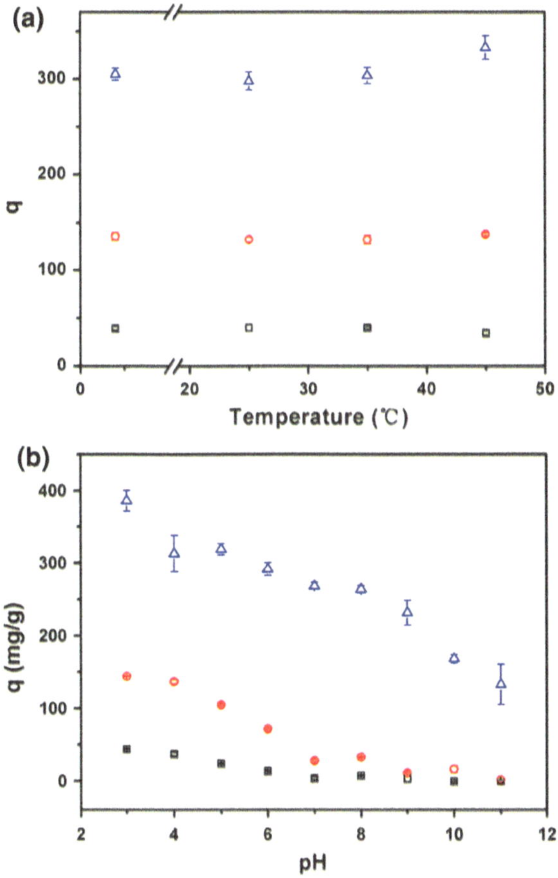

Fig. 27 **a** Effect of temperature on the adsorption capacity of tetracycline (Experiment condition: pH 3.6; incubation for overnight; GO 0.181 mg mL^{-1}; initial concentration of tetracycline is 8.33 mg L^{-1} (open square), 33.33 mg L^{-1} (open circle), and 166.67 mg L^{-1} (open triangle), respectively). **b** Effect of pH on adsorption capacity of tetracycline (Experiment condition: incubation for overnight; 25 °C; GO 0.181 mg mL^{-1}; initial concentration of tetracycline is 8.33 mg L^{-1} (open square), 33.33 mg L^{-1} (open circle), 166.67 mg L^{-1} (open triangle), respectively) (Gao et al. 2012). Reprinted (adapted) with permission from Gao et al. (2012).

rate constant of sorption (k), 0.065 g mg^{-1} h^{-1}, than other adsorbents. The adsorption capacities of tetracycline on GO decreases with the increasing pH or Na^{+} concentration. Figure 27 gives the details of dependence of adsorption capacity of tetracycline on temperature and pH values.

Peng et al. (2016a) also demonstrated the π–π Interactions induced adsorption of antibiotics on graphene and biochar in aqueous solutions. The adsorption rate of graphene was stronger than other two biochar. The adsorption energy markedly increases with increasing number of the π rings by using the density functional

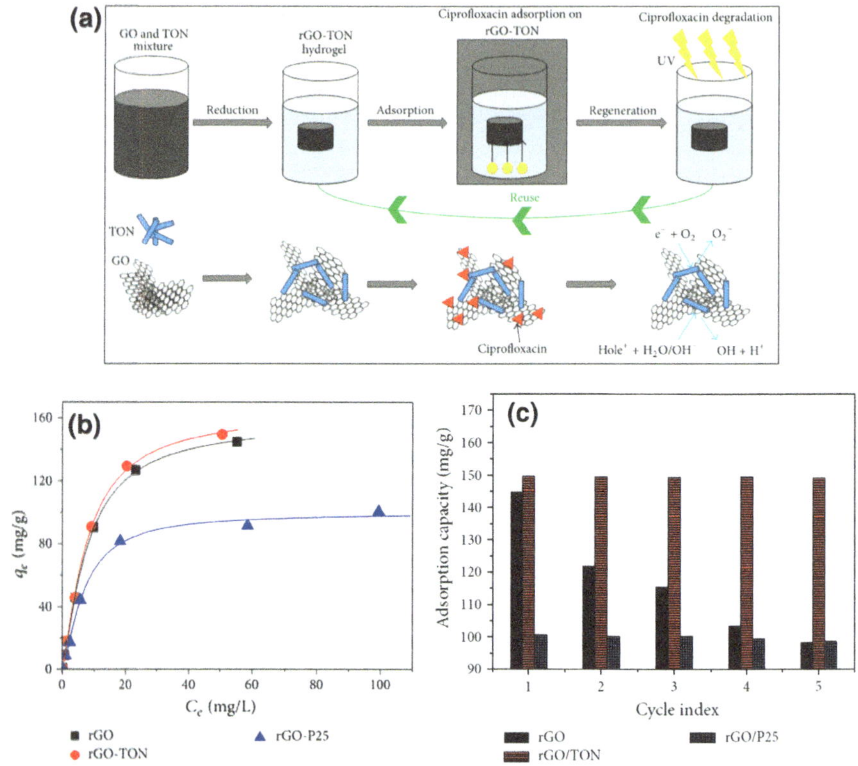

Fig. 28 **a** Scheme for preparation, adsorption, and regeneration process of rGO–TON. Langmuir isotherm model (**b**) and regeneration properties (**c**) (initial concentration 200 mg L^{-1}) of rGO, rGO–TON, and rGO-P25 (Zhuang et al. 2015). Reprinted (adapted) with permission from Zhuang et al. (2015). Copyright 2015; Hindawi Publishing Corporation

theory (DFT), showing the importance of π–π interactions in the adsorption process. Liu et al. (2016b) reported nitrogen-doped graphene nanosheets (N–rGO) as reactive water purification membranes. N–rGO catalyst was prepared via a facile hydrothermal method. An optimized phenol oxidative flux of 0.036 ± 0.002 mmol h^{-1} was achieved by metal-free catalytic activation of persulfate at an influent persulfate concentration of 1.0 mmol L^{-1} and filter weight of 15 mg while N-free rGO filter demonstrated negligible phenol oxidation capability under similar conditions. Compared to a conventional batch system, the flow-through designed batch demonstrated enhanced oxidation kinetics (0.036 vs. 0.010 mmol h^{-1}), mainly due to the liquid flow through the filter leading to convection-enhanced transfer of the target molecule to the filter active sites.

Regenerable long TiO_2 nanotube/GO (rGO–TON) hydrogel adsorbents show enhanced and removal of ciprofloxacin based on adsorption (Zhuang et al. 2015). To improve the adsorption performance and regeneration ability of adsorbent, a simple method was designed to synthesize long TiO_2 nanotube/rGO (rGO–TON) hydrogel (Fig. 28a), which has good adsorption and regeneration capacity toward ciprofloxacin. The rGO–TON attain excellent adsorption capacity, and the maximum adsorption capacities of rGO–TON for ciprofloxacin calculated from Langmuir model are 178.6 mg g^{-1} ($R^2 = 0.9929$), 181.8 mg g^{-1} ($R^2 = 0.9954$), and 108.7 mg g^{-1} ($R^2 = 0.9964$) for GO, GO–TON, and GO-P25, respectively. In regeneration, the adsorption capacity of rGO–TON and rGO-P25 reduced slightly after 5 cycles, while the adsorption capacity of rGO decreases to below 100 mg g^{-1}. Figure 28b, c shows the details information about Langmuir isotherm and the regeneration of rGO, rGO–TON, and rGO-P25 (all the samples were operated in 3 duplicates with presented data equally ($R^2 > 0.98$).

Ghadim et al. (2013) reported the adsorption properties of tetracycline (TC) onto GO. The TC is a broad-spectrum antibiotic produced, by use against many bacterial infections. The studies showed that TC strongly loads on the GO surface via π–π interaction and cation–π bonding. The TC adsorption kinetics showed that the equilibrium was reached within 15 min following the pseudo-second-order model with observed rate constants of $k_2 = 0.2742–0.5362$ g mg^{-1} min (at different temperatures). The adsorption data showed maximum adsorption of 323 mg g^{-1} (298 K). The mean energy of adsorption was 1.83 kJ mol^{-1} (298 K) based on the Dubinin-Radushkevich adsorption isotherm. Magnetic GO doped with strontium titanium trioxide ($SrTiO_3$) nanoparticles were used as a nanocomposite for the removal of antibiotics as tetracycline and cefotaxime from aqueous media (Rashidi Nodeh and Sereshti 2016). The $SrTiO_3$ nanoparticles were synthesized and doped onto GO based magnetic nanoparticles (MNPs) simply via ultrasound. The synthesized GO/MNPs–$SrTiO_3$ magnetic nanocomposite was applied for the removal/adsorption of antibiotics namely tetracycline and cefotaxime from water samples. The GO/MNPs–$SrTiO_3$ nanocomposite showed high adsorption capacities of 65.78 and 18.21 mg g^{-1} toward tetracycline and cefotaxime, respectively. The comparison study of Langmuir, Freundlich and Dubinin-Radushkevich isotherm models along with a kinetic study showed that the adsorption experimental data were well fitted to the Langmuir isotherm ($R^2 > 0.996$) and pseudo-second-order rate model ($R^2 > 0.995$). The thermodynamic studies (+ΔH) of the antibiotics adsorption onto GO/MNPs–$SrTiO_3$ revealed an endothermic nature.

Zhang et al. (2012b) reported the synthesis and environmental implications of graphene-coated biochar for the removal of MB as schematically illustrated in Fig. 29a. Figure 29b, c shows the adsorption isotherms of MB on the original biochar and graphene-coated biochar. Batch sorption experimental results indicate that the graphene-coated biochar has outstanding adsorption ability for polycyclic aromatic hydrocarbons (PAHs) with a maximum MB adsorption capacity of 174 mg g^{-1}, which is 20 times higher than that of the unmodified cotton wood biochar and comparable to those of some physically or chemically activated carbons. The enhanced adsorption of MB on the graphene-coated biochar was mainly controlled by the strong π–π interactions between aromatic molecules and the graphene sheets on biochar surface.

Yu et al. (2016) demonstrated the experimental and theoretical studies on competitive adsorption of aromatic compounds on rGO. The individual and

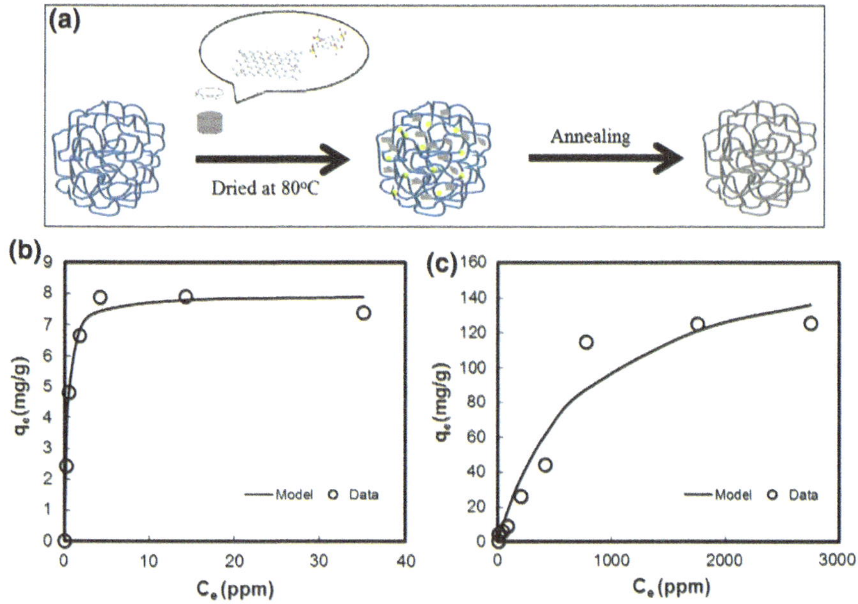

Fig. 29 **a** Illustration of procedures to produce graphene-coated biochar. Adsorption isotherms of MB on the **b** biochar and **c** graphene-coated biochar (Zhang et al. 2012b). Reprinted (adapted) with permission from Zhang et al. (2012b).

competitive adsorption studies of benzene, aniline and naphthylamine on rGO were investigated by batch experiments and theoretically by density functional theory (DFT). Experimental results indicated that (i) in all the single, binary, and ternary aromatic compound systems, the sequence of maximum adsorption capacity is naphthylamine > aniline > benzene on rGOs; (ii) the overall adsorption capacity of rGOs is in the order of ternary > binary > single system. The DFT calculations indicate that (i) the adsorption energy (E_{ad}) follows the order of E_{ad} (benzene) < E_{ad} (aniline) < E_{ad} (naphthylamine); (ii) the binding energy (E_{bd}) values of aromatic mixtures indicate that the intramolecular interactions between the aromatic compounds themselves have an important influence on their adsorption on rGOs. Sulfonated graphene sheetswas used for persistent aromatic pollutant management (Zhao et al. 2011a). It adsorbs persistent organic aromatic pollutants effectively from aqueous solutions. The adsorption capability of the prepared sulfonated graphene nanomaterials approaches ~2.3–2.4 mmol g^{-1} for naphthalene and 1-naphthol.

Graphene-coated materials (GCMs) having SiO_2 particles as a framework can be used for efficient removal of aromatic pollutants from water (Yang et al. 2015). The GCMs were prepared by loading graphene onto SiO_2 to expose the powerful adsorption sites in the interlayers. The adsorption of phenanthrene, a model aromatic pollutant, onto the loaded graphene nanosheets increases up to 100 fold compared with pristine graphene at the same level. The adsorption of GCMs

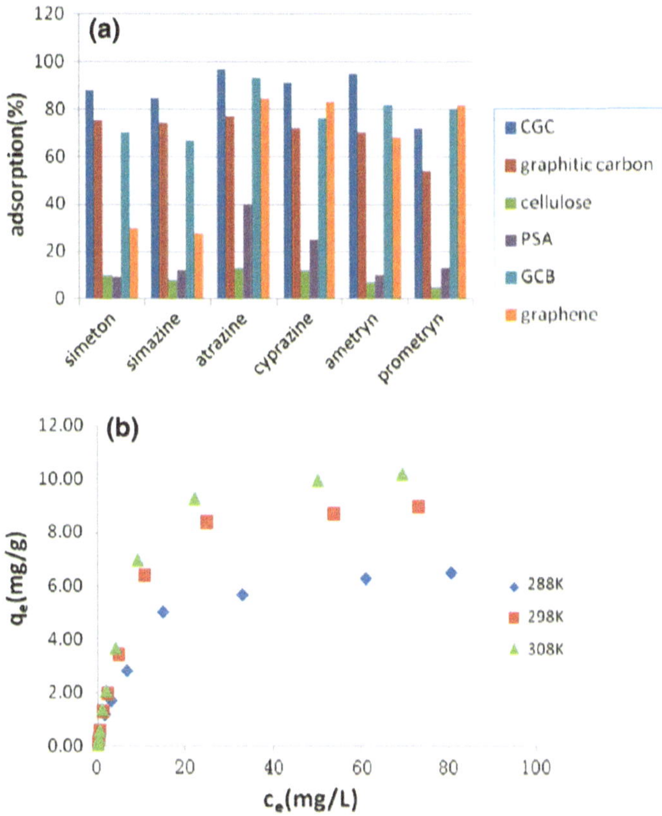

Fig. 30 **a** Comparison of the adsorption capacity of CGC, graphitic carbon, cellulose, PSA, GCB and graphene toward six triazine pesticides and **b** Adsorption isotherms of ametryn on CGC (Zhang et al. 2015). Reprinted (adapted) with permission from Zhang et al. (2015). Copyright (2015) American Chemical Society

increases with the loading amount of the graphene nanosheets and dramatically decreases with the introduction of oxygen-containing groups in the graphene nanosheets. Zhang et al. (2015) prepared cellulose/graphene composite (CGC) and applied it for triazine pesticides (simeton, simazine, atrazine, cyprazine, ametryn, prometryn) adsorption from water. The CGC composite particles achieve higher adsorption levels than five other sorbents (graphite carbons, cellulose, primary secondary amine (PSA), graphite carbon black (GCB), and graphene) for six triazine pesticides (Fig. 30a). Figure 30b shows the adsorption isotherms of ametryn on CGC over the concentration from 1 to 15 mg L^{-1} at 298, 313, and 323 K. The CGC was very stable and can easily be recycled employing simple organic solvent.

Graphene can be used as a reusable substrate for unprecedented adsorption of pesticides (Maliyekkal et al. 2013). The authors reported unmatched adsorption of chlorpyrifos (CP), endosulfan (ES), and malathion (ML) onto GO and rGO from water. The observed adsorption capacities of CP, ES, and ML were ~1200, 1100, and 800 mg g^{-1}, respectively. The first-principles pseudopotential-based density functional analysis of graphene-water-pesticide interactions showed that the adsorption is mediated through water, while direct interactions between graphene and the pesticides are rather weak or unlikely. Boruah et al. (Boruah et al. 2017) studied the use of magnetically recoverable Fe_3O_4/graphene nanocomposite towards efficient removal of triazine pesticides from aqueous solution. The Fe_3O_4/rGO nanocomposite was utilized for the adsorption of five harmful pesticides namely ametryn, prometryn, simazine, simeton and atrazine in an aqueous medium. Electrostatic interaction between the pesticides and Fe_3O_4/rGO nanocomposite was analyzed by the zeta potential analysis. The adsorption isotherm studies show that, the maximum adsorption capacity of 54.8 mg g^{-1} was achieved at pH 5 and it enhances in the presence of different ions (Mg^{2+}, Ca^{2+}, Na^{+} and SO_4^{2-}) and maximum (63.7 mg g^{-1}) adsorption for ametryn was found in seawater medium. Efficient adsorption (93.61%) of pesticides was observed due to electrostatic, hydrophobic and π–π interactions of composite towards the heterocyclic conjugation of pesticide molecules. The Fe_3O_4/rGO nanocomposite can be rapidly separated from an aqueous medium using the external magnet for reuse and 88.66% adsorption efficiency was observed up to seven cycles.

Nazal et al. (2015) reported the kinetics of the adsorption of dibenzothiophene (DBT) in model diesel fuel on carbonaceous materials loaded with aluminum oxide particles. The results showed that modification of the AC, CNT and GO with Al_2O_3 resulted in enhancing the surface chemistry of adsorbents and in turn increases their adsorption capacities and selectivity for DBT from model diesel. Three different types of carbonaceous adsorbents namely coconut activated carbon (AC), CNTs and GO loaded with two loadings of Al (5% and 10.9%) in the form of Al_2O_3 were used for the removal of DBT from n-hexane as a stimulant of diesel fuel. The results shown in Fig. 31a represent all the adsorbents in this study, the adsorption rates of DBT reaches equilibrium within 90 min. The results presented in Fig. 31b shows that the percentage removal of DBT increases with the increase in the dose of AC loaded with Al_2O_3. High removal percentage of DBT (around 98%) appeared at an adsorbent dosage of 500 mg for ACAL5 and ACAL10, while the maximum adsorption of DBT (100% removal) was found at an adsorbent dosage of 1500 mg.

Li et al. (2012b) studied the equilibrium, kinetic and thermodynamics of the adsorption of phenol from aqueous solution onto graphene. The effect of various parameters including pH, dosage, contact time, and temperature on the adsorption properties of phenol onto graphene has also been investigated. The maximum adsorption capacity reaches 28.26 mg g^{-1} at the conditions of initial phenol concentration of 50 mg L^{-1}, pH 6.3 and 285 K. The thermodynamic parameters indicated that the adsorption of phenol onto graphene was endothermic and spontaneous. The percent removal of phenol increases with the adsorbent dosage from 0.2 to 1.7 g L^{-1} and the adsorption capacity reaches to the maximum value of

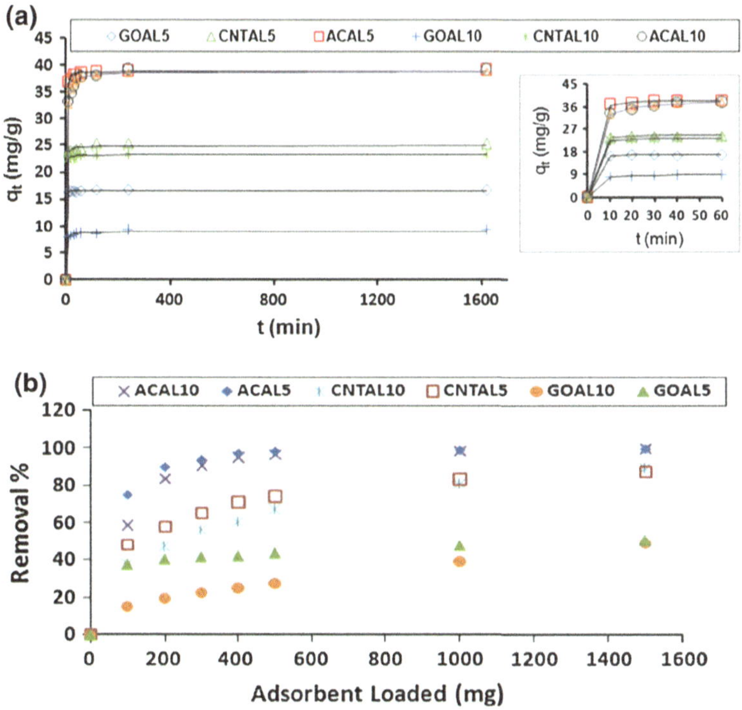

Fig. 31 **a** The effect of agitation time on the adsorption capacity of dibenzothiophene (DBT), using different adsorbents impregnated with different percentages of Al_2O_3 (insets show the effect of agitation time in the first 60 min) and **b** Effect of adsorbents loaded on the removal efficiency of dibenzothiophene (DBT) on AC, CNT and GO impregnated by Al_2O_3 (Nazal et al. 2015). Reprinted (adapted) with permission from Nazal et al. (2015).

28.26 mg g^{-1} at the dosage of 0.5 g L^{-1}. Xu et al. (2012) studied the decontamination of bisphenol A (BPA) from aqueous solution by graphene adsorption. The maximum adsorption capacity of graphene for BPA obtained from Langmuir isotherm was 182 mg g^{-1} at 302.15 K. The authors reported that the π–π interactions and hydrogen bonds might be responsible for the adsorption of BPA on graphene, and the outstanding adsorption capacity of graphene was due to its unique sp^2-hybridized single-atom-layer structure and thus graphene can be regarded as a promising adsorbent for BPA removal in water treatment. The effect of contact time on the adsorption of BPA by graphene are shown in Fig. 32a and the adsorption capacity of graphene increases quickly in the first 0.5 h and then rises slowly until the adsorption reached equilibrium within 5 h. Figure 32b shows the adsorption isotherms of BPA on graphene at three different temperatures. The adsorption capacity of graphene increased with the increasing equilibrium concentration of BPA and reached saturation progressively.

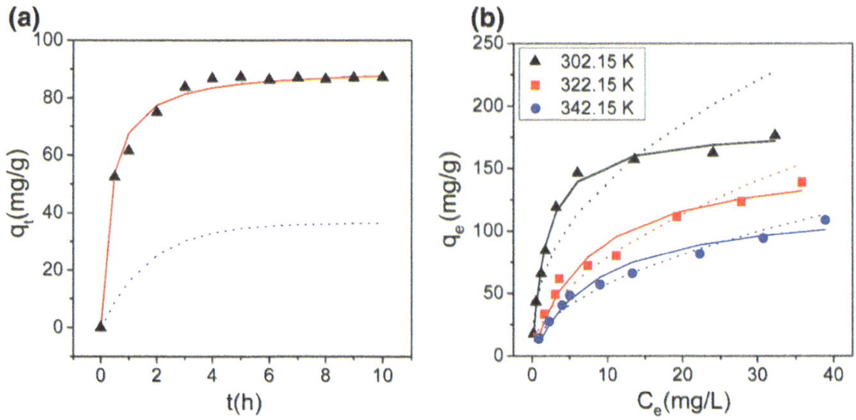

Fig. 32 **a** Effect of contact time on the adsorption of BPA by graphene (10 mg of graphene and 100 mL of 10 mg L^{-1} BPA at 302.15 K, pH 6.0) (dotted line: pseudo-first-order model simulation; solid line: pseudo-second-order model simulation). **b** Adsorption isotherms of BPA by graphene at three different temperatures (10 mg of graphene and 100 mL of BPA reacted for 6 h at pH 6.0) (solid lines: Langmuir model simulation; dotted lines: Freundlich model simulation) (Xu et al. 2012). Reprinted (adapted) with permission from Xu et al. (2012). Copyright (2012) American Chemical Society

8 Conclusions and Outlook

It is well known that among the researches based on novel carbon nanomaterials, graphene and its derivatives have gained countless attention and interest in many areas of physics, chemistry, biomedicine, industrial related and nanotechnology. Recent progress in the development of graphene and its derivatives based adsorbents has generated enormous potential for water treatment technologies. This chapter summarizes the application of GO and its composites as an advanced adsorbent for the removal of heavy metal ions from water. The GO and its composites with metal oxide have attracted widespread attentions and utilized efficiently as adsorbents to remove various heavy metal contaminants in water. Due to unique physicochemical characteristics of GO it becomes an outstanding and most prospective adsorbent material. GO and its composites are broadly explored as superior adsorbent carbon based materials for the exclusion of heavy metal ions from waste water due to their fast kinetics, high efficiency and strong affinity to various heavy metal ions. The π–π interaction between the adsorbate and GO plays a critical role for adsorption on many different organic contaminants. Also, the π–π interaction reveals the major attractive force between organic contaminations and the electron rich surface of GO. The GO portrays negative charges due to attached oxygen containing functional groups and able to exhibit high adsorption for inorganic ions including heavy metals ions. Electrostatic attraction, ion exchange and surface complexation are the promising mechanisms that mostly responsible to the adsorption of various metal ions. Specially, the formation of composite with GO and magnetic nanoparticles promote convenient magnetic separation along with the reusability of the adsorbents. There are still several challenges that must be overcome before the commercial relevance of these materials for the waste water treatment and removal of toxic metal. The cost effective, scalable and simple approaches are the major factors to bring it for the commercial application. GO have attracted considerable attention as a potential candidate for new filtration technologies. More efforts and research is needed on GO and its derivatives to be used for different water treatment, and nanotechnology has great potential for elimination of toxic materials and other contaminants. GO materials have been explored so far for water treatment applications, this emerging field of research is set to attract a great deal of attention in the near future.

References

2017 Eliminating water contamination. Filtration + Separation 54(1):16–17

Aboubaraka AE, Aboelfetoh EF, Ebeid E-ZM (2017) Coagulation effectiveness of graphene oxide for the removal of turbidity from raw surface water. Chemosphere 181:738–746

Ai L, Zhang C, Chen Z (2011) Removal of methylene blue from aqueous solution by a solvothermal-synthesized graphene/magnetite composite. J Hazard Mater 192(3):1515–1524

AlOthman ZA, Alam MM, Naushad M (2013) Heavy toxic metal ion exchange kinetics: validation of ion exchange process on composite cation exchanger nylon 6,6 Zr(IV) phosphate. J Ind Eng Chem 19:956–960. https://doi.org/10.1016/j.jiec.2012.11.016

An S, Joshi BN, Lee J-G, Lee MW, Kim YI, M-w Kim, Jo HS, Yoon SS (2017) A comprehensive review on wettability, desalination, and purification using graphene-based materials at water interfaces. Catal Today 295:14–25

Anirudhan TS, Ramachandran M (2008) Synthesis and characterization of amidoximated polyacrylonitrile/organobentonite composite for Cu(II), Zn(II), and Cd(II) adsorption from aqueous solutions and industry wastewaters. Ind Eng Chem Res 47(16):6175–6184

Arshad A, Iqbal J, Siddiq M, Ali MU, Ali A, Shabbir H, Nazeer UB, Saleem MS (2017) Solar light triggered catalytic performance of graphene-CuO nanocomposite for waste water treatment. Ceram Int 43(14):10654–10660

Atieh MA, Bakather OY, Tawabini BS, Bukhari AA, Khaled M, Alharthi M, Fettouhi M, Abuilaiwi FA (2010) Removal of chromium (III) from water by using modified and nonmodified carbon nanotubes. J Nanomater 2010:9

Balandin AA (2011) Thermal properties of graphene and nanostructured carbon materials. Nat Mater 10(8):569–581

Balandin AA, Ghosh S, Bao W, Calizo I, Teweldebrhan D, Miao F, Lau CN (2008) Superior thermal conductivity of single-layer graphene. Nano Lett 8(3):902–907

Berg M, Tran HC, Nguyen TC, Pham HV, Schertenleib R, Giger W (2001) Arsenic contamination of groundwater and drinking water in Vietnam: a human health threat. Environ Sci Technol 35 (13):2621–2626

Bhaumik M, Agarwal S, Gupta VK, Maity A (2016) Enhanced removal of Cr(VI) from aqueous solutions using polypyrrole wrapped oxidized MWCNTs nanocomposites adsorbent. J Colloid Interface Sci 470:257–267

Bhowmik K, Mukherjee A, Mishra MK, De G (2014) Stable Ni nanoparticle-reduced graphene oxide composites for the reduction of highly toxic aqueous Cr(VI) at room temperature. Langmuir 30(11):3209–3216

Bolotin KI, Sikes KJ, Jiang Z, Klima M, Fudenberg G, Hone J, Kim P, Stormer HL (2008) Ultrahigh electron mobility in suspended graphene. Solid State Commun 146(9):351–355

Boruah PK, Sharma B, Hussain N, Das MR (2017) Magnetically recoverable Fe3O4/graphene nanocomposite towards efficient removal of triazine pesticides from aqueous solution: investigation of the adsorption phenomenon and specific ion effect. Chemosphere 168:1058–1067

Bradder P, Ling SK, Wang S, Liu S (2011) Dye adsorption on layered graphite oxide. J Chem Eng Data 56(1):138–141

Busetti F, Badoer S, Cuomo M, Rubino B, Traverso P (2005) Occurrence and removal of potentially toxic metals and heavy metals in the wastewater treatment plant of Fusina (Venice, Italy). Ind Eng Chem Res 44(24):9264–9272

Cao N, Lyu Q, Li J, Wang Y, Yang B, Szunerits S, Boukherroub R (2017) Facile synthesis of fluorinated polydopamine/chitosan/reduced graphene oxide composite aerogel for efficient oil/water separation. Chem Eng J 326:17–28

Carmalin Sophia A, Lima EC, Allaudeen N, Rajan S (2016) Application of graphene based materials for adsorption of pharmaceutical traces from water and wastewater—a review. Desalination Water Treat 57(57):27573–27586

Castillo M, Oubiña A, Barceló D (1998) Evaluation of ELISA kits followed by liquid chromatography-atmospheric pressure chemical ionization-mass spectrometry for the determination of organic pollutants in industrial effluents. Environ Sci Technol 32(14):2180–2184

Chandra V, Park J, Chun Y, Lee JW, Hwang I-C, Kim KS (2010) Water-dispersible magnetite-reduced graphene oxide composites for arsenic removal. ACS Nano 4(7):3979–3986

Chen M, Huo C, Li Y, Wang J (2016a) Selective adsorption and efficient removal of phosphate from aqueous medium with graphene-lanthanum composite. ACS Sustain Chem Eng 4 (3):1296–1302

Chen X, Zhou S, Zhang L, You T, Xu F (2016b) Adsorption of heavy metals by graphene oxide/cellulose hydrogel prepared from NaOH/urea aqueous solution. Materials 9(7):582

Chen F, Gong AS, Zhu M, Chen G, Lacey SD, Jiang F, Li Y, Wang Y, Dai J, Yao Y, Song J, Liu B, Fu K, Das S, Hu L (2017) Mesoporous, three-dimensional wood membrane decorated with nanoparticles for highly efficient water treatment. ACS Nano 11(4):4275–4282

Cheng J-S, Du J, Zhu W (2012) Facile synthesis of three-dimensional chitosan–graphene mesostructures for reactive black 5 removal. Carbohyd Polym 88(1):61–67

Cohen DB (1986) Ground water contamination by toxic substances. Evaluation of Pesticides in Ground Water: American Chemical Society. pp 499–529

Cong H-P, Ren X-C, Wang P, Yu S-H (2012) Macroscopic multifunctional graphene-based hydrogels and aerogels by a metal ion induced self-assembly process. ACS Nano 6(3):2693–2703

Du Q, Zhou Y, Pan X, Zhang J, Zhuo Q, Chen S, Chen G, Liu T, Xu F, Yan C (2016) A graphene-melamine-sponge for efficient and recyclable dye adsorption. RSC Advances 6 (59):54589–54596

Dubey R, Bajpai J, Bajpai AK (2015) Green synthesis of graphene sand composite (GSC) as novel adsorbent for efficient removal of Cr (VI) ions from aqueous solution. J Water Process Eng 5:83–94

El-Sherif IY, Tolani S, Ofosu K, Mohamed OA, Wanekaya AK (2013) Polymeric nanofibers for the removal of Cr(III) from tannery waste water. J Environ Manag 129:410–413

El-Taweel YA, Nassef EM, Elkheriany I, Sayed D (2015) Removal of Cr(VI) ions from waste water by electrocoagulation using iron electrode. Egypt J Petrol 24(2):183–192

Fan L, Luo C, Li X, Lu F, Qiu H, Sun M (2012) Fabrication of novel magnetic chitosan grafted with graphene oxide to enhance adsorption properties for methyl blue. J Hazard Mater 215:272–279

Fu F, Wang Q (2011) Removal of heavy metal ions from wastewaters: a review. J Environ Manag 92(3):407–418

Gadipelli S, Guo ZX (2015) Graphene-based materials: synthesis and gas sorption, storage and separation. Prog Mater Sci 69:1–60

Ganesan P, Kamaraj R, Vasudevan S (2013) Application of isotherm, kinetic and thermodynamic models for the adsorption of nitrate ions on graphene from aqueous solution. J Taiwan Inst Chem Eng 44(5):808–814

Gao Y, Li Y, Zhang L, Huang H, Hu J, Shah SM, Su X (2012) Adsorption and removal of tetracycline antibiotics from aqueous solution by graphene oxide. J Colloid Interface Sci 368 (1):540–546

Georgakilas V, Tiwari JN, Kemp KC, Perman JA, Bourlinos AB, Kim KS, Zboril R. (2016) Noncovalent Functionalization of Graphene and Graphene Oxide for Energy Materials, Biosensing, Catalytic, and Biomedical Applications. Chem Rev 116(9):5464–5519

Geim AK, Novoselov KS (2007) The rise of graphene. Nat Mater 6(3):183–191

Ghadim EE, Manouchehri F, Soleimani G, Hosseini H, Kimiagar S, Nafisi S (2013) Adsorption properties of tetracycline onto graphene oxide: equilibrium, kinetic and thermodynamic studies. PLOS ONE 8(11):e79254

Gong J, Gao X, Li M, Nie Q, Pan W, Liu R (2017) Dye adsorption on electrochemical exfoliated graphene oxide nanosheets: pH influence, kinetics and equilibrium in aqueous solution. Int J Environ Sci Technol 14(2):305–314

Guo J, Wang R, Tjiu WW, Pan J, Liu T (2012) Synthesis of Fe nanoparticles@graphene composites for environmental applications. J Hazard Mater 225:63–73

Hao L, Song H, Zhang L, Wan X, Tang Y, Lv Y (2012) SiO2/graphene composite for highly selective adsorption of Pb(II) ion. J Colloid Interface Sci 369(1):381–387

Harijan DKL, Chandra V (2017) Akaganeite nanorods decorated graphene oxide sheets for removal and recovery of aqueous phosphate. J Water Process Eng 19:120–125

Hartono T, Wang S, Ma Q, Zhu Z (2009) Layer structured graphite oxide as a novel adsorbent for humic acid removal from aqueous solution. J Colloid Interface Sci 333(1):114–119

Hu X-j, Y-g Liu, Wang H, A-w Chen, G-m Zeng, S-m Liu, Guo Y-m HuX, T-t Li, Y-q Wang, Zhou L, S-h Liu (2013) Removal of Cu(II) ions from aqueous solution using sulfonated magnetic graphene oxide composite. Sep Purif Technol 108:189–195

Hu X-j, Liu Y-g, Zeng G-m, You S-h, Wang H, Hu X, Guo Y-m, Tan X-f, Guo F-y (2014) Effects of background electrolytes and ionic strength on enrichment of Cd(II) ions with magnetic graphene oxide–supported sulfanilic acid. J Colloid Interface Sci 435:138–144

Hur J, Shin J, Yoo J, Seo Y-S (2015) Competitive adsorption of metals onto magnetic graphene oxide: comparison with other carbonaceous adsorbents. Sci World J 2015:11

Jabeen H, Chandra V, Jung S, Lee JW, Kim KS, Kim SB (2011) Enhanced Cr(vi) removal using iron nanoparticle decorated graphene. Nanoscale 3(9):3583–3585

Javadian H, Angaji MT, Naushad M (2014) Synthesis and characterization of polyaniline/γ-alumina nanocomposite: a comparative study for the adsorption of three different anionic dyes. J Ind Eng Chem 20:3890–3900. https://doi.org/10.1016/j.jiec.2013.12.095

Jayanthi S, KrishnaRao Eswar N, Singh SA, Chatterjee K, Madras G, Sood AK (2016) Macroporous three-dimensional graphene oxide foams for dye adsorption and antibacterial applications. RSC Adv 6(2):1231–1242

Jiao T, Liu Y, Wu Y, Zhang Q, Yan X, Gao F, Bauer AJP, Liu J, Zeng T, Li B (2015) Facile and scalable preparation of graphene oxide-based magnetic hybrids for fast and highly efficient removal of organic dyes 5:12451

Kemp KC, Seema H, Saleh M, Le NH, Mahesh K, Chandra V, Kim KS (2013) Environmental applications using graphene composites: water remediation and gas adsorption. Nanoscale 5 (8):3149–3171

Kharismadewi D, Haldorai Y, Nguyen VH, Tuma D, Shim J-J (2016) Synthesis of graphene oxide-poly(2-hydroxyethyl methacrylate) composite by dispersion polymerization in supercritical CO_2: adsorption behavior for the removal of organic dye. Compos Interfaces 23 (7):719–739

Kireeti KVMK, Chandrakanth G, Kadam MM, Jha N (2016) A sodium modified reduced graphene oxide-Fe_3O_4 nanocomposite for efficient lead(ii) adsorption. RSC Adv 6(88):84825–84836

Köhler A, Hellweg S, Escher BI, Hungerbühler K (2006) Organic pollutant removal versus toxicity reduction in industrial wastewater treatment: the example of wastewater from fluorescent whitening agent production. Environ Sci Technol 40(10):3395–3401

Konicki W, Aleksandrzak M, Moszyński D, Mijowska E (2017) Adsorption of anionic azo-dyes from aqueous solutions onto graphene oxide: equilibrium, kinetic and thermodynamic studies. J Colloid Interface Sci 496:188–200

Kumar R, Singh RK, Dubey PK, Singh DP, Yadav RM, Tiwari RS (2015a) Freestanding 3D graphene–nickel encapsulated nitrogen-rich aligned bamboo like carbon nanotubes for high-performance supercapacitors with robust cycle stability. Adv Mater Interfaces 2 (15):1500191-n/a

Kumar R, Singh RK, Vaz AR, Moshkalev SA (2015b) Microwave-assisted synthesis and deposition of a thin ZnO layer on microwave-exfoliated graphene: optical and electrochemical evaluations. RSC Adv 5(83):67988–67995

Kumar R, Savu R, Singh RK, Joanni E, Singh DP, Tiwari VS, Vaz AR, da Silva ETSG, Maluta JR, Kubota LT, Moshkalev SA (2017a) Controlled density of defects assisted perforated structure in reduced graphene oxide nanosheets-palladium hybrids for enhanced ethanol electro-oxidation. Carbon 117:137–146

Kumar R, Singh RK, Singh DP, Joanni E, Yadav RM, Moshkalev SA (2017b) Laser-assisted synthesis, reduction and micro-patterning of graphene: recent progress and applications. Coord Chem Rev 342:34–79

Kumar R, Singh RK, Vaz AR, Savu R, Moshkalev SA (2017c) Self-assembled and one-step synthesis of interconnected 3D network of Fe_3O_4/reduced graphene oxide nanosheets hybrid for high-performance supercapacitor electrode. ACS Appl Mater Interfaces 9(10):8880–8890

Lakshmi J, Vasudevan S (2013) Graphene—a promising material for removal of perchlorate (ClO_4^-) from water. Environ Sci Pollut Res 20(8):5114–5124

LeClair V (1997) MTBE water contamination raises health concerns, research questions. Environ Sci Technol 31(4):176A–177A

Lee Y-C, Yang J-W (2012) Self-assembled flower-like TiO_2 on exfoliated graphite oxide for heavy metal removal. J Ind Eng Chem 18(3):1178–1185

Lee J-W, Jung J, Cho YH, Yadav SK, Baek KY, Park HB, Hong SM, Koo CM (2014) Fouling-tolerant nanofibrous polymer membranes for water treatment. ACS Appl Mater Interfaces 6(16):14600–14607

Li B, Cao H, Yin G (2011a) Mg(OH)2@reduced graphene oxide composite for removal of dyes from water. J Mater Chem 21(36):13765–13768

Li Y, Zhang P, Du Q, Peng X, Liu T, Wang Z, Xia Y, Zhang W, Wang K, Zhu H, Wu D (2011b) Adsorption of fluoride from aqueous solution by graphene. J Colloid Interface Sci 363(1):348–354

Li S, Lu X, Xue Y, Lei J, Zheng T, Wang C (2012a) Fabrication of polypyrrole/graphene oxide composite nanosheets and their applications for Cr(VI) removal in aqueous solution. PLoS ONE 7(8):e43328

Li Y, Du Q, Liu T, Sun J, Jiao Y, Xia Y, Xia L, Wang Z, Zhang W, Wang K, Zhu H, Wu D (2012b) Equilibrium, kinetic and thermodynamic studies on the adsorption of phenol onto graphene. Mater Res Bull 47(8):1898–1904

Liu M, Chen C, Hu J, Wu X, Wang X (2011) Synthesis of magnetite/graphene oxide composite and application for cobalt(II) removal. J Physical Chem C 115(51):25234–25240

Liu F, Chung S, Oh G, Seo TS (2012) Three-dimensional graphene oxide nanostructure for fast and efficient water-soluble dye removal. ACS Appl Mater Interfaces 4(2):922–927

Liu Z, Yang J, Li C, Li J, Jiang Y, Dong Y, Li Y (2014) Adsorption of Co(II), Ni(II), Pb(II) and U (VI) from aqueous solutions using polyaniline/graphene oxide composites. Korean Chem Eng Res 52(6):781–788

Liu J-N, Chen Z, Wu Q-Y, Li A, Hu H-Y, Yang C (2016a) Ozone/graphene oxide catalytic oxidation: a novel method to degrade emerging organic contaminant N, N-diethyl-m-toluamide (DEET) 6:31405

Liu Y, Yu L, Ong CN, Xie J (2016b) Nitrogen-doped graphene nanosheets as reactive water purification membranes. Nano Res 9(7):1983–1993

Luh J, Royster S, Sebastian D, Ojomo E, Bartram J (2017) Expert assessment of the resilience of drinking water and sanitation systems to climate-related hazards. Sci Total Environ 592:334–344

Luo X, Wang X, Bao S, Liu X, Zhang W, Fang T (2016) Adsorption of phosphate in water using one-step synthesized zirconium-loaded reduced graphene oxide 6:39108

Ma H-L, Zhang Y, Hu Q-H, Yan D, Yu Z-Z, Zhai M (2012) Chemical reduction and removal of Cr (vi) from acidic aqueous solution by ethylenediamine-reduced graphene oxide. J Mater Chem 22(13):5914–5916

Maliyekkal SM, Sreeprasad TS, Krishnan D, Kouser S, Mishra AK, Waghmare UV, Pradeep T (2013) Graphene: a reusable substrate for unprecedented adsorption of pesticides. Small 9 (2):273–283

McCoy M (1998) Different routes to water treatment. Chem Eng News Arch 76(51):21

Mehta D, Mazumdar S, Singh SK (2015) Magnetic adsorbents for the treatment of water/wastewater—a review. J Water Process Eng 7:244–265

Mishra AK, Ramaprabhu S (2011) Functionalized graphene sheets for arsenic removal and desalination of sea water. Desalination 282:39–45

Mohan D, Pittman CU (2007) Arsenic removal from water/wastewater using adsorbents—a critical review. J Hazard Mater 142(1):1–53

Motamedi E, Talebi Atouei M, Kassaee MZ (2014) Comparison of nitrate removal from water via graphene oxide coated Fe, Ni and Co nanoparticles. Mater Res Bull 54:34–40

Mudassir MA, Hussain SZ, Rehman A, Zaheer W, Asma ST, Jilani A, Aslam M, Zhang H, Ansari TM, Hussain I (2017) Development of silver-nanoparticle-decorated emulsion-templated hierarchically porous poly(1-vinylimidazole) beads for water treatment. ACS Appl Mater Interfaces 9(28):24190–24197

Mukherjee M, Ghorai UK, Samanta M, Santra A, Das GP, Chattopadhyay KK (2017) Graphene wrapped copper phthalocyanine nanotube: enhanced photocatalytic activity for industrial waste water treatment. Appl Surf Sci 418:156–162

Nair RR, Blake P, Grigorenko AN, Novoselov KS, Booth TJ, Stauber T, Peres NMR, Geim AK (2008) Fine structure constant defines visual transparency of graphene. Science 320 (5881):1308

Nandi D, Basu T, Debnath S, Ghosh AK, De A, Ghosh UC (2013) Mechanistic insight for the sorption of Cd(II) and Cu(II) from aqueous solution on magnetic Mn-Doped Fe(III) oxide nanoparticle implanted graphene. J Chem Eng Data 58(10):2809–2818

Naushad M, ALOthman ZA, Awual MR et al (2015) Adsorption kinetics, isotherms, and thermodynamic studies for the adsorption of Pb_2 + and Hg_2 + metal ions from aqueous medium using Ti(IV) iodovanadate cation exchanger. Ionics (Kiel) 21:2237–2245. https://doi.org/10.1007/s11581-015-1401-7

Nazal MK, Khaled M, Atieh MA, Aljundi IH, Oweimreen GA, Abulkibash AM (2015) The nature and kinetics of the adsorption of dibenzothiophene in model diesel fuel on carbonaceous materials loaded with aluminum oxide particles. Arab J Chem

Nguyen DD, Tai N-H, Lee S-B, Kuo W-S (2012) Superhydrophobic and superoleophilic properties of graphene-based sponges fabricated using a facile dip coating method. Energy Environ Sci 5(7):7908–7912

Nguyen-Phan T-D, Pham VH, Kim EJ, Oh E-S, Hur SH, Chung JS, Lee B, Shin EW (2012) Reduced graphene oxide–titanate hybrids: morphologic evolution by alkali-solvothermal treatment and applications in water purification. Appl Surf Sci 258(10):4551–4557

Nicomel N, Leus K, Folens K, Van Der Voort P, Du Laing G (2016) Technologies for arsenic removal from water: current status and future perspectives. Int J Environ Res Public Health 13 (1):62

Pan S, Liu X (2012) CdS-Graphene nanocomposite: synthesis, adsorption kinetics and high photocatalytic performance under visible light irradiation. New J Chem 36(9):1781–1787

Pathania D, Sharma G, Kumar A et al (2015) Combined sorptional–photocatalytic remediation of dyes by polyaniline Zr(IV) selenotungstophosphate nanocomposite. Toxicol Environ Chem 97:526–537. https://doi.org/10.1080/02772248.2015.1050024

Peng B, Chen L, Que C, Yang K, Deng F, Deng X, Shi G, Xu G, Wu M (2016a) Adsorption of antibiotics on graphene and biochar in aqueous solutions induced by π–π interactions 6:31920

Peng S, Jin G, Li L, Li K, Srinivasan M, Ramakrishna S, Chen J (2016b) Multi-functional electrospun nanofibres for advances in tissue regeneration, energy conversion & storage, and water treatment. Chem Soc Rev 45(5):1225–1241

Peng W, Li H, Liu Y, Song S (2017) A review on heavy metal ions adsorption from water by graphene oxide and its composites. J Mol Liq 230:496–504

Perreault F, Fonseca de Faria A, Elimelech M (2015) Environmental applications of graphene-based nanomaterials. Chem Soc Rev 44(16):5861–5896

Poornima Parvathi V, Umadevi M, Bhaviya Raj R (2015) Improved waste water treatment by bio-synthesized graphene sand composite. J Environ Manag 162:299–305

Pous N, Casentini B, Rossetti S, Fazi S, Puig S, Aulenta F (2015) Anaerobic arsenite oxidation with an electrode serving as the sole electron acceptor: a novel approach to the bioremediation of arsenic-polluted groundwater. J Hazard Mater 283:617–622

Radojčić Redovniković I, De Marco A, Proietti C, Hanousek K, Sedak M, Bilandžić N, Jakovljević T (2017) Poplar response to cadmium and lead soil contamination. Ecotoxicol Environ Saf 144:482–489

Raghu MS, Yogesh Kumar K, Prashanth MK, Prasanna BP, Vinuth R, Pradeep Kumar CB (2017) Adsorption and antimicrobial studies of chemically bonded magnetic graphene oxide-Fe_3O_4 nanocomposite for water purification. J Water Process Eng 17:22–31

Raghubanshi H, Ngobeni SM, Osikoya AO, Shooto ND, Dikio CW, Naidoo EB, Dikio ED, Pandey RK, Prakash R (2017) Synthesis of graphene oxide and its application for the adsorption of Pb+2 from aqueous solution. J Ind Eng Chem 47:169–178

Rashidi Nodeh H, Sereshti H (2016) Synthesis of magnetic graphene oxide doped with strontium titanium trioxide nanoparticles as a nanocomposite for the removal of antibiotics from aqueous media. RSC Adv 6(92):89953–89965

Ren Y, Yan N, Wen Q, Fan Z, Wei T, Zhang M, Ma J (2011) Graphene/δ-MnO_2 composite as adsorbent for the removal of nickel ions from wastewater. Chem Eng J 175:1–7

Samadi N, Ansari R, Khodavirdilo B (2017) Removal of copper ions from aqueous solutions using polymer derivations of poly (styrene-alt-maleic anhydride). Egypt J Petrol 26(2):375–389

Sharma SK, Petrusevski B, Amy G (2008) Chromium removal from water: a review. J Water Supply: Res Technol Aqua 57(8):541–553

Sharma G, Naushad M, Pathania D et al (2015) Modification of *Hibiscus cannabinus* fiber by graft copolymerization: application for dye removal. Desalin Water Treat 54:3114–3121. https://doi.org/10.1080/19443994.2014.904822

Singh V, Joung D, Zhai L, Das S, Khondaker SI, Seal S (2011) Graphene based materials: past, present and future. Prog Mater Sci 56(8):1178–1271

Singh RK, Kumar R, Singh DP (2016) Graphene oxide: strategies for synthesis, reduction and frontier applications. RSC Adv 6(69):64993–65011

Singh DK, Kumar V, Mohan S, Bano D, Hasan SH (2017a) Breakthrough curve modeling of graphene oxide aerogel packed fixed bed column for the removal of Cr(VI) from water. J Water Process Eng 18:150–158

Singh G, Kaushik SK, Mukherji S (2017b) Revelations of an overt water contamination. Med J Armed Forces India 73(3):250–255

Sitko R, Turek E, Zawisza B, Malicka E, Talik E, Heimann J, Gagor A, Feist B, Wrzalik R (2013) Adsorption of divalent metal ions from aqueous solutions using graphene oxide. Dalton Trans 42(16):5682–5689

Sreeprasad TS, Maliyekkal SM, Lisha KP, Pradeep T (2011) Reduced graphene oxide–metal/metal oxide composites: facile synthesis and application in water purification. J Hazard Mater 186 (1):921–931

Stoller MD, Park S, Zhu Y, An J, Ruoff RS (2008) Graphene-based ultracapacitors. Nano Lett 8 (10):3498–3502

Suárez-Iglesias O, Collado S, Oulego P, Díaz M (2017) Graphene-family nanomaterials in wastewater treatment plants. Chem Eng J 313:121–135

Sui Z, Meng Q, Zhang X, Ma R, Cao B (2012) Green synthesis of carbon nanotube-graphene hybrid aerogels and their use as versatile agents for water purification. J Mater Chem 22 (18):8767–8771

Sun H, Cao L, Lu L (2011) Magnetite/reduced graphene oxide nanocomposites: one step solvothermal synthesis and use as a novel platform for removal of dye pollutants. Nano Res 4 (6):550–562

Sun L, Yu H, Fugetsu B (2012) Graphene oxide adsorption enhanced by in situ reduction with sodium hydrosulfite to remove acridine orange from aqueous solution. J Hazard Mater 203:101–110

Sun J-Z, Liao Z-H, Si R-W, Kingori GP, Chang F-X, Gao L, Shen Y, Xiao X, Wu X-Y, Yong Y-C (2014) Adsorption and removal of triphenylmethane dyes from water by magnetic reduced graphene oxide. Water Sci Technol 70(10):1663–1669

Taseidifar M, Makavipour F, Pashley RM, Rahman AFMM (2017) Removal of heavy metal ions from water using ion flotation. Environ Technol Innovation 8:182–190

Ting H, Lingyu K, Xiaoyu G, Yiping W, Feng W, Ying W, Haifeng Y (2016) Magnetic ferrous-doped graphene for improving Cr(VI) removal. Mater Res Express 3(4):045006

Tiwari JN, Mahesh K, Le NH, Kemp KC, Timilsina R, Tiwari RN, Kim KS (2013) Reduced graphene oxide-based hydrogels for the efficient capture of dye pollutants from aqueous solutions. Carbon 56:173–182

Tran DNH, Kabiri S, Wang L, Losic D (2015) Engineered graphene-nanoparticle aerogel composites for efficient removal of phosphate from water. J Mater Chem A 3(13):6844–6852

Unuabonah EI, Günter C, Weber J, Lubahn S, Taubert A (2013) Hybrid clay: a new highly efficient adsorbent for water treatment. ACS Sustain Chem Eng 1(8):966–973

Upadhyay RK, Soin N, Bhattacharya G, Saha S, Barman A, Roy SS (2015) Grape extract assisted green synthesis of reduced graphene oxide for water treatment application. Mater Lett 160:355–358

Vadahanambi S, Lee S-H, Kim W-J, Oh I-K (2013) Arsenic removal from contaminated water using three-dimensional graphene-carbon nanotube-iron oxide nanostructures. Environ Sci Technol 47(18):10510–10517

Van Hoa N, Khong TT, Thi Hoang Quyen T, Si Trung T (2016) One-step facile synthesis of mesoporous graphene/Fe_3O_4/chitosan nanocomposite and its adsorption capacity for a textile dye. J Water Process Eng 9:170–178

Vasudevan S, Lakshmi J (2012) The adsorption of phosphate by graphene from aqueous solution. RSC Adv 2(12):5234–5242

Vilela D, Parmar J, Zeng Y, Zhao Y, Sánchez S (2016) Graphene-based microbots for toxic heavy metal removal and recovery from water. Nano Lett 16(4):2860–2866

Vuong Hoan NT, Anh Thu NT, Duc HV, Cuong ND, Quang Khieu D, Vo V (2016) Fe_3O_4/reduced graphene oxide nanocomposite: synthesis and its application for toxic metal ion removal. J Chem 2016:10

Wang C, Feng C, Gao Y, Ma X, Wu Q, Wang Z (2011) Preparation of a graphene-based magnetic nanocomposite for the removal of an organic dye from aqueous solution. Chem Eng J 173 (1):92–97

Wang S, Sun H, Ang HM, Tadé MO (2013a) Adsorptive remediation of environmental pollutants using novel graphene-based nanomaterials. Chem Eng J 226:336–347

Wang Y, Liang S, Chen B, Guo F, Yu S, Tang Y (2013b) Synergistic removal of Pb(II), Cd(II) and humic acid by Fe_3O_4@mesoporous silica-graphene oxide composites. PLoS ONE 8(6):e65634

Wang H, Wang E, Liu Z, Gao D, Yuan R, Sun L, Zhu Y (2015) A novel carbon nanotubes reinforced superhydrophobic and superoleophilic polyurethane sponge for selective oil-water separation through a chemical fabrication. J Mater Chem A 3(1):266–273

Wei Z, Xing R, Zhang X, Liu S, Yu H, Li P (2013) Facile template-free fabrication of hollow nestlike α-Fe_2O_3 nanostructures for water treatment. ACS Appl Mater Interfaces 5(3):598–604
Whelton AJ, McMillan L, Connell M, Kelley KM, Gill JP, White KD, Gupta R, Dey R, Novy C (2015) Residential tap water contamination following the freedom industries chemical spill: perceptions, water quality, and health impacts. Environ Sci Technol 49(2):813–823
Wu S, Kong L, Liu J (2016a) Removal of mercury and fluoride from aqueous solutions by three-dimensional reduced-graphene oxide aerogel. Res Chem Intermed 42(5):4513–4530
Wu Z, Yuan X, Zhong H, Wang H, Zeng G, Chen X, Wang H, Zhang L, Shao J (2016b) Enhanced adsorptive removal of p-nitrophenol from water by aluminum metal–organic framework/ reduced graphene oxide composite 6:25638
Xie G, Xi P, Liu H, Chen F, Huang L, Shi Y, Hou F, Zeng Z, Shao C, Wang J (2012) A facile chemical method to produce superparamagnetic graphene oxide-Fe_3O_4 hybrid composite and its application in the removal of dyes from aqueous solution. J Mater Chem 22(3):1033–1039
Xu J, Wang L, Zhu Y (2012) Decontamination of bisphenol A from aqueous solution by graphene adsorption. Langmuir 28(22):8418–8425
Yang S-T, Chang Y, Wang H, Liu G, Chen S, Wang Y, Liu Y, Cao A (2010) Folding/aggregation of graphene oxide and its application in Cu_2+ removal. J Colloid Interface Sci 351(1):122–127
Yang S-T, Chen S, Chang Y, Cao A, Liu Y, Wang H (2011) Removal of methylene blue from aqueous solution by graphene oxide. J Colloid Interface Sci 359(1):24–29
Yang X, Chen C, Li J, Zhao G, Ren X, Wang X (2012) Graphene oxide-iron oxide and reduced graphene oxide-iron oxide hybrid materials for the removal of organic and inorganic pollutants. RSC Adv 2(23):8821–8826
Yang K, Chen B, Zhu L (2015) Graphene-coated materials using silica particles as a framework for highly efficient removal of aromatic pollutants in water 5:11641
Yao Y, Miao S, Liu S, Ma LP, Sun H, Wang S (2012) Synthesis, characterization, and adsorption properties of magnetic Fe_3O_4@graphene nanocomposite. Chem Eng J 184:326–332
Ye Y, Yin D, Wang B, Zhang Q (2015) Synthesis of three-dimensional Fe_3O_4/graphene aerogels for the removal of arsenic ions from water. J Nanomater 2015:6
Ye S, Liu Y, Feng J (2017) Low-density, mechanical compressible, water-induced self-recoverable graphene aerogels for water treatment. ACS Appl Mater Interfaces 9(27):22456–22464
Yu S, Wang X, Ai Y, Tan X, Hayat T, Hu W, Wang X (2016) Experimental and theoretical studies on competitive adsorption of aromatic compounds on reduced graphene oxides. J Mater Chem A 4(15):5654–5662
Zaaba NI, Foo KL, Hashim U, Tan SJ, Liu W-W, Voon CH (2017) Synthesis of graphene oxide using modified hummers method: solvent influence. Procedia Eng 184:469–477
Zhang K, Dwivedi V, Chi C, Wu J (2010) Graphene oxide/ferric hydroxide composites for efficient arsenate removal from drinking water. J Hazard Mater 182(1):162–168
Zhang S, Shao Y, Liu J, Aksay IA, Lin Y (2011) Graphene-polypyrrole nanocomposite as a highly efficient and low cost electrically switched ion exchanger for removing ClO_4^- from wastewater. ACS Appl Mater Interfaces 3(9):3633–3637
Zhang K, Kemp KC, Chandra V (2012a) Homogeneous anchoring of TiO_2 nanoparticles on graphene sheets for waste water treatment. Mater Lett 81:127–130
Zhang M, Gao B, Yao Y, Xue Y, Inyang M (2012b) Synthesis, characterization, and environmental implications of graphene-coated biochar. Sci Total Environ 435:567–572
Zhang W, Shi X, Zhang Y, Gu W, Li B, Xian Y (2013a) Synthesis of water-soluble magnetic graphene nanocomposites for recyclable removal of heavy metal ions. J Mater Chem A 1 (5):1745–1753
Zhang Y, Ma H-L, Peng J, Zhai M, Yu Z-Z (2013b) Cr(VI) removal from aqueous solution using chemically reduced and functionalized graphene oxide. J Mater Sci 48(5):1883–1889
Zhang L, Gao Y, Zhou Q, Kan J, Wang Y (2014a) High-performance removal of phosphate from water by graphene nanosheets supported lanthanum hydroxide nanoparticles. Water Air Soil Pollut 225(6):1967

Zhang Y, Yan L, Xu W, Guo X, Cui L, Gao L, Wei Q, Du B (2014b) Adsorption of Pb(II) and Hg (II) from aqueous solution using magnetic $CoFe_2O_4$-reduced graphene oxide. J Mol Liq 191:177–182

Zhang C, Zhang RZ, Ma YQ, Guan WB, Wu XL, Liu X, Li H, Du YL, Pan CP (2015) Preparation of cellulose/graphene composite and its applications for triazine pesticides adsorption from water. ACS Sustain Chem Eng 3(3):396–405

Zhang R, Liu Y, He M, Su Y, Zhao X, Elimelech M, Jiang Z (2016) Antifouling membranes for sustainable water purification: strategies and mechanisms. Chem Soc Rev 45(21):5888–5924

Zhao G, Jiang L, He Y, Li J, Dong H, Wang X, Hu W (2011a) Sulfonated graphene for persistent aromatic pollutant management. Adv Mater 23(34):3959–3963

Zhao G, Li J, Ren X, Chen C, Wang X (2011b) Few-layered graphene oxide nanosheets as superior sorbents for heavy metal ion pollution management. Environ Sci Technol 45 (24):10454–10462

Zhao G, Li X, Huang M, Zhen Z, Zhong Y, Chen Q, Zhao X, He Y, Hu R, Yang T, Zhang R, Li C, Kong J, Xu J-B, Ruoff RS, Zhu H (2017) The physics and chemistry of graphene-on-surfaces. Chem Soc Rev 46(15):4417–4449

Zhong Y, Zhen Z, Zhu H (2017) Graphene: fundamental research and potential applications. FlatChem 4:20–32

Zhu J, Sadu R, Wei S, Chen DH, Haldolaarachchige N, Luo Z, Gomes JA, Young DP, Guo Z (2012a) Magnetic graphene nanoplatelet composites toward arsenic removal. ECS J Solid State Sci Technol 1(1):M1–M5

Zhu J, Wei S, Gu H, Rapole SB, Wang Q, Luo Z, Haldolaarachchige N, Young DP, Guo Z (2012b) One-pot synthesis of magnetic graphene nanocomposites decorated with core@double-shell nanoparticles for fast chromium removal. Environ Sci Technol 46(2):977–985

Zhu H, Yang S, Chen D, Li N, Xu Q, Li H, He J, Lu J (2016) A robust absorbent material based on light-responsive superhydrophobic melamine sponge for oil recovery. Adv Mater Interfaces 3 (5):1500683-n/a

Zhuang Y, Yu F, Ma J (2015) Enhanced adsorption and removal of ciprofloxacin on regenerable long TiO_2 nanotube/graphene oxide hydrogel adsorbents. J Nanomater 2015:8

Zong P, Wang S, Zhao Y, Wang H, Pan H, He C (2013) Synthesis and application of magnetic graphene/iron oxides composite for the removal of U(VI) from aqueous solutions. Chem Eng J 220:45–52

Zou Y, Wang X, Khan A, Wang P, Liu Y, Alsaedi A, Hayat T, Wang X (2016) Environmental remediation and application of nanoscale zero-valent iron and its composites for the removal of heavy metal ions: a review. Environ Sci Technol 50(14):7290–7304

Chapter 7
Photocatalytic Degradation of Pharmaceuticals Using Graphene Based Materials

William W. Anku, Ephraim M. Kiarii, Rama Sharma, Girish M. Joshi, Sudheesh K. Shukla and Penny P. Govender

Abstract Pharmaceutical products are produced purposely for the treatment of diseases with the aim of improving human health. Despite their usefulness to human and animal health, pharmaceuticals are now being regarded as emerging environmental pollutants. This is due to their increased use and the fact that they are indiscriminately discharged into the aquatic environment from hospitals, households, industries, pharmacies, as well as leakages and leachates from municipal wastewater treatment plants and landfill sites. Moreover, the conventional methods of wastewater treatment were not designed with these emerging pollutants in mind resulting in the discharge of untreated or incomplete treated wastewater into water bodies. Pharmaceuticals in water are believed to exert deleterious effects on humans

W. W. Anku · E. M. Kiarii · S. K. Shukla (✉) · P. P. Govender
Department of Applied Chemistry, University of Johannesburg,
Doornfontein Campus, P. O. Box 17011, Johannesburg 2028, South Africa
e-mail: sudheeshkshukla010@gmail.com

W. W. Anku
e-mail: williamanku85@gmail.com

E. M. Kiarii
e-mail: kiariiem@gmail.com

P. P. Govender
e-mail: pennyg@uj.ac.za

R. Sharma
Department Agriculture Science, AKS University, Satna 485001, Madhya Pradesh, India
e-mail: rama81093@gmail.com

G. M. Joshi
Polymer Nanocomposite Laboratory, Center for Crystal Growth, VIT University, Vellore 632014, Tamilnadu, India
e-mail: varadgm@gmail.com

S. K. Shukla
Department of Biomedical Engineering, Ben-Gurion University of the Negev, Beer Sheva, Israel

M. Naushad (ed.), *A New Generation Material Graphene: Applications in Water Technology*, https://doi.org/10.1007/978-3-319-75484-0_7

and aquatic organisms. The concern to remove these pharmaceutical wastes and their metabolites from wastewater before their final discharge into water bodies has culminated in the development of a wide variety of other treatment technologies such as adsorption, chemical oxidation, liquid extraction, biodegradation, and so on. However, because these pharmaceuticals are mostly water soluble and non-biodegradable, most of the treatment techniques are inappropriate for their effective removal. The deployment of an appropriate technique for effective degradation of pharmaceutical wastes in water has therefore become a necessary requirement. This chapter therefore provides a detailed discussion on pharmaceuticals in general, their occurrence in water and their health consequences. It also delved into the photocatalytic degradation of these chemicals in water with emphasis on the use of graphene based materials.

Keywords Pharmaceuticals · Photocatalytic degradation · Emerging environmental pollutants · Water treatment methods · Graphene pollutants

1 Introduction

The world is faced with a number of challenges. Among these changes is the difficulty in guaranteeing access to potable water. This problem is worsened by the increasing world population, and proliferation of many emerging pollutants which are difficult to remove (Qu et al. 2012). It is estimated that about 884 million people across the world require access to potable water; and about 1.8 million children die annually from water-related illnesses (Qu et al. 2012). These figures are likely to double in the not so distant future due to increase in the quantity and diversity of contaminants being discharged into the natural water bodies driven by a quest for better human life. This scenario presents a big challenge for the current and future generations, posing a significant threat to the well-being of mankind (Dosch 2009).

Environmental and water pollution advocates have raised a big concern over the high rate of water pollution, particularly with pharmaceutical wastes. Pharmaceutical products, despite their usefulness to human and animal health, are now being regarded as emerging environmental pollutants as many of these chemicals are now being detected in water bodies. The main sources of these pollutants into the environment, particularly water bodies, include their indiscriminately discharged through production processes or industrial activities, leakages and leachates from municipal wastewater treatment plants and landfill sites (White and de Groot 2006). In addition, only a few of the drugs consumed get wholly metabolized. The remainders are excreted from the body though the skin as sweat, as urine and feces, and finally find their way into the wastewater (White and de Groot 2006). Some of the drugs identified in water included antidepressants, antibiotics, heart medications such as ACE inhibitors, digoxin, calcium-channel blockers. Others are blood thinners, hormones, painkillers, caffeine, carbamazepine, antiseizures and some fragrance chemicals (tonalide and galaxolide) (Mompelat et al. 2009; Wang et al. 1993).

Even though the health effects of consuming pharmaceutical waste contaminated water is not well documented, toxicity emanating from the prolonged exposure to lower concentrations of mixtures of these pollutants can be devastating (Rodriguez-Mozaz and Weinberg 2010). In addition, many of these pollutants interact with other components of the water and transform into different intermediate products which may exert toxic ecological effects (Tijani et al. 2013). Fishes and some other aquatic organisms have been identified to exhibit both male and female sex features in heavily polluted waters owing to the increase in disposal of estrogen and chemicals that affect the male or female fish species (Woodling et al. 2006). Traces of antibiotics in the water can induce drug resistance tendencies of microorganism resulting in more rampant hospitalization (Guardabassi et al. 2002). Thus, the quality of water for domestic, industrial, and agricultural use is of paramount concern. The deployment of an appropriate technique for effective degradation of pharmaceutical wastes in water has therefore become a necessary requirement.

Because these pharmaceuticals are mostly water soluble and non-biodegradable, most of the conventional treatment techniques are inappropriate for their effective removal. The conventional methods of wastewater treatment were not designed with these emerging pollutants in mind resulting in the discharge of untreated or partially treated wastewater into water bodies (Tambosi et al. 2010). The concern to effectively remove these pharmaceutical wastes and their metabolites from wastewater before their final discharge into water bodies has culminated in the development of a wide variety of other treatment technologies such as membrane separation, chemical oxidation/precipitation, electrochemical separation, liquid coagulation/extraction, biodegradation, and photocatalysis which is the main focus of this chapter.

Photocatalysis has become a powerful way to mitigate the effects of environmental pollution owing to its simple and comprehensive pollutants decomposition tendency (Chen and Ye 2008; Liu et al. 2013; Xu et al. 2011). Metal oxides photocatalysts have attracted interest in research due to their tailor-made bandgap and high photodegradation efficiencies for many organic and inorganic water pollutants (Liu et al. 2014). Combination of graphene materials with metal oxide photocatalysts are known to improve the performance of the catalysts by helping to overcome their inherent problems of ultraviolet light activity, and rapid electron-hole recombination rates (Khan et al. 2017). Numerous favourable properties of graphene makes it a potential promising material for applications in composites for water treatment. It's large surface area, chemical purity, and the possibility for its easy functionalization which makes it link easily with other inorganic and organic materials (Yao et al. 2014, 2012) allow graphene to provide opportunities for water treatment.

Several graphene material based nanocomposites have been synthesized and used efficiently in the photocatalytic degradation of pharmaceuticals in water. The aim of this chapter is to provide a general overview on the use of graphene material based composite for the degradation of pharmaceuticals in water. The chapter also takes into consideration the sources of occurrence of these pharmaceuticals in water, their potential health effects and the unique properties of graphene materials that make them appropriate components of such composites.

2 Pharmaceuticals

Pharmaceuticals are a specific class of synthetic and natural chemicals with bioactive characters that are designed and manufactured purposely to exert therapeutic effects against diseases, or to prevent the occurrence and spreading of disease in humans and animals (Tijani et al. 2013). They may be available as over-the-counter, prescription or veterinary drugs. Increasing number of these chemicals are distributed or dispensed and consumed each year throughout the world due to the emergence of new diseases and discovery of new drugs to cure these emerging diseases. Other factors that cause increasing distribution of pharmaceutical products include increasing population, accessibility/availability of inexpensive drugs as more patents reach their expirations (Daughton 2003). Most of these chemicals are now being regarded as water pollutants since their presence is increasingly being determined in water bodies due to industrial, agricultural, domestic, and veterinary practices; and their likelihood of producing toxic effects on humans, non-target organisms, and microorganisms in the water bodies (Segura et al. 2009).

2.1 Occurrence of Pharmaceuticals in Water

The presence of pharmaceutical products in the aquatic environment has been mainly attributed to anthropogenic activities as research indicates that all pharmaceutical products detected in water bodies, apart from caffeine, are synthetic (Marques et al. 2013; Seiler et al. 1999). Though pharmaceutical products have played significant roles in the prevention, and the treatment of diseases in humans, as well as animals, their presence in the aquatic environment has been a great concert to various organization across the world; including environmental and health organizations (Benotti et al. 2008; Xagoraraki and Kuo 2008). This is because since these chemicals are design specifically to be toxic to disease-causing microorganisms, and interact effectively with receptors in both humans and animals, they have the potential to cause harm to microorganisms and other organisms in the water bodies (Jelić et al. 2012). It is therefore a necessity to understand the modes/routes of entry of pharmaceutical products into water bodies so as to be able to devise appropriate means of avoiding their likely adverse effects on humans, animals, and aquatic lives. Among these routes of entry of pharmaceuticals into water bodies are industrial activities, municipal waste or sewage treatment outlets, landfill leachates and wastewater from hospitals and pharmacies.

2.1.1 Industrial Activities

Industrial activities is a major route of entry of pharmaceutical products into water bodies. Studies conducted in recent times have recognized direct discharges or emissions from manufacturing industries as a major source of pharmaceutical wastes into the environment. Though pollution through industrial activities is not widespread, discharges with the potential to enhance drug-resistant tendencies of microorganisms can create global problems (Larsson 2014). Wastewater emanating from pharmaceutical industries contain a number of these chemicals and their mixtures. These pollutants, either non-degraded or partially degraded (because the treatment plants are ineffective against these micro-pollutants) is finally released into the receiving water bodies. Possible leakages from the sewer system from the industry to the treatment site may also contaminate the land. Runoffs during rain then wash these pollutants into the water bodies. Gaseous or volatile pharmaceuticals may also escape into the atmosphere and bound to air borne particles. They can then directly contaminate surface water bodies through aerial deposition during rain, or settle on land and washed by flood into the water bodies (Burkhardt-Holm 2011). Research conducted in *Patancheru*, an important area for drugs manufacturing in India, confirmed the accession that drugs production activities contribute largely to the presence of pharmaceutical wastes in the environment. The concentration of pharmaceuticals, particularly ciprofloxacin, in the effluent of the plant treating wastewater from these industries was identified to be much higher than those present in patients' blood (Fick et al. 2009; Larsson et al. 2007).

2.1.2 Municipal Waste/Sewage Treatment Plant Outlets

Municipal waste treatment plant effluent is documented as one of the main sources of entry of many pharmaceuticals, and their intermediate products of metabolic reaction into the aquatic environment (Kümmerer 2009). The conventional municipal waste treatment plants were not designed with pharmaceutical wastes in mind, as these immerging pollutants were not an issue of environmental concern at the time these plants were being designed. Some of these pollutants are therefore, either not affected or degraded by the treatment processes. Others may also be converted into other intermediate products which may even be more harmful than the parent chemical (Ellis 2006; Ternes et al. 2004). They then become part of the effluent, which finally is released into the receiving water bodies. According to Ellis (2006), Sewer leakages from drainage systems resulted in the contamination of water bodies with some pharmaceuticals such as diclofenac, clofibric acid, ibuprofen, and iopamidol. One of the ways to handle the sludge generated through the sewage treatment is its application as manure in agricultural fields.

This approach, apart from managing the generated sludge, reduces the cost and the quantity of synthetic fertilizers used on agricultural fields. However, the pharmaceutical waste components of this sludge are washed by agricultural runoff into the nearby water bodies with their subsequent pollution with these chemicals (Passuello et al. 2010). In their study to assess the possible occurrence of pharmaceuticals in wastewater treatment plant streams in Dublin, Lacey et al. (2012) identified propranolol, mefenamic acid, carbamazepine, nimesulide, diclofenac, furosemide, gemfibrozil, clotrimazole and metoprolol as pharmaceuticals present in the three streams assessed.

2.1.3 Landfill Sites/Leachates

In situations where expired or unused pharmaceuticals or medication are not flushed down the toilet, they normally get thrown into the waste bin or household waste and disposed of to landfill sites. At the landfill site, these chemicals are likely to undergo degradation and/or reaction with other reactive pollutants resulting in the generation of other intermediate compounds, which may equally be injurious to human health and the environment. These chemicals then form part of the landfill leachate and enter the ground or nearby surface water (Boxall et al. 2004; Walker et al. 2012).

2.1.4 Wastewater from Hospitals and Pharmacies

Hospitals and pharmacies are major sources of occurrence of pharmaceutical in the aquatic environment. In most cases, drugs do not completely metabolized in the human body. The remaining portions of these drugs and their metabolites are then excreted through feaces and urine into the sewer system (Aksu 2005; Rivera-Utrilla et al. 2013). In addition, some expired and unused/left-over drugs are discarded into the water closet. All these pharmaceutical products then get flushed down through the sewer system into water treatment plants and finally into the surface and groundwater bodies. They can also be directly discharged into the nearby water bodies by floodwater when the unused or expired drugs and urines are indiscriminately discharged into drains. A study conducted in Taiwan by Lin and Tsai (2009) confirmed the occurrence of pharmaceuticals in receiving rivers and wastewaters from hospitals and pharmacies. The study, which was conducted in three rivers and waste streams of six hospitals identified acetaminophen, erythromycin-H_2O, sulfamethoxazole and gemfibrozil as contaminants of the streams and water bodies, with acetaminophen being the most ubiquitous with higher concentrations.

2.2 Health Effects of Pharmaceutical Drugs in Water

Pharmaceutical products, upon entry into the environment, can undergo degradation through biotic and/or abiotic means and adsorb onto suspended particles and sediment. Some can also accumulate into aquatic organism (Ramirez et al. 2009) and become part of the food chain. As far as humans are concerned, drinking of contaminated water is acknowledged as the main route of exposure to pharmaceutical products (Daughton 2001).

There is currently no documented evidence of human health effect in relation to the exposure to pharmaceuticals in water (Jones et al. 2005). However, the likelihood of a potential health effect upon continuous, and low dose chronic exposure to these pollutants is not ruled out (Schwab et al. 2005). This is because individual pharmaceutical products are intended to exert specific effects on targeted parts of the body over a specific period. Therefore, continuous exposure to mixtures of these chemicals over a long period can gradually induce effects that could be noticeable with time. In addition, some of these drugs are gender or age specific. Therefore, the situation where the exposure cuts across age and gender is suspected to create health problem with prolonged exposure. Sensitive groups such as children, the elderly, and pregnant women are at high risks (Kolpin et al. 2002).

The possible effects of these chemicals on aquatic lives is also not clearly understood. The probable effects of these chemicals on the organisms in field situations are often based on laboratory toxicity data. These data may, however, not be the exact representation of what happens in the field (Versteeg et al. 2005). These experiments normally consider the effects of a single pharmaceutical product on one organism. Meanwhile, real-life situations involved simultaneous exposure to mixtures of the pollutants across generations (Trudeau et al. 2005). Unlike humans who experience limited exposure to these compounds usually through drinking of contaminated water, aquatic organisms experience life-time exposure and are therefore more at risk compared to humans. Laboratory studies have indicated that these pollutants have the potential to inflict adverse health effects on aquatic organisms (Brooks et al. 2003). Research has shown that effluents emanating from pharmaceuticals wastewater treatment plants with traces of birth control drugs induced reproduction problems in fish upon exposure. Effluents containing endocrine disrupting drugs, particularly hormones and synthetic steroids, have also been identified as having the tendency of altering sex ratios in aquatic organism, including fish (Schultz et al. 2003). They can also cause male fish to exhibit feminine features (feminization) and other changes that affect the overall health of the aquatic organisms (Schultz et al. 2003; Zillioux et al. 2001). Another concern is the possibility that pathogens may develop resistance to antibiotics making these drugs ineffective in curing several diseases.

3 Graphene and Graphene Materials

Graphene is a modern from of carbon nanomaterials derived from graphite. Its two dimensional structure coupled with its auspicious structural, electrical, optical, and mechanical characteristics make it favorable for use in many fields. It is used as a nano-electric constituent, a component of nano-electrochemical products, energy storage devices, sensors among others (Dosch 2009).

Graphene can be described as the backbone of carbon nanomaterials (Castro Neto et al. 2009). Its structure is described as a hexagonal lattice and is viewed as two interleaving lattices as illustrated in Fig. 1.

Graphene is a very stable carbon material due to the closely packed atoms of carbon and their SP^2 hybridized orbitals in addition to the S, P_X, and P_Y orbitals constituting the α-bond (Cooper et al. 2012). The π bond is made up of the final P_Z electrons. This π bond forms the π band and π* bands that are accountable for graphene's distinguished electronic properties as there is free movement of electrons from the half-filled band (Cooper et al. 2012).

Graphene oxide (GO) (Fig. 2a), is the functionalization product of graphene sheet. Presently, Hummers' method ($KMnO_4$, $NaNO_3$, H_2SO_4) is the technique mostly used for the preparation of graphene oxide (Marcano et al. 2010). (Xiang et al. 2010). Graphene oxide is an electrical insulator; it has the sp^2 bonding of graphene disrupted. To improve the electrical conductivity and retain the honeycomb hexagonal lattice, reduction of the graphene oxide is done to obtain reduced graphene oxide (rGO) (Fig. 2b). The reduced graphene oxide obtained is indispensable. It restores the sp^2 bonding present in pure graphene and produces two dimensional carbon material that allows free movement of charges across con-

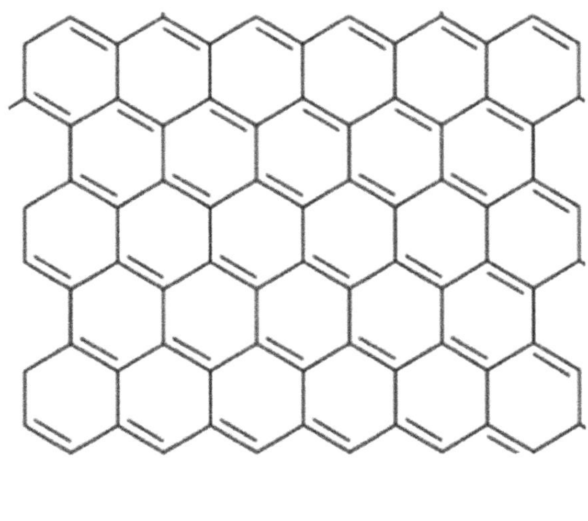

Fig. 1 Lattice structure of graphene

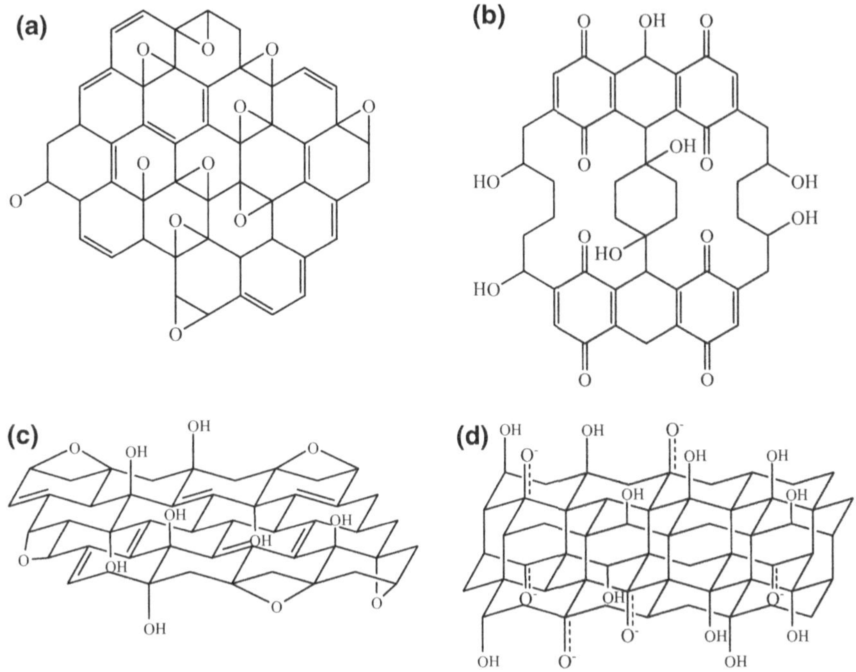

Fig. 2 Structural models of GO proposed by **a** Hofmann, **b** Scholz–Boehm, **c** Ruess and **d** Nakajima-Matsuo (Redrawn from Szabó et al. 2006)

ductive nanomaterials (Lightcap et al. 2010). Graphene oxide has been used in water purification, optoelectronics, coating, and battery electrode and so on. The main advantages of graphene oxide over graphene is its ease of dispersion in organic solvents, water and other inorganic solvents because of the oxygen functionalities (Xiang et al. 2010).

Graphene further yields nanostructures which can be grouped into; quantum dots (GQDs) and nanoribbons (GNR). Graphene nanoribbons can be cut into different widths. This is achieved through electron-beam lithography and an etching mask (Chen et al. 2007; Han et al. 2007). GQDs can be described as small fragments of graphene with all the properties of graphene. However, GQDs normally have dimensions 20 nm in diameter and below. The theoretical study of graphene quantum dots has followed the work on graphene very closely shedding light of the experimentally obtained results. Quantum dots can be made through cutting graphene sheets using a top-down method (Simpson et al. 2002; Wu et al. 2004). GQDs have been used in energy conversion (Zhang et al. 2012), sensors production (Dong et al. 2010) and bioanalysis (Dong et al. 2010).

3.1 Properties of Graphene Materials for Use in Environmental Applications

3.1.1 Optical Properties

Graphene absorbs maximally in the ultraviolet (UV) wavelength (Fig. 3). However, there is a significant absorbance in the visible region of the solar spectrum as shown in the wavelength between 320 and 800 nm. This shows graphene's ability to make maximum utilization of the solar energy which span a whole range of spectrum (Falkovsky 2008). Thus, the tendency of graphene to absorb light across a wide range of the electromagnetic spectrum enables it to enhance the degradation properties of photocatalyst when combined with them.

3.1.2 Electrical Properties

Perfect graphene has a zero bandgap in band structure (Fig. 4). The zero band gap of pristine graphene makes it a perfect conductor with high conductivity. Electron arrangement in the graphene lattice is dominated by the S and P orbitals. Much research has been done on how to tune the band gap in graphene-based semiconductor materials (Gupta et al. 2013) in order to convert graphene from a perfect conductor into a semiconductor. The band gap tuning can be achieved in pristine graphene via nanopatterning (the fabrication of a nanoscale pattern, especially as part of an electronic component) (Giovannetti et al. 2007; Han et al. 2007),

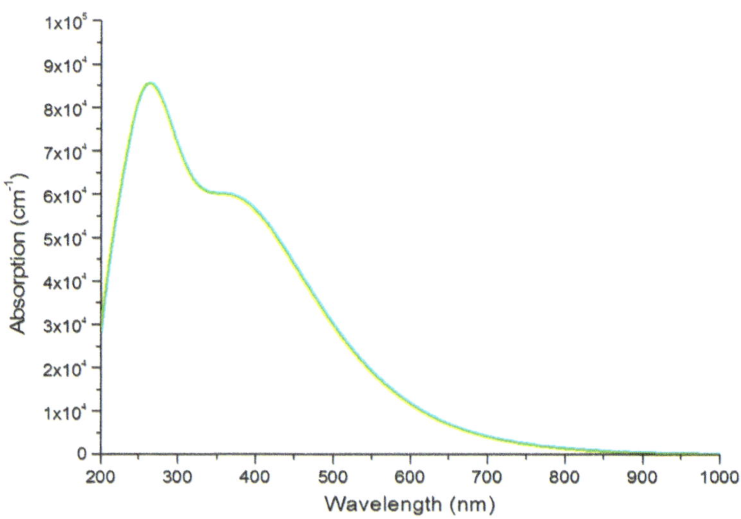

Fig. 3 Graphene optical properties

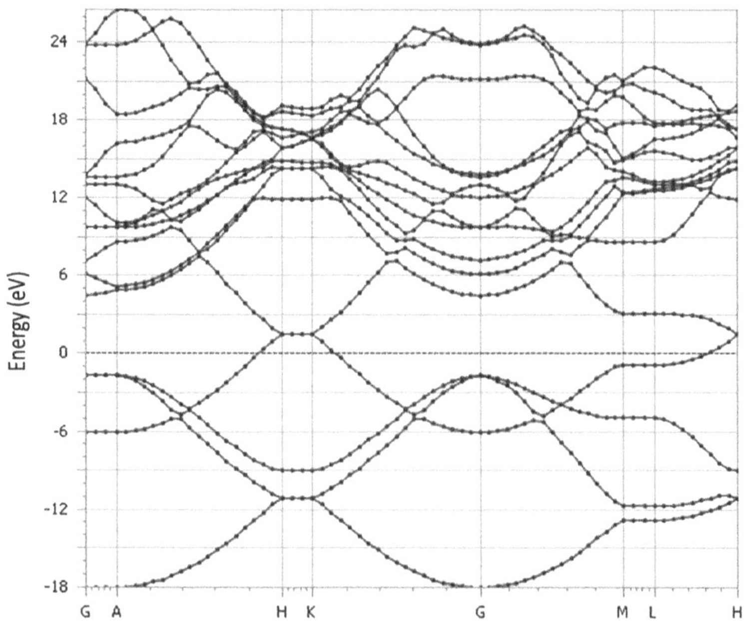

Fig. 4 Graphene zero bandgap

application of gate voltage (Avetisyan et al. 2009; Dragoman et al. 2010; Model 2009; Molitor et al. 2008), or chemical functionalization (Elias et al. 2009; Jung et al. 2008; Wu et al. 2008).

The electronic properties of GO depend on its chemical structure (He et al. 1998; Kumar et al. 2014; Mermoux et al. 1991). The epoxide functional groups on GO significantly induce the local distortion of graphene with a new bond formed by graphene and oxygen atoms. This affects the bonding characteristics of carbon, changing from planar sp^2 to partial sp^3 hybridization. Considering an arrangement of epoxy functional groups in fully oxidized graphene sheet and the effect of epoxy arrangement on electronic properties, there was significant induction of bandgap of 0.529 eV (Lin-Hui et al. 2001). Hybridization of graphene materials with photocatalysts will therefore result in high photodegradation process as the graphene materials exhibit efficient electron conductivity coupled with the ability to minimize charge carriers recombination rate through acceptance of photogenerated electrons.

3.1.3 Mechanical Properties

Graphene is the thinnest and lightest compound known to man (one atom thin) and thus grows into paper-like sheets. It forms the strongest compound discovered to date (graphene's ultimate tensile strength is 130 billion Pascal compared to 400 million for A36 steel or 375.7 million for Aramid (kelwar)) (Dhiman and Ashutosh

2014). Graphene is the lightest at 0.77 mg per square sheet (Rodriguez-Mozaz and Weinberg 2010). This can be qualified into three key features which determines the general mechanical properties of graphene and its nanocomposites: (1) interlayer crosslinks involving mostly the functionalised graphene where the side atoms are involved in covalent bonding mostly with other side atoms in the sheet, or some weak Van Der Waals forces between the carbon-carbon graphene in two separate sheets forming layers, and (2) intralayer mechanical properties, defined by sp^2 carbon-carbon covalent bonds and crosslinks at graphene edges (Liu et al. 2012). Graphene materials are therefore regarded as appropriate materials for catalyst support as a result of their outstanding mechanical properties and high specific surface area which promotes their propensity to adsorb and disperse pollutants for resounding degradation process.

4 Photocatalysis

Photocatalysis refers to the occurrence of a chemical reaction through the use of light as the source of energy, and metal oxide catalysts (photocatalysts) as absorbents of photons of the light to initiate and speed up the reaction (Divya et al. 2014). Photocatalysis is known to have been deployed in solving many environmental problems including the degradation of organic and inorganic pollutant in water as well as disinfection of water (Saggioro et al. 2011). This technique is hailed as the appropriate method for the degradation of many water pollutants because of its ability to completely degrade the pollutant to less harmful products without the generation of any secondary pollutants such as sludge. Other advantages of photocatalysis include its environmental friendliness, cost effectiveness and the fact that it can be operated in areas without electricity (Yu et al. 2011).

Metal oxide photocatalysts normally have filled valence band and empty conduction bands. When these photocatalysts are irradiated with photons of appropriate energy, electrons on the valence band acquire enough energy and become exited and promoted to the conduction band. This occurrence leads to the creation of positive charges (holes) on the valence band and the accumulation of electrons on the conductions band (Fig. 5) (Tryk et al. 2000). It is the creation of the electron-hole pairs that initiates the photocatalytic degradation process. The excited electrons on the conduction band have the tendency to fall back to the valence band and recombine with the holes causing the reaction to cease. Otherwise, they participate in reactions with oxygen molecules (O_2) and reduce them to superoxides ($^{\cdot}O_2^{-}$).

$$O_2 + e^- \rightarrow O_2^{\cdot -}$$

On the other hand, the holes on the valence band also attacks water molecules and oxidize them to hydroxyl radicals ($^{\cdot}OH$).

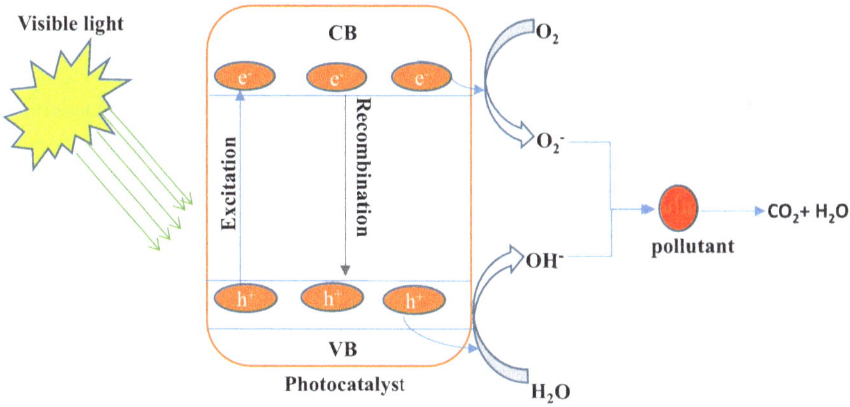

Fig. 5 Mechanism of photocatalytic degradation of pollutants in water

$$H_2O + h^+ \rightarrow HO^{\cdot} + H^+$$

The resultant $^{\cdot}O_2^-$ and $^{\cdot}OH$ (excellent oxidizing agents) then oxidize the pollutants or degrade them into various products. The end products of the photodegradation process is usually carbon dioxide and water provided the reaction is allowed to occur over a long enough period of time (Dung et al. 2005).

4.1 Photocatalytic Degradation of Pharmaceutical Products

As long as various domestic activities, recreation, food supply, and so on continue to rely on the use of surface and ground water bodies, the accumulation of pharmaceuticals in these water bodies will remain a matter of high public health concern. Conventional methods of wastewater treatment, most of which use activated sludge procedure, have been identified to exhibit limited efficiency in the removal of pharmaceutical wastes from polluted wastewater (Li and Randak 2009). Thus, the development of innovative wastewater treatment techniques that are environmentally benign and capable of complete degradation of these pollutants has become a paramount issue in the scientific domain. Heterogeneous semiconductor photocatalysis has been aggressively tested in recent years for its ability to completely irradiate these pollutants from wastewater dues to its numerous advantages as discussed in the previous section.

4.1.1 Graphene Based Materials for Pharmaceutical Products Photocatalytic Degradation

Despite the numerous advantages of photocatalytic degradation of pharmaceuticals in water, the effectiveness of the process in hindered by certain factors. Among these factors are the fast rate at which the photogenerated electrons and holes recombine. Another problem is the fact that most of the photocatalysts are only active under ultraviolet (UV) light rather than visible light (which is more abundant). Thus, in order to attain an efficient photocatalytic degradation process, the photocatalyst must be active under visible light with reduced electron-hole recombination rate. It should be non-toxic with excellent pollutants adsorption property (Doll and Frimmel 2005). It has been identified that one of the best ways to produce an effective photocatalyst is to modify the traditional photocatalyst with graphene materials. Graphene has excellent pollutants adsorption property due its large specific surface area coupled with delocalized pi-electron system that enhances its interaction with the pollutants (Wang et al. 2013). Graphene materials based photocatalysts have been identified to exhibit higher photocatalytic degradation efficiencies compared to the non-graphene based ones due to fact that the later possess large specific surface area with profound pollutants adsorption property, ability to minimize the rate of recombination of the photogenerated electron-hole pairs, and outstanding visible light activity (Oppong et al. 2016). Different types of graphene based materials modified metal oxide photocatalysts have been employed in the degradation of pharmaceuticals in water. The subsequent sections give detail discussions on the use of graphene based TiO_2 photocatalysts, coupled graphene based metal oxide photocatalysts, and other graphene materials based photocatalysts in photocatalytic degradation of pharmaceutical in wastewater.

Use of Graphene Based TiO_2 Photocatalysts

Titanium dioxide (TiO_2) is undoubtedly the most widely used photocatalyst for the degradation of many pollutants in water as a result of fact that this catalyst has high photocatalytic activity, is less toxic and relatively cheap. Modification of TiO_2 with graphene materials is known to improve its visible light activity and reduce the photogenerated electron-hole recombination rate thereby improving its photocatalytic activity. The use of graphene based titanium dioxide photocatalyst in the degradation of pharmaceuticals in water is regarded as one of the appropriate approaches for the effective irradiation of these pollutants. Lin et al. (2017) conducted a study by comparing the effectiveness of TiO_2-rGO and TiO_2-Fe in the degradation of ibuprofen, carbamazepine and sulfamethoxazole in aqueous solution under visible light. TiO_2-rGO was noted to demonstrate higher photodegradation activity against all the pharmaceutical products. The comparatively higher performance of the TiO_2-rGO was attributed to the presence of rGO in the composite. The rGO accepted the photogenerated electrons and thus reduced the recombination rate of the electron-hole pairs. Photocatalytic degradation of 10 ppm carbamazepine aqueous solution with rGO-TiO_2 was also observed to achieve 99% efficiency within 90 min. This value was twice that achieved by the bare TiO_2 due to the availability of abundant surface area for adsorption of the pollutants and effective charge carriers separation and transportation (Nawaz et al. 2017). A graphene oxide modified TiO_2 composite (GO-TiO_2) was synthesis through liquid phase deposition and employed to photocatalytically degrade diphenhydramine under ultraviolet and visible light irradiation (Pastrana-Martínez et al. 2012). The GO content was noted to be responsible for the composite's large surface area, and its porous nature. The GO was also noted to improve the photocatalytic activity of the GO-TiO_2 compare to that of the bare P25. This was due to the ability of GO to effectively capture the photogenerated electrons from the TiO_2 and prevent them from recombining with the holes, enhancing the adsorption of the pollutants due to its large surface area, and by extending the visible light absorption of the composite to longer wavelengths. Similar research work involving the use of GO-TiO_2 in the degradation of diphenhydramine has also been conducted by Morales-Torres et al. (2013) with the GO-TiO_2 demonstrating higher photocatalytic activity against the pollutant.

Use of Coupled Graphene Based Metal Oxide Photocatalysts

Coupled metal oxide photocatalysts immobilized on graphene materials have also been effectively used for the photocatalytic degradation of pharmaceuticals in water. An example is the $ZnFe_2O_4$ coupled Ag nanocomposites synthesized by Khadgi et al. (2016). The coupled nanocomposites were immobilized on reduced graphene oxide (rGO) to form $ZnFe_2O_4$-Ag/rGO nanocomposite, which was effectively deployed in the photocatalytic degradation of 17a-ethinylestradiol, an endocrine disrupting compound. The $ZnFe_2O_4$-Ag/rGO was observed to be photocatalytically effective than the $ZnFe_2O_4$ alone and the $ZnFe_2O_4$-Ag. They attributed the effectiveness of the $ZnFe_2O_4$-Ag/rGO nanocomposite to the presence of the rGO, which minimized agglomeration of the nanocomposites. The rGO also enhanced the composite's visible light utilization, its effective charge carriers generation and transfer/separation, its specific surface area with the accompanying higher pollutant adsorption capacity. Comparatively, the $ZnFe_2O_4$-Ag/rGO demonstrated about 5.6 time higher photocatalytic activity than the $ZnFe_2O_4$-Ag. Similarly, oxytetracycline and ampicillin have been successfully degraded in aqueous solution through visible light photocatalysis using a nanocomposite consisting of Bi_2WO_6/Fe_3O_4 immobilized on graphene and sand (Bi_2WO_6/Fe_3O_4/GSC) (Raizada et al. 2017). The Bi_2WO_6/Fe_3O_4/GSC was noted to degrade 95 and 94% of ampicillin and oxytetracycline respectively. On the other hand, the nanocomposite without graphene (Bi_2WO_6/Fe_3O_4) degraded 74 and 71% respectively, of ampicillin and oxytetracycline, all within 60 min. The results pointed out the significant role played by graphene in enhancing the pollutant adsorption capacity of the composite. The effectiveness of coupled graphene based photocatalysts in the photocatalytic degradation of pharmaceuticals in water has also been demonstrated by the use of a silica coated Fe_3O_4-TiO_2 for efficient degradation of caffeine and carbamazepine in aqueous solution. The composite exhibited higher photocatalytic activity compared to commercial P25 with additional advantages of high recoverability and reusability (Linley et al. 2014). Fe_3O_4/Mn_3O_4-rGO nanocomposite was also used for a comprehensive photocatalytic degradation of aqueous sulfamethazine solution. The results revealed 99% sulfamethazine degradation efficiency at optimum conditions of 0.07 mm/L sulfamethazine concentration, 0.5 g/L of Fe_3O_4/Mn_3O_4-rGO nanocomposites, 35 °C, pH3 and hydrogen peroxide concentration of 6 mM (Wan and Wang 2017). Other studies carried out includes tracycline photodegradation with rGO-CdS/ZnS (Tang et al. 2015) and quinolone photodegradation with Fe_3O_4@Bi_2O_3-rGO (Zhu et al. 2017).

Use of Other Graphene Materials Based Photocatalysts

A nanocomposite consisting of graphene oxide decorated cerium molybdate nanocubes (Ce (MoO_4)$_2$/GO)) was synthesized and utilized in the photocatalytic degradation of chloramphenicol, an antibiotic, under visible light irradiation (Karthik et al. 2017). The Ce(MoO_4)$_2$/GO composite displayed excellent photodegradation potential against the drug and showed higher degradation efficiency (99% within 50 min) than the pure Ce(MoO_4)$_2$ nanocubes. The impressive performance of the composites was assigned to the excellent separation of the photogenerated electrons and holes. With the same intension of photocatalytic degradation of pharmaceuticals in mind, (Anirudhan et al. 2017) fabricated a graphene oxide modified nanohydroxyapatite composite through facile chemical modification of graphene oxide by triethyltetramine and applied it in the removal of aureomycine hydrochloride through adsorption and subsequent photocatalytic degradation. The band gap of GO was observed to improve to 2.4 eV after its modification with the triethyltetramine and enhanced the composites interaction with the pollutant. A sample of the real aureomycine hydrochloride polluted wastewater from the nearby poultry and animal farms was used as the source of aureomycine hydrochloride. The photodegradation experiment was performed under visible light irradiation with successful results. The aureomycine hydrochloride photodegradation was noted to have followed the pseudo first-order kinetics.

Similarly, Anirudhan et al. (2017) degraded ciprofloxacin hydrochloride aqueous solution through visible light photocatalysis using a nano zinc oxide incorporated graphene oxide/nanocellulose composites (ZnO-GO/NC). They observed that the band gap of the GO was tunable to 2.4 eV as a result of incorporation of the ZnO into the GO and further increased to 2.8 eV upon the introduction of the nanocellulose. Their analysis revealed that the composite displayed a maximum photocatalytic degradation efficiency of 98% at optimum pH of 6. The composite exhibited the tendency of being recycled and reused five consecutive times without significantly losing its efficacy. The improved photocatalytic efficiency of the

ZnO-GO/NC was attributed to the improvement in its excited electrons generation and conduction resulting in minimized charge carriers recombination rate due to the presence of GO. Tang et al. (2017) examined the effectiveness of $BiVO_4$ sensitized graphene quantum dots (GQD/$BiVO_4$) in the photocatalytic degradation of carbamazepine while Aalicanoglu and Sponza (2015) used graphene oxide magnetite (GO-Fe_3O_4) in the photocatalytic degradation of ciprofloxacin. In both cases, the GQD/$BiVO_4$ and GO-Fe_3O_4 showed improved optical properties and were identified to be more effective than the bare $BiVO_4$ and Fe_3O_4 due to the presence of the graphene materials. Inorganic salts assisted hydrothermal method was also deployed in the synthesis of $Bi_{3.84}W_{0.16}O_{6.24}$-graphene oxide (BWO-GO). The composite showed successful results when used for the degradation of tetracycline (Song et al. 2016).

5 Conclusion

The need for the development of more effective methods for eradication of pharmaceuticals from water has become imperative as these pollutants possess the tendency to exert devastating health effects on humans as well as aquatic organisms; and due to the inefficiency of the traditional water treatment techniques to effectively remove these pollutants from polluted water. The routes of entry of pharmaceuticals in water, and their health effects on both human and wildlife have been discussed in this chapter. Detailed discussion on graphene and graphene materials, and the importance of using graphene materials based photocatalysts have also been discussed. Our analysis revealed that application of graphene materials based photocatalyst for the degradation of pharmaceuticals in water is an effective approach as most of the available data indicate outstanding results. However, we observed that limited data is available in this regard. We therefore recommend the synthesis and application of other graphene materials based metal oxide photocatalyst, apart from TiO_2, in the photocatalytic degradation of pharmaceuticals in water.

References

Aalicanoglu P, Sponza D (2015) Removal of ciprofloxacin antibiotic with nano graphene oxide magnetite: comparison of adsorption and photooxidation processes. In: Proceedings of the 14th international conference on environmental science and technology

Aksu Z (2005) Application of biosorption for the removal of organic pollutants: a review. Process Biochem 40:997–1026

Anirudhan T, Deepa J, Nair AS (2017) Fabrication of chemically modified graphene oxide/nano hydroxyapatite composite for adsorption and subsequent photocatalytic degradation of aureomycine hydrochloride. J Ind Eng Chem 47:415–430

Avetisyan AA, Partoens B, Peeters FM (2009) Electric-field control of the band gap and Fermi energy in graphene multilayers by top and back gates. Phys Rev B 80:1–11

Benotti MJ, Trenholm RA, Vanderford BJ, Holady JC, Stanford BD, Snyder SA (2008) Pharmaceuticals and endocrine disrupting compounds in US drinking water. Environ Sci Technol 43:597–603

Boxall AB, Sinclair CJ, Fenner K, Kolpin D, Maund SJ (2004) When synthetic chemicals degrade in the environment-what are the absolute fate, effects, and potential risks to humans and the ecosystem? Environ Sci Technol 38:368A–375A

Brooks B, Smith W, Blank C, Weston J, Slattery M, Foran C (2003) Pharmaceutical neuromodulation of teleost catecholamines. In: 24th Annual meeting, SETAC North America. Presented on November 12th Austin

Burkhardt-Holm P (2011) Linking water quality to human health and environment: the fate of micropollutants. Inst Water Policy Natl Univ Singapore 1–62

Castro Neto AH, Guinea F, Peres NMR, Novoselov KS, Geim AK (2009) The electronic properties of graphene. Rev Mod Phys 81:109–162 https://doi.org/10.1103/revmodphys.81.109

Chen D, Ye J (2008) Hierarchical WO_3 hollow shells: dendrite, sphere, dumbbell, and their photocatalytic properties. Adv Funct Mater 18:1922–1928

Chen Z, Lin Y-M, Rooks MJ, Avouris P (2007) Graphene nano-ribbon electronics. Phys E: Low Dimens Syst Nanostruct 40:228–232

Cooper DR, D'Anjou B, Ghattamaneni N, Harack B, Hilke M, Horth A, Majlis N, Massicotte M, Vandsburger L, Whiteway E, Yu V (2012) Experimental review of graphene. ISRN Condens Matter Phys 2012:1–56

Daughton CG (2001) Pharmaceuticals and personal care products in the environment: overarching issues and overview. In: ACS symposium series, vol 791. ACS Publications, Washington, DC, pp 2–38

Daughton CG (2003) Cradle-to-cradle stewardship of drugs for minimizing their environmental disposition while promoting human health. I. Rationale for and avenues toward a green pharmacy. Environ Health Perspect 111:757

Dhiman L, Ashutosh D (2014) Multifaceted graphene: novelty in electronics. Int J Adv Res Electr Electron Instrum Eng 3:11807–11811

Divya K, Umadevi T, Mathew S (2014) Graphene-based semiconductor nanocomposites for photocatalytic applications. J Nanosci Lett 4:21–55

Doll TE, Frimmel FH (2005) Removal of selected persistent organic pollutants by heterogeneous photocatalysis in water. Catal Today 101:195–202

Dong H, Gao W, Yan F, Ji H, Ju H (2010) Fluorescence resonance energy transfer between quantum dots and graphene oxide for sensing biomolecules. Anal Chem 82:5511–5517

Dosch H (2009) Gennesys white paper: a new European partnership between nanomaterials science & nanotechnology and synchrotron radiation and neutron facilities. Max-Planck-Institut für Metallforschung, Stuttgart

Dragoman D, Dragoman M, Plana R, Dragoman D, Dragoman M, Plana R (2010) Tunable electrical superlattices in periodically gated bilayer graphene. J Appl Phys 107:044312

Dung NT, Van Khoa N, Herrmann J-M (2005) Photocatalytic degradation of reactive dye RED-3BA in aqueous TiO_2 suspension under UV-visible light. Int J Photoenergy 7:11–15

Elias DC, Nair RR, Mohiuddin TM, Morozov SV, Blake P, Halsall MP, Ferrari AC, Boukhvalov DW, Katsnelson MI, Geim AK, Novoselov KS (2009) Control of graphene's properties by reversible hydrogenation: evidence for graphane. Science 323:610–613. https://doi.org/10.1126/science.1167130

Ellis JB (2006) Pharmaceutical and personal care products (PPCPs) in urban receiving waters. Environ Pollut 144:184–189

Falkovsky L (2008) Optical properties of graphene. In: Journal of physics: conference series, vol 1. IOP Publishing, p 012004

Fick J, Söderström H, Lindberg RH, Phan C, Tysklind M, Larsson D (2009) Contamination of surface, ground, and drinking water from pharmaceutical production. Environ Toxicol Chem 28:2522–2527

Giovannetti G, Khomyakov PA, Brocks G, Kelly PJ, Van Den Brink J (2007) Substrate-induced band gap in graphene on hexagonal boron nitride: *ab initio* density functional calculations. Phys Rev B Condens Matter Mater Phys 76:2–5. https://doi.org/10.1103/PhysRevB.76.073103
Guardabassi L, Wong DMLF, Dalsgaard A (2002) The effects of tertiary wastewater treatment on the prevalence of antimicrobial resistant bacteria. Water Res 36:1955–1964
Gupta SK, Soni HR, Jha PK (2013) Electronic and phonon bandstructures of pristine few layer and metal doped graphene using first principles calculations. AIP Adv 3:032117
Han MY, Özyilmaz B, Zhang Y, Kim P (2007) Energy band-gap engineering of graphene nanoribbons. Phys Rev Lett 98:206805
He H, Klinowski J, Forster M, Lerf A (1998) A new structural model for graphite oxide. Chem Phys Lett 287:53–56
Jelić A, Gros M, Petrović M, Ginebreda A, Barceló D (2012) Occurrence and elimination of pharmaceuticals during conventional wastewater treatment. In: Emerging and priority pollutants in rivers. Springer, pp 1–23
Jones OA, Lester JN, Voulvoulis N (2005) Pharmaceuticals: a threat to drinking water? Trends Biotechnol 23:163–167
Jung I, Dikin DA, Piner RD, Ruoff RS (2008) Tunable electrical conductivity of individual graphene oxide sheets reduced at "Low" temperatures. Nano Lett 8:4283–4287
Karthik R, Vinoth Kumar J, Chen S-M, Karuppiah C, Cheng Y-H, Muthuraj V (2017) A study of electrocatalytic and photocatalytic activity of cerium molybdate nanocubes decorated graphene oxide for the sensing and degradation of antibiotic drug chloramphenicol. ACS Appl Mater Interfaces 9:6547–6559
Khadgi N, Li Y, Upreti AR, Zhang C, Zhang W, Wang Y, Wang D (2016) Enhanced photocatalytic degradation of 17α-ethinylestradiol exhibited by multifunctional $ZnFe_2O_4$–Ag/rGO nanocomposite under visible light. Photochem Photobiol 92:238–246
Khan A, Wang J, Li J, Wang X, Chen Z, Alsaedi A, Hayat T, Chen Y, Wang X (2017) The role of graphene oxide and graphene oxide-based nanomaterials in the removal of pharmaceuticals from aqueous media: a review. Environ Sci Pollut Res 24:7938–7958
Kolpin DW, Furlong ET, Meyer MT, Thurman EM, Zaugg SD, Barber LB, Buxton HT (2002) Pharmaceuticals, hormones, and other organic wastewater contaminants in US streams, 1999–2000: a national reconnaissance. Environ Sci Technol 36:1202–1211
Kumar PV, Bardhan NM, Tongay S, Wu J, Belcher AM, Grossman JC (2014) Scalable enhancement of graphene oxide properties by thermally driven phase transformation. Nat Chem 6:151–158
Kümmerer K (2009) Antibiotics in the aquatic environment–a review–part I. Chemosphere 75:417–434
Lacey C, Basha S, Morrissey A, Tobin JM (2012) Occurrence of pharmaceutical compounds in wastewater process streams in Dublin, Ireland. Environ Monit Assess 184:1049
Larsson DJ (2014) Pollution from drug manufacturing: review and perspectives. Phil Trans R Soc B 369:20130571
Larsson DJ, de Pedro C, Paxeus N (2007) Effluent from drug manufactures contains extremely high levels of pharmaceuticals. J Hazard Mater 148:751–755
Li Z, Randak T (2009) Residual pharmaceutically active compounds (PhACs) in aquatic environment–status, toxicity and kinetics: a review. Vet Med (Praha) 52:295–314
Lightcap IV, Kosel TH, Kamat PV (2010) Anchoring semiconductor and metal nanoparticles on a two-dimensional catalyst mat. Storing and shuttling electrons with reduced graphene oxide. Nano Lett 10:577–583
Lin-Hui Y, Bang-Gui L, Ding-Sheng W (2001) *Ab initio* molecular dynamics study on small carbon nanotubes. Chin Phys Lett 18:1496
Lin AY-C, Tsai Y-T (2009) Occurrence of pharmaceuticals in Taiwan's surface waters: impact of waste streams from hospitals and pharmaceutical production facilities. Sci Total Environ 407:3793–3802

Lin L, Wang H, Jiang W, Mkaouar AR, Xu P (2017) Comparison study on photocatalytic oxidation of pharmaceuticals by TiO_2-Fe and TiO_2-reduced graphene oxide nanocomposites immobilized on optical fibers. J Hazard Mater 333:162–168
Linley S, Liu Y, Ptacek CJ, Blowes DW, Gu FX (2014) Recyclable graphene oxide-supported titanium dioxide photocatalysts with tunable properties. ACS Appl Mater Interfaces 6:4658–4668
Liu Y, Xie B, Zhang Z, Zheng Q, Xu Z (2012) Mechanical properties of graphene papers. J Mech Phys Solids 60:591–605
Liu Y, Jiao Y, Yin B, Zhang S, Qu F, Wu X (2013) Hierarchical semiconductor oxide photocatalyst: a case of the SnO_2 microflower. Nano-Micro Lett 5:234–241
Liu Y, Jiao Y, Zhang Z, Qu F, Umar A, Wu X (2014) Hierarchical SnO_2 nanostructures made of intermingled ultrathin nanosheets for environmental remediation, smart gas sensor, and supercapacitor applications. ACS Appl Mater Interfaces 6:2174–2184
Marcano DC, Kosynkin DV, Berlin JM, Sinitskii A, Sun Z, Slesarev A, Alemany LB, Lu W, Tour JM (2010) Improved synthesis of graphene oxide. ACS Nano 4:4806–4814
Marques RR, Sampaio MJ, Carrapiço PM, Silva CG, Morales-Torres S, Dražić G, Faria JL, Silva AM (2013) Photocatalytic degradation of caffeine: developing solutions for emerging pollutants. Catal Today 209:108–115
Mermoux M, Chabre Y, Rousseau A (1991) FTIR and ^{13}C NMR study of graphite oxide. Carbon 29:469–474
Model II (2009) Graphene bilayer field-effect phototransistor for terahertz and infrared detection. Phys Rev B 79:245311
Molitor F, Jacobsen A, Stampfer C, Güttinger J, Ihn T, Ensslin K (2008) Transport gap in side-gated graphene constrictions. Phys Rev B 79:1–5
Mompelat S, Le Bot B, Thomas O (2009) Occurrence and fate of pharmaceutical products and by-products, from resource to drinking water. Environ Int 35:803–814
Morales-Torres S, Pastrana-Martínez LM, Figueiredo JL, Faria JL, Silva AM (2013) Graphene oxide-P25 photocatalysts for degradation of diphenhydramine pharmaceutical and methyl orange dye. Appl Surf Sci 275:361–368
Nawaz M, Miran W, Jang J, Lee DS (2017) One-step hydrothermal synthesis of porous 3D reduced graphene oxide/TiO_2 aerogel for carbamazepine photodegradation in aqueous solution. Appl Catal B 203:85–95
Oppong SO-B, Anku WW, Shukla SK, Agorku ES, Govender PP (2016) Photocatalytic degradation of indigo carmine using Nd-doped TiO_2-decorated graphene oxide nanocomposites. J Sol-Gel Sci Technol 80:38–49. https://doi.org/10.1007/s10971-016-4062-8
Passuello A, Mari M, Nadal M, Schuhmacher M, Domingo JL (2010) POP accumulation in the food chain: integrated risk model for sewage sludge application in agricultural soils. Environ Int 36:577–583
Pastrana-Martínez LM, Morales-Torres S, Likodimos V, Figueiredo JL, Faria JL, Falaras P, Silva AM (2012) Advanced nanostructured photocatalysts based on reduced graphene oxide–TiO_2 composites for degradation of diphenhydramine pharmaceutical and methyl orange dye. Appl Catal B 123:241–256
Qu X, Brame J, Li Q, Alvarez PJ (2012) Nanotechnology for a safe and sustainable water supply: enabling integrated water treatment and reuse. Acc Chem Res 46:834–843
Raizada P, Kumari J, Shandilya P, Dhiman R, Singh VP, Singh P (2017) Magnetically retrievable Bi_2WO_6/Fe_3O_4 immobilized on graphene sand composite for investigation of photocatalytic mineralization of oxytetracycline and ampicillin. Process Saf Environ Prot 106:104–116
Ramirez AJ, Brain RA, Usenko S, Mottaleb MA, O'Donnell JG, Stahl LL, Wathen JB, Snyder BD, Pitt JL, Perez-Hurtado P (2009) Occurrence of pharmaceuticals and personal care products in fish: results of a national pilot study in the United States. Environ Toxicol Chem 28:2587–2597
Rivera-Utrilla J, Sánchez-Polo M, Ferro-García MÁ, Prados-Joya G, Ocampo-Pérez R (2013) Pharmaceuticals as emerging contaminants and their removal from water. A review. Chemosphere 93:1268–1287

Rodriguez-Mozaz S, Weinberg HS (2010) Meeting report: pharmaceuticals in water–an interdisciplinary approach to a public health challenge. Environ Health Perspect 118:1016–1020

Saggioro EM, Oliveira AS, Pavesi T, Maia CG, Ferreira LFV, Moreira JC (2011) Use of titanium dioxide photocatalysis on the remediation of model textile wastewaters containing azo dyes. Molecules 16:10370–10386

Schultz IR, Skillman A, Nicolas JM, Cyr DG, Nagler JJ (2003) Short-term exposure to 17α-ethynylestradiol decreases the fertility of sexually maturing male rainbow trout (*Oncorhynchus mykiss*). Environ Toxicol Chem 22:1272–1280

Schwab BW, Hayes EP, Fiori JM, Mastrocco FJ, Roden NM, Cragin D, Meyerhoff RD, Vincent J, Anderson PD (2005) Human pharmaceuticals in US surface waters: a human health risk assessment. Regul Toxicol Pharm 42:296–312

Segura PA, François M, Gagnon C, Sauvé S (2009) Review of the occurrence of anti-infectives in contaminated wastewaters and natural and drinking waters. Environ Health Perspect 117:675

Seiler RL, Zaugg SD, Thomas JM, Howcroft DL (1999) Caffeine and pharmaceuticals as indicators of waste water contamination in wells. Ground Water 37:405–410

Simpson CD, Brand JD, Berresheim AJ, Przybilla L, Räder HJ, Müllen K (2002) Synthesis of a giant 222 carbon graphite sheet. Chem Eur J 8:1424–1429

Song C, Li X, Wang L, Shi W (2016) Fabrication, characterization and response surface method (RSM) optimization for tetracycline photodegration by $Bi_{3.84}W_{0.16}O_{6.24}$-graphene oxide (BWO-GO). Sci Rep 6:37466

Szabó T, Berkesi O, Forgó P, Josepovits K, Sanakis Y, Petridis D, Dékány I (2006) Evolution of surface functional groups in a series of progressively oxidized graphite oxides. Chem Mater 18:2740–2749. https://doi.org/10.1021/cm060258+

Tambosi JL, Yamanaka LY, José HJ, Moreira RFPM, Schröder HF (2010) Recent research data on the removal of pharmaceuticals from sewage treatment plants (STP). Quim Nova 33:411–420

Tang Y, Liu X, Ma C, Zhou M, Huo P, Yu L, Pan J, Shi W, Yan Y (2015) Enhanced photocatalytic degradation of tetracycline antibiotics by reduced graphene oxide–CdS/ZnS heterostructure photocatalysts. New J Chem 39:5150–5160

Tang L, Wang J-j, Jia C-t, Lv G-x, Xu G, Li W-t, Wang L, Zhang J-y, Wu M-h (2017) Simulated solar driven catalytic degradation of psychiatric drug carbamazepine with binary $BiVO_4$ heterostructures sensitized by graphene quantum dots. Appl Catal B 205:587–596

Ternes TA, Joss A, Siegrist H (2004) Scrutinizing pharmaceutical and personal care products in wastewater treatment. Environ Sci Technol 38:393A–399A

Tijani JO, Fatoba OO, Petrik LF (2013) A review of pharmaceuticals and endocrine-disrupting compounds: sources, effects, removal, and detections. Water Air Soil Pollut 224:1770

Trudeau VL, Metcalfe CD, Mimeault C, Moon TW (2005) Pharmaceuticals in the environment: drugged fish? Biochem Mol Biol Fishes 6:475–493

Tryk D, Fujishima A, Honda K (2000) Recent topics in photoelectrochemistry: achievements and future prospects. Electrochim Acta 45:2363–2376

Versteeg DJ, Alder AC, Cunningham VL, Kolpin DW, Murray-Smith R, Ternes T (2005) Environmental exposure modeling and monitoring of human pharmaceutical concentrations in the. In: Human pharmaceuticals: assessing the impacts on aquatic ecosystems. SETAC, Pensacola, Florida, USA, pp 71–110

Walker CH, Sibly R, Hopkin S, Peakall DB (2012) Principles of ecotoxicology, 2nd edn. Taylor & Francis, London

Wan Z, Wang J (2017) Degradation of sulfamethazine using Fe_3O_4-Mn_3O_4/reduced graphene oxide hybrid as Fenton-like catalyst. J Hazard Mater 324:653–664

Wang J, Lin Y, Chen L (1993) Organic-phase biosensors for monitoring phenol and hydrogen peroxide in pharmaceutical antibacterial products. Analyst 118:277–280

Wang S, Sun H, Ang H-M, Tadé M (2013) Adsorptive remediation of environmental pollutants using novel graphene-based nanomaterials. Chem Eng J 226:336–347

White IR, de Groot AC (2006) Cosmetics and skin care products. In: Frosch P, Menné T, Lepoitevin J (eds) Contact dermatitis. Springer, Berlin-Heidelberg, pp 493–506

Woodling JD, Lopez EM, Maldonado TA, Norris DO, Vajda AM (2006) Intersex and other reproductive disruption of fish in wastewater effluent dominated Colorado streams. Comp Biochem Physiol Part C: Toxicol Pharmacol 144:10–15

Wu J, Tomovic Z, Enkelmann V, Müllen K (2004) From branched hydrocarbon propellers to C3-symmetric graphite disks. J Org Chem 69:5179–5186

Wu X, Sprinkle M, Li X, Ming F, Berger C, De Heer WA (2008) Epitaxial-graphene/graphene-oxide junction: an essential step towards epitaxial graphene electronics. Phys Rev Lett 101:26801

Xagoraraki I, Kuo D (2008) Water pollution: emerging contaminants associated with drinking water. In: Heggenhougen K, Quah S (eds) International encyclopedia of public health. Academic Press, San Diego, pp 539–550

Xiang H, Wei S-H, Gong X (2010) Structural motifs in oxidized graphene: a genetic algorithm study based on density functional theory. Phys Rev B 82:1–5

Xu J, Zhu Y, Huang H, Xie Z, Chen D, Shen G (2011) Zinc-oleate complex as efficient precursor for 1-D ZnO nanostructures: synthesis and properties. CrystEngComm 13:2629–2635

Yao Y, Miao S, Liu S, Ma LP, Sun H, Wang S (2012) Synthesis, characterization, and adsorption properties of magnetic Fe_3O_4@graphene nanocomposite. Chem Eng J 184:326–332

Yao Y, Cai Y, Lu F, Wei F, Wang X, Wang S (2014) Magnetic recoverable $MnFe_2O_4$ and $MnFe_2O_4$-graphene hybrid as heterogeneous catalysts of peroxymonosulfate activation for efficient degradation of aqueous organic pollutants. J Hazard Mater 270:61–70

Yu J, Ma T, Liu S (2011) Enhanced photocatalytic activity of mesoporous TiO_2 aggregates by embedding carbon nanotubes as electron-transfer channel. Phys Chem Chem Phys 13:3491–3501

Zhang Z, Zhang J, Chen N, Qu L (2012) Graphene quantum dots: an emerging material for energy-related applications and beyond. Energy Environ Sci 5:8869–8890

Zhu Y, Xue J, Xu T, He G, Chen H (2017) Enhanced photocatalytic activity of magnetic core–shell Fe_3O_4@Bi_2O_3–RGO heterojunctions for quinolone antibiotics degradation under visible light. J Mater Sci Mater Electron 28:8519–8528

Zillioux EJ, Johnson IC, Kiparissis Y, Metcalfe CD, Wheat JV, Ward SG, Liu H (2001) The sheepshead minnow as an in vivo model for endocrine disruption in marine teleosts: a partial life-cycle test with 17α-ethynylestradiol. Environ Toxicol Chem 20:1968–1978

Chapter 8
Catalytic Ozonation of Aromatics in Aqueous Solutions Over Graphene and Their Derivatives

Qi Bao

Abstract Catalytic ozonation is progressively becoming an attractive technique for quick mineralization of aromatic compounds in water, yet efficient and stable heterogeneous catalysts remain elusive. Graphene (G) and related materials have attracted growing interests as carbocatalysts given their superior specific surface area, facile decoration, and high adsorption capacity. They could not only function as a support for nanocatalysts but also behave as a co-catalyst for the enhancement in ozonation reaction. Some G-based catalysts have been synthesized and reported with unprecedented adsorption-ozonation synergistic effect. In this chapter, the pros and cons of the ozonation reaction catalyzed by G and their derivatives have been discussed tentatively. We focus our attention on the unique properties of G that are of relevance to catalysis, with emphasis on the adsorption, electrostatic interaction, active sites that have been proposed to be responsible for the catalytic activity. Moreover, some challenging issues of G based carbocatalysts have been proposed to be resolved for the future development in this field.

Keywords Graphene · Heterogeneous catalytic ozonation (HCO)
Advanced oxidation processes (AOPs) · Reactive oxygen species (ROS)
Polycyclic aromatic hydrocarbons (PAHs)

1 Introduction

Aromatic compounds are quite common be present in a large variety of synthetic dyes, antibiotics, pesticides, crude oil, and natural organic matter. Alarmingly, many of them like polycyclic aromatic hydrocarbons (PAHs) are of considerable concern due to their well-recognized carcinogenicity, teratogenicity and genotoxi-

Q. Bao (✉)
Hong Kong Applied Science and Technology Research Institute Company Limited (ASTRI), 5/F, Photonics Centre, 2 Science Park East Avenue, Hong Kong Science Park, Shatin, Hong Kong SAR, China
e-mail: qibao@mx.nthu.edu.tw; baoqi23@126.com

M. Naushad (ed.), *A New Generation Material Graphene: Applications in Water Technology*, https://doi.org/10.1007/978-3-319-75484-0_8

city. Such concerns have motivated tremendous efforts in developing advanced industrial processes to remove aromatic compounds. Among them, ozonation has attracted increasing attention, whose intrinsic characteristics were listed as follows (Nawrocki 2013).

(1) Ozone (O_3) is a powerful oxidant with a standard redox potential (E^0) of 2.07 V versus normal hydrogen electrode (NHE) at 25 °C as shown in Eq. (1). At the simplest level, O_3 can function as oxidant in a single electron transfer reaction, for example, ozonation with iodide to form iodine.

$$O_3 + 2H^+ + 2e^- \rightarrow O_2 + H_2O \tag{1}$$

(2) Molecular O_3 shows an electrophilic nature, which can be attributed to the positive formal charges on the oxygen terminal atoms as shown in Fig. 1 (Youmin et al. 2012). This accounts for its high reactivity with unsaturated bonds and aromatic rings. Ozone prefers to react at the points of highest electronic density in organic substrates firstly. Such reactivity would be enhanced by the presence of electron-donating groups and suppressed by the effect of electron-withdrawing substituents. Additionally, Sun et al. calculated that the orbital energy of LUMO (lowest unoccupied molecular orbital) of ozone is −0.2044 Hartree based on the frontier orbital analysis. The orbital energy of HOMO (highest occupied molecular orbital) of phenol is −0.2344 Hatree. The huge difference (0.030 Hartree) between them makes the reaction easily.
(3) The stability and reactivity of ozone in water is strongly dependent on the pH of the aqueous solution (Table 1) and temperature. Ozone decomposes faster in increasingly alkaline solution and at higher solution temperature.
(4) At room temperature, molecular ozone can decompose organic substances through a direct route or a chain reaction mechanism as shown in Fig. 2 (Hoigné and Bader 1976). Molecular ozone reacts directly with organic

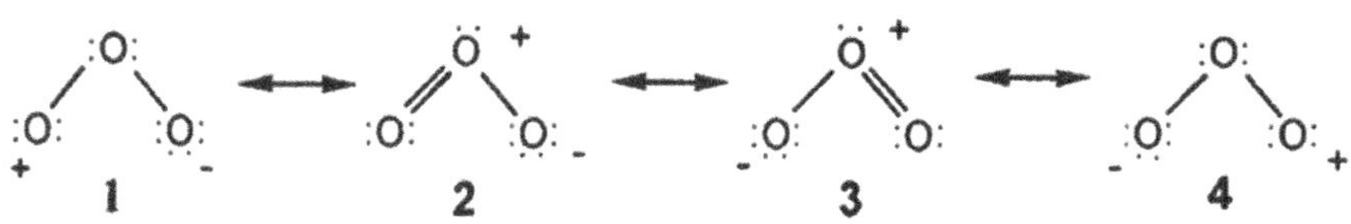

Fig. 1 Four possible resonance structure of ozone molecule

Table 1 Influence of pH on half-life of ozone in water

pH	Half-life (min)
7.6	41
8.5	11
8.9	7
9.2	4
9.7	2
10.4	0.5

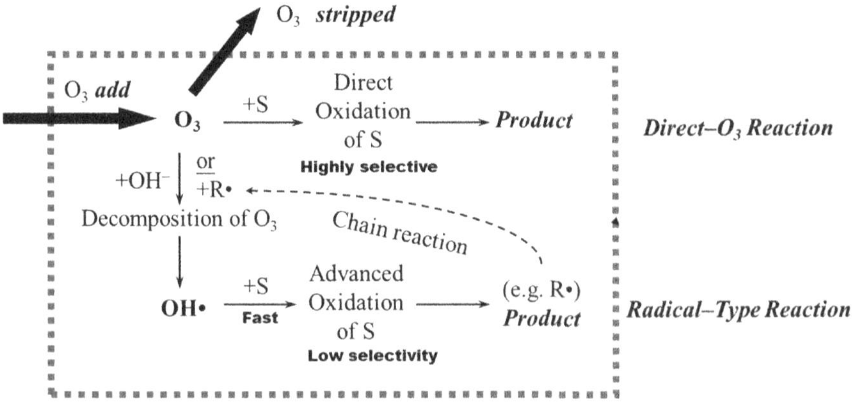

Fig. 2 Reactions of substrates (S) with O_3 and the O_3-decomposition reaction complete for O_3 consumption

pollutants in aqueous solutions when pH < 7. For pH > 7, ozone can also follow an indirect pathway involving generation of $^{\cdot}OH$ radicals. The direct radical-free route suffers from substrate selection and the according reaction rate constants vary significantly with different substrates, for instance, $k \sim 10^{-5}$ for acetic acid and $k \sim 10^{9}$ for phenolate ion (Hoigné and Bader 1976, 1983a, b; Hoigné et al. 1985).

(5) Last but not the least, ozonation could serve multifunction simultaneously like disinfection, precipitation of metal ions, removal of turbidity, destruction of volatile organic compounds, oxidation of taste and odors. Ozonation could also be enhanced by the combination with UV photolysis and H_2O_2 oxidation.

However, ozonation suffered from high operational cost, low gas–liquid mass transfer efficiency, low oxidation efficiency, and toxicity. Normally, ozonation products mainly consist of short-chain (C < 5) carboxylic acids, such as formic, acetic and oxalic acid as well as aldehydes. These saturated end products are refractory towards O_3 attack and only small mineralization is achieved in most cases even after longer oxidation time.

Introducing catalysts to ozonation is one of the applicable approaches to increase ozonation efficiency and ozone utilization degree. Thereafter, catalytic ozonation is highly expected to generate reactive oxygen species (ROS) (e.g., $^{\cdot}OH$) particularly at lower pHs. Due to its higher standard potentials of 2.8 V versus NHE in acidic media and 1.55 V versus NHE in basic media, hydroxyl radicals are highly reactive and nonselective towards substrates, particularly, ozonation by-products (Andreozzi et al. 1999).

Although hydroxyl radicals are highly expected in advanced oxidation processes (AOPs), catalytic ozonation is confirmed once the efficiency of ozonation at the presence of a catalyst is higher than without it at the same pHs. The catalysts can be categorized as homogenous or heterogeneous depending on whether reactions involve reactive species in a fluid medium or solid substance that are phase separated from the fluid. Notably, when investigating catalytic activity of a solid catalyst, also another condition must be fulfilled, the total effect of catalytic ozonation must be higher than a combined effect of adsorption on the catalyst surface and ozonation alone at the same pHs.

2 Advantages of Graphene as Ozonation Catalysts

As a new member of carbonaceous materials, graphene (denoted as G) has been attracting increasing interests as a metal-free heterogeneous carbocatalyst taken account of its high electrical conductivity, large surface area, high mechanical strength, and structural flexibility. Particularly, given its relative cytocompatibility and cost effectiveness, G seems much more advantageous in environmental remediation compared to CNTs. Its millimeter-scale lateral dimension also facilitated their post-separation process from the aqueous solutions by the filtration systems. Meanwhile, the oxy-functional groups located on its basal planes and at the sheet edges make G a versatile platform for in situ growing/anchoring numerous highly active nanocatalysts. Thereafter, G also represents an excellent support to anchor chemical functionalities or nanomaterials and, thus, G-based nanocomposites have become an active area of research for novel catalysts.

2.1 Interactions Between Catalysts and Substrates

The adsorption properties of catalysts are of significance when discussing the catalytic activity of heterogeneous surfaces, which is also essential in understanding overall catalytic reactions. Adsorption and activation of the reactants on the surface of catalysts is a critical stage in catalytic oxidation. The highest surface area of G with a theoretical value of 2630 $m^2\ g^{-1}$ makes it an ideal candidate for such processes involving adsorption or heterogeneous surface reactions. Like CNTs, the mechanisms involved in the adsorption of organic compounds on the surface of G could be associated with five different molecular interactions, that is, electrostatic interaction, van der Waals interactions, hydrophobic effect, π–π bonding, hydrogen bonding, and covalent bonding. Particularly, π–π interactions of G with condensed PAHs are known to be very strong and intermolecular complexes have been reported between G and pyrenes, among other aromatic compounds (Chu et al. 2010). In other cases, this preassociation of substrates with the G basal plane has been proposed as the cause of a synergy with other active sites resulting in high activity of G.

Besides the specific surface area, the superficial charged state of G and solution pH determines the possible electrostatic interaction between adsorbate and adsorbent and the efficiency of the catalytic ozonation process. Like metallic ions in aqueous solutions, the point of zero charge (pH_{pzc}, the pH at which the net surface charge is zero, which was measured typically by potentiometric mass titration method) affected the charged nature of G and its derivative through the presence or absence of different functional groups ($-NH_2$, –OH, –COOH, etc.). When solution $pH > pH_{pzc}$, the G surface is negatively charged due to the deprotonation of carboxyl and hydroxyl groups. Then, the electrostatic interaction with cations is favored. In contrast, when $pH < pH_{pzc}$, G becomes positively charged and electrostatic interactions are weakened due to charge repulsion and the adsorption of anions will take place. Similarly, pH also influences the charged nature of the adsorbates.

The large π-conjugation system also makes G an excellent absorbent to ozone molecule. Based on ab initio density functional theory (DFT) simulations and experimental validation, Lee et al. recently reported that ozone molecules adsorb on the basal plane of G with a physisorption energy of 0.25 eV, indicating that ozone adsorption on G is gentle and reversible (Lee et al. 2009).

2.2 *Electron Charge Transfer*

The electronic properties of G also make it attractive for environmental applications. Graphene has a high electrical conductivity ($10^3 - 10^4$ S m^{-1}) and electron mobility. It has been found that electrons have high mobility reaching 10,000–50,000 cm^2 V^{-1} s^{-1} at room temperature, with an intrinsic mobility limit of 4,200,000 cm^2 V^{-1} s^{-1} (Geim 2009). In addition, G represents an excellent support to anchor chemical functionalities for in situ growth of nanocatalysts (e.g., multivalent metal and their oxide).

As pointed out by Legube and Karpel Vel Leitner, the two mechanisms of heterogeneous catalytic ozonation in water are as follows (Legube and Leitner 1999). On the surface of the reduced Me-catalyst (Me_{red}, Fig. 3 mechanism 2) ozone oxidizes metal with the generation of $^{\cdot}OH$ radicals. Aromatics after adsorption on the catalyst surface are subsequently oxidized by an electron-transfer reaction through a typical the cycle-addition and ring-opening reaction pathway to regenerate a reduced catalyst (Me_{red}-$A^{\cdot}$). Then, the organic radical species $A^{\cdot}$ are subsequently desorbed from the catalyst and oxidized by $^{\cdot}OH$ or O_3 either in bulk solution, or more probably, within the thickness of an electric double layer.

Electron transfer always plays an important role in determining reaction mechanisms and rates. Both the nature of support and the kind of metal determine the reaction rate. Lin et al. evidenced that the average rates of aqueous ozone decomposition strongly depend on metal-support interaction. In this regard, an increase of electron transfer reaction rate and, furthermore, an increase in the rate of redox reaction that takes place at the presence of Me-metal oxide catalysts can have a promotional effect in the catalytic ozonation reaction. The DFT calculations indicating that the HOMO and LUMO energies are concentrated around the edge plane sites of the G sheet. Experimental results reveal that G with a variety of layer

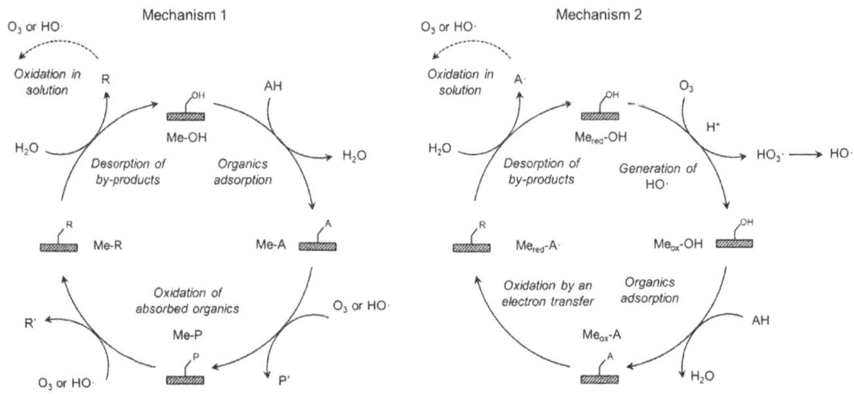

Fig. 3 Scheme illustrating two possible mechanisms for ozonation process in the presence of Me-support (metal oxide) catalysts for ozonation process including (1) adsorption on catalyst and oxidation by ozone or hydroxyl radical of adsorbed organics and (2) hydroxyl radical or other radical species generated by reaction of ozone with reduced metal of catalyst, and oxidation of organics by oxidized metal and/or homogeneous solution (AH = organic acid; P, R = adsorbed primary and final by-products; P′, R′ = primary and final by-products in solutions respectively; modified based on (Legube and Leitner 1999))

numbers and increased edge plane sites for fast electron transfer and thus an improvement in the electrochemical response.

2.3 Active Sites

The unique property of G also offers a joint, delocalized electron pool for further modification by the presence of oxygenated functional groups, vacancies, holes, edges, and dopant elements. It has been reported that these distant heteroatoms atoms/chemical functionalities on the G layers were responsible for serving as active sites for promoting catalyze ozonation reactions. In other words, it is possible to further enhance the catalytic activity by adequate and selective fine tuning of the density of the active sites by increasing the knowledge on the structure of the active sites combined with well-designed methodologies. However, mechanistic and systematic research over G catalysts remains a tough challenge and some conclusions seem inconclusive to date. This is because of their complex surface structure and the co-existence of various kinds of functional groups makes the identification and determination of discrete active sites difficult.

Hydroquinone/quinone is a redox pair that has similar potential to that of Fe^{2+}/Fe^{3+} and, therefore, which has the greatest potential of promoting the generation of more aggressive oxidant species such as hydroxyl radicals or superoxide at the presence of H_2O_2 and/or O_3. These quinone/hydroquinone-like moieties should be located in holes or at the periphery of the G sheet and, in this regard, terminal OH groups can behave as the redox centers. Garcia and co-workers evidenced that hydroquinone/quinone-like functional groups presented in rGO can act as catalytic sites, making rGO efficient carbocatalysts for the (photo)Fenton reaction. The DFT calculations have also revealed that the carbonyl oxygen of quinone-like groups is the most nucleophilic site compared to oxygen atoms of carboxyl, 1,2- and 1,3-diketones, isolated ketone or lactone. The Ag^+-Binding energy calculations were employed to establish theoretically the relative oxygen nucleophilicity order based on electron density parameters, showing again that the quinone group was the most reactive site among the various possible oxygens for the electrophilic attack of Ag^+. More importantly, DFT calculations suggest that hydroquinone/quinone-like substructures could be the active sites for the decomposition of H_2O_2 to $^{\cdot}OH$ radicals in a similar way to the homogeneous metal-promoted Fenton reaction like zero-valent nano Fe.

The preparation and post-treatment of G determined a certain degree of control on their functional groups. The prevalence of carbonyl/quinone groups could be realized by thermal annealing treatment or HNO_3 activation. Yoon et al. evidenced that rGO was more effective than GO for the transformation yields of O_3 into $^{\cdot}OH$ (Yoon et al. 2017).

Regarding oxidation reactions catalyzed by carbonaceous materials, like activated carbons, carbon nanotubes, the introduction of Lewis basic centers produced by nitration and/or amination can play a positive role of active sites. Gonçalves et al. demonstrated that the Lewis basic nature further enhanced the catalytic activity compared with the acid nature through chemical oxidation for carbon nanotubes (Gonçalves et al. 2013). Cao et al. also pointed out activated carbons with high basicity and large surface areas were the most active catalyst (Cao et al. 2014). Recently, we found that compositional modification with nitrogen atoms doped into G can produce more significant enhancement for catalytically oxidative phenol degradation by comparative studies (Bao et al. 2016). Generally, nitrogen dopant atoms could be divided into four categories: pyridinic N, pyrrolic/pyridone N, quaternary or graphitic N, and pyridine-N-oxide. Specially, pyridinic and graphitic N was believed to be responsible for the ORR activity, while pyridinic and pyrrolic/pyridine N was considered representing the pseudo-capacitance effect. Graphitic N was evidenced to be electron deficient and act as active sites by increasing oxygen adsorption and favoring its reduction by electron transfer (Machado and Serp 2012). The introduction of nitrogen influenced the collective properties of G such as surface polarity and enhanced the electron-donor properties of the carbon matrix, thus increasing the chemical reactivity for acid-base or redox chemistry. Nitrogen atoms substituted for carbon atoms of the G layers have an excess electron that must go to a state of higher energy, i.e. an antibonding π^* state. It is then easily transferred to an adsorbed molecule with a lower-lying unoccupied orbital, e.g. O_2. This can explain the increased catalytic activity of nitrogen-doped carbons in oxidation reactions. It is known that dopant nitrogen made the HOMO of electrons more positive, i.e. more electron poor and thereby more electron accepting.

Meanwhile, some ROS (e.g., superoxide anion, $O_2^{\cdot-}$) could be generated in the bulk solutions at the presence of pyridinic N and pyrrolic N somewhat like TEMPO-mediated oxidation reaction (Buffle and von Gunten 2006):

$$R_2NH + O_3 \rightarrow R_2NO^{\cdot} + H^+ + O_2^{\cdot-} \quad (2)$$

Then, $^{\cdot}OH$ radicals can be produced by the disproportionation of $O_2^{\cdot-}$ ions:

$$O_2^{\cdot-} + H_2O \rightarrow HO_2^{\cdot} + OH^- \quad (3)$$

$$HO_2^{\cdot} \rightarrow O_2 + H_2O_2 \rightarrow O_2 + 2^{\cdot}OH \quad (4)$$

Additionally, some highly active species such as O-radical could be formed on the surface of carbocatalysts. The π electron-rich region could also form electron donor–acceptor complexes with H_2O molecules as shown in Eq. (5), yielding large amount of hydroxyl ions, and subsequently initiated the adsorbed ozone decompose into surface bound O-radicals, thereby catalyzing the ozonation reaction for pH = 2–6 solutions as proposed in Eqs. (6)–(8) (Kasprzyk-Hordern et al. 2003)

$$C_\pi + 2H_2O \rightleftharpoons C_\pi{-}H_3O^+ + OH^- \quad (5)$$

$$O_3 + G \rightleftharpoons O_3{-}C_\pi \quad (6)$$

$$O_3{-}C_\pi \rightleftharpoons O{-}G + O_2 \quad (7)$$

$$O_3 + O{-}C_\pi \rightleftharpoons 2O_2 + C_\pi \quad (8)$$

In addition, the presence of nitrogen dopant heteroatoms increased the pH_{pzc} value of the G surface, enhancing the electrostatic interaction between G and acid aromatic substrates and its intermediate products (small molecular weight organic acid) via acid–base interactions.

3 Disadvantages of Graphene as Ozonation Catalysts

Although G nanomaterials can be easily produced by chemical exfoliation, the cost-effective and large-scale manufacturing of G nanomaterials is not yet well-established. On the other hand, according to the principles of green chemistry towards sustainable energy and environment, the recovery and reuse of catalysts is an important factor because of stringent ecological and economical demands. Unfortunately, the poor reusability of G makes more efforts are still necessary to increase G stability. Wang et al. suggested that loss of specific surface area and changes in the surface chemistry were suggested to be responsible for catalyst deactivation for ozonation reaction. It's known that graphite is not attacked by molecular oxygen at ambient or mildly elevated temperatures although the thermodynamic data imply that it should burn to CO_2. This kinetic stability can be explained with the different orbital symmetries of dioxygen molecules and graphite. It has been shown that at higher temperatures oxidation proceeds from the edges of graphite crystals and from defects in the (0001) surfaces. However, the stability of graphite was quite challenging at the presence of ozone. The ab initio density functional theory reveals that the physisorption–chemisorption transition energy is 0.72 eV for the interaction of G and ozone. The calculated chemisorption has the binding energy of 0.33 eV to form epoxide groups on G and a gas phase oxygen molecule (Lee et al. 2009). In other words, the physisorbed molecule can chemically react with G to form an epoxide group and an oxygen molecule. It also shows that the epoxide group has the C–O bonding distance of 1.44 Å and that the desorption energy of an epoxide group into an atomic oxygen is 3.23 eV. This means chemisorption was also plausible and might involve in the adsorption of ozone on the π-electron-rich structure of G. Such theoretical finding of ozone-induced epoxide formation on G basal plane was validated by the experimental alteration of the surface property of HOPG after ozone exposure of different duration (Tao et al. 2011). Then, the active sites as well as pH_{pzc} have been changed.

Generally, the catalytic ozonation process has been recognized an eco-friendly technique for water remediation since carbocatalysts do not produce any secondary contaminants compared with metal-based catalysts. However, such a viewpoint isn't applicable once Br containing compounds appeared in the water treatment system. Bromate formation should be avoided at the presence of ozone due to its hazards toxicity and carcinogenicity.

4 Conclusion

The ozonation process is an eco-friendly technique favorable for the decomposition of aromatic compounds for water remediation because it does not produce any secondary contaminants. However, it can hardly treat saturated aliphatic compounds and will be only effective in the first-step decomposition of aromatics and incapable for some intermediate removal. Introduction of G and their compositing catalysts could promote the decomposition of ozone molecules to generate more ROS to increase the high efficiency and oxidation degree. By exploring the plausible mechanism, it could be explained by the enhancement of adsorption, electrostatic interactions and charge transfer reaction. To develop the potential of G in ozonation catalysis, one major issue in the current state of the art is the limited understanding of the nature of the active sites responsible for the observed catalytic activity and how this activity and selectivity can be further increased. Combining the current knowledge of the advantages and disadvantages of G in catalytic ozonation reaction, one of applicable approaches was composting multi-component and multi-phase metal/metal oxides such as metallic glass/high entropy alloy based nanoparticles/nanoclusters and 2D/3D N doped G materials to pursue exceptional activity for catalytic ozonation reaction. The key role of G lies in that it can incorporate into their structure centers that can exhibit similar catalytic properties, by introducing Lewis basic and electrophilic centers, redox pairs and/or by charge transfer interactions. There are three main barriers that must be overcome to develop a successful new business from a innovative technology, commonly referred to as the "Devil's River," the "Valley of Death," and the "Darwinian Sea." With an attempt to cross the "Devil's River", numerous efforts have to be spent to swap away many demons lurking at the research and development stage.

References

Andreozzi R, Caprio V, Insola A, Marotta R (1999) Advanced oxidation processes (AOP) for water purification and recovery. Catal Today 53(1):51–59. https://doi.org/10.1016/S0920-5861(99)00102-9

Bao Q, Hui KS, Duh JG (2016) Promoting catalytic ozonation of phenol over graphene through nitrogenation and Co_3O_4 compositing. J Environ Sci 50:38–48. https://doi.org/10.1016/j.jes.2016.03.029

Buffle M-O, von Gunten U (2006) Phenols and amine induced $HO^{\cdot}$ generation during the initial phase of natural water ozonation. Environ Sci Technol 40(9):3057–3063. https://doi.org/10.1021/es052020c

Cao H, Xing L, Wu G, Xie Y, Shi S, Zhang Y, Minakata D, Crittenden JC (2014) Promoting effect of nitration modification on activated carbon in the catalytic ozonation of oxalic acid. Appl Catal B 146:169–176. https://doi.org/10.1016/j.apcatb.2013.05.006

Chu SN, Sands S, Tomasik MR, Lee PS, McNeill VF (2010) Ozone oxidation of surface-adsorbed polycyclic aromatic hydrocarbons: role of PAH^- surface interaction. J Am Chem Soc 132(45):15968–15975. https://doi.org/10.1021/ja1014772

Geim AK (2009) Graphene: status and prospects. Science 324(5934):1530–1534. https://doi.org/10.1126/science.1158877

Gonçalves AG, Órfão JJM, Pereira MFR (2013) Ozonation of sulfamethoxazole promoted by MWCNT. Catal Commun 35:82–87. https://doi.org/10.1016/j.catcom.2013.02.012

Hoigné J, Bader H (1976) The role of hydroxyl radical reactions in ozonation processes in aqueous solutions. Water Res 10(5):377–386. https://doi.org/10.1016/0043-1354(76)90055-5

Hoigné J, Bader H (1983a) Rate constants of reactions of ozone with organic and inorganic compounds in water—I. Water Res 17(2):173–183. https://doi.org/10.1016/0043-1354(83)90098-2

Hoigné J, Bader H (1983b) Rate constants of reactions of ozone with organic and inorganic compounds in water—II. Water Res 17(2):185–194. https://doi.org/10.1016/0043-1354(83)90099-4

Hoigné J, Bader H, Haag WR, Staehelin J (1985) Rate constants of reactions of ozone with organic and inorganic compounds in water—III. Inorganic compounds and radicals. Water Res 19(8):993–1004. https://doi.org/10.1016/0043-1354(85)90368-9

Kasprzyk-Hordern B, Ziółek M, Nawrocki J (2003) Catalytic ozonation and methods of enhancing molecular ozone reactions in water treatment. Appl Catal B 46(4):639–669. https://doi.org/10.1016/S0926-3373(03)00326-6

Lee G, Lee B, Kim J, Cho K (2009) Ozone adsorption on graphene: ab initio study and experimental validation. J Phys Chem C 113(32):14225–14229. https://doi.org/10.1021/jp904321n

Legube B, Leitner NKV (1999) Catalytic ozonation: a promising advanced oxidation technology for water treatment. Catal Today 53(1):61–72. https://doi.org/10.1016/S0920-5861(99)00103-0

Machado BF, Serp P (2012) Graphene-based materials for catalysis. Catal Sci Technol 2(1):54–75. https://doi.org/10.1039/C1CY00361E

Nawrocki J (2013) Catalytic ozonation in water: controversies and questions. Discussion paper. Appl Catal B: Environ 142:465–471. http://dx.doi.org/10.1016/j.apcatb.2013.05.061

Tao H, Moser J, Alzina F, Wang Q, Sotomayor-Torres CM (2011) The morphology of graphene sheets treated in an ozone generator. J Phys Chem C 115(37):18257–18260. https://doi.org/10.1021/jp2050756

Yoon Y, Oh H, Ahn YT, Kwon M, Jung Y, Park WK, Hwang TM, Yang WS, Kang JW (2017) Evaluation of the O_3/graphene-based materials catalytic process: pH effect and iopromide removal. Catal Today 282:77–85. https://doi.org/10.1016/j.cattod.2016.03.014

Youmin S, Xiaohua R, Zhaojie C, Guiqin Z (2012) The degradation mechanism of phenol induced by ozone in wastes system. J Mol Model 18(8):3821–3830. https://doi.org/10.1007/s00894-012-1376-5

Chapter 9
Removal of Arsenic from Water Using Graphene Oxide Nano-hybrids

Sharf Ilahi Siddiqui, Rangnath Ravi and Saif Ali Chaudhry

Abstract In 21st century, providing the fresh and affordable water through protects and purifying the water source from pollutants is biggest and most concern environmental challenges. Toxic element particularly arsenic in water is serious matter of threat for human from many developing countries, and long exposure of arsenic is generally associated with skin lesions and hyperkeratosis like adverse effects. Graphene oxide (GO) and its composites have attracted widespread attentions as novel adsorbents for the adsorption of various water pollutants due to their unique physicochemical characteristics. This chapter presents advances made in the synthesis of graphene oxides and their composites, and summarizes the application of these materials as a superior adsorbent for the removal of arsenic from water. The adsorption affinity in terms of contact time, pH, and temperature has been discussed. Competitive ion effect and regeneration are included within the text. Moreover, the challenges for the commercial uses are discussed.

Keywords Heavy metal · Arsenic · Adsorption · Graphene oxide
Magnetic graphene oxide · Remediation of arsenic

1 Introduction

Human life and lives of other animals depend on water; therefore, it is a vital concern for mankind. 71% of the earth's surface is covered up with water but only 2.5% water is fresh. Fresh water in the form of glacial is unusable, thus only ground

S. I. Siddiqui · R. Ravi · S. A. Chaudhry (✉)
Department of Chemistry, Jamia Millia Islamia University,
New Delhi 110025, India
e-mail: saifchaudhry09@gmail.com

S. I. Siddiqui
e-mail: sharf_9793@rediff.com

R. Ravi
e-mail: rangnathravi@gmail.com

M. Naushad (ed.), *A New Generation Material Graphene: Applications in Water Technology*, https://doi.org/10.1007/978-3-319-75484-0_9

Table 1 Some heavy metals and their toxic effect

Heavy metal	Major activities	Effect	References
Lead	Auto mobile, coal burning, smoking, paint and pesticides	Mental retardation in children, liver and kidney failure and cancer	Kumar et al. (2014), Dikilitas et al. (2016)
Mercury	Pesticides, batteries, paper industries	Neurological effect	Ha et al. (2017), Li et al. (2015)
Cadmium	Welding, pesticides, nuclear fusion plant, electroplating	Cancer, kidney damages and gastrointestinal disorder	Fristachi and Chaudhry (2017), Rodríguez and Mandalunis (2016)
Arsenic	Mining, smelting and fertilizers	Renal, dermal, mutagenic, carcinogenic, cardiovascular, and neurological effect	Siddiqui and Chaudhry (2017a, b, c), Rasheed et al. (2017)

and surface water is available for civilization. Major population developed over the river line systems due to rapid availability of sufficient and fresh amount of water but reliable and sustainable supply of fresh water in this era is a challenge due to rapid industrialization and human activities (Abdul et al. 2015; Carolin et al. 2017). The human population and their activities are increasing dramatically; consequently a large amount of fresh water would be required for the newly added people (Siddiqui and Chaudhry 2017a). In addition, contaminant released in water due to rapid industrialization, agricultural activities, geological activities and other human activities are deteriorating the water quality continuously. WHO, environmental agencies, government authorities, scientists, and academicians all over the world are worried and serious over the issue of water contamination. Thousands of organic, inorganic and biological species have been reported as water contaminants (Gao and Zhou 2000; Naushad 2014; Sang et al. 2003; Siddiqui et al. 2018). Heavy metals are among them which show serious adverse effects and toxicities; with lethal and carcinogenic effect, and cause the large damage to ecosystem and human health (Al-Othman et al. 2012; Carolin et al. 2017; Chaudhry et al. 2017a, b, c; Robinson 2017). Heavy metals having high specific gravity are found in earth crust and are non-biodegradable. They accumulate in food chain and soft and hard tissues of human body and get stored for long term, and affect the growth and development of the target organisms (Carolin et al. 2017; Robinson 2017; Siddiqui et al. 2017). The toxicities of various heavy metals are given in Table 1.

2 Arsenic Contamination and Toxicity

Arsenic contamination in water is becoming a severe problem for many countries particularly Asian countries such as Bangladesh, China, India, Pakistan, and Taiwan. The people of these countries are living in threat of arsenic (Bowell et al. 2014; Chaudhry et al. 2017b, c; Kao et al. 2013; Ng et al. 2003). Bangladesh and West Bengal province of India is the most affected zone of arsenic.

Arsenic inhalation through water is generally associated with skin lesions and hyperkeratosis like adverse effects (Matschullat 2000). Long-term exposure of arsenic contamination through water, affects the functioning of nervous and cardiovascular systems, and also causes cancers of many types (Abdul et al. 2015; Biswas et al. 2008; Siddiqui and Chaudhry 2017a, b, c, d; Watanabe et al. 2017). Moreover, arsenic alters the cell calcium signals; induce the oxidative stress, affect the cell mitochondrial function, and cell cycle progression (Abdul et al. 2015; Flora 2011; Kulshrestha et al. 2014).

Arsenic is found frequently in soil and rocks in the minerals form, which gets mobilized into ground water by natural weathering, geochemical reactions, biological activities, volcanic emissions and anthropogenic activities like mineralization, mining and smelting (Chaudhry et al. 2016a; Siddiqui and Chaudhry 2017a, b). Therefore, to control the arsenic effect, removal of arsenic from wastewater or drinking water is the best option to safe millions of people across the world. Various treatment techniques such as oxidation-coagulation, electro-coagulation and co-precipitation, oxidation-precipitation, reverse osmosis, electro dialysis, and ion exchange are being utilized. However, these techniques are inconvenient, require large space and are very costly (Anastopoulos et al. 2017; Chaudhry et al. 2016b; Devi and Saroha 2017; Mohan and Pittman 2007).

Adsorption being inexpensive; does not involve sophisticated instrumentation and do not require long procedure (Alqadami et al. 2016; Khan et al. 2015; Sharma et al. 2017). The process is simple, safe to handle and effectively work at low and high arsenic concentration in water (Gupta et al. 2009; Han et al. 2013; Mondal et al. 2007; Maliyekkal et al. 2009). Therefore, adsorption can be the better option to clean the arsenic contaminated water at different scales ranging from household module to community plants.

3 Adsorption Technology

Adsorption is a well-known water purification technology which involves the chemical or physical interaction between pollutant and solid surface, where solid surface is known as adsorbent and should have low particle size, high surface area, high active sites for higher removal capacity for pollutants (Bai et al. 2010; Fendorf et al. 1997; Johnston et al. 2016; Khan et al. 2013; Mohan and Pittman 2007; Sansone et al. 2013; Sherman and Randall 2003; Su et al. 2017). Large number of adsorbents have been utilized for removal of arsenic species (arsenite, As^{3+} and arsenate, As^{5+}) but the arsenite, As^{3+} removal requires pre-oxidation of As^{3+} to As^{+5} using oxidizing agents, which makes the process costly and sometimes produce unhealthy by-products (Siddiqui and Chaudhry 2017a; Zhang et al. 2007). Therefore, to avoid the pre-oxidation step using costly oxidizing agents, various solid materials with oxidative properties have been developed (Siddiqui and Chaudhry 2017a, c).

Among the various materials graphene oxide provides high surface area and porosity, and strong active sites which can easily trap arsenic from water

(Peng et al. 2017; Platero et al. 2017; Kumar et al. 2014; Sheshmani et al. 2015). Graphene contains hydrophobic surface while large acidic groups are attached on the graphene oxide surface which results greater attraction for cations (Zhou et al. 2014). Abundant oxygenic groups such as epoxy, hydroxyl, and carboxyl are attached on the graphitic backbone of GO which protrude from its surface and can easily interact with the pollutants via coordination, electrostatic, and covalent interactions (Ray et al. 2017; Peng et al. 2017; Machida et al. 2006). Therefore, GO based adsorbents can be the best choice for researchers for arsenic removal. To ease the further study we have incorporated valuable literature here in this chapter. The objective of this chapter is to discuss the latest status of GO and its modified forms. Further, to inspire the environmental community, various GO based adsorbents, their adsorption capacity for arsenic, and comparative study has been incorporated. This chapter will help scientist community for the commercial application of the GO.

4 Graphene Oxide

Graphite oxide, also known as graphitic oxide or graphitic acid, is the compound of carbon, oxygen, and hydrogen. Graphite on oxidation with strong oxidizing agents like $KMnO_4$ gives the Graphite oxide (GrO). GrO is normally yellow colour bulk solid having layer structure of graphite. GrO dispersed in alkaline solutions and produce the monolayer sheets which are known as graphene oxide (Wang et al. 2013; Bian et al. 2015; Gao et al. 2011).

Three approaches have been made for the preparation of GrO, such as Brodie, Staudenmaier and Hummers process. Brodie (1859) suggested that GrO can be prepared by oxidation of graphite powder with mixture of potassium per chlorate and concentrated nitric acid in water. Staudenmaier (1898) reported the preparation of GrO through the oxidation of graphite with mixture of sulfuric acid, nitric acid and potassium per chlorate. A similar approach was attempt by the Hummers and Offeman (1958) using solution of concentrated sulfuric acid, sodium nitrate and potassium permanganate. Obtained layer structure of GrO can easily be converted into the GO through mechanical stirring or ultrasonic process.

Actually, distance between the graphitic layers of GrO increases with increase in the oxygenous functional groups, and therefore, the interaction between the layers decreases which enhances the probability of formation of monolayer GO (Brodie 1859; Hummers and Offeman 1958; McAllister et al. 2007; Schniepp et al. 2006; Staudenmaier 1898).

4.1 Preparation of GO and It's Composite

Recently, Hummers method have been modified which is frequently being used for the preparation of GO. In brief, certain amount of graphite flaks was dissolved into the solution of concentrated H_2SO_4:H_3PO_4 (360:40 mL) under continuous stirring. Further, $KMnO_4$ was added slowly to the suspension under vigorous stirring for 36 h at 50 °C, thus reddish suspension was obtained, cooled to room temperature, and added ~400 mL ice water containing 3.0 mL of 30% H_2O_2 into the suspension under continues stirring until uniform mixture was obtained. Afterward, suspension was centrifuged, washed and dried (Kumar et al. 2014).

Due to the presence of oxygenous groups onto the surface, GO shows high adsorption capacity for water pollutants, but recovery of exhausted GO is not easy and require costly process such as centrifugation and filtration. This problem can be resolved by the modification of GO with magnetic particles. Large number magnetic GO have been utilized, that permit the use of bare GO in large-scale water treatment. Large numbers of magnetic GO have been developed by the simple co-precipitation method (Kumar et al. 2014).

Kumar et al. (2014) reported the preparation of magnetic GO-$MnFe_2O_4$ nano-hybrid. In brief, 3.0 g GO was ultrasonically dispersed in 400 mL of water, then certain amount of ferric chloride ($FeCl_3 \cdot 6H_2O$) and manganese sulfate ($MnSO_4 \cdot H_2O$) were added to the GO solution and stirred for 0.5 h. Afterwards, the temperature of the reaction was increased up to 80 °C and pH of solution was adjusted to 10.5 by addition of 8 M NaOH solution, under continuous stirring. The reaction was continued for 5 min, and then cooled to room temperature and resulting precipitate was magnetically separated, washed and dried. Similarly, approaches were adopted for FeO_x-GO (Su et al. 2017) and a-FeOOH@GCA (Fu et al. 2017). One more attempt was made to improve the adsorption capacity of GO by the functionalization of GO. The functionalized GO surface have large number of oxygen atoms on the graphitic backbone, which provide the large number of active sites for charged ions (Fig. 1) (License No. 4147561017088) (Kumar and Jiang 2016).

In brief, 1.0 g of GO was dispersed in 50 mL thionyl chloride ($SOCl_2$)-dimethylformamide (DMF) solutions and then refluxed for 24 h at 60 °C. The obtained precipitate of GO-COCl was centrifuged, washed with tetrahydrofuran (THF) and

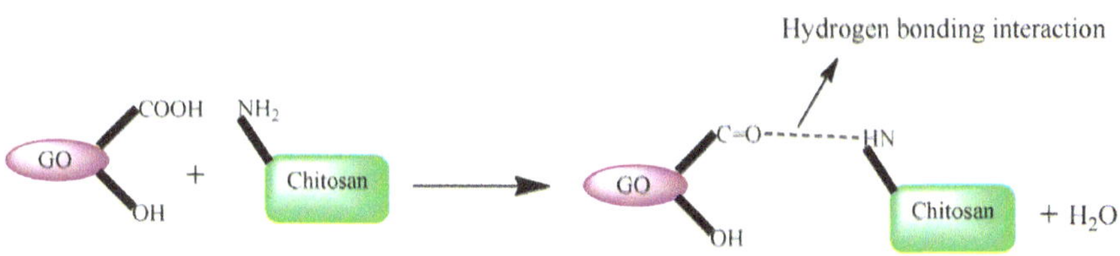

Fig. 1 Hydrogen-bonding interaction between GO and chitosan. Reprinted with permission from Kumar and Jiang (2016) Copyright (2016) Elsevier (License No. 4147561017088)

dried under the vacuum. Afterwards, 1.0 g of GO-COCl was slowly added into the chitosan solutions (0.5 g of chitosan in 10 mL of 1% (v/v) acetic acid solutions) under the continues stirring for 10 min to get uniform mixture. The obtained mixture was slowly heated up to 50 °C and stirred for 3 h, and the resulting mixture was washed with acetone several times, filtered, and dried under vacuum (Kumar and Jiang 2016).

4.2 Characterization of GO

Preparation of GO and its composites was confirmed by FT-IR, Raman, and XRD techniques. Characteristic absorption peak of FT-IR spectrum of GO has shown in Table 2, which is clearly indicating the formation of GO and presence of oxygenated groups on a graphene matrix which grown for further modification.

Kumar et al. (2014) reported the characteristic absorption peaks at 1729, 1620, 1415, 1046, and 1236 cm^{-1} for FTIR spectrum of GO (Fig. 2) (License No. 4147600394062). The C=O group is confirmed by the appearance of peak at 1729 cm^{-1} in IR spectrum and C=C stretching was assigned by the appearance of peak at 1620 cm^{-1}.

The absorption frequency at 1415 cm^{-1} confirms C–H deformation bond. A band at 1046 and 1236 cm^{-1} was appeared due to C–O stretching vibration for epoxy and alkoxy groups, respectively. Similar result was observed by (Khatamian et al. 2017; Kumar and Jiang 2016; Peng et al. 2017). Su et al. (2017) confirmed the distribution of iron oxide species onto GO matrix by assigning the absorption peaks for Fe–O, between the 750–400 cm^{-1}. The formation of FeO_x-GO nano-composite was confirmed by the two bands at 552 and 442 cm^{-1} for Fe–O and one band at 1578 cm^{-1} for C=C stretch in GO. FTIR spectrum of GO-$MnFe_2O_4$ (Fig. 2) exhibited absorption peaks at 490 and 577 cm^{-1} due to M–O stretching vibrations of manganese ferrite (Kumar et al. 2014). Similar result was observed for (Huang et al. 2011; Kumar et al. 2014; Marcano et al. 2010) studies. The XRD spectrum of GO gives the characteristic peak at $2\theta = 9.4–10.50°$ corresponding to (001) plane of GO.

Table 2 Characteristics absorption peaks for GO

Stretching frequency	Information (Khatamian et al. 2017; Kumar et al. 2014; Kumar and Jiang 2016; Peng et al. 2017)
3420–3200	V (O–H) stretching
2850–2950	Hydrogen bonding with oxygenous groups
1710–1730	V (C=O) stretching
1620–1640	V (C=C) in-plane stretching
1415–1550	V (O–H) deformation
1230–1400	V (C–O) Alkoxy stretching
1020–1070	V (C–O) Epoxy stretching

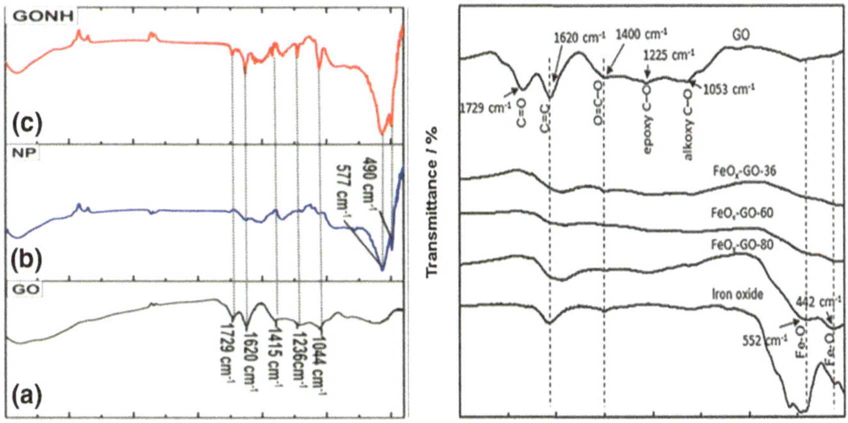

Fig. 2 FTIR spectrum of GO and its composites. Reprinted with permission from **a** Kumar et al. (2014) Copyright (2014) American Chemical Society, **b** Su et al. (2017) Copyright (2017) Elsevier (License No. 4147600394062)

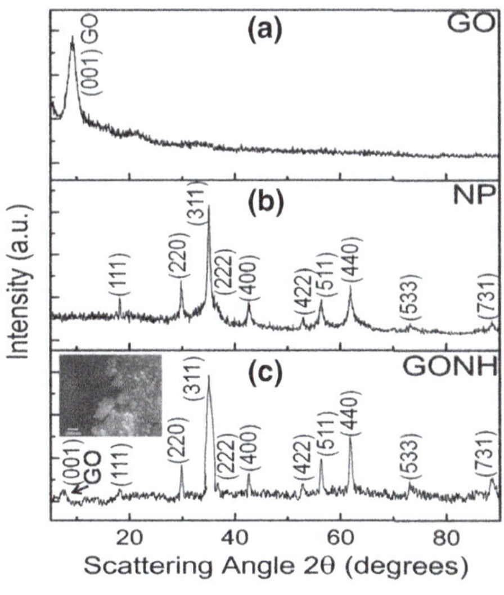

Fig. 3 XRD spectra of GO, $MnFe_2O_4$ and its composites (GO-$MnFe_2O_4$). Reprinted with permission from **a** Kumar et al. (2014) Copyright (2014) American Chemical Society

After treatment of GO with metal, the diffraction intensity decreased which indicates the decline in crystalline structure. This was due to the bonding between the metal and GO. The similar result was also obtained for intramolecular or intermolecular hydrogen bonding interaction between the polymers and GO (Huang et al. 2011; Kumar et al. 2014; Kumar and Jiang 2016; Marcano et al. 2010). XRD pattern of GO and its composite is given in Fig. 3.

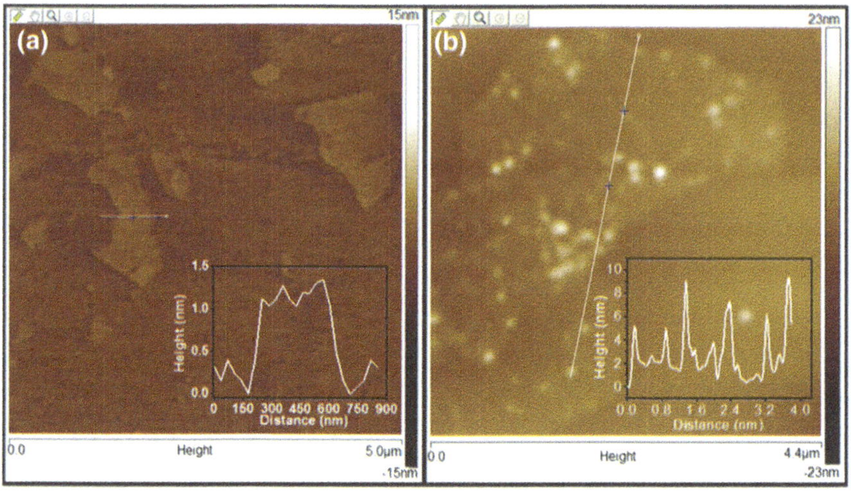

Fig. 4 Atomic force microscopy (AFM) of **a** GO and **b** its composites (GO-$MnFe_2O_4$). Reprinted with permission from **a** Kumar et al. (2014) Copyright (2014) American Chemical Society

The average flake size and average thickness of GO flake was measured ~ 2 μm and $\sim$ 1–2 nm, respectively, using AFM (Fig. 4) (Huang et al. 2011; Kumar et al. 2014; Marcano et al. 2010).

Raman spectra has been widely used to characterize GO and GO based materials. Raman spectrum of GOs showed two prominent peaks at 1580–1620 and 1340–1365 cm^{-1} corresponding to the first order E_2g mode from sp2 carbon domains (G-band) and disorder mode (D-band), respectively. Kumar and Jiang (2016) of GO which showed strong peaks at 1610 and 2445 cm^{-1} related to G band and D′ band, and one broad peak at 1360 cm^{-1} corresponding to D band.

The functionalization of GO with chitosan shows two new peaks which emerged at 718 and 1047 cm^{-1} due to the interaction of COOH of the GO and OH group of chitosan (Fig. 5) (License No. 4147561017088). Similar results were reported by Chen et al. (2013), Kumar et al. (2014), Yang et al. (2013). The morphology of GO and their composites has been characterized from the SEM and TEM images by Chen et al. (2013), Kumar et al. (2014), Su et al. (2017), Yang et al. (2013).

5 GO as Adsorbent for Arsenic

Iron oxide loaded graphene oxide, magnetite (Fe_3O_4)-graphene oxide, magnetite (Fe_3O_4)-reduced graphene oxide, magnetite (Fe_3O_4)-reduced graphite oxide-MnO_2, graphene-$FeMnO_x$, and hydrous cerium oxide-graphene nano-composite are well known graphene based nano-composites that have been used for arsenic remediation (Kumar et al. 2014; Yu et al. 2015). These composites are magnetic in nature,

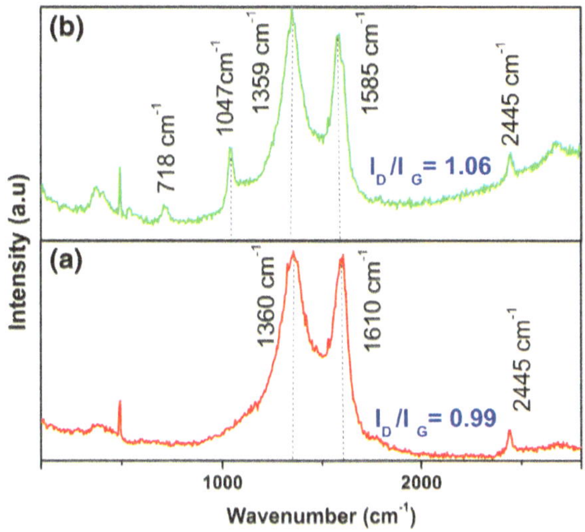

Fig. 5 Raman spectrum of **a** GO and **b** its composites (GO-Chitosan). Reprinted with permission from Kumar and Jiang (2016) Copyright (2016) Elsevier (License No. 4147561017088)

show large sorption capacity towards arsenic and highly effective at various pHs, low concentration, and in the presence of competitive ions.

In comparison to virgin $MnFe_2O_4$ nanoparticles, graphene oxide-$MnFe_2O_4$ nano-composite removed much higher As(III) and As(V). The maximum sorption capacity of graphene oxide-$MnFe_2O_4$ for As(III) and As(V) was recorded as 146 and 207 mg/g, respectively (Kumar et al. 2014). In addition, magnetite (Fe_3O_4)-reduced graphene oxide-MnO_2 nano-composite removed 14 mg/g of As(III) and 12 mg/g of As(V) (Xubiao et al. 2012), and graphene-$FeMnO_x$ removed 10.20 mg/g of As(III) and 11.50 mg/g of As(V) due to electrostatic interaction and pre-oxidation step (Jin et al. 2015). Yoon et al. (2016) reported the preparation of Fe_3O_4-graphene oxide composite (M-GO) and Fe_3O_4-reduced graphene oxide composite (M-rGO) (Fig. 6), investigated its arsenic removal capacity, and reached to the result that M-GO showed higher removal capacity for both As(III) and As(V) than M-rGO.

β-FeOOH incorporated carboxylic graphene oxide nano-composite β-FeOOH@GO-COOH nanocomposite has removed 100% arsenic ions from water. The composite has shown tremendous adsorption efficiency even after 20 successive adsorption-desorption cycles, and removed >80% of both As(III) and As (V) from given initial concentration (Chen et al. 2015). The bonding between the β-FeOOH@GO-COOH and As(III), and As(V) has shown in Fig. 7. β-FeOOH@GO-COOH was also effective in the presence of 2000-fold of SO_4^{2-}, NO_3^-, Cl^- and Mg^{2+} ions, and provided 90% removal efficiency for 5 successive cycles. Maximum adsorption capacity for β-FeOOH@GO-COOH was found to be 77.5 and 45.7 mg/g for As(III) and As(V) ions, respectively (Chen et al. 2015).

Khatamian et al. (2017) reported the preparation of different composites of GO and RGO viz magnetite (Fe_3O_4)/RGO, magnetite (Fe_3O_4)/RGO/Cu-ZEA, GO/Cu-ZEA, and magnetite (Fe_3O_4)/GO/Cu-ZEA using solvothermal method with Cu-exchanged zeolite A (Cu-ZEA) and magnetite nanoparticles (Fe_3O_4). This work was done for the

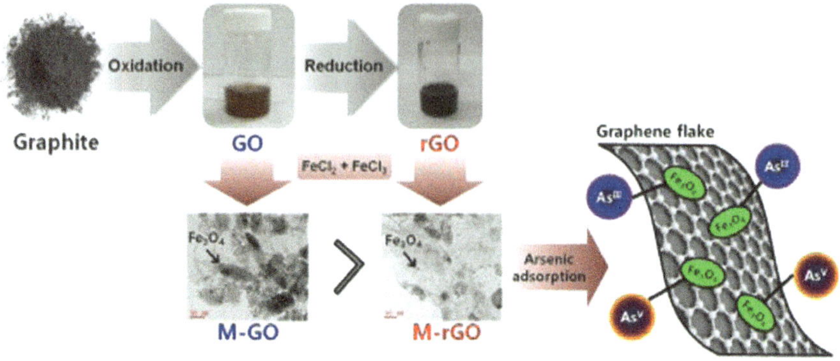

Fig. 6 Preparative steps of magnetite-GO and magnetite-rGO nano-composite and their adsorption activity for As(III) and As(V). Reprinted with permission from Yoon et al. (2016) Copyright (2016) Elsevier (License No. 4147590561897)

improvement in the adsorption capacity of GO, RGO, magnetite (Fe_3O_4) nanoparticles and Cu-ZEA. Among them magnetite (Fe_3O_4)/RGO/Cu-ZEA had the high specific surface area thus they have shown the highest removal capacity for arsenic. The adsorption kinetics data followed the pseudo-second-order kinetic model. Similarly, Su et al. (2017) reported the synthesis FeOx-GO nano-composites with different wt% (36–80 wt%) iron oxide content and further utilized for arsenic removal. GO sheets were prepared by an improved Hummers method and then iron oxide was incorporated on the GO through Co-precipitation reaction. With increased in the wt% of FeOx on GO, the surface area for FeO_x-GO also increased, and iron oxide content of 80 wt% (FeO_x-GO-80) having predominant mesopore structures has highest surface area and consequently higher adsorption sites. Therefore, FeO_x-GO-80 showed maximum adsorption capacity of 147 and 113 mg/g for As(III) and As(V), respectively, which was highest among all the reported iron oxide-GO/reduced GO composite adsorbents. FeO_x-GO-80 removed ~100% of arsenic from initial concentration of 118 and 108 μg/L, of As(III) and As(V).

The mass production of non-oxidative graphene and magnetite/non-oxidative graphene (M-nOG) composite has been reported by Yoon et al. (2017) for arsenic removal. The M-nOG showed greater adsorption capacity for arsenic. The batch experiment was performed to evaluate the adsorption capacity of M-nOG for arsenic in terms of pH, temperature, competing anions, and humic acid. As(III) was largely influenced by the surface complexation while As(V) showed intraparticle diffusion mechanism. The repetitive reuse and regeneration of M-nOG were performed.

Further, three dimension (3D)-magnetite (Fe_3O_4)-graphene macroscopic composites have been utilized as adsorbents for the removal of arsenic from low level (Guo et al. 2015). 3D-graphene macroscopic composite was synthesized through the self-assembly of GO nanosheets with polydopamine and magnetite (Fe_3O_4) nanoparticles under the basic conditions at low temperature. Polydopamine strengthen the 3 Dimension-graphene based macroscopic architect as well as enhanced the load and binding ability of magnetite (Fe_3O_4) nanoparticles.

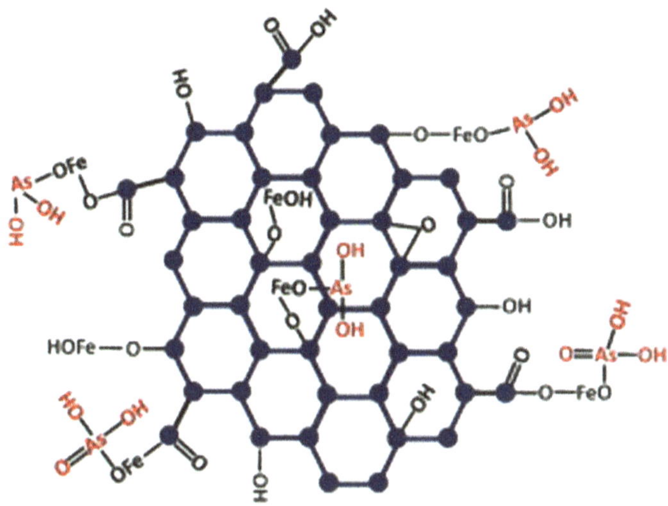

Fig. 7 The bonding between β-FeOOH@GO-COOH and As(III), and As(V). Reprinted with permission from Chen et al. (2015) Copyright (2015) Elsevier (License No. 4147590927555)

Graphene based metal and metal oxide composites such as single metal Ag-RGO, Cu_2O-RGO, and magnetite (Fe_3O_4)-RGO and bimetallic Ag-Cu_2O-RGO, Ag-magnetite (Fe_3O_4)-RGO, and Cu_2O-magnetite (Fe_3O_4)-RGO have been utilized for the adsorptive removal of As(V) (Dubey et al. 2015). These adsorbents were synthesized hydrothermally using sodium borohydride as a reducing agent and sodium sorbate as a stabilizer. Similarly, Al and Fe-doped graphene was used for the removal of methylated As(III) and As(V) and their adsorption capacity was studied by quantum chemistry computations (Arriagada and Labbe 2016).

Kumar and Jiang (2016) reported the preparation of chitosan functionalized GOs which act as interactive sites due to the presence of one amino group and two hydroxyl groups on each glucosamine monomer of chitosan. The formation of chitosan functionalized GO was due to the electrostatic affinity between the amino group (NH_2), primary and secondary hydroxyl groups (CH_2OH, CHOH) of the chitosan and surface of carboxyl (COOH) and hydroxyl groups (OH) on the GO. The chitosan functionalized GO worked as good host for incoming arsenic oxy-anion. Various interaction mechanism such as cation–π interaction (RNH_3^+—aromatic π moiety), electrostatic interaction ($H_2AsO_4^-$, $HAsO_4^{2-}$—$^+NH_3R$), and inter and intermolecular hydrogen bonding as well as anion–π interaction (R–COO^-—aromatic π moiety), (R–O^-—aromatic π moiety) defined the interaction between arsenic and GO moieties. Mishra and Ramaprabhu (2011) prepared the graphene sheets by hydrogen induced exfoliation of graphitic oxide followed by functionalization and used for the removal of high concentration of inorganic species of arsenic As(III) and As(V) from aqueous solution using super capacitor based water filter. The maximum monolayer adsorption capacities for As(III) and As(V) were 142, and 139 mg/g, respectively, which were higher than other, reported adsorbents. Graphene based super capacitor provides a solution for commercially feasible water filter.

Similarly, β-Cyclodextrins (CDs) functionalized GO (β-CDs@GO hybrid), magnetite (Fe_3O_4) nanoparticles decorated with β-CDs-functionalized GO (Fe_3O_4-β-CDs@GO hybrid), β-CDs-GO@ magnetite (Fe_3O_4) nanoparticles and polyelectrolyte of poly (diallyldimethylammonium chloride) (pDADMAC) with paramagnetic anions based on $FeCl_4$-GO (Fe-polyDADMAC@GO) have been synthesized for the removal of arsenic from water. Kumar and Jiang (2017) reported the magnetically separable and recyclable magnetic nanoparticles decorated with β-cyclodextrin functionalized graphene oxide for As(III) and As(V) cleanup from water. Due to hydroxyl and carboxyl groups present on the surface of β-CDs-GO@ magnetite (Fe_3O_4) nanoparticles and their superior magnetic property, this adsorbent showed excellent adsorption capacity for As(III) and As(V). Batch adsorption experiments were performed to demonstrate the maximum adsorption capacity of prepared adsorbent in terms of various parameters such as pH, temperature, and contact time. Adsorption isotherms and kinetics data were well described by Freundlich isotherm and the pseudo-second order kinetic models, respectively. Thermodynamics studies suggested that the reaction was feasible, spontaneous and endothermic. Adsorbent was regenerated using sodium hydroxide (NaOH) and was reutilized for two adsorption/desorption cycles.

Graphene-doped titanium nano tube coated to super paramagnetic nanoparticles (GN-MNP-TNT) was used to remove As(III) from their solutions (Lin et al. 2017). GO layers served as precursor and stabilizer for GN-MNP-TNT due to their excellent electron transfer capacities. TNT-doped graphene (GN-TNT) was prepared by hydrothermal method. GN-MNP-TNT could be tested for 4 consecutive adsorption/desorption cycle without major loss. After 4 cycle of regeneration adsorbent efficiency remained 83% which was much higher than other reported value in literature.

5.1 GO Coated Carbon Nanotubes

In addition to this, carbons nanotubes have also been incorporated with GO but limited number of literature is available. Goethite impregnated graphene oxide (GO)-carbon nanotubes (CNTs) aerogel (α-FeOOH@GCA) was prepared for adsorptive removal of arsenic from water (Fu et al. 2017). α-FeOOH@GCA was prepared by a facile self-assembly of GO-CNTs by in situ Fe^{2+} reduction method. Batch experiment was performed to investigate the adsorption capacity of prepared α-FeOOH@GCA for different arsenic species like As(V), DMA, and *p*-ASA. The maximum adsorption capacities for As(V), DMA and *p*-ASA was obtained as 56.43, 24.43 and 102.11 mg/g which was much higher than (25.71, 8.03 and 14.52 mg/g) pristine α-FeOOH, respectively. This was due to the incorporation of GO-CNTs, which not only hindered the aggregation of GO-CNTs but also inhibited the growth of α-FeOOH nanoparticles and facilitated the diffusion and adsorption.

α-FeOOH@GCA showed excellent reusability with adsorption capacity, however, adsorption rate was affected by the presence of phosphate and silicate anions due to the similar anionic structure. The complex prepared between the arsenic species and the surface of α-FeOOH@GCA was inner sphere complex.

Park et al. (2016) developed a feasible water flow filter with facilely functionalized magnetite (Fe_3O_4)-non-oxidative graphene/CNT composites for arsenic removal in household use for continuous purification of water. Feasible water flow filter filled with magnetite (Fe_3O_4)-functionalized non-oxidative graphene/CNT was fabricated through facile functionalization in a Couette-Taylor flow reactor. Couette-Taylor flow method allows the fast production of filters. magnetite (Fe_3O_4)-functionalized non-oxidative graphene/CNT composites as flow filter showed higher arsenic removal efficiency than when it was used in the batch experiment. This is due to its 3D-structure which enhanced the water flow pathway and the contact area with magnetite (Fe_3O_4). Similarly, Roy et al. (2016) reported a Europium doped magnetic graphene oxide-MWCNT nanohybrid adsorbent for effective and rapid removal of As(III) and As(V) from real water samples. This adsorbent could be prepared by a combination of Eu-doped magnetic graphene oxide and gold nanoparticle functionalized multiwalled carbon nanotubes. Prepared nano-hybrid adsorbent having magnetic property (15,000 emu/g) showed extraordinary adsorption capacity as 320.0 and 298.0 mg/g for As(III) and As(V), respectively, and also showed excellent photo catalytic activity for oxidation of As (III) to As(V). Therefore, this adsorbent could be utilized for the conversion of highly toxic As(III) to less toxic As(V) as well as adsorptive removal of As(V). These were recently utilized GO based adsorbents, which could be effective for arsenic removal, however, more study is needed to make the process more convenient.

6 Conclusion and Future Prospects

The excellent adsorption performance, along with their low cost and convenient synthesis, makes GO and it composites highly promising for commercial applications in drinking water purification and wastewater treatment. Reusability, ease of magnetic separation, high removal efficiency, and fast kinetics make them very attractive and smart materials for the effective removal of heavy metals particularly arsenic from contaminated water. The experimental conditions and results for GO for arsenic remediation provide a way for further research and development of GO based adsorbent for others charged pollutants.

Acknowledgements The financial support from the University Grant Commission, UGC, India and Department of Chemistry, Jamia Millia Islamia, New Delhi, India, is gratefully acknowledged.

References

Abdul KS, Jayasinghe SS, Chandana EP, Jayasumana C, De-Silva PM (2015) Arsenic and human health effects: a review. Environ Toxicol Pharmacol 40:828–846

AL-Othman ZA, Ali R, Naushad M, (2012) Hexavalent chromium removal from aqueous medium by activated carbon prepared from peanut shell: Adsorption kinetics, equilibrium and thermodynamic studies. Chem Eng J 184:238–247

Alqadami AA, Naushad M, Abdalla MA et al (2016) Synthesis and characterization of Fe_3O_4 @TSC nanocomposite: highly efficient removal of toxic metal ions from aqueous medium. RSC Adv 6:22679–22689

Anastopoulos I, Karamesouti M, Mitropoulos AC, Kyzas GZ (2017) A review for coffee adsorbents. J Mol Liq 229:555–565

Arriagada DC, Labbe AT (2016) Aluminum and iron doped graphene for adsorption of methylated arsenic pollutants. Appl Surf Sci 386:84–95

Bai L, Ma XJ, Liu JF, Sun XM, Zhao DY, Evans DG (2010) Rapid separation and purification of nanoparticles in organic density gradients. J Am Chem Soc 132:2333–2337

Bian Y, Bian ZY, Zhang JX, Ding AZ, Liu SL, Wang H (2015) Effect of the oxygen-containing functional group of graphene oxide on the aqueous cadmium ions removal. Appl Surf Sci 329:269–275

Biswas BK, Inoue JI, Inoue K, Ghimire KN, Harada H, Ohto K, Kawakita H (2008) Adsorptive removal of As(V) and As(III) from water by a Zr(IV)-loaded orange waste gel. J Hazard Mater 154:1066–1074

Bowell RJ, Alpers CN, Jamieson HE, Nordstrom DK, Majzlan J (2014) The environmental geochemistry of arsenic—an overview. Rev Mineral Geochem 79:1–16

Brodie BC (1859) On atomic weight of graphite. Philos Trans R Soc Lond 149:249–259

Carolin F, Kumar PS, Saravanan A, Joshiba GJ, Naushad M (2017) Efficient techniques for the removal of toxic heavy metals from aquatic environment: a review. J Environ Chem Eng 5:2782–2799

Chaudhry SA, Khan TA, Ali I (2017a) Equilibrium, kinetic and thermodynamic studies of Cr(VI) adsorption from aqueous solution onto manganese oxide coated sand grain (MOCSG). J Mol Liq 236:320–330

Chaudhry SA, Khan TA, Ali I (2017b) Zirconium oxide-coated sand based batch and column adsorptive removal of arsenic from water: Isotherm, kinetic and thermodynamic studies. Egypt J Petrol 26:553–563

Chaudhry SA, Ahmed M, Siddiqui SI, Ahmed S (2016a) Fe(III)-Sn(IV) mixed binary oxide-coated sand preparation and its use for the removal of As(III) and As(V) from water: application of isotherm, kinetic and thermodynamics. J Mol Liq 224:431–441

Chaudhry SA, Khan TA, Ali I (2016b) Adsorptive removal of Pb(II) and Zn(II) from water onto manganese oxide-coated sand: Isotherm, thermodynamic and kinetic studies. Egypt J Basic App Sci 3:287–300

Chaudhry SA, Zaidi Z, Siddiqui SI (2017c) Isotherm, kinetic and thermodynamics of arsenic adsorption onto Iron-Zirconium Binary Oxide-Coated Sand (IZBOCS): modelling and process optimization. J Mol Liq 229:230–240

Chaudhry SA, Siddiqui SI (2017) Arsenic removal from water using nano-composites: a review. Cur Environ Eng. https://doi.org/10.2174/2212717804666161214143715

Chen J, Yao B, Li C, Shi G (2013) An improved Hummers method for eco-friendly synthesis of graphene oxide. Carbon 64:225–229

Chen ML, Sun Y, Huo CB, Liu C, Wang JH (2015) Akaganeite decorated graphene oxide composite for arsenic adsorption/removal and its pre-concentration at ultra-trace level. Chemosphere 130:52–58

Devi P, Saroha AK (2017) Utilization of sludge based adsorbents for the removal of various pollutants: a review. Sci Total Environ 578:16–33

Dikilitas M, Karakas S, Ahmad P (2016) Chapter 3: effect of lead on plant and human DNA damages and its impact on the environment. Plant Metal Interact 41–67

Dubey SP, Nguyen TTM, Kwon YN, Lee C (2015) Synthesis and characterization of metal-doped reduced graphene oxide composites, and their application in removal of *Escherichia coli*, arsenic and 4-nitrophenol. J Indus Eng Chem 29:282–288

Fendorf SE, Grossl MJP, Sparks DL (1997) Arsenate and chromate retention mechanisms on goethite surface structure. Environ Technol 31:315–320

Flora SJS (2011) Arsenic-induced oxidative stress and its reversibility. Free Rad Bio Med 51:257–281

Fristachi A, Chaudhry H (2017) Cadmium. In: International encyclopedia of public health (vol 5, 2nd edn), pp 316–319

Fu D, He Z, Su S, Xu B, Liu Y, Zhao Y (2017) Fabrication of α-FeOOH decorated graphene oxide-carbon nanotubes aerogel and its application in adsorption of arsenic species. J Colloid Interface Sci 505:105–114

Gao TY, Zhou ZY (2000) The simple denitrification and phosphate removal transformation for municipal waste water treatment plant. J Tongji Uni (Sci) 28:324–327

Gao W, Majumder M, Alemany LB, Narayanan TN, Ibarra MA, Pradhan BK, Ajayan PM (2011) Engineered graphite oxide materials for application in water purification. ACS Appl Mater Interfaces 3:1821–1826

Guo L, Ye P, Wang J, Fu F, Wu Z (2015) Three-dimensional Fe_3O_4-graphene macroscopic composites for arsenic and arsenate removal. J Hazard Mater 298:28–35

Gupta A, Chauhan VS, Sankararamakrishnan N (2009) Preparation and evaluation of iron-chitosan composites for removal of As(III) and As(V) from arsenic contaminated real life groundwater. Water Res 43:3862–3870

Ha E, Basu N, O'Reilly SB, Dórea JG, Chan HM (2017) Current progress on understanding the impact of mercury on human health. Environ Res 152:419–433

Han C, Li H, Pu H, Yu H, Deng L, Huang S, Luo Y (2013) Synthesis and characterization of mesoporous alumina and their performances for removing arsenic (V). Chem Eng J 217:1–9

Huang NM, Lim HN, Chia CH, Yarmo MA, Muhamad MR (2011) Simple room-temperature preparation of high-yield large-area graphene oxide. Int J Nanomed 6:3443–3448

Hummers WS Jr, Offeman RE (1958) Preparation of graphitic oxide. J Am Chem Soc 80:1339

Jin Z, Zimo L, Yu L, Ruiqi F, Shams AB, Xinhua X (2015) Adsorption behavior and removal mechanism of arsenic on graphene modified by iron-manganese binary oxide ($FeMnO_x$/RGO) from aqueous solutions. RSC Adv 5:67951–67961

Johnston SG, Burton ED, Moon EM (2016) Arsenic mobilization is enhanced by thermal transformation of schwertmannite. Environ Sci Technol 50:8010–8019

Kao AC, Chu YJ, Hsu FL, Liao VHC (2013) Removal of arsenic from groundwater by using a native isolated arsenite-oxidizing bacterium. J Contam Hydrol 155:1–8

Khan TA, Chaudhry SA, Ali I (2013) Thermodynamic and kinetic studies of As(V) removal from water by zirconium oxide coated marine sand. Environ Sci Pollut Res 20:5425–5440

Khan TA, Chaudhry SA, Ali I (2015) Equilibrium uptake, isotherm and kinetic studies of Cd(II) adsorption onto iron oxide activated red mud from aqueous solution. J Mol Liq 202:165–175

Khatamian M, Khodakarampoor N, Oskoui MS (2017) Efficient removal of arsenic using graphene-zeolite based composites. J Colloid Interface Sci 498:433–441

Kulshrestha A, Jarouliya U, Prasad GBKS, Flora SJS, Bisen PS (2014) Arsenic-induced abnormalities in glucose metabolism: biochemical basis and potential therapeutic and nutritional interventions. World J Trans Med 3:96–111

Kumar S, Nair RR, Pillai PB, Gupta SN, Iyengar MAR, Sood AK (2014) Graphene oxide–$MnFe_2O_4$ magnetic nano-hybrids for efficient removal of lead and arsenic from water. ACS Appl Mater Interfaces 6:17426–17436

Kumar SK, Jiang SJ (2016) Chitosan-functionalized graphene oxide: a novel adsorbent an efficient adsorption of arsenic from aqueous solution. J Environ Chem Eng 4:1698–1713

Kumar SK, Jiang SJ (2017) Synthesis of magnetically separable and recyclable magnetic nanoparticles decorated with β-cyclodextrin functionalized graphene oxide an excellent adsorption of As(V)/(III). J Mol Liq 237:387–401

Li P, Du B, Chan HM, Feng X (2015) Human inorganic mercury exposure, renal effects and possible pathways in Wanshan mercury mining area China. Environ Res 140:198–204

Lin YJ, Cao WZ, Ouyang T, Chen BY, Chang CT (2017) Developing sustainable graphene-doped titanium nano tube coated to super paramagnetic nanoparticles for arsenic recovery. J Taiwan Inst Chem Eng 70:311–318

Machida M, Mochimaru T, Tatsumoto H (2006) Lead(II) adsorption onto the graphene layer of carbonaceous materials in aqueous solution. Carbon 44:2681–2688

Maliyekkal SM, Philip L, Pradeep T (2009) As(III) removal from drinking water using manganese oxide-coated-alumina: performance evaluation and mechanistic details of surface binding. Chem Eng J 153:101–107

Marcano DC, Kosynkin DV, Berlin JM, Sinitskii A, Sun Z, Slesarev A, Alemany LB, Lu W, Tour JM (2010) Improved synthesis of graphene oxide. ACS Nano 4:4806–4814

Matschullat J (2000) Arsenic in the geosphere—a review. Sci Total Environ 249:297–312

McAllister MJ, Li JL, Adamson DH, Schniepp HC, Abdala AA, Liu J, Alonso MH, Milius DL, Car R, Prud'homme V (2007) Chem Mater 19:4396

Mishra AK, Ramaprabhu S (2011) Functionalized graphene sheets for arsenic removal and desalination of sea water. Desalination 282:39–45

Mohan D, Pittman CU Jr (2007) Arsenic removal from water/wastewater using adsorbents-a critical review. J Hazard Mater 142:1–53

Mondal P, Balomajumder C, Mohanty B (2007) A laboratory study for the treatment of arsenic, iron, and manganese bearing ground water using Fe^{3+} impregnated activated carbon: effects of shaking time, pH and temperature. J Hazard Mater 144:420–426

Naushad M (2014) Surfactant assisted nano-composite cation exchanger: Development, characterization and applications for the removal of toxic Pb2+ from aqueous medium. Chem Eng J 235:100–108

Ng JC, Wang J, Shraim A (2003) Global health problems caused by arsenic from natural sources. Chemosphere 52:1353–1359

Park WK, Yoon Y, Kim S, Yoo S, Do Y, Kang JW, Yoon DH, Yang WS (2016) Feasible water flow filter with facilely functionalized Fe_3O_4-non-oxidative graphene/CNT composites for arsenic removal. J Environ Chem Eng 4:3246–3252

Peng W, Li H, Liu Y, Song S (2017) A review on heavy metal ions adsorption from water by graphene oxide and its composites. J Mol Liq 230:496–504

Platero E, Fernandez ME, Bonelli PR, Cukierman AL (2017) Graphene oxide/alginate beads as adsorbents: influence of the load and the drying method on their physicochemical-mechanical properties and adsorptive performance. J Colloid Interface Sci 491:1–12

Rasheed H, Kay P, Slack R, Gong YY, Carter A (2017) Human exposure assessment of different arsenic species in household water sources in a high risk arsenic area. Sci Total Environ 584–585:631–641

Ray SK, Majumdera C, Saha P (2017) Functionalized reduced graphene oxide (RGO) for removal of fulvic acid contaminant. RSC Adv 7:21768–21779

Robinson T (2017) Removal of toxic metals during biological treatment of landfill leachates. Waste Manage 63:299–309

Rodríguez J, Mandalunis PM (2016) Effect of cadmium on bone tissue in growing animals. Exp Toxicol Pathol 68:391–397

Roy E, Patra S, Madhuri R, Sharma PK (2016) Europium doped magnetic graphene oxide-MWCNT nanohybrid for estimation and removal of arsenate and arsenite from real water samples. Chem Eng J 299:244–254

Sang JQ, Zhang XH, Wang ZS (2003) Improvement of organics removal by bio-ceramic filtration of raw water with addition of phosphorus. Water Res 37:4711–4718

Sansone V, Pagani D, Melato M (2013) Chronic arsenicals dermatoses from tube-well water in West Bengal during 1983–87. Clin Cases Miner Bone Metab 10:34–40

Schniepp HC, Li JL, McAllister MJ, Sai H, Alonso MH, Adamson DH, Prud'homme RK, Car R, Saville DA, Aksay IA (2006) Functionalized graphene sheets derived from splitting graphite oxide. J Phys Chem B 110:8535–8539

Sharma G, Naushad M, Al-Muhtaseb AH, Kumar A, Khan MR, Kalia SS, Bala M, Sharma A (2017) Fabrication and characterization of chitosan-crosslinked-poly(alginic acid) nanohydrogel for adsorptive removal of Cr(VI) metal ion from aqueous medium. Int J Biol Macromol 95:484–493

Sherman DM, Randall SR (2003) Surface complexation of arsenic (V) to iron(III) (hydr)oxides: structural mechanism from ab initio molecular geometries and EXAFS spectroscopy. Geochim Cosmochim Acta 67:575–580

Sheshmani S, Nematzadeh MA, Shokrollahzadeh S, Ashori A (2015) Preparation of grapheme oxide/chitosan/FeOOH nanocomposite for the removal of Pb (II) from aqueous solution. Int J Biol Macromol 80:475–480

Siddiqui SI, Chaudhry SA, Islam SU (2017) Green adsorbents from plant sources for the removal of arsenic: an emerging wastewater treatment technology. In plant-based natural products: derivatives and applications, Ed. Islam SU, John Wiley & Sons, Inc, pp. 193–215

Siddiqui SI, Chaudhry SA (2017a) Arsenic removal from water using nano-composites: a review. Cur Environ Eng 4:81–102

Siddiqui SI, Chaudhry SA (2017b) Arsenic: toxic effects and remediation. In advanced materials for wastewater treatment, Ed. Islam SU, John Wiley & Sons, Inc, pp. 1–27

Siddiqui SI, Chaudhry SA (2017c) Iron oxide and its modified forms as an adsorbent for arsenic removal: a comprehensive recent advancement. Process Saf Environ Prot 111:592–626

Siddiqui SI, Chaudhry SA (2017d) Removal of arsenic from water through adsorption onto metal oxide-coated material. Mater Res Found 15:227–276

Siddiqui SI, Ravi R, Rathi G, Tara N, Islam SU, Chaudhry SA (2018) Decolorization of textile wastewater using composite materials. In nano materials in the wet processing of textiles, Ed. Islam SU, Butola BS, John Wiley & Sons, Inc, pp. 187–218

Staudenmaier L (1898) Verfahren zur darstellung der graphitsaure. Ber Dtsch Chem Ges 31:1481–1487

Su H, Ye Z, Hmidi N (2017) High-performance iron oxide–graphene oxide nanocomposite adsorbents for arsenic removal efficient removal of arsenic using graphene-zeolite based composites. Colloids Surf A: Physicochem Eng Asp 522:161–172

Wang H, Yuan X, Wu Y, Huang H, Zeng G, Liu Y, Wang X, Lin N, Qi Y (2013) Adsorption characteristics and behaviors of graphene oxide for Zn (II) removal from aqueous solution. Appl Surf Sci 279:432–440

Watanabe CH, Monteiro ASC, Gontijo ESJ, Lira VS, Bueno CDC, Kumar NT, Fracácio R Rosa AH (2017) Toxicity assessment of arsenic and cobalt in the presence of aquatic humic substances of different molecular sizes. Ecotoxicol Environ Saf 139:1–8

Xubiao L, Cheng W, Shenglian L, Ruizhi D, Xinman T, Guisheng Z (2012) Adsorption of As(III) and As(V) from water using magnetite Fe_3O_4-reduced graphite oxide-MnO_2 nano composites. Chem Eng J 187:45–52

Yang G, Cao J, Li L, Rana RK, Zhu JJ (2013) Carboxymethyl chitosan-functionalized graphene for label free electrochemical cytosensing. Carbon 51:124–133

Yoon Y, Park WK, Hwang TM, Yoon DH, Yang WS, Kang JW (2016) Comparative evaluation of magnetite–graphene oxide and magnetite-reduced graphene oxide composite for As(III) and As (V) removal. J Hazard Mater 304:196–204

Yoon Y, Zheng M, Ahn YT, Park WK, Yang WS, Kang JW (2017) Synthesis of magnetite/non-oxidative graphene composites and their application for arsenic removal. Sep Pure Technol 178:40–48

Yu L, Ma Y, Ong CN, Xie J, Liu Y (2015) Rapid adsorption removal of arsenate by hydrous cerium oxide–graphene composite. RSC Adv 5:64983–64990

Zhang G, Qu J, Liu H, Liu R, Wu R (2007) Preparation and evaluation of a novel Fe-Mn binary oxide adsorbent for effective arsenite removal. Water Res 41:1921–1928

Zhou Q, Zhong YH, Chen X, Liu JH, Huang XJ, Wu YC (2014) Adsorption and photo catalysis removal of fulvic acid by TiO_2-graphene composites. J Mater Sci 49:1066–1075

Chapter 10
Bromate Formation in Drinking Water and Its Control Using Graphene Based Materials

Mu. Naushad, P. Senthil Kumar and S. Suganya

Abstract The increasing interest in the use of ozone in drinking water treatment has led to concern over the formation of ozonation by-products. Disinfection with various hypochlorite solution results in the bromate formation which is being carcinogenic tested against laboratory animal. Looking at near future, the bromate concentration level is beyond the limit of provisional water guidelines drawn by the World Health Organization. From the analysis of low dose bromate extrapolation models at an upper 95% confidence limit risk of 1 in 10^5 of the population, the results are life threatening. Possible control options originate from 'design-for-purpose' to control bromate formation or reduction scientifically. In this chapter, the potential grapheme based materials in creating next-generation and ground-breaking solutions to the water challenges of our times have been discussed. It is believed that grapheme based materials can meet water challenges in a sustainable ways.

Keywords Bromate · Bromide · Ozone · Chlorination · Disinfection

1 Introduction

Water containing various heavy metals, radionuclides and carcinogenic agents such as arsenic, cadmium, cobalt, chromium, copper, mercury, uranium, thorium, nitrate and bromate are released by several sectors like mining industry, sewage irrigation,

Mu. Naushad
Department of Chemistry, College of Science, King Saud University, Riyadh, Saudi Arabia
e-mail: mnaushad@ksu.edu.sa

P. Senthil Kumar (✉) · S. Suganya
Department of Chemical Engineering, SSN College of Engineering, Chennai 603110, India
e-mail: psk8582@gmail.com; senthilkumarp@ssn.edu.in

S. Suganya
e-mail: suganyas@ssn.edu.in

M. Naushad (ed.), *A New Generation Material Graphene: Applications in Water Technology*, https://doi.org/10.1007/978-3-319-75484-0_10

metal plating, electronic and nuclear industry, including more specifically wastewater treatment industry leads to water contamination (Oliveira et al. 2012). The toxicity of those compounds highly depend on several factors, including concentration, exposure time, oxidation states and chemical species, as well as the age, gender, genetics and nutritional status of exposed individuals. Toxic metals, i.e., arsenic, cadmium, chromium, lead and mercury, sodium and potassium bromate have high degree of toxicity, which rank among the priority agents causing cancer and are also classified as human carcinogens (known or probable) according to the U.S. Environmental Protection Agency, and the International Agency for Research on Cancer Potentially-toxic metals, e.g., Cu, Fe, Mg, Co, Ni, Zn and Br, are essential for microbial growth and metabolism at low concentrations; however, they are toxic in excess amounts, which produce cellular and tissue damage leading to a variety of adverse effects and human diseases.

Bromate is well-known oxyhalide obtained as byproduct of disinfection of drinking water via ozonation (Naushad et al. 2016a; Khan et al. 2014). It can easily contaminate drinking water. Bromate is also classified as one of a carcinogenic and genotoxic compound by the International Agency of Research on Cancer (Zhang and Yang 2013). In addition, the US Environmental Protection Agency (EPA), the European Council (EC) and the World Health Organization (WHO) are stated that a maximum permissible level of bromate in drinking water is 10 μg/L. The excess level of bromate in drinking water leads to the damage of kidneys, nervous system and renal cell tumors (Naushad et al. 2013, 2016b). Hence, the accurate and low level detection of bromate is of great importance in real samples for human health. So far, the different analytical methods have been developed for sensitive detection of bromate, including spectrophotometry, chromatographic methods, florescent and electrochemical methods.

Various active physical and chemical processes have been extensively employed to treat wastewater containing these inorganic toxic species, such as adsorption, coagulation, catalysis, ion exchange, and membrane technology. Among these approaches, adsorption technique has been used as an attractive alternative for removing these toxic metals due to its easy operation, low-cost, and environmental friendly (Romele 1998). Recently, numerous researches have been studied through various technologies using a variety of adsorbent materials, including biosorbents (e.g., *Saccharomyces cerevisiae*), agricultural wastes (e.g., sweet sorghum stalk), natural and modified minerals (e.g., zeolites), bentonite and vermiculite, polymers (e.g., chitosan), carbon-based materials (e.g., activated carbon, carbon nanotube, graphene, biochar and carbon soot), nanoparticles (e.g., CuO, CdS, gold, iron oxide, TiO_2 and ZnO). Among them, graphene-based materials are becoming promising option (Cavalli et al. 2005).

Graphene, a single-atom-thick planar sheet of sp^2-bonded carbon atoms arranged in a hexagonal crystalline structure, possesses unique physicochemical properties, notably the exceptionally large specific surface area (theoretically 2630 m^2/g), rapid heterogeneous electron transfer, high thermal conductivity, and outstanding mechanical strength (Ali 2012). In the environmental field, graphene and graphene-based materials have been applied to develop novel adsorbents, membranes, photocatalysts, electrode materials, or disinfectants for contaminant monitoring or removal. Such graphene based materials show consistent role in simultaneous bromate detection and reduction in various forms for real-time-applications.

2 Bromate Formation

Bromate is a contaminant usually found in drinking water (Haag and Hoigné 1983) as a result of water treatment using commercially produced solutions of sodium hypochlorite used for disinfection of drinking water, rather than through source water contamination. The presence of bromate in drinking water is primarily occurred during the reaction between naturally arising bromide in source water and ozone (Naushad et al. 2015). This is due to the presence of bromide in the raw materials such as chlorine and sodium hydroxide used while manufacturing of sodium hypochlorite. Indeed, disinfection of drinking water with chlorine produces a number of undesirable and potentially cancer-causing organic disinfection byproducts (DBPs) such as trihalomethanes and haloacetic acids.

Water treatment facilities eventually employ ozone to minimize the production of such harmful compounds, followed by chloramination for disinfection. Though, ozone is a powerful oxidant but unstable in water, requiring disinfectant for the purpose of maintaining a disinfectant residual in water distribution systems. Nevertheless, bromate is a major DBP caused from bromide-containing source waters that undergo ozonation. Inspite, bromide is not oxidized by chlorine dioxide (ClO_2), so the use of ClO_2 do not negatively generate hypobromous acid, hypobromite ion or bromate (US EPA 1999). A simultaneous exposure of chlorine dioxide and light can form bromate, leads to thermodynamically unfavourable reaction and treatment conditions. On the other side, electrolytically generated hypochlorous acid solutions are able to form bromate when the presence of bromide is in brine found.

2.1 *Industrial Applications of Bromate*

Sodium and potassium bromate are powerful oxidizers used as a chemical reagent to mature flour during milling and to condition dough during baking. They also have been used as oxidizers mainly in permanent wave neutralizing solutions and the dying of textiles using sulfur dyes. Additionally, treating barley in beer making with potassium bromate for the improvement of the quality of fish paste products in Japan is inappropriate (Ministry of Health and Welfare 1979) leading to the presence of traces of bromate in food as consumed. Bromate has been found in the consumer products like cosmetics in which sodium and potassium bromate are major reagents used. Little bromate is expected to be found in air or soil as well in ozonated bottled water. Bromate is category I group B2 carcinogen, regulated in treated drinking water by Safe Drinking Water Act with a maximum contaminant level (MCL) of 10 g/L. Such guidelines have been drawn based on the 10^{-4} excess lifetime cancer risk from exposure to bromate in drinking water at 5 g/L. The maximum acceptable concentration of 0.01 mg/L (10 μg/L) undergoes for public consultation.

2.2 *Environmental Fate—Exposure of Bromate*

Bromate is being strong oxidant, does not volatilize and is only adsorbed onto soil or sediment. It mostly reacts with organic matter, leading to the formation of bromide ion.

2.2.1 Water

At this time, there are no data for groundwater plants (US EPA 2001). Many researchers conducted survey on the bromate formation and its concentration in different sources based on the treatment processes and conditions involved. Bromate is not normally found in water. Indeed, conversion of bromide to bromate upon ozonation is affected by several factors that are organic matter, pH and temperature and others. The level of bromate presence is compared with (over time or as a function of concentration × time, or CT). The use of CT is a significant factor or indicator to describe the relative rate of bromate formation with effect of disinfection efficiency. The rate of formation of bromate ion may increase with temperature (AWWARF 1991) and increased alkalinity during ozonation. However, ozone characteristics play major role in the rate of bromate formation. As mentioned earlier, ozone becomes unstable with increasing temperature and/or alkalinity. Among all the factors are concerned equal, bromide concentration and ozone dose are the best predictors of bromate formation during ozonation.

World Health Organization (WHO)—International Programme on Chemical Safety (IPCS) establishes the sound management on bromate control in drinking water. A study conducted by European utilities found bromate levels ranging from less than the detection limit (2 μg/L) to 16 μg/L in finished water. Canada reported an average concentration of 3.17 μg/L, with a range of 0.73–8.00 μg/L. In ozonated bottled water, the average level of bromate was found 18 μg/L with a range of 4.3–37.3 μg/L. The range of bromate concentration after with a variety of source water characteristics after ozonation was <2–293 μg/L depending on dissolved organic carbon. American Water Work Association (AWWA) has reported a mean level of bromate in finished waters of 3.06 μg/L and a median level of 3.64 μg/L is acceptable. These reports demonstrate the high rates of conversion of bromide to bromate are pure laboratory studies with very high bromide levels and thus may not be representative of conversion rates at environmentally relevant doses. US

Environmental Protection Agency (EPA) reported annual mean bromated concentration in finished surface water was 2.9 μg/L, with a range of <0.2–25 μg/L; the annual mean concentration of bromate in the same waters determined using the utility method was 2.1 μg/L, with a range of <5–50 μg/L for comparison. As a result, the minimum reporting using the utility method can be 5 μg/L, while for the EPA method is 0.2 μg/L.

Over all, optimal dosage of ozonation is not only responsible or causing a substantial bromate formation. It should be noted that the disinfection target using G. lamblia in the number of water plants with bromate concentrations above 10 mg/L increases according to the optimized biomass implementation which also require high dosage of ozone (Rice and Gomez-Taylor 1986).

2.2.2 Food

Bromate has been added to beer or fish paste for treating barley. Small amounts of bromate (≤ 30 mg/kg) are allowed to add to flour or dough during the preparation of bread, but this is broken down to bromide during baking. However, increasing residual potassium bromate is detected, when bread was made from flour doughs containing >50 mg of potassium bromate per kg of dough, either by bulk fermentation or by mechanical development. The previous acceptable level of treatment of flours for bread making was 0–60 mg of bromate per kg of flour, has been withdrawn later and inappropriate. In the USA, the Code of Federal Regulations allows up to 50 mg/kg in finished bromated flour and up to 75 mg/kg in finished bromated whole wheat flour if this addition improves the baking qualities (US EPA 1999; Gordon and Emmert 1996). However, the US FDA has suggested that bakers voluntarily decrease potassium bromate use, and, if bromate is used, it must be indicated on packages.

2.3 Total Exposure and Relative Contribution of Drinking-Water

For most people, exposure to bromate is unlikely to be practical. If ozone is used to disinfect drinking-water, intake of bromate might range from 120 to 180 μg/day (Hoigné and Bader 1994). In the USA, the amount of residual inorganic bromides in fermented malt beverages may not exceed 25 μg/L (calculated as bromine). The levels in baked goods and products made from bromated flour are expected to be low, since bromates are broken down to bromide during baking (IARC 1986).

2.4 Mechanism on by-Product Formation

Bromide ion (Br) usually does not pose direct public health ramifications in drinking water supplies. The presence of average concentration in drinking water along with Natural Organic Matter (NOM) in some natural systems such as water intrusion or anthropogenic sources, run-off pesticides and industrial waste, ozone (O_3) oxidizes Br to hypobromous acid (HOBr), which can then oxidize NOM to brominated compounds. If in the case of absence of NOM, O_3 can still oxidize Br to HOBr or the ionized form of HOBr, hypobromite (OBr). These compounds can further react to produce bromate ions (BrO_3) and Br. pH, O_3 concentration, contact time that take place upon the ozonation also affect the formation of BrO_3. HOBr may then react with NOM to form brominated organics such as bromoform, brominated acids, bromoacetonitriles, etc. HOBr is primarily responsible for bromination reactions with NOM while its ionized form, or conjugate base, OBr), is the primary intermediate in the oxidation of Br) to BrO_3. Br^- may be oxidized either to BrO_3 or HOBr (Mack 1988). Further HOBr reacts with NOM to form brominated organics such as bromoform, brominated acids, bromoacetonitriles, etc. While at HOBr ionization, they become primarily responsible for bromination reactions with NOM or conjugate base OBr^-, since it is been intermediate in the oxidation of Br^- to BrO_3. The formation of organic bromides is favorable at lower pH while a higher pH favors the intermediate formation of OBr^- which inevitably leads to BrO_3^- formation. By determining the reaction rate can be useful in calculating the effects that O_3 exposure, pH, and influent Br^- concentration and BrO_3 relative to that for NOM-free water. Based on the review, it is concluded that low Br^- concentration and O_3 in drinking water first reacts with NOM, while at high or increasing Br^- concentration at constant organic matter concentrations and O_3 dosage leads to the formation of bromo-substituted organics in increased manner, particularly at a low pH. The occurrence of BrO_3^- in water containing high Total Organic Carbon (TOC) is due to the slow reaction between O_3/Br^- and O_3/OBr^- and the faster reaction between O_3/NOM and OBr^-/NOM. In conclusion, changing water quality parameters affect not only by-product formation but also the efficiency of disinfection.

2.5 Effects on Humans

Accidental or intentional ingestion of human poison bromate is mostly from home permanent wave solutions that contains either 2% potassium bromate or 10% sodium bromated (Fielding and Hutchison 1993). The most bromate poison suffering is children with ingestion of 60–120 ml of 2% potassium bromated (equivalent to 46–92 mg of bromate per kg of body weight per day for a 20-kg child). WHO estimated the lethal doses of potassium bromated to be 200–500 mg/kg of body weight (150–385 mg of bromate per kg of body weight).

Toxic effects of bromate salts include nausea, vomiting, abdominal pain, anuria and diarrhoea, varying degrees of central nervous system depression, haemolytic anaemia and pulmonary oedema. Most of the time, these effects are reversible including renal failure and deafness, observed the ingestion of 240–500 mg of potassium bromate per kg of body weight (185–385 mg of bromate per kg of body weight).

3 Disinfection Chemistry

3.1 Ozonation

Considerable research has been conducted to avail the mechanisms for better understanding of the formation of bromate in ozonated water. Ozone reacts with bromide ion in natural water source in three ways. Direct pathway is the first whereby $HOBr/OBr^-$ forms as a result of reaction between O_3 and Br^-; additional oxidation of OBr^- by O_3 ultimately forms bromate. The second pathway is referred to as a "direct/indirect" pathway whereby $HOBr/OBr^-$ forms followed by additional oxidation by both hydroxyl radicals OH^- and O_3 ultimately forms bromate. In the third pathway, referred to as an "indirect/direct" pathway, bromide is oxidized by OH^- to form bromine radicals; additional oxidation by O_3 ultimately forms bromate. When both O_3 and OH^- cause oxidation process, the reaction becomes extremely complex and nonlinear. Since, both arte powerful oxidants with oxidation potentials of 2.70 electron volts (eV), respectively, compared with chlorine at 1.49 eV contribute to oxidation. Comparatively, OH^- is a powerful oxidant than O_3 because of its high reactivity with both organic and inorganic compounds in source water. Knowledge of the reaction pathways can help water utilities select the most appropriate control strategy. If the dominant pathways are blocked, bromate formation can be minimized (17) (Fig. 1).

3.2 Hypochlorite Treatment Chemicals

Hypochlorite solutions are used as disinfectants in water treatment, produced by electrolyzing sodium or calcium chloride brines. The amount of bromate contamination is dependent on the sources of the salt used to produce brines for electrolysis for both commercially produced bulk hypochlorite. Brines are natural evident for the co-existence bromide with chloride and is oxidized to bromate during the manufacturing process (Upadhyay et al. 2014). As a result, the type of electrolysis cells and the bromide content in the hypochlorite solution extremely impact the process utilities. To manufacture commercial bulk hypochlorite include diaphragm, mercury and membrane cells, each of them contribute 60, 40 and 20% in the

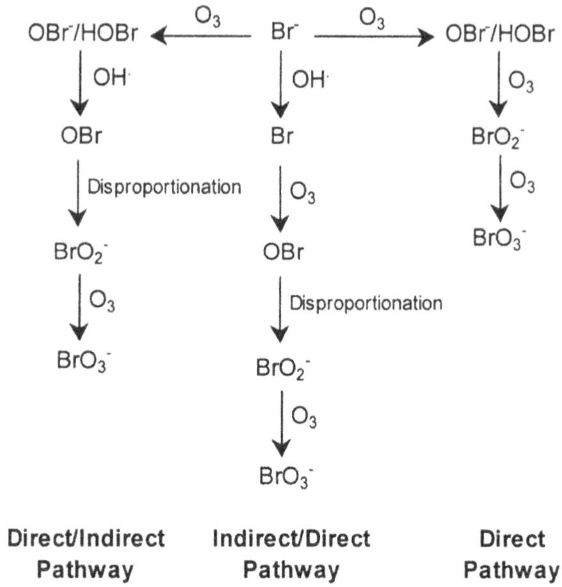

Fig. 1 Mechanisms for bromate formation. Br^-: Bromide; O_3: Ozone; OBr^-: Hypobromite; HOBr: Hypobromous acid; OH^-: Hydroxyl radical; Br: Bromine; radical OBr: Bromine oxide radical; BrO_2^-: Bromite; BrO_3^-: Bromate

formation of sodium or calcium hydroxide when the bromide is originated from the brine; when a mercury or membrane electrolytic cells used; and 80% in the chlorine. Combination of sodium or calcium hydroxide and chlorine in producing the hypochlorite product, 100% efficiency of the conversion of bromide ion in the hydroxide product to bromate can be obtained.

3.3 *Recommendations*

It is recommended that water utilities establish a quality control program to verify product quality and manage solution storage. AWWA Standard B300 also recommends a verification program to confirm that hypochlorite treatment chemicals meet specifications (Wang and Chen 2015).

In summary, to minimize bromate concentrations in drinking water when using hypochlorite solutions, water utilities should:

- specify hypochlorite treatment chemicals that are certified as meeting NSF/ANSI Standard 60 in their purchasing documents; for water utilities using onsite-generated hypochlorite systems, use a low-bromide salt;
- follow the handling and storage recommendations outlined in Appendix B; and
- establish a quality control program to verify product quality and manage solution storage.

4 Factors Influencing Bromate Formation

4.1 Bromide Concentration

Bromide ion is naturally found in drinking water from saltwater intrusion and the dissolution of geological formations, or it can enter water sources by human activities such as the spreading of methyl bromide on crops, the spreading of salt on roads during winter and sewage or industrial effluent discharges. Disinfection-by-product (DBP) concentrations due to bromide discharges (Ding et al. 2016) to the environment by particular human activities include salt mining, textile manufacturing and hydraulic fracturing operations and blowdown wastes from coal power plants.

Bromide is more detected in treated water distribution system than the source water contamination. In concern, the average raw water bromide concentration is fixed by the US approximately 0.1 mg/L in a study undertaken to assess the nationwide occurrence of bromide in preparation for new DBP regulations. Water sources are highly dependent on rainwater as raw water) untreated is of primary concern for bromide concentration. As well the bromide concentrations in rainwater and snow range are found approximately from 5 μg/L to >150 μg/L. Conceptually, increasing bromide ion concentration leads to increase in bromate formation during ozonation. However, because other water quality parameters influence the oxidation process, it is difficult to instinctly predict bromate formation potential. The water source with low bromate concentration is found to contain natural ammonia, which inhibits bromate formation.

Water utilities should have a good understanding of the bromide concentrations in their source water to control the formation of bromate and other brominated DBPs. The maximum amount of bromate occurrence can be calculated if 100% of bromide is converted to bromate. This is useful as quality control measure for water utilities, ensures low Minimum Reporting Level (MRL) for bromide analysis (Gadd 2009). A minimum MDL of 0.01 mg/L (10 μg/L) is recommended for bromide.

4.2 Ozone Dose

Ozone dosage plays an important role in bromate formation. A higher O_3 dose leads to higher bromate formation in general. The water utility's treatment objectives and water quality characteristics may get fluctuated by O_3 fixation and seasonal changes in temperature as result in natural organic matter (NOM) and other water quality characteristics. The locality establishment for O_3 injection must fulfill the treatment objectives (i.e., pre-ozonation or intermediate ozonation). O_3 demand is proportional to Dissolved Oxygen Concentration (DOC) relevant to the drinking water industry. Seasonal changes (Wei et al. 2014) to optimize the applied O_3 dose at all times while minimizing the formation of bromate and other DBPs must be prioritized to meet lowest possible O_3 dose during operation. Water quality, O_3 consumption and energy consumption are secondary factors to ensure that treatment objectives are satisfied. It is critical that the effectiveness of disinfection not be compromised by any method used to control DBP concentrations in drinking water.

4.3 Temperature

Temperature is a rate determining factor on bromate formation is threefold: dissolved O_3 is more stable at lower temperatures, oxidation rate increases with increased temperature and the pKa of the $HOBr/OBr^-$ system decreases as temperature increases.

The influence of temperature from 5 to 35 °C on the reaction rate of O_3 and OH^- reflects to O_3 decay at pH 8.0 with increased temperature. O_3 is considerably having temperature dependency whereas the OH^- reactions have very little temperature dependency. Therefore, reaction pathways involving O_3 will be more affected by temperature than those involving OH^-. Pathogen inactivation during treatment is extremely dependent on temperature and becomes more difficult at low temperature. Typical DBP (Dong et al. 2013) formation increases at higher temperature, yet seasonal changes are again a big question which collapse water quality goals accordingly.

4.4 pH

The effect of pH on bromate formation is important in determining the equilibrium of the $HOBr/OBr^-$ system. A low pH drives the equilibrium towards HOBr, which is less amenable to form bromated, whereas a high pH drives the equilibrium towards OBr^-, which readily reacts with O_3 and OH^- to form bromate. Because at lower pH, the O_3 decay rate is also slower and meet with disinfection targets. At lower pH, the bromate formation is reduced because the HOBr species, predominant below pH 8.8, is less readily oxidized by O_3 or OH^-. The reduction rate from 30–50% can be achieved per unit decrease of pH when the initial pH is above 7.5. Water utilities consider that pH depression seems advantageous to minimize bromate formation, particularly for water with low alkalinity (Xiang et al. 2012). It is applicable towards acid addition prior to enhanced coagulation treatment. The conditions for bromated control are

- a low pH is undesirable with respect to corrosion control; hence, supplementary pH adjustment may be necessary;
- a lower pH increases the formation of HOBr, which can react with organic matter to form brominated DBPs;

- pH adjustment increases the total dissolved solids (TDS) concentration of the water; and
- the feasibility of pH depression may be limited for high-alkalinity sources owing to the amount of acid that can reasonably be added.

In addition, the common chemicals used to depress pH include carbon dioxide (CO_2) and sulphuric acid (H_2SO_4). Carbon dioxide may be preferred as a safer alternative to H_2SO_4 and because it stabilizes O_3 to a greater extent; however, it is a weaker acid, and costs may be prohibitive if a significant pH change is necessary.

4.5 Alkalinity

Alkalinity is an important water quality parameter that is known to affect bromate formation. Carbonate system is the primary source of alkalinity, which is assumed to be at equilibrium in water treatment processes, whereas bicarbonate (HCO_3^-) and carbonate (CO_3^{2-}) act as OH^- scavengers. Thus, increasing alkalinity leads to decrease bromate formation by blocking the "direct/indirect" and "indirect/direct" pathways. However, alkalinity scavenges OH^- and produces a carbonate radical (CO_3^-) which is a mild oxidant. Therefore, bromate formation pathways or O_3 decay may be blocked or reduced when alkalinity scavenges OH^-. As a result, more bromate can be formed ultimately via the "direct" pathway (Perreault et al. 2015), or oxidization of OBr^- by CO_3^-. From survey conducted, reaction kinetics are shown complex which confirms that alkalinity can either inhibit or promote bromate formation. pH variation influences the bromated formation. For example, at pH 7, increased alkalinity reduces bromate formation, whereas at pH 8, it increases bromate formation. Remarkably, an increase in alkalinity results in increased bromate formation as a result Mao of the oxidation of OBr^- by CO_3^-. The confounding effects regarding alkalinity is influenced by source water characteristics with seasonal change but no natural Organic Matter (NOM) is considered. Water utilities should be aware of how their process responds for bromate and other DBP formation so that they can adjust their water quality goals accordingly.

4.6 Organic Matter

As mentioned in previous section, Natural organic matter (NOM) is another important water quality parameter that is known to affect bromate formation. NOM is referred to describe the complex composition of organic material present in source waters. NOM reactivity in water is measured generally by TOC, DOC and UV absorbance at a wavelength of 254 nm (Mao et al. 2015). The chemical fractions of NOM are thrust interest and can be measured to assess hydrophobicity and treatability characteristics using liquid chromatography with organic carbon detection.

Seasonal change of aquifer and its characteristics during rain or by algal blooms can affect the NOM reactivity in the source water area. It is important to know how NOM works

- First, NOM exerts an O_3 demand, which may affect the performance of disinfection.
- Second, NOM reacts with $HOBr/OBr^-$ to form DBPs, such as bromoform and bromoacetic acids.
- Third, NOM scavenges OH^-, which may block the "direct/indirect" and "indirect/direct" bromate formation pathways.
- Fourth, when NOM reacts with O_3, it forms OH^-, hydrogen peroxide (H_2O_2) and superoxide (O_2^-), all of which accelerate O_3 decay.

Bromate concentration becomes lower when NOM is present, because the reaction between O_3 and OH^- with NOM reduces the amount of oxidant available to oxidize bromide that confirms the amount of bromate formation is dependent on the nature and amount of the NOM. Ozone reacts preferentially with certain NOM fractions, whereas OH^- reacts unselectively. Likely, the NOM hydrophobic fraction is less effective at scavenging OH^-.

The influence of NOM compositions on bromate formation for several source waters is relevant to hydrophobic fraction. Such confounding effects can even inhibit bromate formation. Beside, NOM is specific and beneficial to compensate O_3 demand, O_3 decay and disinfection prior to ozonation. This would suggest that intermediate ozonation is preferable to pre-ozonation. Over all, the production of bromate increases as the O_3/DOC ratio increases, suggesting that the ratio of O_3/DOC should be kept low to minimize bromate formation. This can be achieved by reducing the NOM fraction that consumes O_3, thereby lowering the ozone dose. For better understanding, the water source and the nature and generation of NOM according to seasonal change and how it relates to O_3 demand and bromate formation must be thoroughly studied before experimenting. The calculation of Specific UV Absorbance (SUVA) (Li et al. 2014) can provide water utilities with a good measure of NOM reactivity and be an indicator of changes in source water quality; this would require water utilities to monitor for DOC and UV254 as process control tool not a bromate control strategy.

5 Bromate Removal Strategies

5.1 Bromate Minimization Before Formation

Bromate must be minimized or to be eliminated during its formation process in several treatment options including pH depression and ammonia addition. They have turned out to be feasible in drinking water treatment. The depression of pH to below 7 prior to ozonation will result in the reduction on bromate formation. 50–

63% of bromate formation can be reduced for decreasing a one-pH-unit. Likely, HOBr has a pKa of 8.8–9.0, at pH less than 7, the amount of hypobromite (OBr_2) decrease. Consequently, the level of bromate is also reduced. Bromate formation through hydroxyl radical pathway is limited with decreasing pH, the generation of hydroxyl radicals is reduced. Moreover, lower ozone dosage is required to achieve the same level of disinfection than ozone dosage (Guo et al. 2010) required at ambient pH. However the degradation of the micropollutants is getting slow down while the formation of hydroxyl radicals are retarded. Since, pH depression is a significant factor lead to HOBr formation and then reacts with NOM, resulting in the formation of brominated organics such as bromoform and cyanogen bromide.

5.1.1 Addition of Ammonia

Another way to control bromate formation is ammonia addition. To achieve 50% reduction, ammonia reacts with HOBr, contributes to the formation of monbromamine. 200 mg/L ammonia concentration can compensate effective reduction of bromate, later more concentration of ammonia gets saturated in its function. In some cases, ammonia yields inconsistent reduction rate from 0 to 30%. Meanwhile, unstable bromamines react with ozone to release nitrate and bromide, which acts as a catalyst in ammonia oxidation. During this, ammonia is converted to nitrate, released bromide can react with ozone or hydroxyl radicals (HO^-) again to produce bromate. It is concluded that from literature, not all but in some cases ammonia addition and pH depression play a major role in bromate reduction below the current bromate MCL. A bromated minimization takes a step forward to chlorine-ammonia process, including a prechlorination followed by ammonia addition and ozonation. This strategy is capable of reducing bromated formation levels below the standard of 10 mg/L.

5.1.2 Addition of H_2O_2

Advanced Oxidation Process (AOPs) is more powerful forming Peroxone, after the combination of ozone and hydrogen peroxide. H_2O_2 can be added or generated in situ by O_3 photolysis in which H_2O_2 initiate the O_3 decomposition cycle by a single electron transfer process involving its conjugate base, HO_2^- to show the efficiency of O_3 boosted by H_2O_2. During the reaction, H_2O_2 is acting as a scavenger at high concentrations. BrO_3 formation is highly dependent on pH. At a high pH, H_2O_2 addition shows decreased formation and high at a lower pH. Gradual H_2O_2 addition leads to O_3 decomposition and substantially decreases carbon concentration. Another way to control bromated formation is α-irradiation. At pH range 6–8 showing bromide containing solution subject into $HOBr/OBr^-$ radical mechanism. $HOBr/OBr^-$ itself act as intermediate in bromate formation during the process. Then, H_2O_2 plays a major role at its presence, leads to the reduction of $HOBr/OBr^-$ (Vilian et al. 2016) for the purpose of controlling bromated formation.

AOP is failed to produce HO radical concentration which is unable to compensate for this rapid reduction reaction. In fact, if no bromated is formed at the presence of H_2O_2, it is concluded that only HO is oxidizing $HOBr/BrO^-$.

5.1.3 Hypochlorite Solutions

To minimize the potential poison of bromate, handling, operational and storage conditions must be taken extra care by manufacturers. Light, warmer temperatures, organic matter and certain heavy metal cations such as copper, nickel and cobalt, accelerate the decomposition of the chlorine in the hypochlorite solution leads to the lesser addition of bromate in drinking water. However, the bromate concentration does not show fluctuation during storage. As a result, a decrease in chlorine concentration in the hypochlorite solution will lead to an increase in chlorine dose to maintain disinfection targets; the increased chlorine dose will result in a higher bromate concentration in the treated water. Based on the regulations drawn for storage of hypochlorite solutions and real time use in water utilities can minimize the concentrations of contaminants in hypochlorite solutions. As part of the treatment process, the maximum concentration is referred to as the single product allowable concentration (SPAC) for impurities such as bromated can directly be added to drinking water. It also monitors hypochlorite dosage for bromated drinking water to avoid negative effect on the value of the SPAC. To promote drinking water quality, hypochlorite treatment chemicals must meet the NSF/ANSI Standard 60 in their purchasing documents.

The bromate contribution to drinking water during chlorination will therefore depend on the bromate and chlorine concentrations in the sodium hypochlorite solution and the applied chlorine dose. For on-site sodium hypochlorite generators, NSF/ANSI Standard 60 specifies that bromide should not exceed 54 mg/kg in NaCl at chlorine maximum feed concentration of 10 mg/L provided there is no bromide in the water used for brine. NSF/ANSI Standard 60 allows a higher concentration of bromide for generators delivering lower maximum feed concentrations of chlorine but the total concentration of bromate must not exceed 0.003 mg/L (NSF 1999).

5.2 *Removal of Bromate After Formation*

The Best Available Technology (BAT) must offer control of the ozonation process to minimize bromate formation, it is technically possible to remove bromate after formation using several treatment approaches, including activated carbon, ferrous iron addition, UV irradiation, membranes and anion exchange resins. For full scale applications, these methods have been evaluated primarily at bench and pilot scales, and little information is available for researchers. However, all of them are cost effective and technically feasible to remove bromate from treated water.

5.2.1 Adsorption with Activated Carbon

Granular activated carbon (GAC) filters for water utilities has been encourageable for bromate concentrations in both the influent and effluent. As employing O_3 in the removal of bromate, is ineffective at full scale, which has given platform for other technologies. The bromate removal is dependent on the pH of the source water, the type of activated carbon and contact time; On other side, Powdered activated carbon (PAC) is resulting ineffective for a number of reasons (e.g., confounding effects from pH, NOM and sulphate, needs positively charged carbon). Therefore, bromate removal is carbon specific, dependent on source water quality and negatively impacted by the presence of NOM and anions such as bromide, nitrate and sulphate. As part of the removal, pretreatment of contaminated water with prechlorination, PAC adsorption, flocculation–sedimentation and sand filtration (Zhao et al. 2015) can be employed to achieve bromate removal approximately ranging from 50 to 90% for at least 98 days as reported. The occurrence of bromated in natural water systems is very quick due to the presence of NOM and other anions (e.g., chloride, sulphate, bromide and nitrate). Biologically active carbon (BAC) is a breakthrough could reduce an influent bromate concentration of 20 μg/L by up to 40%, but the dissolved oxygen (DO) concentration had to be less than 2 mg/L to support bromate-reducing biomass. It is concluded that the 50–80% bromate concentration can be reduced by GAC filtration only for 3 months of operation.

5.2.2 Ferrous Iron Addition

Using a reducing agent such as ferrous iron (Fe_2^+), chemical reduction of bromate to bromide is possible. It results in 40–80% reduction in bromate concentrations, depending on dose, pH and temperature; but temperature control below 10 °C during the reaction seems less effective. The process was also dependent on alkalinity and NOM, with effectiveness decreasing as alkalinity and NOM increases. Researchers recommend that 1 mg/L Fe_2^+ residual is required to ensure good bromate removal. Unlikely it is unacceptable for treated water quality. In addition, the process required delicate balancing and was not practical at all temperatures.

5.2.3 Ultraviolet Irradiation

Bromate removal by UV irradiation at bench and pilot scales is dependent on the water quality, UV fluence and light wavelength. Medium-pressure lamps (200–600 nm) are more effective than low-pressure lamps (254 nm) in destroying bromate, owing to a higher energy input and a broader range of emitting radiation. However, low-pressure lamp with radiation at wavelengths below 200 nm is more effective than a medium-pressure lamp at similar doses (50 μg/L). It is stated that negligible bromate removal would occur using medium-pressure UV treatment at fluences typically used for drinking water disinfection.

5.2.4 Membrane Technologies

Membrane technologies for bromate removal include reverse osmosis, electro-dialysis reversal (EDR) and anion exchange membranes. The operational condition may vary for each bench, pilot and full-scale applications.

Reverse Osmosis

Reverse osmosis (RO) is a pressure-driven process that forces water through a semi-permeable membrane to remove organic contaminants and salts. Pilot-scale experiments show 96% reduction rate by RO membranes. Notably, membrane performance is not affected by temperature.

Electrodialysis Reversal

Electrodialysis (ED) uses direct current to transfer ionic species through cell pairs of oppositely charged membranes, achieving their removal from source water. EDR poses slight change in working performance of the ED process, whereby electrode polarity is reversed during the treatment process to reduce scaling and clean membrane surfaces. EDR membranes seem to be less effective than RO and decreased performance at lower temperatures. Comparably, RO is preferential to achieve multiple treatment objectives including disinfection, as well as the removal of salts, hardness and organic compounds. EDR neither removes organics nor has disinfection capacity.

Anion Exchange Membranes

Anion exchange membrane process is similar to ED, whereby chemical potential gradient is used between the feedwater and exchange solution to transfer ions (Saraji et al. 2014) across the membrane. Anion exchange membranes could achieve 90% removal efficiencies when bromate is the only anion to be removed. In regard to other anions, such as nitrate and bicarbonate, removal efficiency decreased to approximately 60%, and bromate concentrations could not be reduced below 20 μg/L. A mono-anion-selective membrane is an optional to increase the bromate removal efficiency to 93% which exerts a favourable effect on the removal of bromate, while minimizing the exchange of associated anions, such as sulphate and bicarbonate.

Anion Exchange Resins

The removal of bromate from aqueous solution using a macroporous strong base anion exchange resins with polystyrene matrix shows selective sorption and pH dependency. Quaternary amine functional groups and chloride as exchange ion are having maximum sorption capacity of the resin. The presence of other anions, such as nitrate, sulphate and chloride, reduced resin sorption capacity by 28, 27 and 23%, respectively.

To certify drinking water devices and materials as meeting NSF/ANSI standards:

- CSA Group (www.csagroup.org);
- NSF International (www.nsf.org);
- Water Quality Association (www.wqa.org);
- UL LLC (www.ul.com);
- International Association of Plumbing and Mechanical Officials (www.iapmo.org). An up-to-date list of accredited certification organizations can be obtained directly from the SCC (2014).

Residential RO devices are expected to be effective for decreasing bromate concentrations in drinking water. The use of distillation for bromate removal from drinking water is expected to adequately remove bromate, because it is effective for the reduction of inorganic contaminants. However, this process requires an electrical energy input. Ion exchange may also effectively remove bromate from drinking water based on the municipal-scale. Although no residential treatment devices are certified for bromate removal, it is recommended that drinking water treatment devices certified as meeting NSF/ANSI Standard 58 (Reverse Osmosis Drinking Water Treatment Systems), NSF/ANSI Standard 62 (Drinking Water Distillation Systems) or NSF/ANSI Standard 53 (Drinking Water Treatment Units —Health Effects) be used.

RO treated water or distillation may be corrosive to internal plumbing components. Therefore, these devices should be installed only at the point of use. In addition, these two types of drinking water treatment systems are intended only for point-of-use installation, as large quantities of influent water are needed to obtain the required volume of treated water, and are generally not practical for point-of-entry installation at the residential scale. Periodic testing by an accredited laboratory should be conducted on both the water entering the treatment device and the finished water to verify that the treatment device is effective. Treatment devices may lose their removal capacity through usage and time and need to be maintained or replaced. Consumers should verify the expected longevity of the components in Guidelines Technical Document 29 their treatment device according to the manufacturer's recommendations and service it when required.

6 Graphene Based Materials

6.1 Nanoparticles

Platinum based nanocomposites such as bimetallic Pt-based alloy nanoparticles, Pt nanoparticles dispersed on various substrates such as metal oxides, carbon nanotubes (CNTs), reduced graphene oxide (rGO), and rGO/ionic liquid (IL) composites, may further enhance the oxidation of bromide ions molecules due to their synergistic electronic effects (Xu et al. 2011). Platinum based nanomaterials as sensing electrode materials may be added to the surfaces of electrodes or modified Pt electrodes through numerous strategies, including physical adsorption, chemical covalent bonding, electrochemical reduction, electrochemical deposition, electrochemical polymerization with redox polymers, etc.

6.2 Nanosensor

Platinum nanomaterial modified electrodes may have the ability to selectively adsorb and sense bromide ions in drinking water. The development of electrochemical sensor platforms with unique functionalities and processes that are activated on exposure to bromate, as well as the capacity for enhanced mobility and the real time monitoring of other biomolecules in water utilities. The main focus of this section is to provide insights into electrochemical sensors based on reduce PtO nanomaterials. It will endeavor to describe the utility of Pt nanomaterials in the design of high-performance electrochemical sensor platforms for the detection of bromide ions, which presently assist in the detection of various source of water and facilitate the real-monitoring of biological processes (Geim and Novoselov 2007). It seems hopeful for sustainable advances and concepts articulated in this chapter will inspire new discoveries in this area for the benefit of bromated control in drinking water.

6.3 Nanocomposites

Hemoglobin (Hb) is an ideal redox metalloprotein for the construction of biosensors due to its low cast and availability. In addition, Hb is widely used for the detection of hydrogen peroxide, nitrite, trichloroacetic acid and bromate, due to its high specificity. Recent years, different nano and micro materials have been widely used for immobilization of Hb on the electrode surface. Furthermore, graphene and reduced graphene oxide (RGO) based composites are continuously used for the construction of electrochemical biosensors. Though, the special properties of graphene are highly a bridged in RGO when compared to the properties of pristine graphene (Dong et al. 2013). Hence, the pristine graphene is an alternative choice to prolong its special properties to the biosensors fabrication. However, pristine graphene can be easily agglomerates to graphite in aqueous solutions due to the π stacking of individual grapheme sheets, hence different materials have been used to prevent the agglomeration of graphene sheets and enrich the biocompatibility.

6.4 Nanocatalyst

The effect of reduced graphene oxide (rGO) on the inhibition of BrO_3^- reacting with HOBr to Br^- in thermal activated peroximonosulfate (PMS) treatment process. Synthesis of rGO using hydro-thermal methods are showing better decomposition rate of PMS, the removal kinetics of bromated as target pollutant. It must be understood that the slight changes in performance of thermal/PMS and rGO$^-$ thermal/PMS process under the same dosage. Interestingly, behavior of rGO on targeted contaminant does not create secondary radicals, but bromine species were prone to react with rGO. The quick reduction of HOBr to Br^- is possible at room temperature (35 °C) by applying H_2O_2 and ammonia at typical dosage. In addition, no abatement of BrO_3^- and Br^- is found for rGO adsorption. Thus, it is an evident for the role of rGO on masking HOBr and inhibition effect in rGO/ozonation. It is encourageable to overcome the extra formation of unwanted bromine species in advanced oxidation. In comparison with other HOBr masking measures, i.e. ammonia and H_2O_2, shows some unique benefits with suitable oxidation procedure.

6.5 Thin Films

A highly conductive Reduced Graphene Oxide (RGO) films are extremely attractive for a range of electronic devices, especially for printed macroelectronics. It is highly potential to replacing heavy, metal-based current collectors with thin, light, flexible, and highly conductive films will further improve the energy density of such devices. Films with two-dimensional (2D) building blocks, such as graphene or reduced graphene oxide (RGO) nanosheets, are particularly promising due to their low percolation threshold with a high aspect ratio, excellent flexibility and low cost. However, the electrical conductivity of these films is low, typically less than 1000 S/cm with record-high DC conductivity, up to 3112 S/cm. Whereas, high conductivity in RGO films is achievable through an electrical current induced annealing process at high temperature, up to 2750 K, in less than one minute of anneal time. Further, it is fabricated by a solution based filtration process followed by thermal reduction (773 K) and then a high temperature reduction (2750 K by Joule heating). Through a combination of experimental and computational studies, the fundamental mechanism behind the formation of a highly conductive three-dimensional (3D) structure composed of well-connected RGO layers for the effective removal defects from graphene oxide, improve the crystalline structure of individual RGO nanosheets and densify their stacking in DC conductivity

(Tchounwou et al. 2012). High temperature annealing attributes to the defect-facilitated cross-linking of neighboring RGO layers, as confirmed by our molecular dynamics simulations. A wide range of applications from microelectronics to large-scale printed electronics using RGO film has been noted. It can be applied in Li-ion batteries as current collectors. The simple but effective Joule heating method can provide new tools to study and investigate materials under extreme thermal conditions successfully.

7 Conclusion

From the discussion throughout this chapter, readers can understand that the rational design emphasizes 'design-for-purpose'. Graphene based materials to control bromate formation or reduction during wastewater treatment starts with scientifically, generally, chemically, defining the problem to be solved in details, such as what the barrier is, what the key to the solution with certain conditions are met etc. A conceptual design of nanomaterial-based solution has been proposed after a clear understanding of the problem. The currently available synthesis capability of nanomaterials is advantageous if passing. The designed nanomaterial must be scientifically teated the event of an unsatisfactory performance. "Think-outside-the-box", has a high potential of creating next-generation and ground-breaking solutions to the water challenges of our times. Such inventive material contributes to the already vast library of nanomaterial database as effective in their purpose. Looking at the near future, graphene based materials can meet the water challenges in a sustainable way.

References

Ali I (2012) New generation adsorbents for water treatment. Chem Rev 112:5073–5091

AWWARF (1991) Disinfection by-products database and model project. Denver, CO, American Water Works Association Research Foundation

Cavalli S, Polesello S, Valsecchi S (2005) Chloride interference in the determination of bromate in drinking water by reagent free ion chromatography with mass spectrometry detection. J Chromatogr A 1085:42–46

Ding SY, Yang Y, Li C, Huang HO, Hou LA (2016) The effects of organic fouling on the removal of radionuclides by reverse osmosis membranes. Water Res 95:174–184

Dong J, Hu J, Wang JL (2013) Radiation-induced grafting of sweet sorghum stalk for copper (II) removal from aqueous solution. J Hazard Mater 262:845–852

Fielding M, Hutchison J (1993) Formation of bromate and other ozonation by-products in water treatment. In: Proceedings of the IWSA international workshop on bromate and water treatment, Paris 1993. London, International Water Supply Association, pp 81–84

Gadd GM (2009) Heavy metal pollutants: Environmental and biotechnological aspects reference module in life sciences. In: Encyclopedia of microbiology, 3rd edn., pp 321–334

Geim AK, Novoselov KS (2007) The rise of graphene. Nat Mater 6:183

Gordon G, Emmert GL (1996) Bromate ion formation in water when chlorine dioxide is photolyzed in the presence of bromide ion. In: Proceedings of the water quality technology conference, New Orleans, LA. Denver, CO, American Water Works Association

Guo Y, Guo S, Ren J, Zhai Y, Dong S, Wang E (2010) Cyclodextrin functionalized graphene nanosheets with high supramolecular recognition capability: synthesis and host–guest inclusion for enhanced electrochemical performance. ACS Nano 4:4001–4010

Haag WR, Hoigné J (1983) Ozonation of bromide-containing water: kinetics of formation of hypobromous acid and bromate. Environ Sci Technol 17:261–267

Hoigné J, Bader H (1994) Kinetics of reactions of chlorine dioxide (OClO) in water-I. Rate constants for inorganic and organic compounds. Water Res 28(1):45–55

IARC (1986) Some naturally occurring and synthetic food components, furocoumarins and ultraviolet radiation. Lyon, International Agency for Research on Cancer, pp 207–220 (IARC Monographs on the Evaluation of Carcinogenic Risks to Humans, vol 40)

Khan MR, AlOthman ZA, Alqahtani NJ, Alsohaimi IH, Naushad M (2014) Quantitative analysis of bromate in non-alcoholic beer using ultra performance liquid chromatography-electrospray ionization mass spectrometry. Analy Meth 6:4038–4045

Li J, Mei H, Zheng W, Pan P, Sun XJ, Li F, Guo F, Zhou HM, Ma JY, Xu XX, Zheng YF (2014) Anovel hydrogen peroxide biosensor based on hemoglobin-collagen-CNTs composite nanofibers. Colloids Surf B 118:77–82

Mack RB (1988) Round up the usual suspects. Potassium bromate poisoning. N C Med J 49: 243–245

Mao R, Zhao X, Lan H, Liu H, Qu J (2015) Graphene-modified Pd/C cathode and Pd/GAC particles for enhanced electrocatalytic removal of bromate in a continuous three-dimensional electrochemical reactor. Water Res 77:1–12

Ministry of Health and Welfare (1979) The Japanese standards of food additives, 4th edn. Tokyo, p 367

Naushad M, AlOthman ZA, Khan MR, Wabaidur SM (2013) Removal of bromate from water using De-Acidite FF-IP resin and determination by ultra-performance liquid chromatography-tandem mass spectrometry. Clean-Soil Air Water 41:528–533

Naushad M, Khan MR, AlOthman ZA, AlSohaimi IH, Rodriguez-Reinoso F, Turki TM, Ali R (2015) Removal of Br O_3^- from drinking water samples using newly developed agricultural waste-based activated carbon and its determination by ultra-performance liquid chromatography-mass spectrometry. Environ Sci Pollut Res 22:15853–15865

Naushad M, Khan MR, AlOthman ZA, Awual MR (2016a) Bromate removal from water samples using strongly basic anion exchange resin Amberlite IRA-400: kinetics, isotherms and thermodynamic studies. Desal Water Treat 57:5781–5788

Naushad M, Khan MR, AlOthman ZA, Awual MR, Alqadami AA (2016b) Water purification using cost effective material prepared from agricultural waste: kinetics, isotherms and thermodynamic studies. CLEAN–Soil Air Water 44:1036–1045

NSF (National Sanitation Foundation, International) (1999) Drinking water treatment chemicals-health effects; Document ANSI/NSF 60-1999; NSF: Ann Arbor, MI

Oliveira SM, Oliveira HM, Segundo MA, Rangel AOSS, Lima JLFC, Cerdà V (2012) Automated solid-phase spectrophotometric system for optosensing of bromate in drinking waters. Anal Methods 4:1229–1236

Perreault F, de Faria AF, Elimelech M (2015) Environmental applications of graphene-based nanomaterials. Chem Soc Rev 44:5861–5896

Rice RG, Gomez-Taylor M (1986) Occurrence of by-products of strong oxidants reacting with drinking water contaminants-scope of the problem. Environ Health Perspect 69:31–44

Romele L (1998) Spectrophotometric determination of low levels of bromate in drinking water after reaction with fuchsin. Analyst 123:291–294

Saraji M, Khaje N, Ghani M (2014) Cetyltrimethyl ammonium coated magnetic nanoparticles for the extraction of bromate, followed by its spectrophotometric determination. Microchim Acta 181:925–933

Tchounwou PB, Yedjou CG, Patlolla AK, Sutton DJ (2012) Heavy metal toxicity and the environment. In: Molecular, clinical and environmental toxicology, Experientia Supplementum, pp 133–164

Upadhyay RK, Soin N, Roy SS (2014) Role of graphene/metal oxide composites as photocatalysts, adsorbents and disinfectants in water treatment: a review. RSCB Adv 4:3823–3851

US EPA (1999) Guidelines for carcinogen risk assessment (SAB review draft). Washington, DC, US Environmental Protection Agency, Risk Assessment Forum (NCEA-F-0644)

US EPA (2001a) Stage 2 occurrence and exposure assessment for disinfectants and disinfection byproducts (D/DBPs). US Environmental Protection Agency, Washington, DC

Vilian ATE, Chen SM, Kwak CH, Hwang SK, Huh YS, Han YK (2016) Immobilization of hemoglobin on functionalized multiwalled carbon nanotubes-poly-L-histidine-zinc oxide nanocomposites toward the detection of bromate and H_2O_2. Sens Actuators B 224:607–617

Wang JL, Chen C (2015) The current status of heavy metal pollution and treatment technology development in China. Environ Technol Rev 4:39–53

Wei XC, Wolfe FA, Han YX (2014) Mine drainage: characterization, treatment, modeling, and environmental aspect. Water Environ Res 86:1515–1534

Xiang QJ, Yu JG, Jaroniec M (2012) Graphene-based semiconductor photocatalysts. Chem Soc Rev 41:782–796

Xu J, Liu C, Wu Z (2011) Direct electrochemistry and enhanced electrocatalytic activity of hemoglobin entrapped in graphene and ZnO nanosphere composite film. Microchim Acta 172:425–430

Zhang J, Yang X (2013) A simple yet effective chromogenic reagent for the rapid estimation of bromate and hypochlorite in drinking water. Analyst 138:434–437

Zhao H, Ji X, Wang B, Wang N, Li X, Ni R, Ren J (2015) An ultrasensitive acetylcholine esterase biosensor based on reduced grapheme oxide-Au nanoparticles-β-cyclodextrin/Prussian blue-chitosan nanocomposites for organophosphorus pesticides detection. Biosens Bioelectron 65:23–30

Chapter 11
Efficient Removal of Aqueous Aromatic Pollutants by Various Techniques

Natrayasamy Viswanathan, Soodamani Periyasamy and Ilango Aswin Kumar

Abstract Water plays a vital role in the universe for all the living organisms, but water quality gets affected by a lot of pollutants. Due to the rapid increase in industries, the disposal of numerous aromatic pollutants into the environment leads to different unpredicted effects on human health. The removal of aromatic pollutants from water is very essential for good health. Hence, this chapter focuses on the catalytic ozonation of aqueous aromatic pollutants using graphene/carbon nanotube (CNT) and their derivative materials. The various sources of aqueous aromatic pollutants were demonstrated. Some compatible treatment methods of aromatic pollutants in water with their merits and demerits were debated. The detail account of ozonation and catalytic ozonation using different materials were examined for the removal of toxic aromatic pollutants. In addition, graphene/CNT and their properties on catalytic ozonation process for the removal of aromatic pollutants were discussed. Graphene/CNT with their supported materials towards the application of catalytic ozonation was summarized by providing the operating conditions, degradation efficiency and mineralization of their products. The versatile graphene and CNT supported materials convert the toxic aromatic pollutants into non-toxic products.

Keywords Ozonation · Catalysts · Aromatic pollutants · Graphene
Water treatment

N. Viswanathan (✉) · S. Periyasamy · I. A. Kumar
Department of Chemistry, Anna University, University College of Engineering - Dindigul, Reddiyarchatram, Dindigul 624622, Tamilnadu, India
e-mail: drnviswanathan@gmail.com

S. Periyasamy
e-mail: periyasamy187@gmail.com

I. A. Kumar
e-mail: aswinmariya@gmail.com

M. Naushad (ed.), *A New Generation Material Graphene: Applications in Water Technology*, https://doi.org/10.1007/978-3-319-75484-0_11

1 Introduction

Any physical, chemical or biological changes in water that affects the water quality which leads to various diseases is known as water pollution (Naushad et al. 2016; Viswanathan et al. 2014). Water pollutants can be classified into organic and inorganic pollutants. Inorganic pollutants are nitrate, cyanide, phosphate, etc., Organic pollutants are highly reactive, toxic and it may enter into the environment from various anthropogenic sources. Organic pollutants are further classified into both aliphatic and aromatic pollutants. In general, the aromatic pollutants (compounds) having low molecular mass could easily get soluble in water whereas the aromatic compounds with high molecular mass are insoluble in water (Abdel-Shafy and Mansour 2016). The aromatic pollutant is one of the major environmental pollutants that frequently affects the water system. Aromatic compound plays an essential key role in the biochemistry of all the living organisms and hence it is extensively used in pharmaceutical industries. Indian chemical industry has appreciably contributing their effective research towards the industrial and economic growth of the country. The chemical industry of India placed 12th rank around the world in terms of manufacturing and developing the chemicals and pharmaceutical drugs. About 14% of drugs have been obtained from India (Vinod 2017).

India is a strong and well known country for the development of basic chemicals that are generally used to produce the consumer items like paints, dyes, soaps, medicines, toiletries, cosmetics, etc. Therefore, major organic pollutants such as dyes, pesticides, phenols, aromatic hydrocarbons and their derivatives have been liberated day by day from various industries (Kile et al. 1995; Singh and Shah 2016). In particular, the aromatic pollutants are mainly affects the environment by displaying many actions such as light sensitivity, conductivity, instability and unidentified nature in water system (Kumar et al. 2017; Builee and Hatherill 2004). The polycyclic aromatic hydrocarbons are commonly present in the environment and get contaminated with soil, water and food. Due to the extreme usage of aromatic compounds, their effluents can get mixed with the ground water leads to various human diseases such as liver damage, cardiovascular, thyroid hormone disruption and feminization (Rengarajan et al. 2015). Hence, the removal of aqueous aromatic pollutant is an important for the environmental remediation. The different techniques were used for the remediation of aqueous aromatic pollutants such as adsorption, membrane process, catalytic ozonation, fenton processes and physicochemical processes (Midathana and Moholkar 2009; Zhang et al. 2006; Wu et al. 2016a, b; Daneshvar et al. 2017; Shi et al. 2017). In recent years, catalytic ozonation has been developed and comparatively this process seems to be green, effortless and economical one. This is because most of the organic pollutants can be oxidized, and gets mineralized into CO_2 and H_2O by the catalytic ozonation.

A variety of high-resolution catalytic ozonation materials have been developed for the removal of aqueous aromatic organic pollutants. Currently, many ozonation catalysts were used but each of them having own merits and demerits. So, researchers have paid an increase in attention to improve graphene/CNT and their derivatives towards the removal of aromatic organic pollutants from aqueous medium in the presence of ozone (Ahn et al. 2017). Graphene/CNT based catalytic ozonation materials are new classes of techniques with ultrafine dimensions, and have received considerable attention for waste-water treatment. This present chapter which deals the issues of nature of the catalytic active sites available for catalytic ozonation in graphene/CNT and its supported materials for the removal of toxic aromatic pollutants (Sun et al. 2012).

2 Sources of Aromatic Pollutants in Water

The sources of aromatic pollutants are both natural and man-made sources.

2.1 Natural Sources

Oil streaming, burning woods, volcanoes, chlorophyllous plants, fungi and bacteria are the main natural sources of aromatic pollutants (Fischer et al. 2016).

2.2 Anthropogenic Sources

Pharmaceuticals, chemicals and food industries produce the numerous aromatic pollutants such as polychlorobiphenyls, bisphenyl, trinitrotoluene, chlorodioxins and aromatic azo dyes. These toxic pollutants are directly mixed with water sources (Carter and Blizard 2016). Coal and petrochemical industries are generating maximum aromatic contaminants like polycyclic aromatic hydrocarbons and gasoline. Other aromatic hydrocarbon may enter into the aquatic system from harbour waste-water, coastal refineries and crude oil stations. The use of pesticides is often affects the water system because, it possesses harmful chlorinated aromatic pollutants such as dichloro diphenyl trichloroethene, aldrin, dieldrin, heptachlor, hexachlorobenzene, lindane and chloro methoxy benzene (Lehner et al. 2017). These pollutants are directly mixed with the ground water causes water pollution. The household waste exits many aromatic pollutants such as glycopeptide, macrolide, quinolone, sulfonamides, and tetracycline compounds (Meij and Winkel 2007).

3 Removal Techniques of Aromatic Pollutants

3.1 Adsorption

Adsorption is one of the feasible methods for the removal of various aromatic pollutants from water. Many adsorbents like natural materials, carbon based materials, inorganic, biopolymers and synthetic polymeric materials were demonstrated for the removal of aromatic pollutants. Hu et al. have developed graphene supported polyacrylamide (PAM) composite for the removal of water soluble aromatic pollutants (Hu et al. 2017). Aromatic organoarsenic compounds like p-arsanilic acid and roxarsone were removed from water using ferric and manganese binary oxides and its removal capacities were found to be 0.52 and 0.25 mmol/g respectively (Jung et al. 2015). Llado et al. have explained the application of micro porous activated carbon derived from biocollagenic wastes of the leather industry for the removal of organic pollutants (Llado et al. 2015). However, these methods have some demerits such as low capacity and difficult to regenerate. Hence, adsorption process is not suitable for the removal of all aromatic pollutants.

3.2 Advanced Oxidation Process

Advanced oxidation process (AOPs) is used in various removal technologies. Mostly, AOPs are able to provide the complete mineralization of the organic matters. This method is based on the formation and reaction of hydroxyl free radical. The hydroxyl radical is more effective than other oxidative species. The different AOPs include chemical oxidation process, photochemical degradation process and photocatalysis have been developed for toxic ion removal from water. This processes are explained their complete degradation and exceptional mineralization efficiency for the environmental detoxification. Wang et al. have reported the use of N-doped graphene for the removal of phenol by stimulated peroxymonosulfate. This study has well explained for green catalyst of nitrogen doped graphene and those catalysts have contributed a new technology for waste-water treatment without producing any secondary contamination (Wang and Xu 2012). Carre et al. have evaluated the application of different advanced oxidation process as tertiary treatment for the complete removal of pesticides from water. This study have been demonstrated that the humic acid, lignin-derivative and lauryl sulphate which were affect the efficiency of AOP. The persulphate in AOP process have been proved for the complete efficient removal of micro pollutant from water (Taheran et al. 2016). However, AOPs possess some drawback that it acquires high operational cost.

3.3 Membrane Process

The contribution of membrane technology in waste-water purification has been improved in the recent years. This process offers the additional benefits such as low energy consumption, eco-friendly and compatible. Membrane processes have the potential ability towards the removal of organic pollutants because, they are selectively remove the specific groups of organic compounds mainly cyclic and non-cyclic alcohols, aldehydes, ketones, etc. Membrane processes which include the micro filtration, nano filtration and ultrafine filtration was applied for the pollutants removal. Microfiltration and ultrafiltration are employed to remove both micro particles and macromolecules. But, membrane process is not obeyed to remove aromatic pollutants having low molecular weight, because of its limitations like low selectivity, short life time of membrane and high operational cost (Carra et al. 2014).

3.4 Flocculation and Coagulation

Coagulation and flocculation methods have been utilized for the treatment of primary effluents from the industrial waste-water. This method is used for many applications such as the removal of natural organic matter, pathogen, inorganic and chemical phosphorus from water. Commonly, coagulation is a process where the repulsive charge of electrical double layers of colloids has been reduced in such via that micro particle is formed. These micro particles colloid with every wastes and create the large structure in the flocculation process. Factually, coagulation has been utilized in water treatment in order to reduce color, turbidity, odour and pathogen. This process has been activated by some coagulants such as aluminum sulfate, ferrous sulfate, ferric chloride, aluminum chloride, ferric sulfate and sodium aluminate. However, the use of those coagulants could create many secondary pollutants such as metal residuals, chemical residuals and physico-chemical sludges. Moreover, flocculation and coagulation possess some demerits such as high operational cost, need of various coagulants and the pollutant compounds which have been collected only as the solid sludge could not be regenerate and reuse. Likewise, micro level dissolved organic pollutants are very difficult for the removal from waste-water using this method (Ntampou et al. 2006).

3.5 Electrochemical Oxidation Process

Electrochemical oxidation process is another method for the degradation of toxic aromatic compounds in waste-water because it possesses some benefits like easy operation, control process and no more protection of the secondary pollution. So, this method is not suitable for the removal of dissolved organic pollutants, micro organism and soluble heavy metals (Lv et al. 2009). To overcome these disadvantages of conventional treatment methods, the advanced catalytic ozonation processes have been developed for the remediation of aromatic pollutants from water.

3.6 Ozonation Process

Ozone is a powerful oxidizing agent which selectively oxidizes the numerous aromatic pollutants from aqueous solution. Besides, ozonation is a type of advanced oxidation methods with high efficient numbers of oxygen species to attack the extensive aromatic compounds. The degradation of organic pollutants from waste-water using ozonation processes are highly pH dependent. Because, ozone has been produced by the potential discharge over an oxygen molecule to reacts with aromatic pollutants. The ozonation reaction is accomplished via two pathways as direct ozone oxidation and indirect free hydroxyl radical oxidation. In case of direct ozonation, the organic molecules can be destroyed into various fragments, which includes the breakage of double bonded systems, the formation of carbonyl forms (aldehyde and ketones), the addition of oxygen atom to the aromatic ring and the reaction with alcohols to form organic acids. The indirect type of ozonation occurs between free radical hydroxyl species and organic matter present in waste-water. At the same time, the detoxification and mineralization of the aromatic pollutants using ozonation process is not as efficient as advanced oxidation process, hence the numerous technologies such as microwave, ultrasound, ultra violet, hydrogen peroxide and persulphate were applied to enhance the ozonation efficiency and to increase in degradation rate by the formation of new hydroxyl radicals.

However, the most important drawback of this method as high energy consumption for ozone generation leads to incomplete mineralization of various organic matters. For instance, when ozonation is utilized alone, the saturated intermediates might accumulate in the effluent solution which is more toxic than initial pollutant, which may happen in the earlier stage of ozonation. Consequently, the studies were focused onto improving the ozonation efficiency by overcoming the drawbacks of the single ozonation process. Stockinger et al. have successfully utilized the ozone for complete degradation of chloro and bromo substituted organic compounds by chemical oxidation (Banjoo and Nelson 2005). Moraes et al. have explained the significant improvements could be made using solid catalysts in ozonation process (De Moraes et al. 1996). Therefore, ozone has been activated using some catalyst that activated ozone possesses enhanced mineralization capacities. To further enhance the mineralization, a combination of heterogeneous photocatalysis with ozone has been expected to decompose organic substances more quickly (Agustina et al. 2005). The catalytic ozonation have received an increase in attention due to its high effectiveness in the degradation and mineralization of organic pollutants. Organic compounds are difficult to dissociate by single ozonation and can be oxidized by catalytic ozonation at ambient temperature and pressure.

3.7 Advantages of Catalysts in Waste Water Treatment

Nowadays the water researchers have paid more attention on catalysts for ground/industrial water treatment process. In particular, the advanced oxidation catalysts have been applied for the degradation of organic pollutants due to its advantages like enhanced physical and chemical properties, effective degradation rate and simple method. Consequently, the advanced oxidation catalysts have been extensively utilized in various water treatment technologies due to its high oxidative potential ($E_0 = 2.7$ V), very quick process, efficient and economical one towards the detoxification of aromatic pollutants (Atalay and Ersoz 2016).

4 Catalytic Ozonation

Catalytic ozonation using solid catalyst can potentially enhance the degradation of organic pollutants in drinking/waste-water by different reaction pathway. Catalytic ozonation covers advanced oxidation technology which is expected as efficient technology for the treatment of industrial waste-water. The variety of aromatic pollutants such as sulfamethoxazole, diclofenac, clofibric acid, ibuprofen, bezafibrate, erythromycin, fenofibric acid, ciprofloxacin and aromatic hydrocarbons were degraded by catalytic ozonation. Nowadays different catalysts such as metal oxides, graphene, graphene oxide, carbon nanotube, alumina supported materials, zeolites, magnetic materials and minerals were utilized for the detoxification of aromatic pollutants from industrial and ground waste-water. The catalytic ozonation were classified into types as homogeneous catalytic ozonation and heterogeneous catalytic ozonation. In homogeneous catalytic ozonation, the metal ions were used as catalysts in the reaction system whereas metal oxides and metal oxide supported materials were predominantly used as catalysts in the heterogeneous catalytic ozonation (Guo et al. 2012).

4.1 Homogeneous Catalytic Ozonation

The d-block metal ions like Ag^+, Cd^{2+}, Co^{2+}, Cu^{2+}, Fe^{2+}, Mn^{2+}, Ni^{2+} and Zn^{2+} were commonly utilized in the homogeneous catalytic ozonation process for the decomposition of organic pollutants in aqueous medium. The metal ion plays a vital role in the rate of the reaction system for the selectivity of ozonation process (Pirkanniemi and Sillanpaa 2002). The mechanism of homogeneous catalytic ozonation is based on ozone decomposition followed by the generation of hydroxyl radicals and the possible mechanism is shown in Scheme 1. The mechanism was followed by metal ions as increased for the decomposition of ozone to given $^{\cdot}O_2^-$, and the available free electron of $^{\cdot}O_2^-$ removed to O_3 and give $^{\cdot}O_3$ in order to produce the hydroxyl radical. Homogeneous catalytic ozonation can effectively

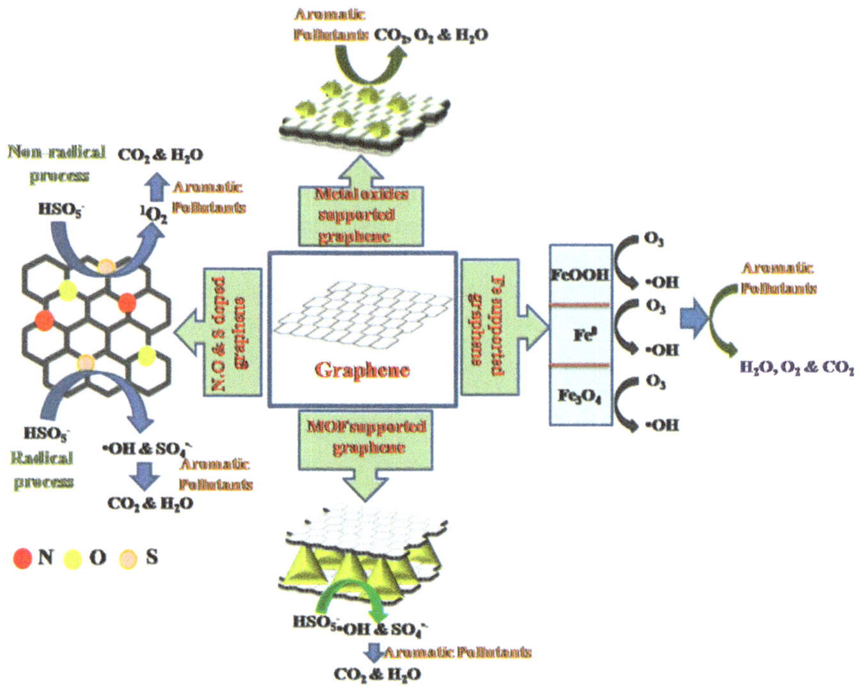

Scheme 1 Catalytic ozonation of aromatic pollutants by graphene supported materials

improve the efficiency in the removal of aromatic pollutants in water. However, the introduction of various metal ions to the ozonation process causes in the formation of secondary pollution, which needs high operational cost for further treatment would considered as the disadvantage of this technology.

4.2 *Heterogeneous Catalytic Ozonation*

First heterogeneous ozonation catalysts were demonstrated for waste-water treatment at 1970s. The heterogeneous ozonation catalysts are preferred for the detoxification of aromatic pollutants, because these systems offer a cleaner process, independent of additional stoichiometric reagents, as controlling the treatment systems, increasing the ozone decomposition and thereby forming the highly reactive free hydroxyl radicals. The various types of heterogeneous catalysts, including metal oxides and carbon based materials were extensively utilized to enhance the activation of ozonation process. Whilst, the several water treatment researchers have been claimed many metal oxides as heterogeneous ozonation catalysts, like FeOOH, CuO, Mn_2O_3, Co_2O_4, Al_2O_3, MnO_2, TiO_2 and ZnO and utilized for the removal of different aromatic pollutants (Valdes and Zaror 2006).

During reaction, the catalysts might given the toxic intermediate during the decomposition of aromatic pollutants and so, these types of troubles were reduced using free catalysts like carbon supported materials are applied for the catalytic ozonation of aromatic pollutants from water.

5 Catalytic Ozonation Materials for Organic Pollutants Removal

5.1 Inorganic Materials

Commonly, catalytic ozonation is focused some physicochemical properties such as surface area, surface charge, pore size, chemical stability and active surface site. Inorganic materials possesses enhance stability, surface area and number of active surface site. Hence, the numerous inorganic materials like metals, metal oxides and their supported materials were utilized for the catalytic ozonation of the degradation of aromatic pollutants in efficient manner. This is because of the decomposition of ozone on the Lewis metal oxides could induce the provision of high hydroxyl radical and this radical probably decompose the aromatic compounds. Metal oxides based catalytic ozonation materials were depended on the highly affected factors such as temperature and pH which lowering their catalytic activity (Shahamat and Sadeghi 2016).

5.2 Carbon Based Materials

Carbon and activated carbon materials are well demonstrated for the removal of numerous aromatic pollutants from water by using various techniques like adsorption, ozonation and catalytic ozonation. Jans et al. have studied the activated carbon can exit in the degradation of ozone in water by the formation of hydroxyl and superoxide radicals (Xing et al. 2016). The result describes that the reactive sites on the surface of activated carbon could make the ability of ozone disintegration was limited but ozone decomposition and formation of both hydroxyl and superoxide radicals was faster even though there was no catalyst added. The decomposition of aquatic zone using activated carbon leads to the formation of active oxygen species causes easiest mineralization of aromatic organic compounds. Aniline is a toxic aromatic compound obtained by the synthesis of dyes and some pesticides and it is also the by-product in paper and textile industries and hence it should remove from water sources. Faria et al., utilized activated carbon for the degradation of aniline. During ozonation, the complete conversion of aniline was attained at 120 min (Faria et al. 2007).

5.3 Mesoporous Materials

Mesoporous alumina, silicates, zirconium and their derivatives were used in the catalytic ozonation of aromatic pollutants. The catalytic ozonation process depends on the size distribution and morphology of catalytic particles, smaller particle possesses numerous active centers in order to enhance the catalytic activity. Besides, the catalytic efficiency has been increased when reducing the catalyst's diameter into nano level (Hu et al. 2016).

5.4 Zeolites Supported Materials

Zeolites and its supportive materials have excellent adsorption behavior and catalytic features. Valdes et al. have studied ozone decay in water by zeolites supported materials and it was observed that the rate of ozone decomposition was increased than the self decomposition of ozone. Zeolites have produced more number of hydroxyl free radical. In addition, hydroxyl amine and metal oxide tailored zeolites also possesses good catalytic property towards aromatic organic pollutants (Ikhlaq and Kasprzyk-Hordern 2017).

5.5 Metal Organic Frameworks

Metal-organic frameworks (MOFs) are a new type of hybrid functional material. MOFs have large surface area, high thermal stability and tailorable chemistry. These properties are considered as advantages for the catalytic ozonation of aromatic pollutants. Moreover, MOFs produced the sulphate free radical from the decomposition of ozone. Sulphate free radical has high redox potential (between 2.5 and 3.1 V) than hydroxyl free radical (between 1.9 and 2.7). Zang et al. have reported the use of yolk-shell Co_3O_4@MOFs nanocomposites for the improved degradation of organic pollutants and almost 100% removal was achieved within 60 min (Zeng et al. 2015).

5.6 Natural Minerals

Natural minerals/materials like brucite, alumina, zirconia, natural zeolites and clays possess good catalytic ozonation capacity, because they are effective in the generation of hydroxyl radicals naturally by ozone decomposition in aqueous solution. Dong et al. have utilized the natural mineral like brucite for the catalytic ozonation of dye waste-water of active brilliant red X-3B (Dong et al. 2007).

6 Graphene Based Materials

Graphene is a 2D material with thin sheets of a layer bonded sp^2 hybridized carbon atom. The graphene materials are used in various applications due to its high specific surface area, good thermal resistance, very good conductivity and ozonation catalytic property. The ability of graphene and its materials like graphene oxide and reduced graphene oxide were performed as best ozonation catalysts for the degradation of aromatic pollutants (Choi et al. 2010). Likewise, graphene and their supportive materials act as metal free catalysts were applied for ozonation process. At the same time, nitrogen was also doping into graphene and metal oxide incorporated graphene materials have shown the superior catalytic ozonation property towards the degradation of aromatic pollutants in aqueous solution. Hu et al. have demonstrated MnO_2 tightly anchored birnessite on graphene nano sheet for catalytic ozonation of gaseous toluene. Their results shows that 64.6 wt% of MnO_2 loaded graphene nano sheet are very aggressive than low and high loading of MnO_2. The catalytic activity of 64.6 wt% MnO_2/graphene shows the degradation rate as 7.89×10^{-6} mol min^{-1} g^{-1} for gaseous toluene (Wang et al. 2016a, b). The synergetic effect of graphene and MnO_2 has been studied due to the strong bonding between the active sites on graphene and MnO_2 for adsorption of toluene. The synthesized 3D sea urchin-like reduced graphene oxide modified with α-MnO_2 nanostructures by hydrothermal method was utilized for the degradation of aromatic pollutants. Graphene and their derivative materials were utilized for catalytic ozonation of aromatic pollutants and various types of possible mechanism for the catalytic ozonation of aromatic pollutants on graphene and their derivatives are shown in Scheme 1.

7 Carbon Nanotube (CNT) Based Materials

Currently, carbon nanotubes (CNTs) acts as novel supportive materials in nano catalytic applications, because CNTs have exceptional physicochemical properties such as high oxidation stability, enhanced specific surface area, impressive mechanical property and outstanding electron conductivity. Multiwalled carbon nanotube (MWCNT) is more efficient material for catalytic ozonation than activated carbon. Further, CNT is converted into multiwall carbon nanotube by oxidizing with nitric acid, which possesses oxygen rich surface (carboxylic acid, hydroxyl and carbonyl groups). CNTs and their derivative materials were utilized for the catalytic ozonation of aromatic pollutants are shown in Scheme 2.

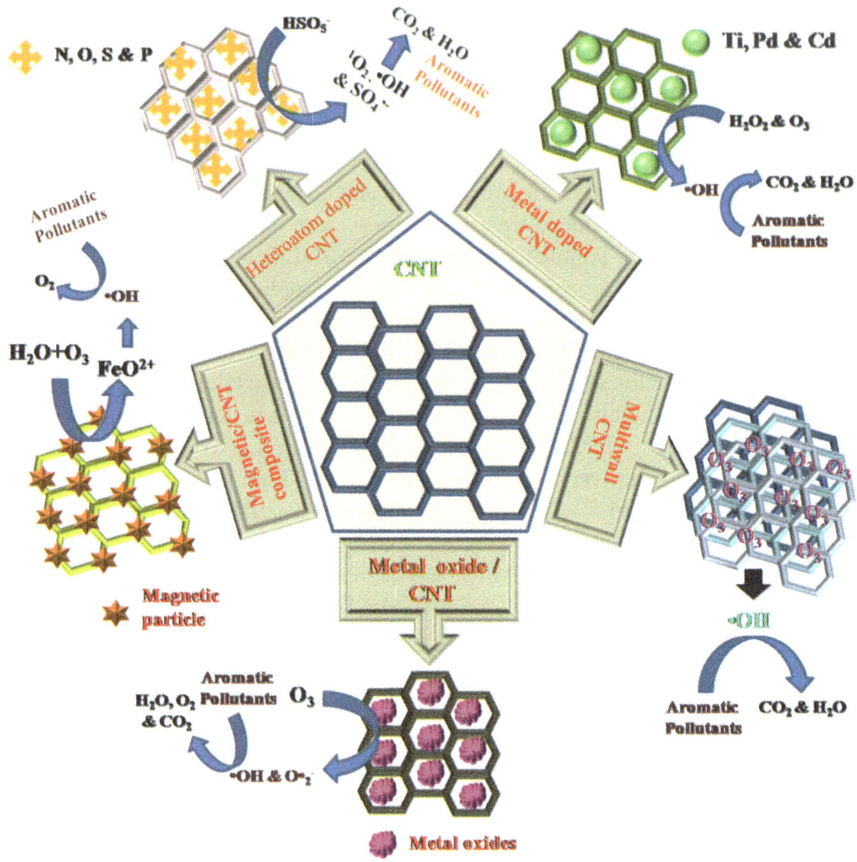

Scheme 2 Catalytic ozonation of aromatic pollutants by CNT supported materials

8 Catalytic Ozonation of Aromatic Pollutants Using Graphene and Their Derivatives

Graphene is thick two dimensional sheets of hexagonally arranged sp^2 bonded carbon atoms have large specific surface area and abundant functional groups. The peculiar physical and chemical properties of graphene which makes as good material for the widespread applications in drug delivery, sensors, super capacitors, nanocomposites, batteries, nano-electronics, solar cells and hydrogen storage and also it paid more attention towards environmental risk and assessment (Shahil and Balandin 2012). Most of the effluents liberated from various industries like textiles, paper, rubber, plastics and printing have aromatic ring in their structures which are highly toxic, non-biodegradable, mutagenic and carcinogenic effects on human beings and aquatic life which are very important to remove for environmental safety (Hou et al. 2012). Du et al. have synthesized graphene oxide/chitosan (GO/CS) fibers by wet-chemical etching off silica nanoparticles for highly enhanced adsorption of congo red dye and about 89% of dye removal was observed (Du et al. 2014). Yao et al. have prepared the magnetic manganese ($MnFe_2O_4$) nanoparticle with reduced graphene oxide (rGO) namely $MnFe_2O_4$-rGO hybrid adsorbent and utilized as catalysts to activate peroxymonosulfate (PMS) to oxidatively degrade various organic methyl orange (MO) dye from water and about 90% of MO removal was attained within 120 min (Yao et al. 2014). Metal-organic frameworks (MOFs) are group of materials possess rapid development and applications in environmental field owing to their excellent properties such as high porosity, enhanced surface area, modifiable surfaces, tunable pore size and regular structure (Furukawa et al. 2010). Liu et al. have developed metal organic framework with GO (MOFs/GO) composite for the adsorption of benzene and it possess adsorption capacities of 72 cm^3/g (Guo-qiang et al. 2015). Bisphenol-A (2,2-bis(4-hydroxyphenyl) propane or BPA), an aromatic compound with two phenol moieties, are wildly used for synthesizing epoxy resins, polycarbonate plastics, automotive lenses, food and drink packaging, medical equipment, toys, compact disks, etc. Due to its slow degradation, BPA was found in various waste-water treatment plants, which could cause liver damage, thyroid hormone disruption, cardiovascular disease, and feminization phenomena to both human beings and animals (Ahmed et al. 2011). To overcome such problem, Li et al. have synthesized the efficient catalyst using reduced graphene oxide modified sea urchin-like α-MnO_2 (RGO/α-MnO_2) architectures for BPA removal and 90.5% removal was achieved up to 60 min (Li et al. 2015).

The heteroatoms like oxygen and nitrogen functionalities on carbon based materials especially in graphene could play a vital role in the environmental applications as catalysts (Figueiredo and Pereira 2010). Rocha et al. have synthesized nitrogen-doped graphene (N-GO) based materials for the advanced oxidation processes towards phenol removal from water which clearly exhibits the best catalytic performance of 50% of phenol removal after 120 min (Rocha et al. 2015). Borthakur et al. have synthesized the semiconductor based copper sulfide (CuS) with reduced graphene oxide (CuS-rGO) nanocomposite materials have drawn a considerable attention towards the photodegradation of diazo congo red (CR) dye. The synthesized CuS-rGO nanocomposite behaved as efficient catalyst for the degradation of diazo CR dye molecule with a maximum degradation efficiency of 98.76% (Boruah et al. 2016). Chen et al. have prepared nitrogen doped graphene (N-GP) and aminated graphene (NH_2-GP) by persulfate (PS) activation for the removal of aromatic pollutant namely sulfamethoxazole (SMX) and the results shows that SMX removal by N-GP and NH_2-GP was observed as >99 and 50% respectively (Chen and Carroll 2016).

Bisphenol A (BPA) is endocrine disrupting compound (EDC) which has been widely used in epoxy food-can linings and polycarbonate baby feeding bottles. Liao et al. have achieved efficient mineralization of BPA by photocatalytic ozonation with titanium oxide (TiO_2) combined reduced graphene oxide (RCO) (TiO_2-RGO) hybrid composite using liquid phase deposition (LPD) method and separation efficiency of BPA was achieved as 98.4% within 45 min (Liao et al. 2016). Nowadays, the various textiles industrial wastes namely methylene blue

(MB) cause large environmental pollution. Nuengmatcha et al. have applied the simple solvothermal approach for the preparation of ZnO/graphene/TiO_2 (ZGT) sonocatalyst towards MB dye removal and the removal efficiency of 67.08% was achieved within 150 min (Nuengmatcha et al. 2016). Semi synthetic penicillin namely ampicillin (AMP) have been widely utilized in human and veterinary medicine. The presence of β-lactam ring in its structure makes it as the good antimicrobial agent.

The various water sources and sediments contains oxytetracycline (OTC) which is one of the tetracycline antibiotic usually discharged from pharmaceutical manufacture, agriculture, livestock and hospital waste leads to water pollution (Elmolla and Chaudhuri 2010). Priya et al. have synthesized nano-sized BiOCl was immobilized onto graphene sand composite and chitosan for the degradation of both AMP and OTC from water and the removal efficiency were achieved as 70 and 84% respectively (Priya et al. 2016). Rosu et al. have synthesized graphene/TiO_2-Ag composites for the photocatalytic performance towards anionic azo dye namely amaranth that is trisodium 2-hydroxy-1-(4-sulfonato 1-naphthylazo) naphthalene-3,6-disulfonate from water. The remarkable photocatalytic activity of about 85.3–98% of amaranth degradation under UV and was observed within 2 h (Marcela-Corina et al. 2016). Umukoro have synthesized Ag–Ag_2O–ZnO nanostuctures anchored graphene oxide (GO) namely Ag–Ag_2O–ZnO/GO adsorbent for the degradation of acid blue 74 in water. As a result, the photocatalytic degradation activity of Ag–Ag_2O–ZnO/GO, Ag–Ag_2O–ZnO and ZnO were observed as 90, 85 and 75% respectively (Umukoro et al. 2016).

The refractory organic pollutants namely dimethyl phthalate (DMP) and the derivatives of phthalate esters (PAEs) were applied in the manufacture of plastics in order to improve their properties like flexibility and durability which causes prevalent environmental problem (Bauer and Herrmann 1997). Wang et al. have prepared graphene and WO_3 co-modified TiO_2 nanotube array (GR-WO_3/TNA) photo electrodes through in situ anodization and electrode position process for the catalytic degradation of DMP with the removal efficiency of 75.9% which is higher than the individual TNA, WO_3/TNA and GR/TNA were observed as 21.5, 12.1 and 15.2% respectively (Wang et al. 2016a, b). Wu et al. have synthesized a high specific surface area, monodisperse and highly active magnetic composite namely reduced graphene oxide (rGO/$CoFe_2O_4$) using one-pot solvothermal reaction for the removal of thiazin dye known as methylene blue (MB) from water and 100% of MB have been degraded within 15 min (Wu et al. 2016a, b). Yin et al. have prepared the reduced graphene oxide (rGO) and heteroatoms (N and P) doped graphene oxides (NGO, PGO) for the application of catalytic ozonation of sulfamethoxazole (SMX). The catalytic cleavage of S–C and S–N bond of SMX has been occurred by NGO and PGO and the removal efficiency of 95 and 99% were observed for the removal of SMX using NGO and PGO respectively (Yin et al. 2017).

The modification of RGO by doping with heteroatom like nitrogen was achieved by Khojasteh et al. and utilized for the removal of organic dyes such as methylene blue (MB) and rhodamine B (RhB) with the removal efficiencies of 100 and 69%

was observed after 30 min (Hojasteh et al. 2017). Alam et al. have reported the one step hydrothermal synthesis of Bi-TiO_2 nanotube/graphene composite for an efficient degradation of organic pollutant namely Dinoseb which are belong to dinitrophenol family were oftenly liberated from the various dye industries leads to environmental trouble. The removal efficiency of Dinoseb was observed as 72% within 180 min (Alam et al. 2017). In order to develop the adsorption and electrosorption techniques for the removal of organic and inorganic pollutants from waste-water, the novel magnetite/porous graphene (Fe_3O_4/porous graphene)-based nanocomposites was synthesized by Bharath et al. The synthesized material was utilized for the catalytic degradation of methyl violet (MV) dye from water and the 90% of removal efficiency was achieved (Bharath et al. 2017). Kim et al. developed 2D reduced graphene oxide-titania nanocomposites (RGO-TiO_2) under different hydrothermal conditions for treatment of hazardous organic pollutants especially for hexane and toluene with the removal efficiencies were observed as 39 and 87% respectively (Kim et al. 2017).

The widespread usage of chlorotetracycline (CTC) has caused the perseverance of antibiotic residues in aquatic environment, resulting in serious threat to human health and ecosystem. To overcome such problem, Li et al. have prepared graphene oxide/titanium dioxide (GO/TiO_2) nanocomposite via in situ hydrolysis of tetra-n-butyl titanate ($Ti(BuO)_4$) to TiO_2 particles on GO sheets were utilized as adsorbent for the efficient adsorptive removal of CTC from aqueous solution and the degradation efficiency was observed as 99–100% (Li et al. 2017). Yang et al. have synthesized metal organic framework (MOF)/graphene oxide namely (MIL-68 (In)-NH_2/GrO) composite as photocatalyst with the enhanced photocatalytic activity toward the degradation of amoxicillin (AMX). The MIL-68(In)-NH_2/GrO composite was prepared by dispersing GrO powder in the well-dissolved indium nitrate/BDC-NH_2 mixture and the percentage degradation of AMX was observed as 93% (Yang et al. 2017). A report shows that about 60% of organic matter affects the surface and subsurface water, that could liberated from the various dye industries, especially the dyes namely acid red 265 (AR265) and acid orange 7 (AO7) cause critical health hazards (Yanhui et al. 2013). To overcome such problem, Yusuf et al. have developed cetyltrimethyl ammonium bromide (CTAB) intercalated graphene (GN-CTAB) for the application of AR265 and AO7 degradation from water and the removal efficiency was achieved as 92.1 and 89.1% respectively (Yusuf et al. 2017).

9 Catalytic Ozonation of Aromatic Pollutants Using CNT and Their Derivatives

In recent years, the potential approach of carbon nanotubes (CNTs) in adsorption technology have been improved for the removal of various organic matters from aqueous system. Unlike many microporous adsorbents, CNTs possess the structural advantages like fibrous shape with high ratio, well developed mesopores and large accessible external surface area, all these contribute to the greater removal capacities towards aqueous aromatic pollutants (Upadhyayula et al. 2009; Chiang and Wu 2010). The aromatic natural dye namely indigo which are extracted from certain plants leaves. The textile industry generates greater amount of waste-water with intense colour of these dyes by various processes like bleaching, printing and finishing. These types of dyes possess mutagenic and carcinogenic could cause cruel difficulties to humans such as dysfunction of liver, kidney, brain, central nervous system and reproductive system (Kadirvelu et al. 2003).

The mineralization of the aromatic antibiotic sulphamethoxazole (SMX) using CNT has been investigated by Goncalves et al. The higher catalytic performance of MWCNTs was justified by surface chemistry and by the higher internal mass transfer resistances. During the catalytic performance, the sulphur present in SMX was almost completely converted into sulphate and nitrogen content in SMX was converted into ammonium ion and nitrate with 60% removal of SMX was observed at 150 min (Goncalves et al. 2012). Phenol and its derivatives were considered as the major organic pollutants appear in waste-waters from pharmaceutical, paper, petrochemical and from many other industries. These compounds are very toxic and difficult to eliminate from waste-water (Azabou et al. 2010). For the purpose of removing phenol from water, Pinho et al. have synthesized carbon nanotubes as catalysts for the catalytic wet peroxide oxidation of highly concentrated phenol (Pinho et al. 2015). The catalytic degradation of phenol were achieved by the presence of $OH^{\cdot}$ radicals, typically through an electrophilic addition to the aromatic ring. Phenol can be degraded by $OH^{\cdot}$ radicals, typically through an electrophilic addition to the aromatic ring and the synthesized catalyst have been removed more than 80% of initial phenol content after 24 h (Oturan et al. 2000).

Photocatalytic ozonation facilitates the degradation and mineralization of aromatic pollutants in a rapid manner. The discharge of untreated direct green 6 (DG6) and reactive red 198 (RR198) dyes colored waste-water into natural streams poses severe problems due to the toxicity of dyes to the aquatic life and damaging to the esthetic nature of the environment. Mahmoodi has utilized MWCNTs for the photocatalytic ozonation (UV/O_3) of direct green 6 (DG6) and reactive red 198 (RR198) dyes (Mahmoodi 2013a, b). In recent years nano-Fe_3O_4 has received increasing attention in ozonation technology due to the easiest separation from the bulk solution using external magnetic field and also has unique electric properties as electron transfer of Fe^{2+} and Fe^{3+} ions in the octahedral vacant sites (Nidheesh et al. 2014). To overcome the effect of solution pH on ozonation of methylene blue (MB), the composite heterogeneous catalyst using virtue of carbon nanotube and zero-valent iron (Fe^0-CNTs) have been developed (Zhang et al. 2013). Though the presence of oxygen-containing active species on the surface of CNTs at pH 9 might contribute to the removal of aromatic compounds, the given XPS band intensity denotes carboxyl groups on CNTs were slightly lower at pH 3 with addition MB than that detected without MB, which demonstrated thorough catalytic ozonation of intermediates adsorbed or enmeshed on the surface region of CNTs and 55% removal of MB was observed (Faria et al. 2008).

The ozonation technique has been used worldwide because of its strong oxidizing capacity and its eco-friendly nature to achieve the maximum degradation of many contaminants during waste-water treatment (Gharbani et al. 2008). p-hydroxybenzoic acid (p-HBA) is obtained from the personal care products in pharmaceuticals namely parabens. p-HBA are biodegradable, they are always appear at minimum concentration levels in effluents of waste-water treatment plants (Haman et al. 2015). For the degradation of p-HBA, Bai et al. was synthesized Fe_3O_4/multi-walled carbon nanotubes which achieved 32% degradation of p-HBA with in 5 min (Bai et al. 2015). Dimethyl phthalate (DMP) is the simplest form of phthalic acid esters (PAEs) which are widely used as plasticizers in polyvinyl chloride plastics which are difficult to biodegrade also found to be widely distributed in natural water, sediment, waste-water and soils (Wang et al. 1996). Sulfamethazine (SMT) is an antibiotic drug are widely used in veterinary and human medicine which is oftenly detected in environments. It causes alterations of the immune system and allergic reactions, which creates the severe health threat (Polesel et al. 2016). Bai et al. have prepared Fe_3O_4/Multiwalled carbon nanotubes for the catalytic ozonation of dimethyl phthalate (DMP) and sulfamethazine (SMT) antibiotics. The redox recycling of Fe (II)/Fe (III) at the surface of catalysts can promote the production of hydroxyl radicals, which facilitate the catalytic efficiency towards both DMP and SMT of 70 and 90% respectively (Bai et al. 2016, 2017).

The use of metal ions and metal nanoparticles such as silver, gold, cerium, lead and iron as catalyst and reagent for the water treatment have been improved due to their unique reactivity towards pesticides and pathogens (Wang et al. 2013). CNT are also possessing the large specific surface area required for adopting metal nano particles easily. m-Nitrophenol (m-NP) are frequently present in the industrial effluents because of they are widely used in many manufactures, such as wood preservatives, rubber chemicals, pharmaceuticals, pigments, dyes, explosives and herbicides which should be removed from the water sources (Sun and Lemley 2011). Duan et al. have prepared Ce co-modified PbO_2 (CNT-Ce-PbO_2) electrode for the electro-catalytic oxidation of m-NP (Duan et al. 2014). The degradation mechanism of m-NP covers that releasing of nitro group from aromatic ring to form 1,2,4-trihydroxybenzene or resorcin and then the aromatic rings were cleaved into the formation of aliphatic carboxylic acid followed by the same were oxidized into carbon dioxide and water, in this ozonation treatment 58% degradation of m-NP was achieved.

Ozone is found to uphold the catalytic oxidation of monochlorobenzene (CB) over CNTs supported copper oxide (CNT-CuO) composite at high reaction temperature of 300 °C. Chen et al. was synthesized CNT-CuO composite for catalytic oxidation of monochlorobenzene (CB). The selectivity of adsorption of CB was achieved by the promotion of Cu^+ to Cu^{2+} oxidation and also it was found that 99% of CB was converted into CO_2 by degradation above 250 °C. Similar way Pt/carbon nanotube (Pt-CNT) catalysts was fabricated by a molecular-level mixing method which was investigated at temperature ranging from 40 to 150 °C by Hea-Jung et al. for the degradation volatile organic compounds (VOCs) from industrial water (Hea-Jung et al. 2014). VOCs are considered to be detrimental to human health and the environment (Atkinson and Arey 2003). The oxidation activity was most probably supported due to the higher surface benzene, toluene, ethylbenzene and oxylene (BTEX) concentration afford by the adsorption ability of MWCNTs. Likewise, Zi-Rong et al. was synthesized one-dimensional titanate nanotubes and carbon nanotubes (TNT-CNT) by hydrothermal treatment for the photocatalytic gas-phase degradation of volatile aromatic pollutant especially for benzene and the removal efficiency was observed as 90% (Zi-Rong et al. 2011).

The aromatic dyes namely direct red 23 and direct red 31 were degraded using CNT and TiO_2 combined single system called (UV/CNT/H_2O_2 and UV/n-TiO_2/H_2O_2) and binary catalyst called (UV/CNT/n-TiO_2 and UV/CNT/n-TiO_2/H_2O_2). This photocatalytic dye degradation was studied using UV-Vis spectrophotometer and ion chromatography and these studies are reported (Mahmoodi 2013a, b). As a result, about 35 and 42% of mineralization of dyes leads to the conversion of harmful organic carbon and hydrogen into less toxic products like CO_2 and H_2O, various heteroatoms like N and S into inorganic anions such as NO_3^- and SO_4^{2-} respectively. Sulfamethoxazole (SMX) is an antibiotic were utilized worldwide for the treatment of urinary infection leads to identify in many waste-water treatment plant effluents. Hence, it is required to develop the effective approaches for SMX removal. Liu et al. was synthesized Zn-Fe-CNTs for the catalytic ozonation of SMX (Liu et al. 2017). The mechanism for SMX degradation are involved by in situ generation of H_2O_2, the in situ generation of $OH^{\cdot}$ radical via the decomposition H_2O_2 by Fe^0/Fe_2O_2 and the oxidation for pollutants by $OH^{\cdot}$. Finally, the complete degradation was achieved within 10 min at acidic pH condition. To enhance the decolorization rate of indigo dye, the ozonation of indigo by CNTs was functionalized with carboxyl groups (CNTs-COOH) as catalysts have been applied by Qu et al. The functional –COOH groups in CNT are considered to play a vital role in the internal mechanism of the catalytic ozonation process and about 55% of indigo concentration was reduced within 20 min (Qu et al. 2015). Chen et al. have synthesized copper oxide-carbon nanotube (CuOX/CNTs) for the ozone promotion of monochlorobenzene (CB) and about 84.0% of CB was degraded (Chen et al. 2013). The potentially higher effectiveness in the degradation and mineralization of aromatic compounds in water and waste-water was observed by the heterogeneous catalytic ozonation. Sui et al. have synthesized the manganese oxide (MnO_2) loaded on multi-walled carbon nanotube (MWCNT) surfaces by simply impregnating the carbon nanotube in order to obtain carbon nanotube-supported manganese oxides (MnO_2/MWCNT) and were used as ozonation catalysts to assist ozone in degrading ciprofloxacin (CF) in water (Sui et al. 2012). The probable catalytic mechanism was proposed for the heterogeneous catalytic ozonation which includes the promotion of hydroxyl radical ($OH^{\cdot}$) generation, the ozone and organic-compound concentration on the catalyst surface and complex catalysis. The removal efficiency of CF in this study was observed as 26.7% (Nawrockia and Kasprzyk-Hordern 2010).

The catalytic activity of carbon materials in oxidation process have been enhanced by the modification of their surface chemistry using the addition of various heteroatoms like N, O and S onto the carbon materials in catalytic ozonation (Cao et al. 2014). Sun et al. have prepared the surface nitrogen-modified

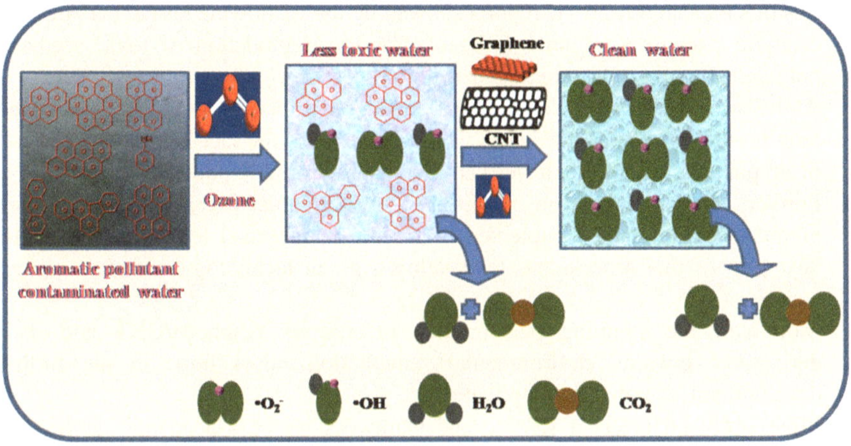

Scheme 3 The conversion of toxic aromatic pollutants into non-toxic products by catalytic ozonation using graphene/CNT supported materials

carbon nanotubes with sulfate radicals for the catalytic oxidation of aromatic pollutants namely phenol and about 76% degradation of phenol was achieved within 150 min (Sun et al. 2014). Restivo et al. have performed a systematic study on O, S, and N heteroatoms modified CNTs (CNTs-O, CNTs-S and CNTs-N) for the catalytic ozonation of phenolic water. The removal efficiency of phenol were observed as 44, 57 and 64% for CNTs-O, CNTs-S and CNTs-N materials respectively (Restivo et al. 2014). As a result conclude that, all the toxic aromatic pollutants could be converted into various less toxic products by the process of catalytic ozonation using graphene/CNT and their derivatives which are shown in Scheme 3.

10 Conclusions

This chapter was discussed on the green catalytic ozonation of the degradation of aqueous aromatic pollutants using graphene/CNT and their derivatives.

- The use of aromatic compounds in certain industries is unavoidable. The effluents from those industries contains aromatic pollutants and it leads to various effects on human and environment.
- The effluents must be treated by the catalytic ozonation in order to convert the toxic pollutants into non-toxic products like CO_2, H_2O, etc., which helps in the remediation of aromatic pollutants.
- From the various reports, the application of catalytic ozonation over graphene/CNT and their derivative materials have been decorated by providing the suitable operating conditions to achieve the enhanced removal of aromatic pollutants from industrial waste-water.

- It is much cleared from the reports showed by the worldwide researchers that the catalytic ozonation is more favorable for the degradation of toxic aromatic pollutant than the other treatment technologies.
- At the same time, factors like incomplete degradation of aromatic pollutants and long period process as the issues which come from the fact that the application of graphene/CNT and their derivatives for the aromatic pollutant removal.
- Further, the researchers are required to concern this important aspect before graphene oxide can be implemented for the commercial applications and the next step of this process has to develop a green technology with low energy consumption.
- Moreover, the stunning performance of various graphene/CNT and their derivatives towards environmental remediation raises hope in the further development of industrial applications.
- More study of foremost science and teamwork between academic circles and manufactures will yield the development of the novel remediation technology based on graphene/CNT and their derivatives towards the removal of aromatic pollutants from water in order to protect and save the environment.

References

Abdel-Shafy HI, Mansour MSM (2016) A review on polycyclic aromatic hydrocarbons: source, environmental impact, effect on human health and remediation. Egypt J Petrol 25(1):107–123

Agustina TE, Ang HM, Vareek VK (2005) A review of synergistic effect of photocatalysis and ozonation on wastewater treatment. J Photochem Photobiol C: Photochem Rev 6(4):264–273

Ahmed S, Rasul MG, Martens WN, Brown R, Hashi MA (2011) Advances in heterogeneous photocatalytic degradation of phenols and dyes in wastewater: a review. Water Air Soil Pollut 215(1–4):3–29

Ahn Y, Oh H, Yoon Y, Park WK, Yang WS, Kang JW (2017) Effect of graphene oxidation degree on the catalytic activity of graphene for ozone catalysis. J Environ Chem Eng 5(4):3882–3894

Alam U, Fleisch M, Kretschmer I, Bahnemann D, Muneer M (2017) One-step hydrothermal synthesis of Bi-TiO_2 nanotube/graphene composites: an efficient photocatalyst for spectacular degradation of organic pollutants under visible light irradiation. Appl Catal B Environ 218:758–769

Atalay S, Ersoz G (2016) Advanced oxidation processes. In: Novel catalysts in advanced oxidation of organic pollutants, pp 23–34

Atkinson R, Arey J (2003) Atmospheric degradation of volatile organic compounds. Chem Rev 103(12):4605–4638

Azabou S, Najjar W, Bouaziz M, Ghorbel A, Sayadi S (2010) A compact process for the treatment of olive mill wastewater by combining wet hydrogen peroxide catalytic oxidation and biological techniques. J Hazard Mater 183(1–3):62–69

Bai ZY, Yang Q, Wang JL (2015) Fe_3O_4/multi-walled carbon nanotubes as an efficient catalyst for catalytic ozonation of p-hydroxybenzoic acid. Int J Environ Sci Technol 13(2):483–492

Bai Z, Yang Q, Wang J (2016) Catalytic ozonation of dimethyl phthalate using Fe_3O_4/multi-wall carbon nanotubes. Environ Technol 38(16):2048–2057

Bai Z, Yang Q, Wang J (2017) Catalytic ozonation of sulfamethazine antibiotics using Fe_3O_4/multiwalled carbon nanotubes. Environ Prog Sustain Energy 37(2):678–685

Banjoo DR, Nelson PK (2005) Improved ultrasonic extraction procedure for the determination of polycyclic aromatic hydrocarbons in sediments. J Chromatogr A 1066(1–2):9–18
Bauer MJ, Herrmann R (1997) Estimation of the environmental contamination by phthalic acid esters leaching from household wastes. Sci Total Environ 208(1–2):49–57
Bharath G, Alhseinat E, Ponpandian N, Khan MA, Siddiqui MR, Ahmed F, Alsharaeh AH (2017) Development of adsorption and electrosorption techniques for removal of organic and inorganic pollutants from wastewater using novel magnetite/porous graphene-based nanocomposites. Sep Purf Technol 188:206–218
Boruah PK, Darabdhara G, Sengupta P, Das MR, Boronin AI, Kibis LS, Kozlova MN, Fedorov VE (2016) Microwave assisted synthesis of CuS-reduced graphene oxide nanocomposite with efficient photocatalytic activity towards azo dye degradation. J Environ Chem Eng 4(4):4600–4611
Builee TL, Hatherill JR (2004) The role of polyhalogenated aromatic hydrocarbons on thyroid hormone disruption and cognitive function: a review. Drug Chem Toxicol 27(4):405–424
Cao H, Xing L, Wu G, Xie Y, Shi S, Zhang Y, Minakata D, Crittenden JC (2014) Promoting effect of nitration modification on activated carbon in the catalytic ozonation of oxalic acid. Appl Catal B Environ 146:169–176
Carra I, Perez JAS, Malato S, Autin O, Jefferson B, Jarvis P (2014) Performance of different advanced oxidation processes for tertiary wastewater treatment to remove the pesticide acetamiprid. J Chem Technol Biotechnol 91(1):72–81
Carter CJ, Blizard RA (2016) Autism genes are selectively targeted by environmental pollutants including pesticides, heavy metals, bisphenyl A, phthalates and many others in food, cosmetics or household products. Neurochem Int 101:83–109
Chen H, Carroll KC (2016) Metal-free catalysis of persulfate activation and organic-pollutant degradation by nitrogen-doped graphene and aminated graphene. Environ Pollut 215:96–102
Chen R, Jin D, Yang H, Ma Z, Liu F, Zhang X (2013) Ozone promotion of mono chlorobenzene catalytic oxidation over carbon nanotubes-supported copper oxide at high temperature. Catal Lett 143(11):1207–1213
Chiang YC, Wu PY (2010) Adsorption equilibrium of sulfur hexafluoride on multiwalled carbon nanotubes. J Hazard Mater 178(1–3):729–738
Choi W, Lahiri I, Seelaboyina R (2010) Synthesis of graphene and its applications: a review. Crit Rev Solid State Mater Sci 35(1):52–71
Daneshvar E, Vazirzadeh A, Niazi A, Kousha M, Naushad M, Bhatnagar A (2017) Desorption of Methylene blue dye from brown macroalga: effects of operating parameters, isotherm study and kinetic modeling. J Clean Prod 152:443–453
De Moraes SG, Freire RS, Duran N (1996) Ozonation of wastewater containing *N*-methylmorpholine-*N*-oxide. Water Res 30(8):1745–1748
Dong Y, He K, Zhao B, Yin Y, Yin L, Zhang A (2007) Catalytic ozonation of azo dye active brilliant red X-3B in water with natural mineral brucite. Catal Commun 8(11):1599–1603
Du Q, Sun J, Li Y, Yang X, Wang X, Wang Z, Xia L (2014) Highly enhanced adsorption of congo red onto graphene oxide/chitosan fibers by wet-chemical etching off silica nanoparticles. Chem Eng J 245:99–106
Duan X, Zhao Y, Liu W, Chang L (2014) Investigation on electro-catalytic oxidation properties of carbon nanotube-Ce-modified PbO_2 electrode and its application for degradation of m-nitrophenol. Arab J Chem. https://doi.org/10.1016/j.arabjc.2014.11.025
Elmolla ES, Chaudhuri M (2010) Effect of photo-fenton operating conditions on the performance of photo-fenton-SBR process for recalcitrant wastewater treatment. J App Sci 10(24):3236–3242
Faria PCC, Orfao JJM, Pereira MFR (2007) Ozonation of aniline promoted by activated carbon. Chemosphere 67(4):809–815
Faria PCC, Orfao JJM, Pereira MFR (2008) Catalytic ozonation of sulfonated aromatic compounds in the presence of activated carbon. Appl Catal B Environ 83(1–2):150–159
Figueiredo JL, Pereira MFR (2010) The role of surface chemistry in catalysis with carbons. Catal Today 150(1–2):2–7

Fischer A, Manefield M, Bombach P (2016) Application of stable isotope tools for evaluating natural and stimulated biodegradation of organic pollutants in field studies. Curr Opin Biotechnol 41:99–107

Furukawa H, Ko N, Go YB (2010) Ultrahigh porosity in metal-organic frameworks. Science 329 (5990):424–428

Gharbani P, Tabatabaii SM, Mehrizad A (2008) Removal of Congo red from textile wastewater by ozonation. Int J Environ Sci Technol 4(5):495–500

Goncalves AG, Orfao JM, Pereira MFR (2012) Catalytic ozonation of sulphamethoxazole in the presence of carbon materials: catalytic performance and reaction pathways. J Hazard Mater 239–240:167–174

Guo Y, Yang L, Cheng X, Wang X (2012) The application and reaction mechanism of catalytic ozonation in water treatment. J Environ Anal Toxicol 2(7):1–7

Guo-qiang L, Ming-xi W, Zheng-hong H, Fei-yu K (2015) Preparation of graphene/metal-organic composites and their adsorption performance for benzene and ethanol. New Carbon Mater 30 (6):566–571

Haman C, Dauchy X, Rosin C, Munoz J (2015) Occurrence, fate and behavior of parabens in aquatic environments: a review. Water Res 68:1–11

Hea-Jung J, Jae-Ha K, Jun-Sik O, Dong-Wook Y, Han-Oh P, Kwang-Woo J (2014) Catalytic oxidation of VOCs over CNT-supported platinum nanoparticles. App Surf Sci 290:267–273

Hojasteh H, Salavati-Niasari M, Safajou H, Safardoust-Hojaghan H (2017) Facile reduction of graphene using urea in solid phase and surface modification by N-doped graphene quantum dots for adsorption of organic dyes. Diam Relat Mater 79:133–144

Hou H, Zhou R, Wu P, Wu L (2012) Removal of congo red dye from aqueous solution with hydroxyapatite/chitosan composite. Chem Eng J 211–212:336–342

Hu E, Wu X, Shang S, Tao X, Jiang S, Gan L (2016) Catalytic ozonation of simulated textile dyeing wastewater using mesoporous carbon aerogel supported copper oxide catalyst. J Clean Prod 112:4710–4718

Hu H, Chang M, Zhang M, Wang X, Chen D (2017) A new insight into PAM/graphene-based adsorption of water-soluble aromatic pollutants. J Mater Sci 52(14):8650–8664

Ikhlaq A, Kasprzyk-Hordern B (2017) Catalytic ozonation of chlorinated VOCs on ZSM zeolites and alumina: formation of chlorides. Appl Catal B 200:274–282

Jung BK, Jun JW, Hasan Z, Jhung SH (2015) Adsorptive removal of p-arsanilic acid from water using mesoporous zeolitic imidazolate framework-8. Chem Eng J 267:9–15

Kadirvelu K, Kavipriya M, Karthika C, Radhika M, Vennilamani N, Pattabhi S (2003) Utilization of various agricultural wastes for activated carbon preparation and application for the removal of dyes and metal ions from aqueous solutions. Bioresour Technol 87(1):129–132

Kile DE, Chiou CT, Zhou H, Li H, Xu Q (1995) Partition of nonpolar organic pollutants from water to soil and sediment organic matters. Environ Sci Technol 29(5):1401–1406

Kim M, Won-Hwa H, Kim W, Park SH, Wan-Kuen J (2018) 2D reduced graphene oxide-titania nanocomposites synthesized under different hydrothermal conditions for treatment of hazardous organic pollutants. Particuology 36:165–173

Kumar A, Naushad M, Rana A, Inammuddin Preeti, Sharma G, Ghfar AA, Stadler FJ, Khan MR (2017) ZnSe-WO_3 nano-hetero-assembly stacked on gum ghatti for photo-degradative removal of Bisphenol A: symbiose of adsorption and photocatalysis. Int J Biol Macromol 104:1172–1184

Lehner AF, Johnson M, Buchweitz J (2017) Veterinary utility of dried blood spots for analysis of toxic chlorinated hydrocarbons. Toxicol Mech Meth 101:1–9

Li G, Lu Y, Lu C, Zhu M, Zhai C, Du Y, Yang P (2015) Efficient catalytic ozonation of bisphenol-A over reduced graphene oxide modified sea urchin-like-MnO_2 architectures. J Hazard Mater 294:201–208

Li Z, Qi M, Tu C, Wang W, Chen J, Ai-Jun W (2017) Highly efficient removal of chlorotetracycline from aqueous solution using graphene oxide/TiO_2 composite: properties and mechanism. Appl Surf Sci 425:765–775

Liao G, Zhu D, Zheng J, Yin J, Lan B, Li L (2016) Efficient mineralization of bisphenol A by photocatalytic ozonation with TiO_2-graphene hybrid. J Taiwan Ins Chem Eng 67:300–305
Liu Y, Fan Q, Wang J (2017) Zn-Fe-CNTs catalytic in situ generation of H_2O_2 for Fenton-like degradation of sulphamethoxazole. J Hazard Mater 342:166–176
Llado J, Lao-Luque C, Ruiz B, Fuente E, Sole-Sardansa S, Doradoa AD (2015) Role of activated carbon properties in atrazine and paracetamol adsorption equilibrium and kinetics. Process Saf Environ Prot 95:51–59
Lv G, Wu D, Fu R (2009) Performance of carbon aerogels particle electrodes for the aqueous phase electro-catalytic oxidation of simulated phenol wastewaters. J Hazard Mater 165(1–3):961–966
Mahmoodi NM (2013a) Photocatalytic ozonation of dyes using multiwalled carbon nanotube. J Mol Catal A: Chem 366:254–260
Mahmoodi NM (2013b) Photocatalytic degradation of dyes using carbon nanotube and titania nanoparticle. Water Air Soil Pollut 224:1–8
Marcela-Corina R, Socaci C, Floare-Avram V, Borodi G, Pogacean F, Coros M, Magerusan L, Pruneanu S (2016) Photocatalytic performance of graphene/TiO_2-Ag composites on amaranth dye degradation. Mater Chem Phys 179:232–241
Meij R, Winkel H (2007) The emissions of heavy metals and persistent organic pollutants from modern coal-fired power stations. Atmos Environ 42(40):9262–9272
Midathana VR, Moholkar VS (2009) Mechanistic studies in ultrasound-assisted adsorption for removal of aromatic pollutants. Ind Eng Chem Res 48(15):7368–7377
Naushad M, Abdullah ALOthman Z, Rabiul Awual M et al (2016) Adsorption of rose Bengal dye from aqueous solution by amberlite Ira-938 resin: kinetics, isotherms, and thermodynamic studies. Desalin Water Treat 57:13527–13533
Nawrockia J, Kasprzyk-Hordern B (2010) The efficiency and mechanisms of catalytic ozonation. Appl Catal B Environ 99(1–2):27–42
Nidheesh PV, Gandhimathi R, Velmathi S, Sanjini NS (2014) Magnetite as a heterogeneous electro fenton catalyst for the removal of rhodamine B from aqueous solution. RSC Adv 4 (11):5698–5708
Ntampou X, Zouboulis AI, Samaras P (2006) Appropriate combination of physico-chemical methods (coagulation/flocculation and ozonation) for the efficient treatment of landfill leachates. Chemosphere 62(5):722–730
Nuengmatcha P, Chanthai S, Mahachai R, Won-Chun O (2016) Sonocatalytic performance of ZnO/graphene/TiO_2 nanocomposite for degradation of dye pollutants (methylene blue, texbrite BAC-L, texbrite BBU-L and texbrite NFW-L) under ultrasonic irradiation. Dyes Pigm 134:487–497
Oturan MA, Peiroten J, Chartrin P, Acher AJ (2000) Complete destruction of *p*-Nitrophenol in aqueous medium by electro-fenton method. Environ Sci Technol 34(16):3474–3479
Pinho MT, Gomes HT, Ribeiro RS, Faria JL, Silva AMT (2015) Carbon nanotubes as catalysts for catalytic wet peroxide oxidation of highly concentrated phenol solutions: towards process intensification. App Catal B Environ 165:706–714
Pirkanniemi K, Sillanpaa M (2002) Heterogeneous water phase catalysis as an environmental application: a review. Chemosphere 48(10):1047–1060
Polesel F, Andersen HR, Trapp S, Plosz BG (2016) Removal of antibiotics in biological wastewater treatment systems—a critical assessment using the activated sludge modeling framework for Xenobiotics (ASMX). Environ Sci Technol 50(19):10316–10334
Priya B, Shandilya P, Raizada P, Thakur P, Singh N, Singh P (2016) Photocatalytic mineralization and degradation kinetics of ampicillin and oxytetracycline antibiotics using graphene sand composite and chitosan supported BiOCl. J Mole Catal A Chem 10:3236–3242
Qu R, Xu B, Meng L, Wang L, Wang Z (2015) Ozonation of indigo enhanced by carboxylated carbon nanotubes: performance optimization, degradation products, reaction mechanism and toxicity evaluation. Water Res 68:316–327

Rengarajan T, Rajendran P, Nandakumar N, Lokeshkumar B, Rajendran P, Nishigaki I (2015) Exposure to polycyclic aromatic hydrocarbons with special focus on cancer. Asian Pac J Trop Biomed 5(3):182–189

Restivo J, Rocha RP, Silva AMT, Orfao JJM, Pereira MFR, Figueiredo JL (2014) Catalytic performance of heteroatom modified carbon nanotubes in advanced oxidation processes. Chin J Catal 35(6):896–905

Rocha RP, Goncalves AG, Pastrana-Martinez LM, Bordoni BC, Soares OSGP, Orfao JJM, Faria JL, Figueiredo JL, Silva AMT, Pereira MFR (2015) Nitrogen-doped graphene-based materials for advanced oxidation processes. Catal Today 249:192–198

Shahamat YD, Sadeghi M (2016) Heterogeneous catalytic ozonation of 2,4-dinitrophenol in aqueous solution by magnetic carbonaceous nanocomposite: catalytic activity and mechanism. Desalin Water Treat 57(43):20447–20456

Shahil KMF, Balandin AA (2012) Graphene-multilayer graphene nanocomposites as highly efficient thermal interface materials. Nano Lett 12(2):861–867

Shi X, Ng KK, Fu C, Low SL, Ng HY (2017) Removal of toxic component of wastewater by anaerobic processes. In: Current developments in biotechnology and bioengineering, 443–467

Singh RP, Shah AB (2016) Monitoring of hazardous inorganic pollutants and heavy metals in potable water at the source of supply and consumers end of a tropical urban municipality. Int J Environ Res 10(1):149–158

Sui M, Xing S, Sheng L, Huang S, Guo H (2012) Heterogeneous catalytic ozonation of ciprofloxacin in water with carbon nanotube supported manganese oxides as catalyst. J Hazard Mater 227–228:227–236

Sun SP, Lemley AT (2011) p-Nitrophenol degradation by a heterogeneous Fenton-like reaction on nano-magnetite: process optimization, kinetics, and degradation pathways. J Mol Catal 349(1–2):71–79

Sun H, Liu S, Zhou G, Ang HM, Tadé MO, Wang S (2012) Reduced graphene oxide for catalytic oxidation of aqueous organic pollutants. ACS Appl Mater Interfaces 4(10):5466–5471

Sun H, Kwan C, Suvorova A, Ang HM, Tadea MO, Wang S (2014) Catalytic oxidation of organic pollutants on pristine and surface nitrogen-modified carbon nanotubes with sulfate radicals. Appl Catal B Environ 154–155:134–141

Taheran M, Brar SK, Verma M, Surampalli RY (2016) Membrane processes for removal of pharmaceutically active compounds (PhACs) from water and wastewaters. Sci Total Environ 547:60–77

Umukoro EH, Peleyejua G, Ngilaa JC, Arotibaa OA (2016) Photocatalytic degradation of acid blue 74 in water using Ag-Ag_2O-ZnO nanostuctures anchored on graphene oxide. Solid State Sci 51:66–73

Upadhyayula VKK, Deng S, Mitchell MC, Smith GB (2009) Application of carbon nanotube technology for removal of contaminants in drinking water: a review. Sci Total Environ 408 (1):1–13

Valdes H, Zaror CA (2006) Heterogeneous and homogeneous catalytic ozonation of benzothiazole promoted by activated carbon: kinetic approach. Chemosphere 65(7):1131–1136

Vinod H (2017) Handbook on Indian chemical industry, Federation of Indian Chambers of Commerce and Industry. https://www.ibef.org

Viswanathan N, Pandi K, Meenakshi S (2014) Synthesis of metal ion entrapped silica gel/chitosan biocomposite for defluoridation studies. Int J Biol Macromol 70:347-353

Wang JL, Xu LJ (2012) Advanced oxidation processes for wastewater treatment: formation of hydroxyl radical and application. Crit Rev Environ Sci Technol 42(3):251–325

Wang JL, Liu P, Qian Y (1996) Biodegradation of phthalic acid esters by immobilized microbial cells. Environ Int 22(2):737–741

Wang H, Dong Z, Na C (2013) Hierarchical carbon nanotube membrane-supported gold nanoparticles for rapid catalytic reduction of p-nitrophenol. ACS Sustain Chem Eng 1(7): 746–752

Wang G, Chen Q, Xin Y, Liu Y, Zang Z, Hu C, Zhang B (2016a) Construction of graphene-WO_3/TiO_2 nanotube array photo electrodes and its enhanced performance for photocatalytic degradation of dimethyl phthalate. Electrochim Acta 222:1903–1913

Wang Y, Xie Y, Sun H, Xiao J, Cao H, Wang S (2016b) Efficient catalytic ozonation over reduced graphene oxide for p-Hydroxylbenzoic acid (PHBA) destruction: active site and mechanism. ACS Appl Mater Interfaces 8(15):9710–9720

Wu J, Ma L, Chen Y, Cheng Y, Liu Y, Zha X (2016a) Catalytic ozonation of organic pollutants from bio-treated dyeing and finishing wastewater using recycled waste iron shavings as a catalyst: removal and pathways. Water Res 92:140–148

Wu Q, Zhang H, Zhou L, Bao C, Zhu H, Zhang Y (2016b) Synthesis and application of rGO/$CoFe_2O_4$ composite for catalytic degradation of methylene blue on heterogeneous Fenton-like oxidation. J Taiwan Ins Chem Eng 67:484–494

Xing S, Lu X, Liu J, Zhu L, Ma Z, Wu Y (2016) Catalytic ozonation of sulfosalicylic acid over manganese oxide supported on mesoporous ceria. Chemosphere 144:7–12

Yang C, You X, Cheng J, Zheng H, Chen Y (2017) A novel visible-light-driven in-based MOF/graphene oxide composite photocatalyst with enhanced photocatalytic activity toward the degradation of amoxicillin. Appl Catal B Environ 200:673–680

Yanhui L, Qiuju D, Tonghao L, Xianjia P, Junjie W, Jiankun S, Yonghao W, Zonghua W, Yanzhi X, Linhua X (2013) Comparative study of methylene blue dye adsorption onto activated carbon, graphene oxide, and carbon nanotubes. Chem Eng Res Des 91(1):361–368

Yao Y, Cai Y, Lu F, Wei F, Wang X, Wang S (2014) Magnetic recoverable $MnFe_2O_4$ and $MnFe_2O_4$-graphene hybrid as heterogeneous catalysts of peroxymonosulfate activation for efficient degradation of aqueous organic pollutants. J Hazard Mater 270:61–70

Yin R, Guo W, Du J, Zhou X, Zheng H, Wu Q, Chang J, Ren N (2017) Heteroatoms doped graphene for catalytic ozonation of sulfamethoxazole by metal-free catalysis: performances and mechanisms. Chem Eng J 317:632–639

Yusuf M, Khan MA, Otero M, Abdullah EC, Hosomi M, Terada A, Riya S (2017) Synthesis of CTAB intercalated graphene and its application for the adsorption of AR265 and AO7 dyes from water. J Colloid Interf Sci 493:51–61

Zeng T, Zhang X, Wang S, Niu H (2015) Spatial confinement of a Co_3O_4 catalyst in hollow metal-organic frameworks as a nano reactor for improved degradation of organic pollutants. Environ Sci Technol 49(4):2350–2357

Zhang H, Quan X, Chen S, Zhao H, Zhao Y (2006) Fabrication of photocatalytic membrane and evaluation its efficiency in removal of organic pollutants from water. Sep Purif Technol 50(2):147–155

Zhang S, Wang D, Quan X, Zhou L, Zhang X (2013) Multi-walled carbon nanotubes immobilized on zero-valent iron plates (Fe^0-CNTs) for catalytic ozonation of methylene blue as model compound in a bubbling reactor. Sep Purif Technol 116:351–359

Zi-Rong T, Li F, Zhang Y, Fu X, Yi-Jun X (2011) Composites of titanate nanotube and carbon nanotube as photocatalyst with high mineralization ratio for gas-phase degradation of volatile aromatic pollutant. J Phys Chem C 115(16):7880–7886

Chapter 12
Efficient Removal of Nitrate and Phosphate Using Graphene Nanocomposites

P. Senthil Kumar, P. R. Yaashikaa and S. Ramalingam

Abstract The increase in demand for food products has forced production in agricultural sector at faster rate. Application of chemical fertilizers consisting of phosphorous, nitrates and potassium increased the yield with more negative impacts on environment. The run-off water from agriculture field with high concentration of nitrates and phosphates contaminate the water bodies such as rivers and lakes thus affecting the aquatic system and creatures. Algal growth and degradation occurs at faster rate and the microbes utilize the dissolved oxygen present in water. Lack of oxygen in water causes suffocation and death of living creatures resulting in eutrophication. Adsorption technique has been widely used to overcome these problems. Graphene nanocomposites find promising application in adsorbing and removing nitrates and phosphates from the water bodies. Graphene nanocomposites can be obtained from its derivatives such as graphene oxide, chemically and thermally reduced graphene oxides. They act as nanosorbents and are widely used owing to their properties such as biocompatibility, stability, high surface area to volume ratio, good conductivity and low-cost synthesis. It can be synthesised by solvent processing, melt processing or in situ polymerization methods. Studies reveal that graphene nanocomposites prove to be ideally potent in removing pollutants such as nitrates and phosphates.

Keywords Eutrophication · Graphene nanocomposites · GO synthesis
Nitrates · Phosphates

P. Senthil Kumar (✉) · P. R. Yaashikaa
Department of Chemical Engineering, SSN College of Engineering,
Chennai 603110, India
e-mail: psk8582@gmail.com; senthilkumarp@ssn.edu.in

P. R. Yaashikaa
e-mail: pryaashikaa@gmail.com; yaashikaapr@ssn.edu.in

S. Ramalingam
Department of Chemical Engineering, University of Louisiana at Lafayette,
Lafayette, LA 70504, USA
e-mail: raams@louisiana.edu

M. Naushad (ed.), *A New Generation Material Graphene: Applications in Water Technology*, https://doi.org/10.1007/978-3-319-75484-0_12

1 Introduction

Removal of pollutants from environment such as soil, ground and surface water, air paved the way for remediation (Naushad et al. 2015). Organic toxins and pollutants speak to a risk to human, creature, and ecological wellbeing (Naushad et al. 2013). If unmanaged, these contaminations could cause concern. Numerous researchers have ventured up endeavours to discover more supportable and inexpensive methods to utilizing dangerous chemicals and medicines to evacuate existing destructive pollutants. Ecological biotechnology, for example, bioremediation and phytoremediation, is a promising field that uses common assets including microorganisms and plants to wipe out poisonous natural contaminants (Kang 2014). This innovation offer an appealing contrasting option to other traditional remediation forms as a result of its moderately advantages like low cost and eco-friendly approach. There is restricted probability of an expansion in the supply of fresh water because of contending requests of expanding population all through the world; additionally, water-related issues are relied upon to increment assist because of atmosphere changes and because of populace development throughout the following two decades (Bruggen and Vandecasteele 2003). Deficiency of new water supply is additionally a consequence of the misuse of water assets for residential, industry, and water system purposes in many parts of the world. The weight on freshwater assets because of the expanding requirement of food, vitality, etc. is expanding increasingly because of population development and dangers of environmental climate change. Dirtying surface/ground water sources is another reason for diminished new water supplies. Aquifers around the globe are exhausting and being dirtied because of numerous issues of saltwater interruption, soil disintegration, deficient sanitation, sullying of ground/surface waters by algal blossoms, cleansers, manures, pesticides, chemicals, substantial metals, etc. The runoff water from farming area is fundamental source for deposition of toxic pollutants like nitrate and phosphate into ground water and other water bodies. Presence of overabundance of phosphorus in water causes undesirable eutrophication, rare of non-sustainable asset that is crucial for human life and also cause to type of destructive algal blooms. Henceforth, most extreme level of phosphorus in water ought not to exceed 10 mg/L, set by United States Environmental Protection Agency (USEPA). High amount of nitrate in water assets causes baby methemoglobinemia and teratogenic impacts and furthermore antagonistically influence the oceanic plants (development of green algal growth), fish and creatures. In this way, most extreme consent level of nitrate in drinking water has set 50 mg/L by World Health Association in 2008 (Naushad et al. 2014). Conventional techniques for removal of nitrate and phosphate include reverse osmosis, ion-exchange and electro dialysis. At present, Adsorption technique is commonly used because of its advantages including easy, effective, eco-friendly, renewable and low cost (Rashed 2013).

2 Why Nanotechnology for Pollutants Removal

Nanotechnology has been viewed as powerful in taking care of water issues identified with quality and amount in expelling nitrates and phosphates from water. Nanomaterials (e.g., carbon nanotubes (CNTs) and dendrimers) are adding to the improvement of more effective treatment forms among the propelled water frameworks (Amin et al. 2014). There are numerous approaches of nanotechnology to address the various issues of water quality keeping in mind the end goal to guarantee the environmental security and stability. The propelled techniques utilization of nano-materials made of activated carbon, carbon tubes, graphene and graphene oxide carbon nanotubes, biopolymers, single-compound nanoparticles, self-assembled monolayer on mesoporous supportive network, zeolites, nanoparticles of zero valent iron are utilized for water remediation with improved fondness, limit, selectivity for overwhelming metals and different contaminants (Gautam and Chattopathyaya 2016). Higher reactivity, bigger surface contact and better transfer ability are the upsides of utilizing nanomaterials in the evacuation of water contaminations. In addition, higher surface area to volume ratio in nanoparticles enhances the uptake and removal of nitrates, phosphates and other pollutants in water (Wang et al. 2013). Nano-materials used as adsorbents can be modified on the surface using various chemical and biological methods for effective binding of ions and nanomaterials thus enhancing removal percentage (Loganathan et al. 2014).

3 Graphene

Graphene is one of the allotropes (precious stone, carbon nanotube, and fullerene) of carbon, two-dimensional monolayer of lattice of carbon and is the building block of all forms of graphite. Graphene is used to describe single layer of graphite which is a constituent of graphite intercalation compounds. It is made up of four bonds with one σ bond at its three outer neighbours and one π bond present out of the plane. It can be easily modified using nitrate and oxygen containing functional groups characterised using infrared spectroscopy. Graphene is commonly employed mainly for its properties including electrical quantum Hall Effect and high electron mobility at room temperature, stiffest and strongest material and high thermal conductivity. Owing to these properties, graphene finds application in various fields including sensors, electrodes, nanocomposites, solar cells, etc. (Zhu et al. 2010).

4 Morphology

According to Mermin-Wagner theorem it is known that long-range orders do not exist. At any finite temperature dislocation occurs at two-dimensional (2D) crystals. Recent research proved that long-range orders do exist because of the coupling between stretching and bending modes. As a result the 2D form exists but still wrinkled. Graphene expresses tow different morphologies such as intrinsic and extrinsic classified based on their geometric feature (Wang et al. 2012).

4.1 Intrinsic Morphology

Random intrinsic arrangements influence the 2D configuration of graphene in 3D space resulting in increased strain energy and stabilization of random thermal fluctuation. Intrinsic morphology can be viewed using transmission electron microscopy (TEM) and atomic force microscopy (AFM). TEM results showed that the graphene sheets are not flat and roughly surfaced. AFM observations depict the nanometre high buckles present in a single layer of graphene. These buckles in multi layer graphene penetrate through the single layers of graphene. In order to confirm, more simulations have been conducted and most importantly Atomistic Monte Carlo simulations demonstrated that ripples can be formed by the thermal fluctuations. The factors affecting the intrinsic morphology of graphene include structural defect, aspect ratio, sample size, etc. Increase in aspect ratio results in change in morphology structure from planar membrane forming another structural phase. Thus in electronic applications low aspect ratio is suitable in graphene composites avoiding self-folding. For regular and reconstruction of free edges in graphene, it is subjected to compressive stress. The wrinkles are restricted to the edge areas itself. The edge stress results in folding of graphene sheets into cylindrical with their edges facing inside. Morphology of chemical groups in graphene also changes resulting in bond rearrangement from sp^2 to sp^3 type.

4.2 Extrinsic Morphology

In extrinsic morphology, the substrate-supported graphene is controlled by graphene-substrate interactions in contrast to intrinsic morphology. When fabricated on to a substrate, graphene appears to be striated because of the unnecessary photo-resist residue present under the graphene. On removal of these residues, atomic resolution images of graphene depict the corrugations. Research activities showed that the single and multi-layer graphene sheets follow the surface morphology of substrates. The extrinsic morphology of graphene is regulated by three types of energies:

- Graphene strain energy
- Substrate strain energy
- Graphene-substrate interaction energy

When graphene follows up a substrate, graphene and substrate strain energy increases with decrease in graphene-substrate interaction energy. Thus equilibrium energy can be obtained by minimizing the total energy of the system. The substrate present below the graphene layer can be designed with different characteristics such as single dimensioned nanowires and nanotubes or with dimensionless nanoparticles.

5 Synthesis Approaches

Graphene synthesis approach can be classified into two types as top-down and bottom-up approaches (Young et al. 2012).

5.1 Top-Down

This approach deals with processing the existing form of bulk material to produce the final product. The cost involved in this process material varies depending on the type of material used. This method finds application at small scale only and finds only limited quality control during the process. Single layer graphene or graphene sheets are produced by various processes like separation, cleaving, peeling and exfoliation. The capital investment is high to produce minimum yield of product. Few mechanical functions involved in this approach are sonication, exfoliation, thermal and electrochemical exfoliation, alkylation and chemical reduction of graphene oxide.

5.2 Bottom-up

This approach utilizes standard techniques such as epitaxial growth or organic synthesis using metallic substrates as catalysts by means of chemical vapour deposition (CVD) process depending on the choice of precursor materials involved in the process. Other methods involved include arc discharge, carbon nanotube unzipping, chemical conversion and CO reduction. CVD and epitaxial growth methods produces more quantities of graphene sheets with larger size compared to other methods and can be used for large scale applications. Though bottom-up approach shows less deformity, the mode of operation and techniques are quite harder making this unfit for mass production due to its expensive issue.

6 Synthesis Methods

Different methods have been used for synthesis of graphene and its oxides. The first synthesis of graphene was performed using Scotch Tape method (Novoselov et al. 2004). This method involves separating single layers of graphene from graphite and

Table 1 Advantages and disadvantages of methods of graphene synthesis

Methods	Advantages	Disadvantages
Chemical vapour deposition	Bulk quantities of graphene	Time consuming
Epitaxial growth	Bulk quantities of graphene	Requires high temperature
Unzipping of carbon nanotubes	High quality	Expensive

offers good quality of graphene for electrical applications but the main drawback is low yield. The different methods of graphene synthesis are

- Chemical Vapour Deposition
- Arc discharge
- Carbon nanotubes
- Reduction of graphite oxides
- Epitaxial growth on silicon carbide
- Pyrolysis
- Solvothermal

Each of these methods possesses their own advantages and disadvantages (Table 1).

Though these methods have more advantages, they are not suitable for polymer nanocomposites because it fails to produce huge quantities as nanofills. These methods are suitable for synthesis graphene for electronic applications only (Tang et al. 2012).

6.1 Chemical Vapour Deposition (CVD)

Graphene layers are grown on the surface of substrates. The substrates are coated with catalytic metals such as copper, nickel, gold, platinum, cobalt and silver and are exposed to precursor molecular flux (Sun et al. 2011). During this exposure, the metallic coat helps to produce carbon species as graphene layers. This method can be carried out at high and low temperatures. This method also employs thermal decomposition of hydrocarbon gases like methane at high temperatures and low pressure. This method is quite complicated because removal of metallic coat after graphene formation is complex and thus limits its application in electronic devices only (Rümmeli et al. 2010).

6.2 Exfoliation and Reduction of Graphite Oxide

Exfoliation and reduction are two methods widely available for bulk production of graphene for nanofill applications in polymer nanocomposites. This method involves oxidation of graphite. Brodie in 1859 was the first to synthesis graphene using this method. Graphene was also oxidised to its oxide using strong oxidising agents like $KMnO_4$, $NaNO_3$, etc. in presence of concentrated acids like H_2SO_4 and HNO_3 or mixture of both by a process termed Staudenmaier and Hummers methods. Of all these methods, Hummers method is more commonly used. Experimental studies reveal that graphite oxide structure is composed of functional groups like carboxide, epoxide, hydroxyl, etc. Large scale production of graphite oxide also called functionalized graphene is synthesised by converting graphite to graphite oxide followed by exfoliation and this can be easily dispersed in water and organic solvents. Graphite oxide is thermally unstable and electrically insulating so the conversion of graphite oxide to graphene oxide is necessary. Complete conversion of graphite oxide to graphene oxide is not possible so the formed graphene is termed as the "reduced graphene". This reduction can be performed chemically or thermally.

(1) Brodie method (Brodie 1859)

$$\text{Graphite} + KClO_4 \xrightarrow[60\,^\circ C]{+HNO_3} \xrightarrow{H_2O} \xrightarrow{\text{Exfoliation}} \text{GO}$$

(2) Staudenmaier method (Staudenmaier 1898)

$$\text{Graphite} \xrightarrow[5\,^\circ C]{+H_2SO_4/HNO_3} \xrightarrow{KClO_4} \xrightarrow{H_2O} \xrightarrow{\text{Exfoliation}} \text{GO}$$

(3) Hummers-Offeman method (Hummers and Offeman 1958)

$$\text{Graphite} \xrightarrow[0\,^\circ C]{+H_2SO_4} \xrightarrow[<20\,^\circ C]{KMnO_4} \xrightarrow[90\,^\circ C]{H_2O_4} \xrightarrow{H_2O_4} \xrightarrow{\text{Exfoliation}} \text{GO}$$

6.3 Chemical Reduction of Graphite Oxide

This method employs stable colloidal dispersion of graphite oxide followed by exfoliation using reducing agents like dimethyl hydrazine, hydrazine hydrate, etc. with hydroquinone, hydrazine, etc. (Park et al. 2011). The stable dispersion was attained by ultrasonication in water or alcohol. The reduction of graphite oxide to graphene oxide increases the electrical conductivity of the restored graphene oxide by restoring sp^2 character of carbon.

6.4 Thermal Reduction of Graphite Oxide

This method involved continuous heating of graphite oxide for about 30–45 min at high temperature of around 1000 °C at inert pressure (Schniepp et al. 2006). CO_2 is evolved generating high pressure resulting in exfoliation to reduced graphene oxide. Degradation of hydroxyl, carboxyl and epoxide functional groups occurs and the Van der waals force holds the layers of graphene oxide together. The major advantage of thermal reduction over chemical reduction is that no solvent is required for dispersion of graphite oxide (Steurer et al. 2009).

7 Properties

Graphene possess high material strength with strong electrical, mechanical and thermal properties. Graphene has high surface area and gas impermeability useful for improving gas barriers, thermal, mechanical and electrical properties of polymers nanocomposites (Potts et al. 2011).

7.1 Mechanical Properties

Graphenes are stiffer and stronger than carbon nanotubes. Simulations of atoms determine the size and chirality elastic properties in graphenes whereas the effect of size of graphene is negligible on Young's modulus. Higher values of Cauchy stress and fracture strain was found in zigzag direction compared to arm-chair direction. The effect of temperature also influences the mechanical properties of graphene. There is no significant change in Young's modulus with temperature 1200 K after which graphene becomes softer. Temperature is indirectly proportional to fracture strength and strain of graphene. Graphene is formed as monolayer with several layers formed after continuous exfoliation. The number of layers formed also influences mechanical properties of graphene. Monolayer of graphene contains carbon atoms arranged in hexagonal order. Interactions between graphene and other materials occur in the form of non-bonded Van der Waals attraction.

7.2 Electrical Properties

Graphene when acts as a semi-conductor with zero band gaps contain unusual charge carriers behaves as massive relative particles. When subjected to magnetic field, it varies with other electrons and has unique quantum Hall effect which was observed at room temperature also. The band structure of monolayer graphene shows two bands at two equilibrium points K and K_0. Now these points resemble the relativistic electrons. K and K_0 are termed as Dirac points resembling graphene as zero band gap semiconductors. Monolayer graphene also exhibits ambipolar electric field effect at room temperatures by inducing gate voltage. The result showed that at positive gate bias, the Fermi level raises above the Dirac point thus allowing the electrons into conduction band and at the negative gate bias, Fermi level falls below the Dirac point allowing the holes in valence band.

7.3 Thermal Properties

Thermal conductivity of graphene is usually isotropic. Photons determine the specific heat and thermal conductivity of graphenes. These are the electric waves generated in graphene lattice. The maximum activity of graphene as a perfect thermal conductor was found at 5000 W m^{-1} K^{-1} which is found to be higher than thermal conductivity of other carbon materials like carbon nanotubes, diamonds, etc. Graphite formed from several layers of graphene shows thermal conductivity of around 1000 W m^{-1} K^{-1} which is five times lesser than graphene. High thermal conductivity of graphene widens its vast application in electronic devices.

8 Graphene Nanocomposites

The success of polymer nanocomposites synthesis greatly depends on the compatible dispersion of nanofills in polymer matrix. Vast endeavours have been focused on accomplishing a homogeneous and all around scattered framework by growing either covalent or non-covalent functionalization of the filler surface. Polymer nanocomposites with graphenes find its applications in solar cells, electronic devices, green energy, automobile industries, etc. Two dimensional graphene possess better mechanical, thermal and electrical properties with high surface area and aspect ratio. Graphene nanocomposites find application in various biomedical, nanoelectro mechanical systems and electrochemical applications (Hazra and Basu 2016). Few important commercial applications of graphene nanocomposites are given in Fig. 1.

9 Method of Graphene Nanocomposite Preparation

The viable usage of graphene composites firmly relies upon the homogeneous scattering of graphene and the interfacial cooperation between the polymer grid and graphene. Consequently, to accomplish high level of graphene scattering in polymers amid handling without influencing the properties, three unique strategies might be utilized: in situ polymerization, solvent processing and melt processing (Tripathi et al. 2017). Along with polymers, different sorts of networks, for example, poly(methyl methacrylate), polyethylene glycol, polyamide, epoxy saps, poly(ethylene terephthalate), polystyrene, etc. have additionally been utilized to get ready polymer/graphene nanocomposites for enhanced electrical, thermal and mechanical properties.

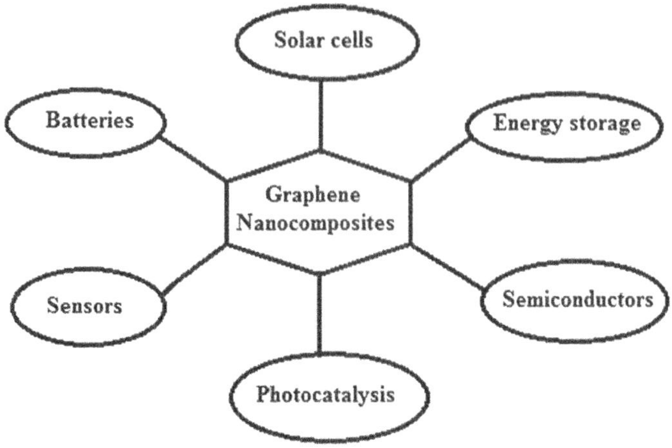

Fig. 1 Applications of graphene nanocomposites

9.1 In Situ Polymerization

This method is the most promising technique for preparation of graphene nanocomposites based on derivatives such as graphene oxide, reduced graphene oxide and thermally reduced graphene oxide. This method allows synthesis of thermally unstable and insoluble graphene that cannot be prepared using other methods. The monolayer graphenes are usually mixed with polymers or monomers in presence of solvent. Polymerization reaction occurs by altering parameters such as temperature and time period. Thermally reduced graphene nanocomposites were prepared in an aqueous solution in presence of surfactant using in situ polymerization method by post polymerization mixing approach. The synthesised graphene nanocomposites were compared with those of commercially available graphene and was found that percolation threshold in synthesised thermally reduced graphene nanocomposites were lower than that of commercial ones (Tchernook et al. 2014). In situ polymerization method suits more for preparation of ultra high molecular weight polyethylene graphene nanocomposites. This method allows preparation of graphene nanocomposites with high compactibility and well dispersed monolayer graphene sheets with improved properties (Fig. 2).

9.2 Solvent Processing

This method is suitable for preparing graphene composites using high molecular weight polymers by blending of graphene with suitable polymer matrix in presence of solvent. Basically three steps are involved in solvent processing namely

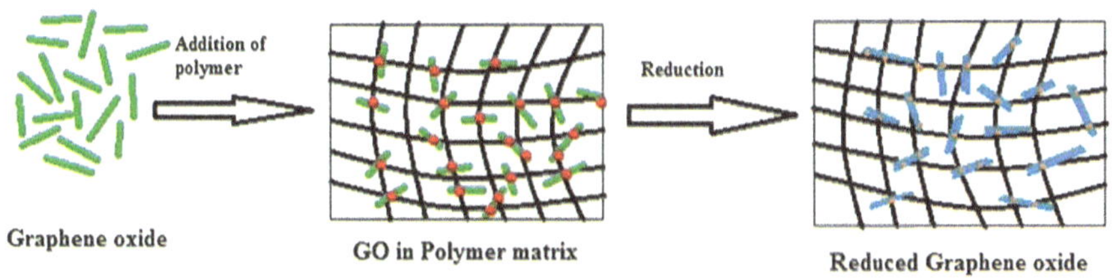

Fig. 2 In situ method of conversion of graphite oxide to graphene oxide

- Dispersing graphene in a suitable solvent solution
- Blending of graphene along with suitable polymer solution at room temperature
- Recovery of nanocomposite either by precipitating or as a film
  Continuous blending of graphene along with polymer in presence of solvent promotes uniform dispersion of graphene in the polymer matrix (Suner et al. 2015). This method provides advantages of low viscosity and even mixing and dispersion of graphene. The major drawback of this method is use of huge quantities of solvent and its evaporation. Ultra High Molecular Weight graphene oxides can be synthesised by dispersing graphene oxide in solvent like ethanol followed by blending in the same solution using ball milling technique. It was found that thermal and mechanical properties of synthesised graphene nancomposite improved. The composite film can be prepared using two methods
- Dispersion of graphene oxide followed by thermal reduction to reduced graphene oxide before addition of solvent
- Dispersion and reduction of graphene oxide after addition if solvent.

9.3 Melt Processing

This method is the common choice for industrial application because of its advantages including low costs and simplicity. This method provides large scale synthesis of graphene nanocomposites for commercial applications. This method involves melting of graphene in the form of pellets to a viscous liquid followed by application of high shear force for the dispersion of nanofillers (Kim et al. 2011). This method does not require any use of solvent for dispersion. It is an eco-friendly approach involving the application of commercially available polymer and conventional mixing techniques like twin screen extrusion by altering parameters like screw speed, temperature and time period. The disadvantages of this strategy are the low mass thickness of thermally shed graphene that makes extruder nourishing a troublesome assignment and the lower level of scattering contrasted with dissolvable mixing. This lessened scattering degree would then be able to bring about poorer properties, suggesting higher permeation limits and lower conductivities.

10 Eutrophication

Presence of excess concentration of nitrates and phosphates causes' death of aquatic organisms by a process termed as Eutrophication. It is defined as the procedure by which a water body gains a high amount of nutrient supplements, particularly phosphates and nitrates. These regularly advance over the more development of algal growth. The death and decomposition of algae results in deposition of high levels of organic matter and other microbial growth thus depleting the available oxygen in the marine water causing lethal to marine organisms like fish (Serediak et al. 2014). Eutrophication seems to be a problem due to following reasons:

- Light penetration into water is reduced due to mat formation by algae.
- Oxygen gets completely depleted from water. Algal decomposition causes much consumption of oxygen leading to decreased or no oxygen.
- Death of aquatic organisms since they require dissolved oxygen for their growth.
- Toxins may be produced by the algal species thus making the water unfit.

Anoxic event results in absence of oxygen caused by high nutrient supplements in water resources which affects algal growth development. At the point when the plants die and decompose, oxygen is cleared from the water, which at that point turns green or smooth white and emits a solid strong spoiled egg scent. The absence of oxygen is frequently lethal for invertebrates, fish and shellfish (Fig. 3).

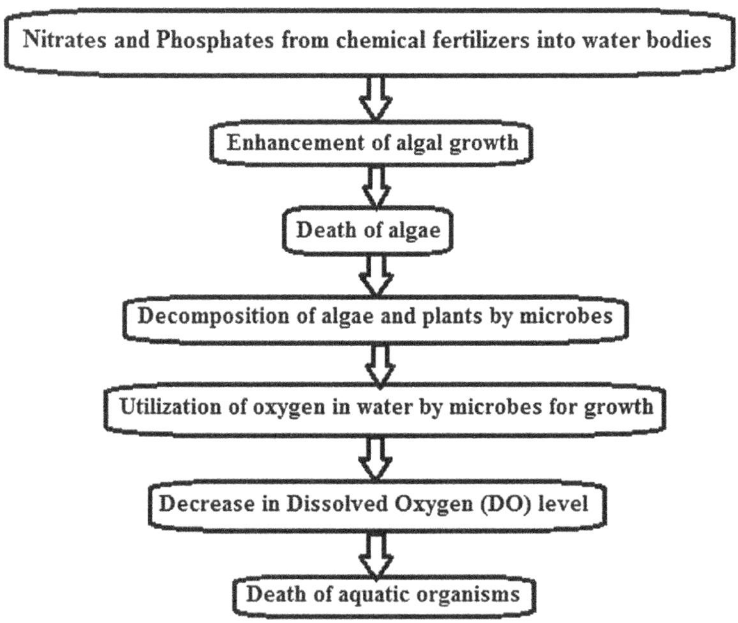

Fig. 3 Effect of eutrophication in water bodies

11 Nitrates

Nitrates are a polyatomic ion and are important source of nitrogen for plants. The major source of nitrate contamination in river resources is through application of nitrogen fertilizer carried away by rain or irrigation. Wastes of humans and animals also results in nitrogen contamination. The quality of drinking water is determined by the concentration of nitrate ions present in it. Presence of high amount of nitrates causes growth of various disease-causing microbes and finally affecting human health. The permissible level of nitrate present in drinking water is 10 mg/L according to Environmental Protection Agency (EPA). Nitrate level if present above this permissible limit may cause "methemoglobinemia or blue-baby syndrome" in infants resulting in fatal blood disorder. These days, customary strategies for denitrification of ground water are produced in light of physico-chemical forms, including ion exchange, reverse osmosis and electrodialysis. For specific lessening of nitrate to nitrogen and reactant decrease, natural assimilation is utilized. The natural biological denitrification procedure is as of now the most broadly utilized process, however, conceivable bacterial contamination, the nearness of remaining organics and additionally conceivable increment in chlorine in refined water limits the use of this procedure. Reactant denitrification has been created to get specific decrease of nitrate into nitrogen with the undesired development of ammonia salts (McAleer et al. 2017). As of late, the catalytic lessening of nitrate by platinum and palladium metals has pulled in impressive consideration. In reducing nitrates, it is important to actuate the valuable metal by the expansion of a promoter, for example, tin or copper. This yields unsuitable measures of ammonia salts as the response by product (maximum level 0.1 mg/L). Different metals and aggravates that have been examined with the point of applying a superior selectivity control are Titania, zirconia, SnO_2 and hydrotalcite. Utilization of bimetallic Pd–Sn catalysts for denitrification of water, upheld on polyaniline and polypyrrole was additionally proposed. Regardless of the way that the catalysts indicate high proficiency in nitrate evacuation, they are not helpful for modern application attributable to the utilization of costly metals and creation of harmful side-effects.

12 Nitrate Removal

Recent studies revealed that graphene, graphene oxide and its composites showed better removal of nitrate and served as good source of materials for efficient removal. Presence of catalyst such as palladium and copper held by graphene with iron as reductants resulted in removal of 82% of nitrate and showed 66% of nitrogen selectivity. Magnetic graphene nanoparticles showed 86.4% removal of nitrate at optimum conditions such as pH-3, temperature-50 °C, contact time-60 min, adsorbent dosage-2 g/L and initial concentration of 50 mg/L. This method of employing magnetic graphene nanocomposites proved to be easy, efficient with high removal rate of nitrate. Nanocomposites showed maximum removal efficiency of nitrate up to 94.3% than nano-scale zero valent iron. Nitrates present in water can be removed by coating nickel and cobalt nanoparticles with graphene oxide. Graphene based nanocomposites with specific properties along with iron, cobalt and nickel proved to be efficient for removal of nitrates present in water. Nanomembrane papers can also be used for efficient removal of nitrates by means of adsorption process (Mautner et al. 2017).

13 Synthesis of Graphene Based Magnetic Nanocomposites

Studies revealed that graphene nanocomposites enhanced the nitrate removal from water (Nodeh et al. 2017). The process of synthesis of graphene based nanocomposites is shown in Fig. 4.

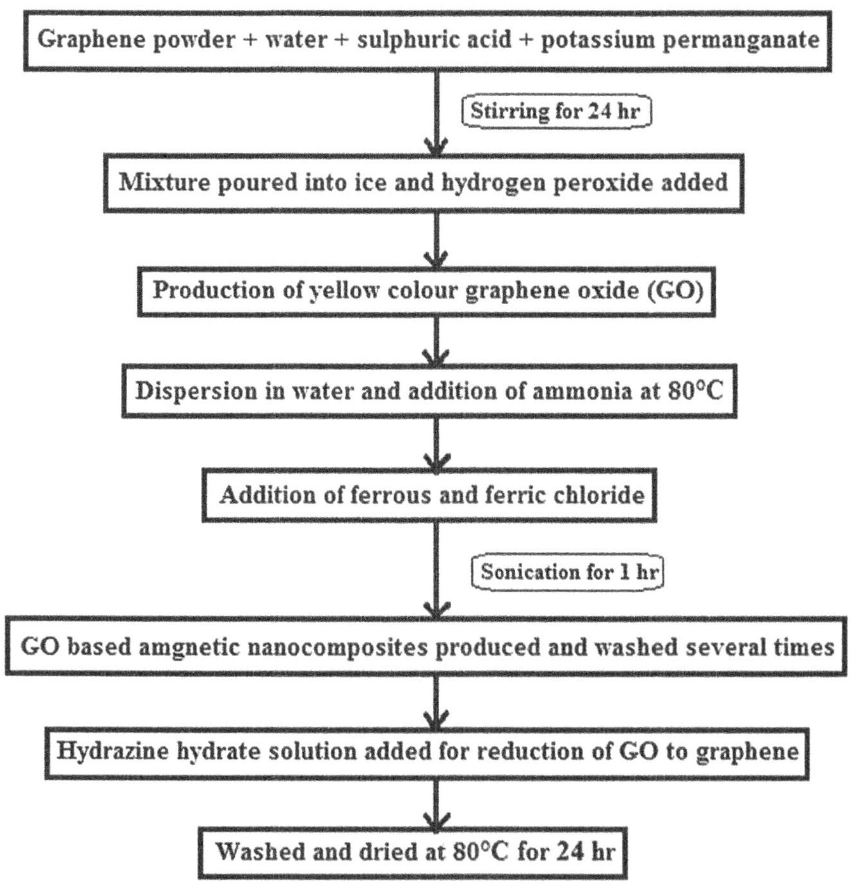

Fig. 4 Process of synthesis of magnetic graphene nanocomposites

14 Factors Affecting the Rate of Nitrate Removal

The rate of nitrate removal is greatly influenced by various parameters like solution pH, temperature, nanocomposite (adsorbent) dosage and time of contact.

14.1 Effect of Solution pH

pH of the solution plays a major role in determining the rate of removal. Efficient removal of nitrate ions was observed at acidic pH 2–6. Nearly 90% of nitrate ions get eliminated at acidic pH. Further increasing the pH to basic condition a gradual decrease in sorption was found. Electrostatic attractive forces between negatively charged nitrate ions and positively charged nanocomposites functions maximum at low pH only.

14.2 Effect of Temperature

Temperature also has a predominant effect on adsorption rate. Research activities reveal that most of the removal process occurs at temperature between 20 and 50 °C. Increase in temperature increases the force of the ions to bind with the active site of the graphene nanocomposite thus enhancing the removal efficiency.

14.3 Effect of Ionic Strength

The influence of ionic strength was analysed by addition of NaCl to the solution. The results showed that there was no effect on the removal rate when NaCl was maintained at concentration 10%. If the concentration was further increased removal rate decreased due to the Na^+ ions that accumulates in the nanocomposites and leaves no space for binding of nitrate ions.

14.4 Effect of Initial Concentration

Removal rate is influenced greatly by the effect of initial concentration of the ions. If the accumulations of nitrate ions are more, rate of removal decreases since the ions compete with one another for binding. So at low concentration more number of sites is available for binding thus enhancing the removal percentage. Increasing the initial concentration of nitrate ions increases the adsorption capacity until equilibrium was achieved. After that the adsorption rate gradually decreased.

14.5 Effect of Adsorbent Dosage

Removal rate was found to be high until equilibrium was reached. This is due presence of number of sites for binding of nitrate ions in the nanocomposites. A decrease in concentration of adsorbent dosage results in less number of active sites for binding. Adsorption rate also decreases with low removal percentage.

14.6 Effect of Contact Time

Increasing the reaction time between the nanocomposites and nitrate ions increased the adsorption rate. Contact time is most important parameter since it determines the time of process to occur. Phosphate ions are more readily adsorbed than nitrate ions due to driving forces for mass transfer during the initial stage of adsorption process. The contact time for nitrate was found to be 90 min.

15 Phosphates

Phosphates are salts of phosphoric acid causes water contamination by promoting excessive growth of algae. These ions serve as feed for algal growth creating imbalances in water and destructing other aquatic life forms like fish by producing lethal toxins. Abundance phosphates make water shady and low in oxygen. Bringing extra phosphates into water brings about an enormous development of green growth, which are oceanic plants including many single-celled, free-skimming plants. Unreasonable measures of green algal growth cloud the water in an impact called an algal bloom, which decreases the daylight accessible to different plants and some of the time kills them. At the point when the green growth die, the microbes that split them down go through broke down oxygen in the water, denying and then suffocating other oceanic life. Fertilizers containing phosphates contaminate surface water. Phosphates enter water frameworks actually by dissolving out of rock; however phosphates are additionally mined and made into substance composts to develop crops. Applying chemical composts to soil officially soaked with phosphates and spreading unreasonable measures of fertilizers makes phosphates keep running off amid overwhelming precipitation and water pollution. At the point when the measure of aggregate phosphorous crosses 100 parts for every billion (ppb) in streams or 50 ppb in lakes, eutrophication—the impact of algal blooms—is a threat. Abundant phosphate levels likewise influence the procedures in drinking water treatment plants (Zhang et al. 2009). Finally, phosphate cannot be totally expelled from the aquarium since natural phosphate is always changed over into in-natural dissolvable orthophosphate. All things considered, phosphates can be controlled with a decent upkeep plan gone for keeping natural phosphates at the very least measure.

16 Factors Influencing the Rate of Phosphate Removal

16.1 Effect of pH

The most important parameter affecting the adsorption rate is pH. The optimum pH for maximum phosphate adsorption by graphene nanocomposites was found to be 6–8. Above pH 8, the adsorption rate of phosphate decreases. This is due to electrostatic repulsion between the negatively charged graphene nanocomposites and phosphate ions. The ability of graphene to adsorb phosphate ions was found maximum at pH 8 to about 99% removal.

16.2 Effect of Temperature

As temperature increases the phosphate adsorption also increases depicting that removal rate increases with increasing temperatures. As temperature rises, the interactive forces between solute and solvent in the solution become weak. This in turn increases the interaction between solute and adsorbent thus resulting in more active sites for binding. This indicates the process is endothermic.

16.3 Effect of Initial Concentration

At the initial concentration, adsorption of phosphates was found to be high till equilibrium is reached. If the concentration increases, there will be no active sites for the phosphate ions for binding thus reducing the removal percentage.

16.4 Effect of Contact Time

The contact time required for maximum phosphate removal is less compared to nitrate ions. This is because of the driving forces that exist between the phosphate ions for binding. The maximum contact time for phosphate adsorption was found to be 10 min.

17 Process of Nitrate and Phosphate Removal by Polymer Nanocomposites

The nitrate removal process of each prepared nanocomposites powder was surveyed as follows: Solution containing sodium nitrate ($NaNO_3$) and sodium phosphate (NaH_2PO_4) with a concentration of 100 ppm at neutral pH was set up as an initial preparation solution. 0.01 g of each nanocomposite powder was completely ground in a glass vessel and added into a 100 mL nitrate and phosphate solution. The suspension containing nitrate and phosphate particles and nanocomposite particles was sonicated for 1 min to accomplish a steady suspension. The suspension was kept at room temperature and consistently stirred amid the procedure. In this manner, the nanocomposite was isolated from the reactor (after 1 h) by means of centrifugation. Finally the concentration of nitrate and phosphate in the supernatant was determined by measuring the absorbance in the UV Spectrophotometer.

18 Mechanism of Adsorption of Nitrate and Phosphate by Graphene Based Magnetic Nanocomposites

Electrostatic attraction is the important parameter in the removal process of nitrate using graphene based magnetic nanocomposites. The opposite charges between nitrate and phosphate and graphene nanocomposites enhance the adsorption rate of ions rapidly. The attractive opposite forces brings about efficient binding of ions at the active site hence competition occurs among the nitrate and phosphate ions in binding with the nanocomposites (Xie et al. 2014) (Fig. 5).

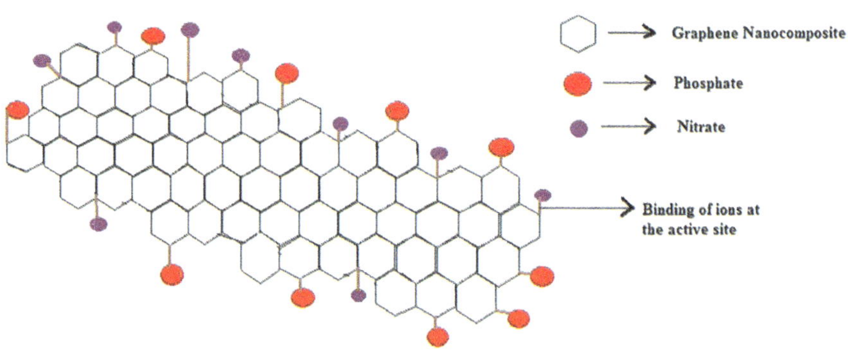

Fig. 5 Adsorption of phosphate and nitrate by graphene nanocomposites

19 Conclusion

The revelation of graphene and its nanocomposites is viewed as an incredible turning point ever. It is presently imagined in view of various research endeavours that the utilization of graphene and its nanocomposites can offer an extensive variety of potential advantages for natural applications. Inspite of their advantages (e.g., surface area, recyclability, stability, and presence of functional groups on graphene-based materials), the prevalence of these materials has been exhibited in the degradation of natural pollutants, for example, nitrates and phosphates from aquatic bodies through adsorption, removal, or a mix of both. Nanocomposites of graphene with various photocatalysts can be utilized to build the efficiency through improvement of the photocatalytic properties. Although the removal procedures of nitrate and phosphate in view of graphene methods are going up against with two difficulties like stability and toxicity, the extraordinary properties of graphene are relied upon to offer significant choices in the advancement of wastewater treatment strategies. Graphene and graphene-based nanocomposite removal process exhibited to have a potential in the removal of nitrate and phosphates. Practicability of utilizing graphene, graphene oxide and nanocomposites is tremendous in denitrification and phosphate expulsion of groundwater because of extraordinary properties including and furthermore ultra substantial particular surface range which contributes to a great extent in upgrading the adsorption limit in expulsion of nitrate and phosphate from groundwater.

References

Amin MT, Alazba AA, Manzoor U (2014) A review of removal of pollutants from water/wastewater using different types of nanomaterials. Adv Mater Sci Eng 24. Article ID 825910

Brodie BC (1859) On the atomic weight of graphite. Philos Trans R Soc London 149:249–259

Bruggen BV, Vandecasteele C (2003) Removal of pollutants from surface water and ground water by nanofiltration: overview of possible applications in drinking water industry. Environ Pollut 122(3):435–445

Gautam RK, Chattopathyaya MC (2016) Graphene-based nanocomposites as nanosorbents (Chap. 4). Nanomaterials for wastewater remediation, pp 49–78

Hazra SK, Basu S (2016) Graphene-oxide nanocomposites for chemical sensor applications. J Carbon Res 2(12). https://doi.org/10.3390/c2020012

Hummers WS, Offeman RE (1958) Preparation of graphitic oxide. J Am Chem Soc 80:1339

Kang JW (2014) Removing environmental organic pollutants with bioremediation and phytoremediation. Biotechnol Lett 36(6):1129–1139

Kim H, Kobayashi S, Rahim MAA et al (2011) Graphene/polyethylene nanocomposites: effect of polyethylene functionalization and blending methods. Polymer 52:1837–1846

Loganathan P, Vigneshwaran S, Kandasamy J (2014) Enhanced removal of nitrate from water using surface modifications of adsorbents—a review. J Environ Manag 131:363–374

Mautner A, Kobkeatthawin T, Bismarck A (2017) Efficient continuous removal of nitrates from water with cationic cellulosic nanopaper membranes. Resour Effic Technol 3(1):22–28

McAleer EB, Coxon CE, Richards KG et al (2017) Groundwater nitrate reduction versus dissolved gas production: a tale of two catchments. Sci Total Environ 586:372–389

Naushad M, Alothman ZA, Khan MR (2013) Removal of malathion from aqueous solution using De-Acidite FF-IP resin and determination by UPLC-MS/MS: equilibrium, kinetics and thermodynamics studies. Talanta 115:15–23. https://doi.org/10.1016/j.talanta.2013.04.015

Naushad M, Khan MA, ALOthman ZA, Khan MR (2014) Adsorptive removal of nitrate from synthetic and commercially available bottled water samples using De-Acidite FF-IP resin. J Ind Eng Chem 20:3400–3407. https://doi.org/10.1016/j.jiec.2013.12.026

Naushad M, ALOthman ZA, Awual MR et al (2015) Adsorption kinetics, isotherms, and thermodynamic studies for the adsorption of Pb^{2+} and Hg^{2+} metal ions from aqueous medium using Ti(IV) iodovanadate cation exchanger. Ionics (Kiel) 21:2237–2245. https://doi.org/10.1007/s11581-015-1401-7

Nodeh HR, Sereshti H, Afsharian EZ et al (2017) Enhanced removal of nitrate and phosphate ions from aqueous media using nanosized lanthanum hydrous doped on magnetic graphene nanocomposite. J Environ Manag 197:265–274

Novoselov KS, Geim AK, Morozov SV et al (2004) Electric field effect in atomically thin carbon sheets. Science 306:666–669

Park S, An J, Potts JR et al (2011) Hydrazine-reduction of graphite and graphene oxide. Carbon 49:3019–3023

Potts JR, Dreyer DR, Bielawski CW et al (2011) Graphene based polymer nanocomposites. Polymers 52:5–25

Rashed MN (2013) Adsorption technique for the removal of organic pollutants from water and wastewater. In: Nageeb Rashed M (ed) Organic pollutants—monitoring, risk and treatment. InTech. https://doi.org/10.5772/54048

Rümmeli MH, Bachmatiuk A, Scott A et al (2010) Direct low-temperature nanographene CVD synthesis over a dielectric insulator. ACS Nano 4:4206–4210

Schniepp HC, Li JL, McAllister MJ et al (2006) Functionalized single graphene sheets derived from splitting graphite oxide. J Phys Chem B 110:8535–8539

Serediak NA, Prepas EE, Putz GJ (2014) Eutrophication of fresh water systems. In: Reference module in earth system and environmental sciences treatise on geochemistry, 2nd edn., vol 11, pp 305–323

Staudenmaier L (1898) Verfahren zur Darstellung der Graphitsäure. Eur J Inorg Chem 31: 1481–1487

Steurer P, Wissert R, Thomann R et al (2009) Functionalized graphene and thermoplastic nanocomposites based upon expanded graphite oxide. Macromol Rapid Commun 30:316–327

Sun NLJ, Cole MT, Teo BK et al (2011) Large-area uniform graphene-like thin films grown by chemical vapour deposition directly on silicon nitride. Appl Phys Lett 98:252107

Suner S, Joffe R, Tipper JL et al (2015) Ultra high molecular weight polyethylene/graphene oxide nanocomposites: thermal, mechanical and wettability characterisation. Composites 78:185–191

Tang H, Ehlert GJ, Lin Y et al (2012) High efficient synthesis of graphene nanocomposites. NanoLetters 12:84–90

Tchernook A, Krumova M, Tolle FJ et al (2014) Composites from aqueous polyethylene nanocrystal/graphene dispersions. Macromolecules 47:3017–3021

Tripathi SN, Rao GSS, Mathur AB et al (2017) Polyolefin/graphene nanocomposites: a review. R Soc Chem 7:23615–23632

Wang M, Yan C, Ma L (2012) Graphene nanocomposites. In: Hu N (ed) Composites and their properties. InTech, Rijeka, pp 17–36

Wang S, Sun H, Ang HM et al (2013) Adsorptive remediation of environmental pollutions using novel graphene-based nanomaterials. Chem Eng J 226:336–347

Xie J, Wang Z, Fang D et al (2014) Green synthesis of a novel hybrid sorbent of zeolite/lanthanum hydroxide and its application in the removal and recovery of phosphate from water. J Colloid Interface Sci 423:13–19

Young RJ, Kinloch IA, Gong L et al (2012) The mechanics of graphene nanocomposites: a review. Compos Sci Technol 72:1459–1476

Zhang GS, Liu HJ, Liu RP, Qu JH (2009) Removal of phosphate from water by a Fe-Mn binary oxide adsorbent. Colloid Interface Sci 335(2):168–174

Zhu Y, Murali S, Cai W et al (2010) Graphene and graphene oxide: synthesis, properties and applications. Adv Mater 22:3906–3924

Chapter 13
Graphene Family Materials for the Removal of Pesticides from Water

T. Paramasivan, N. Sivarajasekar, S. Muthusaravanan, R. Subashini, J. Prakashmaran, S. Sivamani and P. Ajmal Koya

Abstract Graphene, its composites and its modified forms have attracted the attention due to its novel structure and unique properties. They are widely employed in the treatment of organic and inorganic contaminants. One of the organic contaminants class—pesticides present in the aqueous environment is the threat to human and animal biota due to their carcinogenic effects. Graphene-based materials hold great potential in decontaminating pesticide bearing effluents such as adsorbents, photo-catalyst and membranes and are the current research trend. In this chapter, we reviewed the preparation, characterization and application of graphene-based materials in water purification. From the literature, it is known that graphene-based materials are widely used as adsorbents for pesticide removal. Therefore the optimum

T. Paramasivan · N. Sivarajasekar (✉) · S. Muthusaravanan · R. Subashini
Laboratory for Bioremediation Research, Unit Operations Laboratory, Department of Biotechnology, Kumaraguru College of Technology, Coimbatore, TN, India
e-mail: sivarajasekar@gmail.com

T. Paramasivan
e-mail: paramasivan@gmail.com

S. Muthusaravanan
e-mail: muthusaravanan.ind@gmail.com

R. Subashini
e-mail: rameshsubashini159@gmail.com

J. Prakashmaran
Department of Food Science and Nutrition, Periyar University, Salem, TN, India
e-mail: prakashmaran@gmail.com

S. Sivamani
Chemical and Petrochemical Engineering Section, Engineering Department, Salalah College of Technology, Sultanate, Oman
e-mail: sivman.sel@gmail.com

P. Ajmal Koya
Department of Chemistry, National Institute of Technology Mizoram, Aizawal 796012, Mizoram, India
e-mail: ajmalkoya@gmail.com

M. Naushad (ed.), *A New Generation Material Graphene: Applications in Water Technology*, https://doi.org/10.1007/978-3-319-75484-0_13

parameters affecting the adsorption process and a comparison of graphene-based adsorbents with other adsorbents are also discussed.

Keywords Graphene · Adsorption · Water · Pesticide

1 Introduction

A pesticide is a substance or mixture of substances which can be a naturally derived or synthetically produced and used in destroying the life cycle of pest. Pesticides include bactericides, fungicides, herbicides and insecticides. The development of carbamates and pyrethroids lead to manufacture of persistent pesticides (Edwards 1977; Chaudhry et al. 2002). These pesticides are harmful and when released into the environment, they disperse through volatilization, leaching, run-offs and drainage. Most pesticides used on land end up in aquatic environments (Edwards 1977). Once in the aquatic environment, the persistence of a pesticide is dependent on its chemical stability, degradability by microorganisms and uptake by aquatic/ terrestrial species including plants (Cunningham et al. 1997). Various pathological effects of low doses of pesticides in animals and man are immune-pathological effects and carcinogenic effects (Chauhan and Singhal 2006).

There are many methods available for the deactivation/removal of pesticides in aquatic environments. Depending on the chemical nature of a pesticide, treatment with a chemical reagent such as a strong acid or alkali (Chen et al. 2007), oxidants such as hydrogen peroxide or ozone (Sun and Pignatello 1992), catalytic oxidation (Chen et al. 2007), photocatalytic degradation (Gupta et al. 2015), Fenton's reagent (Chen et al. 2007) is usually sufficient to cause deactivation. But most of the chemical methods are expensive or even impracticable. The two main biological methods for the removal or decomposition of pesticides include bioremediation using micro-organisms (Gavrilescu 2005; Hunter 2002; Newcombe and Crowley 1999) and phyto-remediation using plants (Cunningham et al. 1997; Rice et al. 1997; Xia and Ma 2006). Over the years both the fields have emerged as low-cost and eco-friendly technologies (Chatterjee et al. 2010). However, these methods are limited to selected compounds and they require the particular conditions for action (Gavrilescu 2005; Akhtar et al. 2009; Sivarajasekar et al. 2016).

Adsorption is one of the eco-friendly and effective method for micro-pollutant removal (Singh 2009; Sivarajasekar et al. 2017a, b, c) provided that an efficent adsorbent is available (Sivarajasekar et al. 2017d, e, f, g). Graphene is found unique because of its structure, properties and wide applications (Wu et al. 2011). Due to low cost, high flexibility and strength of graphene, different graphene composites can be fabricated according to the type of application (Wang et al. 2014a, b). Amongst, graphene based materials have been identified as the potential adsorbents in decontaminating water because of the larger scale surface area and their adsorption capacity (Liu et al. 2011). For the applications in water purification, graphene and its derivatives have several natural advantages like high adsorption and lower time consumption (Xu et al. 2013; Zhao et al. 2012).

Keeping the above points in mind the objectives of this review are framed such as

- To study the preparation of graphene/graphene based materials
- To understand the characterization methods of graphene/graphene based materials
- To learn the different applications of graphene/graphene based material
- To examine the optimum parameters studied for graphene/graphene based materials for pesticide contaminated effluent.

2 Methods for the Synthesis of Graphene

There are several methods used in the synthesis of graphene. The methods include mechanical exfoliation (Huang et al. 2011), chemical exfoliation (Stankovich et al. 2007), thermal exfoliation (McAllister et al. 2007; Schniepp et al. 2006), electrostatic deposition (Sidorov et al. 2007), chemical vapor decomposition (Kim et al. 2009; Reina et al. 2009), thermal decomposition on metal surface (Huc et al. 2008) and oxidative exfoliation (Brodie 1859; Hummers and Offeman 1958).

Graphene nano-sheets were first obtained from mechanical exfoliation, but in this method some difficulties like formation of point defect and stone wales defect is frequent. The graphene obtained by this method never lies in a single plane (Huang et al. 2011). Exfoliation is one of the routes for synthesis of graphene sheets with a larger production volume. The most common drawback of this technique is the production of by-products, which contaminate the graphite layers. It is also hard to evade the defects in the basal plane (Stankovich et al. 2007). Various organic solvents have been studied to exfoliate graphene oxide, followed by a thermal reduction. This technique could possibly lead to production in large-scale with an easy one-step process (McAllister et al. 2007; Schniepp et al. 2006). Highly oriented pyrolytic graphite sheets separated into layers using electrostatic attractive force. These layers are removed according to the voltage range. This process consumes much time for the synthesis of graphene layers having high surface area (Sidorov et al. 2007). Thermal decomposition method produces graphene layers that are far from the atomic level (Huc et al. 2008). Graphite materials are prepared from hydrocarbons decomposition on metal surfaces by using chemical vapor deposition. In this method there is a possibility of some foreign atoms getting misplaced in place of carbon (Kim et al. 2009; Reina et al. 2009).

2.1 Hummers' Method

Oxidative exfoliation of graphite is done through the Staudemaier, Brodie and Hummers method (Brodie 1859; Hummers and Offeman 1958). In the Hummers method, oxidation of graphite is achieved the steps as shown in Fig. 1. Hydrazine was used in reducing graphene oxide to synthesize graphene (Li et al. 2008; Shi et al. 2014; Stankovich et al. 2007). Hummers' method has more advantages comparing the Brodie's method (Brodie 1859) previously used in graphene synthesis. Hummers' method has following merits such as

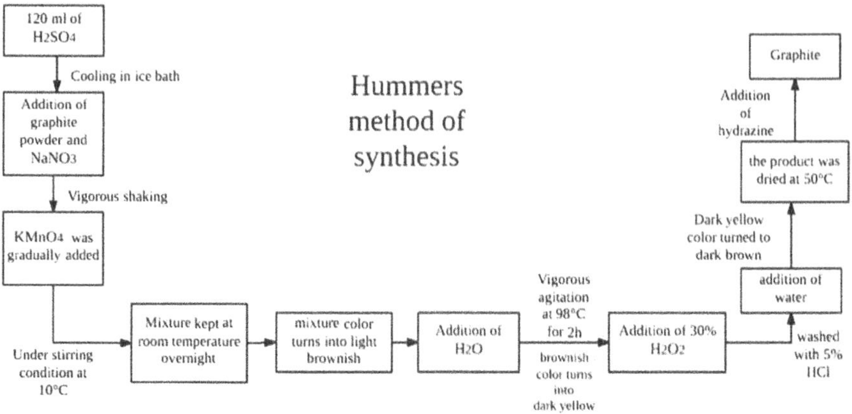

Fig. 1 Synthesis of graphene from Hummers' method

- The hazardous $KClO_3$was replaced with $KMnO_4$
- The use of $NaNO_3$ in place of fuming HNO_3 eliminated the formation of acid fog
- The synthesis of graphite reaction is safer than Brodie's method

The Hummers method possessed some disadvantages like release of toxic gasses such as NO_2 and N_2O_4 which hinders the removal of the residual Na^+ and NO^{3-} ions from the waste water (Chen and Li 2013).

2.2 *Modified Hummers' Method*

There are many methods newly designed and reported as modified Hummers' method. One of which includes, improving the Hummers method by excluding $NaNO_3$ and increasing the amount of $KMnO_4$ used. Thus, they performed the reaction in a 9:1 mixture of H_2SO_4/H_3PO_4 improving the efficiency of the oxidation process. This modification successfully increased the yield and reduced the emission of toxic gas (Marcano et al. 2010). In another work (Chen and Li 2013), Graphene oxide was prepared according to Hummers method using natural graphite powder with a modification of removing $NaNO_3$ as shown in Fig. 2. Graphene oxide prepared by this method was found to be nearly the same in their dispersing ability, chemical structures, thicknesses, and lateral dimensions comparing to the one produced by conventional Hummers' method (Chen and Li 2013).

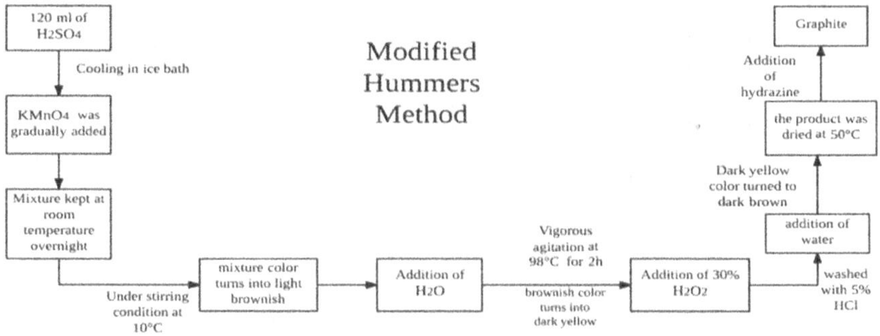

Fig. 2 Synthesis of graphene from modified Hummers' method

3 Synthesis of Graphene Family Material for the Application of Pesticide Removal

After the synthesis of graphene, different methods were followed to prepare graphene derivatives/graphene composites for the application of pesticide removal from water. Table 1 explains the precursor used, treatment and synthesis for graphene materials. The various graphene family materials used for the removal of pesticides from water are listed below.

3.1 Reduced Graphene Oxide Adsorbent

Gupta et al. (2015) have prepared reduced graphene oxide and utilized to adsorb carbofuran pesticides. He reported that graphene oxide was filtered with 0.1 M HCl and then washed with distilled water. Further, graphene oxide was dispersed into 200 mL water under mild ultrasound yielding a yellow-brown suspension, then 4 mL hydrazine hydrate (80 wt%) was added. The solution was heated in an oil bath maintained at 100 °C for 24 h. After the reaction, reduced graphene oxide was collected by vacuum filtration.

3.2 Graphene Sand Composite Adsorbent

Gupta et al. (2012) have used sucrose as the carbon source for production of graphene sand composite for the adsorptive removal of chlorpyrifos. Initially, the sugar was dissolved in water, was mixed with required amount of sand in different loading ratios and was dried at 95 °C in a hot air oven for 6 h. The sugar coated

Table 1 Precursor, treatment and synthesis

Precursor	Treatment	Resulting material	Pesticide	References
Graphite powder	Modified Hummers method	Reduced graphene oxide	Chlorpyrifos, endosulfan, and malathion	Maliyekkal et al. (2013)
	Modified Hummers method	Graphene oxide		
Graphite oxide	Modified Hummers method	Graphene-based magnetic nanocomposite	5 different Carbamates	Wu et al. (2011)
Reduced graphene oxide	Modified Hummers method	$CoFe_2O_4$ + TiO_2/reduced graphene oxide photocatalyst	Chlorpyrifos	Gupta et al. (2015)
Graphene oxide	Hummers method	Magnetite + SiO_2 + TiO_2-reduced graphene	2,4-Dichlorophenoxyacetic acid	Tang et al. (2013)
Graphite oxide	Modified Hummers method	Graphene magnetic nanoparticle	Neonicotinoid insecticides	Wang et al. (2012)
Graphite oxide	Modified Hummers method	Graphene	Six carbamate pesticides	Shi et al. (2014)
Graphene	Modified Hummers method	Graphene coated solid phase micro-extraction fibre	Triazine herbicides	Wu et al. (2012)
Graphene oxide	Modified Hummers method	Graphene coated silica	Organo-phosphorous Pesticides	Liu et al. (2013)
Graphene oxide	Modified Hummers method	Graphene coated solid phase micro-extraction fibre	Six pyrethroid pesticides	Chen et al. (2010)
Common Sugar	Modified Hummers method	Graphene sand composite	Chlorpyrifos	Gupta et al. (2012)
Graphene oxide	Modified Hummers method	Cellulose/graphene composite	Six triazine pesticides	Zhang et al. (2015)

sand was then carbonized in N_2 atmosphere then activated with 10 mL of concentrated sulfuric acid.

3.3 Graphene Coated Silica Adsorbent

Silica (Liu et al. 2013) was mixed with 3 M HCl, filtered and dried at 100 °C in an oven for 3 h. Acid-treated silica and graphene oxide were mixed by ultrasonication. Finally, 85% hydrazine hydrate was added to the solution and the mixture was heated to 80 °C for 12 h. The precipitate was dried and utilized to adsorb organophosphorus pesticides.

3.4 Cellulose Graphene Composite Adsorbent

Zhang et al. (2015) have prepared cellulose graphene composite for triazine pesticides adsorption from water. He has mixed NaOH and urea in water and cellulose was dispersed into the pre-cooled aqueous mixer. The solution was centrifuged, graphene oxide were added, homogenized by ultrasonication, hydrazine hydrate was added and the mixture was heated to 80 °C. The resulting suspension was washed, was frozen at −20 °C and was lyophilized at −50 °C.

3.5 Magnetic Graphene Nano-composite Adsorbent

This composite was synthesized by Wang et al. (2011) and Wu et al. (2011) for the removal of insecticides. The process involves the magnetic graphene nano-composite synthesized by the in situ chemical co-precipitation of Fe^{2+} and Fe^{3+}. The magnetic composite crystal was prepared by suspending Graphene in the solution containing $(NH_4)_2Fe(SO_4)_2 \cdot 6H_2O$ and $NH_4Fe(SO_4)_2 \cdot 12H_2O$ at 50 °C under N_2 atmosphere.

3.6 $CoFe_2O_4@TiO_2$/Reduced Graphene Oxide Photocatalyst

Gupta et al. (2015) have synthesized this graphene composite and used it as photocatalyst. He had mixed $Co(NO_3)_2$ and $Fe(NO_3)_3$ solutions, added gradually 20% NaOH, and further added calcined TiO_2. After that Co-Fe precursor solution was added to the suspension, resultant residue was washed and calcined at 400 °C to obtain $CoFe_2O_4@TiO_2$ nanoparticles. Later, $CoFe_2O_4@TiO_2$ was added to the reduced graphene oxide suspension and was refluxed to procure $CoFe_2O_4@TiO_2$/reduced graphene oxide nano-composite.

3.7 Graphene Coated Fiber as Micro-extraction Medium

This material was used by Chen et al. (2010) and Wu et al. (2012) to remove triazine herbicides. Graphene dispersed in ethanol by ultra-sonication and the fibres were immersed to obtain Graphene coated fibres. They were cured at 150 °C and used as extraction medium.

4 Characterization of Graphene/Graphene Composites for Pesticide Removal

The FT-IR analysis was used to understand the characteristic bonds present in the graphene materials in order to learn the functional interactions of these materials with pesticide pollutants (Liu et al. 2013; Wang et al. 2012; Zhang et al. 2015). The cellulose graphene composite (Zhang et al. 2015) was prepared for the removal of triazine pesticides, graphene coated silica (Liu et al. 2013) was prepared for organo-phosphorous pesticides removal, and graphene–Fe_3O_4 (Wang et al. 2012) was prepared for the removal of neonicotinoid insecticides which are given in this analyses.

XRD patterns of different graphene materials were reported by various workers (Gupta et al. 2012, 2015; Wang et al. 2011; Wu et al. 2012; Zhang et al. 2015). The cellulose/graphene composite (Zhang et al. 2015), graphene–Fe_3O_4 particles (Wang et al. 2012; Wu et al. 2011), silica (Liu et al. 2013), the $CoFe_2O_4@TiO_2$/reduced graphene oxide (Gupta et al. 2015) were examined by the respective authors in order to understand their crystalline nature as well as the degree of graphene formation.

Raman spectroscopy is an effective structural testing instrument for nano-materials. This method is reported by (Gupta et al. 2012; Liu et al. 2013; Zhang et al. 2015) for their materials. Raman patterns of the graphene oxide, cellulose graphene composite (Zhang et al. 2015), graphene coated silica (Liu et al. 2013), Graphene sand Composite (Gupta et al. 2012) were analysed in order to understand the electron bands and the nano-structure of the material.

XPS analysis was done by (Gupta et al. 2015; Zhang et al. 2015). Cellulose graphene composite (Zhang et al. 2015) was analysed to find the elements present in the prepared material. The $CoFe_2O_4@TiO_2$ nanoparticles (Gupta et al. 2015) on graphene oxide were examined to identify the elements present and the bond structures.

The scanning electron microscopy were described by (Gupta et al. 2015; Shi et al. 2014; Tang et al. 2013; Wang et al. 2012; Wu et al. 2012, 2011; Zhang et al. 2015) to examine the surface morphology of the prepared graphene materials. The cellulose graphene composite's (Zhang et al. 2015) rough surface and homogeneous 3D porous structures, Fe_3O_4 nanoparticles' (Wang et al. 2012) silk wave-like carbon sheets of graphenes, graphene-Fe_3O_4 nano-composite's (Wu et al. 2011) nano size, $CoFe_2O_4$ + TiO_2/reduced graphene oxide nanocomposite's (Gupta et al. 2015) uniform dispersion, Magnetite + SiO_2 + TiO_2 particle's (Tang et al. 2013) aggregation, graphene's (Shi et al. 2014) random aggregation, graphene fiber coating's (Wu et al. 2012) homogeneous wrinkled structure were indicated by the scanning electron microscope images.

The Transition electron microscope (TEM) analysis were listed by (Chen et al. 2010; Gupta et al. 2012, 2015; Shi et al. 2014; Zhang et al. 2015). The TEM analysis revealed that $CoFe_2O_4$@TiO_2/reduced graphene oxide nanocomposite (Gupta et al. 2015) were a fine dispersion of dark and light particles, graphene (Shi et al. 2014; Chen et al. 2010), graphene silica composite (Gupta et al. 2012) was made of wrinkled sheets, reduced graphene oxide (Zhang et al. 2015) was ordered with graphite lattices.

For cellulose graphene composite, elemental analysis was done by Zhang et al. (2015) which suggests that oxygen-containing groups of cellulose widely existed in the cellulose graphene composite. The BET surface area was also measured for some of the graphene materials to understand there suitability. The BET surface area of graphene coated silica (Liu et al. 2013), $CoFe_2O_4$ + TiO_2/reduced graphene oxide (Gupta et al. 2015) was measured to be the reduced graphene oxide, $CoFe_2O_4$@TiO_2, TiO_2/reduced graphene oxide and were found to be 328.2; 140.2; 185.9; 210.2; and 305.3 $m^2\ g^{-1}$ respectively. Thermogravimetry analysis was carried out by Chen et al. (2010), Liu et al. (2013) in order to analyze the thermal stability of the prepared materials. Table 2 illustrates the various characterizations used for different graphene materials.

5 Optimization of Process Variables for Pesticide Adsorption

Among the literatures available expect a few all the remaining reports the adsorptive removal pesticides using graphene materials. The adsorptive removal efficiency of pesticide depends on the source of raw material, preparation, and treatment conditions.

Most of the graphene family materials possessed a higher stability in varying pH range (Maliyekkal et al. 2013; Wang et al. 2014a, b; Wu et al. 2011). Few graphene materials had high removal efficiency only at a narrow pH range (Tang et al. 2013;

Table 2 Characterizations of graphene family materials

Material type	Characterizations	BET surface area ($m^2\ g^{-1}$)	References
Reduced graphene oxide	Raman spectroscopy, XPS, TEM and SEM	–	Maliyekkal et al. (2013)
Graphene oxide (GO)	Raman spectroscopy, XPS, TEM and SEM		
Graphene-based magnetic nano-composite	XRD and SEM	–	Wu et al. (2011)
Reduced graphene oxide	TEM, SEM, XPS, BET analysis and XRD	140.2	Gupta et al. (2015)
$CoFe_2O_4$/reduced graphene oxide	TEM, SEM, XPS and XRD	185.9	
TiO_2/reduced graphene oxide	TEM, SEM, XPS and XRD	210.2	
$CoFe_2O_4$@TiO_2/reduced graphene oxide	TEM, SEM, XPS and XRD	305.3	
Magnetic TiO_2-graphene composite	SEM and TEM	–	Tang et al. (2013)
Graphene-Fe_3O_4	FTIR, SEM, TEM and XRD	225	Wang et al. (2012)
Graphene	SEM and TEM	2630	Shi et al. (2014)
Graphene coated solid phase micro-extraction (SPME) fibre	SEM	–	Wu et al. (2012)
Graphene coated silica	Raman spectroscopy, FTIR, XRD, TGA, BET analysis and elemental analysis	328.2	Liu et al. (2013)
Graphene coated solid phase micro-extraction (SPME) fibre	TGA, TEM and SEM	–	Chen et al. (2010)
Graphene sand composite (GSC)	XPS, SEM, TEM and Raman spectroscopy	–	Gupta et al. (2012)
Cellulose/graphene composite	FTIR, XPS, XRD, Raman spectroscopy, elemental analysis, SEM and TEM	–	Zhang et al. (2015)

Wang et al. 2012; Wu et al. 2012). It was reported by Wang et al. (2012) that maximum removal efficiency of the neonicotinoid insecticides was observed at acidic pH. Removal efficiency of pesticides like Isoprocarb, Baycarb, Baygon

(Shi et al. 2014) and 2,4-dichlorophenoxyacetic acid (Tang et al. 2013) remained maximum at neutral pH. Maximum removal of pesticides like Ametryn, Prometryn and Cyprazinewas observed at basic pH (Zhang et al. 2015). No pH change was required for some pesticides like Chlorpyrifos, Endosulfan (Maliyekkal et al. 2013) and Metolcarb (Wu et al. 2011) as they had maximum removal efficiency for entire pH range.

Room temperature was found suitable for pesticide adsorption for many graphene materials whereas higher temperature was reported by (Chen et al. 2010; Liu et al. 2013). The variation in amount of initial concentration of pesticides affected its removal efficiency. Maximum removal of pesticide was reported by Gupta et al. (2012), Liu et al. (2013), Maliyekkal et al. (2013), Zhang et al. (2015) when the initial concentration of pesticides was maximum. In other study, maximum removal efficiency was found for minimum initial concentration of pesticides such as Carbofuran (Wu et al. 2011), Pirimicarb (Shi et al. 2014) and acetamiprid (Wang et al. 2012). The different process parameters optimized during adsorptive removal of pesticides are listed in Table 3.

6 Comparison of Graphene Adsorbents Used for Pesticides Removal

Adsorption process is a surface phenomenon that depends on the number of sites available, porosity and specific surface area of adsorbent as well as various types of interactions (ALOthman et al. 2013; Awual et al. 2015; Naushad et al. 2015; Alqadmi et al. 2016). Adsorbents can be from a carbon sources, agricultural wastes, polymers, industrial wastes, biological sources and inorganic sources (Karthik et al. 2016a, b; Sivarajasekar et al. 2017a, b, c, d). The different adsorbents and their capacity for the selected pesticide were showed in Table 4. Among the adsorption studies only a very few authors reported the isotherms studies (Liu et al. 2013; Zhang et al. 2015) using Langmuir and Freundlich isotherms. The result of chlorfenvinphos on graphene coated silica (Liu et al. 2013) provided a good fit with Freundlich isotherm because of the favourable bonds present for adsorption. For malathion adsorption on graphene coated silica (Liu et al. 2013) well fitted with Langmuir isotherm due to physisorption nature. Langmuir isotherm fitted well for adsorption of ametryn on cellulose graphene composite (Zhang et al. 2015) because of its surface area and pore structures. The adsorption capacity of various adsorbents is compared with graphene family materials and presented in Table 4.

Table 3 Optimization of adsorptive removal of pesticides

Pesticide	pH	Temperature (°C)	Initial concentration (mg L^{-1})	Time (min)	Removal efficiency (%)	References
Chlorpyrifos (CP)	3–9	30 ± 2	2	30	100	Maliyekkal et al. (2013)
Endosulfan (ES)	3–9	30 ± 2	1	45	100	
Malathion (ML)	3–9	30 ± 2	2	60	100	
Carbofuran	2–7	30 ± 2	2×10^{-7}	15	–	Wu et al. (2011)
Metolcarb	2–9	30 ± 2	2×10^{-7}	15	–	
Pirimicarb	2–9	30 ± 2	2×10^{-7}	15	–	
Isoprocarb	2–9	30 ± 2	2×10^{-7}	15	–	
Diethofencarb	2–9	30 ± 2	2×10^{-7}	15	–	
Chlorpyrifos	5.8	30 ± 2	5	60	–	Gupta et al. (2015)
2,4-dichlorophenoxyacetic acid (2,4-D)	7	30 ± 2	20	140	100	Tang et al. (2013)
Thiame-thoxam (TMX)	6	30 ± 2	5×10^{-7}	10	55	Wang et al. (2012)
Imidacloprid (ICL)	6	30 ± 2	5×10^{-7}	10	78	
Acetamiprid (ACT)	6	30 ± 2	5×10^{-7}	10	72	
Thiacloprid (TCL)	6	30 ± 2	5×10^{-7}	10	70	
Pirimicarb	6.8–10	30 ± 2	5×10^{-7}	–	–	Shi et al. (2014)
Diethofencarb	6.8–10	30 ± 2	2.5×10^{-6}	–	–	
Carbaryl	6.8–8.2	30 ± 2	5×10^{-6}	–	–	
Isoprocarb	6.8	30 ± 2	2.5×10^{-6}	–	–	
Baycarb	6.8	30 ± 2	3×10^{-6}	–	–	
Baygon	6.8	30 ± 2	2.5×10^{-6}	–	–	

(continued)

Table 3 (continued)

Pesticide	pH	Temperature (°C)	Initial concentration (mg L^{-1})	Time (min)	Removal efficiency (%)	References
atrazine	6–7	25 ± 2	0.01	30	84–96.4	Wu et al. (2012)
Prometon	6–7	25 ± 2	0.01	30		
Ametryn	6–7	25 ± 2	0.01	30		
Prometryn	6–7	25 ± 2	0.01	30		
Phonamiphos	3–11	260	10	0	<85	Liu et al. (2013)
Dimethoate	3–11	260	10	0	<85	
Phorate	3–11	260	10	0	>85	
Parathion-methyl	3–11	260	10	0	>85	
Pirimiphos-methyl	3–11	260	10	0	>85	
Malathion	3–11	260	10	0	>85	
Fenthion	3–11	260	10	0	>85	
Isocarbophos	3–11	260	10	0	>85	
Chlorfenvinphos	3–11	260	10	0	>85	
Profenofos	3–11	260	10	0	>85	
Methidathion	3–11	260	10	0	>85	
Bifenthrin	–	270	0.01	90	–	Chen et al. (2010)
Cyhalothrin	–	270	0.01	90	–	
Permethrin	–	270	0.01	90	–	
Cypermethrin	–	270	0.01	90	–	
Phenvalerate	–	270	0.01	90	–	
Deltamethrin	–	270	0.01	90	–	
Chlorpyrifos	–	30 ± 2	1	720	–	Gupta et al. (2012)

(continued)

Table 3 (continued)

Pesticide	pH	Temperature (°C)	Initial concentration (mg L^{-1})	Time (min)	Removal efficiency (%)	References
Simeton	9	25–45	1	0	85	Zhang et al. (2015)
Simazine	9	25–45	1	0	82	
Atrazine	9	25–45	1	0	98	
Ametryn	9	25–45	1	0	95	
Prometryn	9	25–45	1	0	72	
Cyprazine	11	25–45	1	0	90	

Table 4 Comparison of adsorption capacity of various adsorbents with graphene family materials

Material	Pesticide	Adsorption capacity (mg/g)	References
Hyper cross-linked polymers of Macronet-150	Methomyl	40	Chang et al. (2008)
Hyper cross-linked polymers of Macronet-500	Methomyl	5.07	
Macro fungi sojarcaju	Endosulfan	1.575	Sudhakar and Dikshit (1999)
Granulated activated carbon	Bifenthrin	0.294	Domingues et al. (2007)
Cork	Bifenthrin	0.260	
Activated carbon fibre	Atrazine	238.1	Faur et al. (2005)
Unmodified maize cob	Copper fungicide	933.7	
Pine Bark	Molinate	10	Silva et al. (2004)
Blast furnace dust	2,4-Dichlorophenoxy-aceticacid	21	Gupta et al. (2006)
Oil shale ash	Deltamethrin	10.74	Al-Qodah et al. (2007)
Rhizopusarrhizus	Pentachloronitrobenzene	4.6	Lièvremont et al. (1998)
Micelle clay Cloisite	Chlorpyrifos	6.63	Suciu and Capri (2009)
Activated clay	Paraquat	58.48	Tsai et al. (2003)
Cellulose graphene composite	Ametryn	9.5877	Zhang et al. (2015)
Graphene coated silica	Malathion	4.878	Liu et al. (2013)
Graphene	Chloropyrifos	48	Gupta et al. (2012)
Reduced graphene oxide	Chlorpyrifos	1200	Maliyekkal et al. (2013)
	Endosulfan	1100	
	Malathion	800	

7 Conclusion

At the whole, this review addressed the issues raised by the pesticides which are present in the aqueous environment and their remediation by graphene-family materials. The synthesis and characterization of the each graphene-based material have been discussed. The wide application of graphenes lies in the adsorption process; therefore the optimal conditions for efficient pesticide removal and a comparison with other adsorbents are also discussed. From this review we came to know that, there is a large scope for the researchers exploring graphene family materials for photo-catalysis treatment and graphene-blended membrane treatment of pesticide bearing effluents.

References

Akhtar M, Iqbal S, Bhanger MI, Moazzam M (2009) Utilization of organic by-products for the removal of organo-phosphorous pesticide from aqueous media. J Hazard Mater 162:703–707. https://doi.org/10.1016/j.jhazmat.2008.05.084

ALOthman ZA, Alam MM, Naushad M (2013) Heavy toxic metal ion exchange kinetics: validation of ion exchange process of composite cation exchanger nylon 6, Zr (IV) phosphate. J Ind Eng Chem 19:956–960

Alqadmi A, Naushad M, Ahamd T, Abdalla MA, ALOthman ZA, Al Shehri SM (2016) Synthesis and characterization of Fe_3O_4@TSC nanocomposite: highly efficient removal of toxic metal ions from aqueous medium. RSC Adv. 6:22679–22689

Al-Qodah Z, Shawaqfeh AT, Lafi WK (2007) Adsorption of pesticides from aqueous solutions using oil shale ash. Desalin 208:294–305. https://doi.org/10.1016/j.desal.2006.06.019

Awual MR, Hasan MM, Naushad M, Shiwaku H, Yaita T (2015) Peparation of new class composite adsorbent for enhanced palladium (II) detection and recovery. Sen Actuat B: Chem 209:790–797

Brodie BC (1859) On the atomic weight of graphite. Philos Trans R Soc London 149:249–259

Chang C, Chang C, Hsu K, Lee S, Wolfgang H (2008) Adsorptive removal of the pesticide methomyl using hyper-crosslinked polymers. J Hazard Mater 155:295–304. https://doi.org/10.1016/j.jhazmat.2007.11.057

Chatterjee S, Das SK, Chakravarty R, Chakrabarti A, Ghosh S, Guha AK (2010) Interaction of Malathion, an Organophosphorus Pesticide with *Rhizopus oryzae* Biomass. J Hazard Mater 174:47–53. https://doi.org/10.1016/j.jhazmat.2009.09.014

Chaudhry Q, Schroder P, Werck-Reichhart D, Grajek W, Marecik R (2002) Prospects and limitations of phytoremediation for the removal of persistent pesticides in the environment. Environ Sci Pollut Res 9:4–17. https://doi.org/10.1007/BF02987313

Chauhan RS, Singhal L (2006) Harmful effects of pesticides and their control through cowpathy. Int J Cow Sci 2:61–70

Chen J, Li C (2013) An improved hummers method for eco-friendly synthesis of graphene oxide synthesis of graphene oxide. Carbon 64:225–229. https://doi.org/10.1016/j.carbon.2013.07.055

Chen S, Sun D, Chung JS (2007) Treatment of pesticide wastewater by moving-bed biofilm reactor combined with Fenton-coagulation pretreatment. J Hazard Mater 144:577–584. https://doi.org/10.1016/j.jhazmat.2006.10.075

Chen J, Zou J, Zeng J, Song X et al (2010) Preparation and evaluation of graphene-coated solid-phase micro-extraction fiber. Anal Chim Acta 678:44–49. https://doi.org/10.1016/j.aca.2010.08.008

Cunningham SD, Shann JR, Crowley DE, Anderson TA (1997) Phytoremediation of contaminated water and soil. ACS Symp Ser 664:2–17. https://doi.org/10.1021/bk-1997-0664.ch001

Domingues VF, Priolo G, Alves AC, Cabral MF, Delerue-matos C (2007) Adsorption behavior of α-cypermethrin on cork and activated carbon. J Environ Sci Heal Part B 42:649–654. https://doi.org/10.1080/03601230701465635

Edwards CA (1977) Environmental aspects of the usage of pesticides in developing countries. In: International symposium over Fytofarmacie en Fytiatrie, Rothamshed experimental station, Harpenden (UK)

Faur C, Métivier-Pignon H, Le Cloirec P (2005) Multi component adsorption of pesticides onto activated carbon fibers. Adsorption 11:479–490. https://doi.org/10.1007/s10450-005-5607-2

Gavrilescu M (2005) Fate of pesticides in the environment. Eng Life Sci 5:497–526. https://doi.org/10.1002/elsc.200520098

Gupta VK, Ali I, Saini VK (2006) Adsorption of 2, 4-D and carbofuran pesticides using fertilizer and steel industry wastes. J Colloid Interface Sci 299:556–563. https://doi.org/10.1016/j.jcis.2006.02.017

Gupta SS, Sreeprasad TS, Maliyekkal SM, Das SK, Pradeep T (2012) Graphene from sugar and its application in water purification. ACS Appl Mater Interfaces 4:4156–4163. https://doi.org/10.1021/am300889u

Gupta VK, Eren T, Atar N, Yola ML, Parlak C, Karimi-Maleh H (2015) $CoFe_2O_4$@TiO_2 decorated reduced graphene oxide nanocomposite for photocatalytic degradation of chlorpyrifos. J Mol Liq 208:122–129. https://doi.org/10.1016/j.molliq.2015.04.032

Huang X, Yin Z, Wu S, Qi X et al (2011) Graphene-based materials: synthesis, characterization, properties, and applications. Small 7:1876–1902. https://doi.org/10.1002/smll.201002009

Huc V, Bendiab N, Bouchiat V, Ebbesen T (2008) Large and flat graphene flakes produced by epoxy bonding and reverse exfoliation of highly oriented pyrolytic graphite. Nanotechnology 19:455601 (6 p). https://doi.org/10.1088/0957-4484/19/45/455601

Hummers WS Jr, Offeman RE (1958) Preparation of graphitic oxide. J Am Chem Soc 80:1339. https://doi.org/10.1021/ja01539a017

Hunter WJ (2002) Bioremediation of chlorate or perchlorate contaminated water using permeable barriers containing vegetable oil. Curr Microbiol 45:287–292. https://doi.org/10.1007/s00284-002-3751-4

Karthik V, Saravanan K, Sivarajasekar N, Suriyanarayanan N (2016a) Bioremediation of dye bearing effluents using microbial biomass. Ecol Environ Conserv 22:S423–S434

Karthik V, Saravanan K, Sivarajasekar N, Suriyanarayanan N (2016b) Utilization of biomass from *Trichoderma harzianum* for the adsorption of reactive red, dye. Ecol Environ Conserv 22: S435–S440

Kim KS, Zhao Y, Jang H et al (2009) Large-scale pattern growth of graphene films for stretchable transparent electrodes. Nature 457:706–710. https://doi.org/10.1038/nature07719

Li D, Müller MB, Gilje S, Kaner RB, Wallace GG (2008) Processable aqueous dispersions of graphene nanosheets. Nat Nanotechnol 3:101–105. https://doi.org/10.1038/nnano.2007.451

Lièvremont D, Seigle-Murandi F, Benoit-Guyod JL (1998) Removal of PCNB from aqueous solution by a fungal adsorption process. Water Res 32:3601–3606. https://doi.org/10.1016/S0043-1354(98)00132-8

Liu Q, Shi J, Zeng L, Wang T, Cai Y, Jiang G (2011) Evaluation of graphene as an advantageous adsorbent for solid-phase extraction with chlorophenols as model analytes. J Chromatogr A 1218:197–204. https://doi.org/10.1016/j.chroma.2010.11.022

Liu X, Zhang H, Ma Y, Wu X et al (2013) Graphene-coated silica as a highly efficient sorbent for residual organophosphorus pesticides in water. J Mater Chem A 1:1875–1884. https://doi.org/10.1039/C2TA00173J

Maliyekkal SM, Sreeprasad TS, Krishnan D et al (2013) Graphene: a reusable substrate for unprecedented adsorption of pesticides. Small 9:273–283. https://doi.org/10.1002/smll.201201125

Marcano DC, Kosynkin DV, Berlin JM et al (2010) Improved synthesis of graphene oxide. ACS Nano 4:4806–4814. https://doi.org/10.1021/nn1006368

McAllister MJ, Li JL, Adamson DH, Schniepp HC et al (2007) Single sheet functionalized graphene by oxidation and thermal expansion of graphite. Chem Mater 19:4396–4404. https://doi.org/10.1021/cm0630800

Naushad M, ALOthman ZA, Sharma G (2015) Inamuddin, Kinetics, isotherm and thermodynamic investigations for the adsorption of Co(II) ion onto crystal violet modified amberlite IR-120 resin. Ionics 21:1453–1459

Newcombe DA, Crowley DE (1999) Bioremediation of atrazine-contaminated soil by repeated applications of atrazine-degrading bacteria. Appl Microbiol Biotechnol 51:877–882. https://doi.org/10.1007/s002530051477

Reina A, Jia X, Ho J, Nezich D et al (2009) Layer area, few-layer graphene films on arbitrary substrates by chemical vapor deposition. Nano Lett 9:3087. https://doi.org/10.1021/nl901829a

Rice PJ, Anderson TA, Coats JR (1997) Phytoremediation of herbicide-contaminated surface water with aquatic plants. ACS Symp Ser 664:133–151. https://doi.org/10.1021/bk-1997-0664.ch010

Schniepp HC, Li J, Mcallister MJ et al (2006) Functionalized single graphene sheets derived from splitting graphite. J Phys Chem B 110:8535–8539. https://doi.org/10.1021/jp060936f

Shi Z, Hu J, Li Q, Zhang S, Liang Y, Zhang H (2014) Graphene based solid phase extraction combined with ultra high performance liquid chromatography-tandem mass spectrometry for carbamate pesticides analysis in environmental water samples. J Chromatogr A 1355:219–227. https://doi.org/10.1016/j.chroma.2014.05.085

Sidorov AN, Yazdanpanah MM, Jalilian R et al (2007) Electrostatic deposition of graphene. Nanotechnology 18:135301 (4 p). https://doi.org/10.1088/0957-4484/18/13/135301

Silva M, Fernandes A, Mendes A, Manaia CM, Nunes OC (2004) Preliminary feasibility study for the use of an adsorption/bio-regeneration system for molinate removal from effluents. Water Res 38:2677–2684. https://doi.org/10.1016/j.watres.2004.03.016

Singh N (2009) Adsorption of herbicides on coal fly ash from aqueous solutions. J Hazard Mater 168:233–237. https://doi.org/10.1016/j.jhazmat.2009.02.016

Sivarajasekar N, Baskar R, Ragu T, Sarika K, Preethi N, Rathika T (2016) Biosorption studies on waste cotton seed for cationic dyes sequestration: equilibrium and thermodynamics. Appl Water Sci. https://doi.org/10.1007/s13201-016-0379-2

Sivarajasekar N, Paramasivan T, Muthusaravanan S, Muthukumaran P, Sivamani S (2017a) Defluoridation of water using adsorbents—a concise review. J Environ Biotechnol Res 6:186–198

Sivarajasekar N, Balasubramani K, Mohanraj N, Prakash Maran J, Sivamani S, Ajmal Koya P, Karthik V (2017b) Fixed-bed adsorption of atrazine onto microwave irradiated Aegle marmelos Correa fruit shell: statistical optimization, process design and breakthrough modelling. J Mol Liq 241:823–830

Sivarajasekar N, Mohanraj N, Baskar R, Sivamani S (2017c) Fixed-bed adsorption of ranitidine hydrochloride onto microwave assisted—activated *Aeglemarmelos Correa* fruit shell: statistical optimization and breakthrough modelling. Arab J Sci Eng. https://doi.org/10.1007/s13369-017-2565-4

Sivarajasekar N, Mohanraj N, Sivamani S, Ganesh Moorthy I (2017d) Response surface methodology approach for optimization of lead (II) adsorptive removal by Spirogyra sp. Biomass. J Environ Biotechnol 6:88–95

Sivarajasekar N, Mohanraj N, Balasubramani K, Prakash Maran J, Moorthy IG, Karthik V, Karthikeyan K (2017e) Optimization, equilibrium and kinetic studies on ibuprofen removal onto microwave assisted - activated Aegle marmelos correa fruit shell. DESALINATION AND WATER TREATMENT 84:48–58

Sivarajasekar N, Mohanraj N, Sivamani S, Moorthy IG, Kothandan R, Muthusaravanan S (2017f) Comparative modeling of fluoride biosorption onto waste Gossypium hirsutum seed microwave-bichar using response surface methodology and artificial neural networks. International Conference on Intelligent Computing, Instrumentation and Control Technologies (ICICICT), IEEE Explore 1631-1635. https://doi.org/10.1109/ICICICT1.2017.8342815

Sivarajasekar N, Paramasivan T, Subashini R, Kandasamy S (2017g) Central composite design optimization of fluoride removal by spirogyra biomass. Asian J Microbiol Biotechnol Environ Sci 19:S130–S137.

Sivarajasekar N, Mohanraj N, Sivamani S, Prakash Maran J, Moorthy IG, Balasubramani K, (2018) Statistical optimization studies on adsorption of ibuprofen onto Albizialebbeck seed pods activated carbon prepared using microwave irradiation. Materials Today: Proceedings 5 (2):7264–7274

Stankovich S, Dikin DA, Piner RD et al (2007) Synthesis of graphene-based nanosheets via chemical reduction of exfoliated graphite oxide. Carbon N Y 45:1558–1565. https://doi.org/10.1016/j.carbon.2007.02.034

Suciu NA, Capri E (2009) Adsorption of chlorpyrifos, penconazole and metalaxyl from aqueous solution by modified clays. J Environ Sci Heal Part B 44:525–532. https://doi.org/10.1080/03601230902997543

Sudhakar Y, Dikshit A (1999) Adsorbent selection for endosulfan removal from water environment. J Environ Sci Heal Part B 34:97–118

Sun Y, Pignatello JJ (1992) Chemical treatment of pesticide wastes. Evaluation of iron (III) chelates for catalytic hydrogen peroxide oxidation of 2,4-D at circumneutral pH. J Agric Food Chem 40:322–327. https://doi.org/10.1021/jf00014a031

Tang Y, Zhang G, Liu C, Luo S, Xu X (2013) Magnetic TiO_2-graphene composite as a high-performance and recyclable platform for efficient photocatalytic removal of herbicides from water. J Hazard Mater 252–253:115–122. https://doi.org/10.1016/j.jhazmat.2013.02.053

Tsai WT, Lai CW, Hsien KJ (2003) Effect of particle size of activated clay on the adsorption of paraquat from aqueous solution. J Colloid Interface Sci 263:29–34. https://doi.org/10.1016/S0021-9797(03)00213-3

Wang C, Feng C, Gao Y, Ma X, Wu Q, Wang Z (2011) Preparation of a graphene-based magnetic nanocomposite for the removal of an organic dye from aqueous solution. Chem Eng J 173:92–97. https://doi.org/10.1016/j.cej.2011.07.041

Wang W, Li Y, Wu Q, Wang C, Zang X, Wang Z (2012) Extraction of neonicotinoid insecticides from environmental water samples with magnetic graphene nanoparticles as adsorbent followed by determination with HPLC. Anal Methods 4:766–772. https://doi.org/10.1039/c2ay05734d

Wang X, Liu B, Lu Q, Qu Q (2014a) Graphene-based materials: fabrication and application for adsorption in analytical chemistry. J Chromatogr A 1362:1–15. https://doi.org/10.1016/j.chroma.2014.08.023

Wang Y, Peng W, Liu X et al (2014b) Study of bilineage differentiation of human-bone-marrow-derived mesenchymal stem cells in oxidized sodium alginate/N-succinyl chitosan hydrogels and synergistic effects of RGD modification and low-intensity pulsed ultrasound. Acta Biomater 10:2518–2528. https://doi.org/10.1016/j.actbio.2013.12.052

Wu Q, Zhao G, Feng C, Wang C, Wang Z (2011) Preparation of a graphene-based magnetic nanocomposite for the extraction of carbamate pesticides from environmental water samples. J Chromatogr A 1218:7936–7942. https://doi.org/10.1016/j.chroma.2011.09.027

Wu Q, Feng C, Zhao G, Wang C, Wang Z (2012) Graphene-coated fiber for solid-phase microextraction of triazine herbicides in water samples. J Sep Sci 35:193–199. https://doi.org/10.1002/jssc.201100740

Xia H, Ma X (2006) Phytoremediation of ethion by water hyacinth (*Eichhorni acrassipes*) from water. Bioresour Technol 97:1050–1054. https://doi.org/10.1016/j.biortech.2005.04.039

Xu J, Lv H, Yang ST, Luo J (2013) Preparation of graphene adsorbents and their applications in water purification. Rev Inorg Chem 33:139–160. https://doi.org/10.1515/revic-2013-0007

Zhang C, Zhang RZ, Ma YQ et al (2015) Preparation of cellulose/graphene composite and its applications for triazine pesticides adsorption from water. ACS Sustain Chem Eng 3:396–405. https://doi.org/10.1021/sc500738k

Zhao G, Wen T, Chen C, Wang X (2012) Synthesis of graphene-based nanomaterials and their application in energy-related and environmental-related areas. RSC Adv 2:9286–9303. https://doi.org/10.1039/c2ra20990j

Chapter 14
Graphene/Graphene Oxide and Carbon Nanotube Based Sensors for the Determination and Removal of Bisphenols

Rajesh Kumar, Rajesh Kumar Singh and Stanislav A. Moshkalev

Abstract In the past decades, the high concerns are increasing about the feasible health threat caused by endocrine disrupting chemicals (EDCs), which could interfere with hormone biosynthesis, metabolism, or activity in a deviation from normal homeostatic control or reproduction. Among the numerous EDCs, bisphenol A (BPA) has strained great attention because of its wide occurrence and toxicity. Graphene oxide (GO) and carbon nanotubes (CNTs) are promising and novel carbon based materials that have high potential for fabrication of toxic detection sensors due to its high chemical stability, large specific surface area, abundant pore size distribution, ease of functionalization and feasibility of mass production. The fabricated sensors using GO and CNTs materials are very useful and primary choice for detection and removal of toxic harmful organic contaminants and toxic pollutants chemicals containing bisphenols. BPA is one such kind of toxic material is well-known as a typical endocrine disruptor which can mimic estrogen and lead to negative health effects on animals, wildlife and human beings. It has been widespread concerned in recent years for detection and removal of BPA due to its health hazard to wildlife and humans. BPA enters in our daily life food and water matrix by depolymerisation, leaching and migration from food packaging materials such as polycarbonates, polysulfone, polyacrylates and epoxy resins used in the industrial manufacture of water bottles, feeding bottles, internal coating in tin cans, microwave oven wares etc. An intensive study reveals that BPA has been also detected in wastewater, groundwater, surface water, and even drinking water. A large number

R. Kumar (✉) · S. A. Moshkalev
Centre for Semiconductor Components and Nanotechnology (CCS Nano),
University of Campinas (UNICAMP), Campinas, Sao Paulo 13083-870, Brazil
e-mail: rajeshbhu1@gmail.com

S. A. Moshkalev
e-mail: stanisla@unicamp.br

R. K. Singh (✉)
School of Physical & Material Sciences, Central University of Himachal Pradesh (CUHP),
Kangra, Dharamshala 176215, India
e-mail: rksbhu@gmail.com

M. Naushad (ed.), *A New Generation Material Graphene: Applications in Water Technology*, https://doi.org/10.1007/978-3-319-75484-0_14

of attentiveness studies have verified that BPA can cause cancerous tumors, birth defects, and other developmental disorders even at very low part-per-trillion doses. These carbon based materials also used to remove low-level environmental EDCs in micro-polluted water with the aid of an investigation about the adsorption of BPA at low levels in a very wide range. The GO and CNTs based sensor exhibited faster response, adequate storage stability, inexpensive, simple fabrication with disposability, satisfactory reproducibility and repeatability, and outstanding selectivity for the determination and removal of BPA. In this chapter we have done selective literature survey concerning the construction of sensors and biosensors based on GO and CNTs related materials for detection and removal of BPA from different kind of solutions and materials. The included literature survey article in this chapter also provides an overview of analytical performance for application in clinical, environmental, and food sciences research, and comments on future and interesting research trends in this field.

Keywords Graphene oxide · Carbon nanotubes · Sensors · Endocrine disruptors
Bisphenols · Health hazard · Polycarbonate · Toxic · Detection
Removal

1 Introduction

In present time, pure drinking water has been on the main focus of international concern and worldwide water scarcity and future demands. The debate on the utilization of recycled wastewater as an alternative water source is gaining extra attention. Mixing of toxic wastages in water due to industrialization and population expansion is a serious environmental concern that requires immediate awareness and solution (Sampaio et al. 2017). The daily routine of domestic life produces wastewater, which contains soaps, shampoos, toothpaste, body care products, shaving waste, detergent powders leading to high concentrations of chemicals (Tsoumachidou et al. 2017). Also, the strengthening of agricultural and industrial activities has led to the contamination of ground water by fertilizers, tannery bi products and other chemicals (Malato et al. 2009; Odukkathil and Vasudevan 2013). A wide variety of phenol and its derivatives can be found in water which generally comes from industrial pollution e.g., olive oil mills, petrochemical, pulp and paper industries and are known as toxic and refractory organic compounds (Mantzavinos and Kalogerakis 2005; Pintor et al. 2011; Saien and Nejati 2007; Zhang et al. 2014a). The rising presence of phenols represents a noteworthy environmental toxicity risk; therefore, the increase of methods for the removal of phenols from industrial wastewater has generated significant interest. The development of effective processes capable of removing persistent organic pollutants from wastewaters has been considered as a serious environmental challenge (Ahamad et al. 2017; Kumar et al. 2016c; Sharma et al. 2014).

Nowadays, diverse methods have been applied for the removal and detection of chemicals/contaminations from polluted water. Water pollution arised due to organic, inorganic or microbial contamination can be removed by adsorption (Bhatnagar and Anastopoulos 2017; Dong et al. 2010; Kim et al. 2011; Kulkarni and Junghare 2013; Roostaei and Tezel 2004), electro-chemical processes or membrane separation (Luo et al. 2015; Mousset et al. 2016; Peiris et al. 2008; Zhang et al. 2006), biological treatment (El-Naas et al. 2009; Marrot et al. 2006), photocatalytic degradation (Bora et al. 2017; Nerger et al. 2015; Nowakowska and Szczubiałka 2017; Sampaio et al. 2017; Tsoumachidou et al. 2017; Wang et al. 2016a), advanced oxidation processes (Chaplin 2014) and other processes. Among these methods, adsorption is a superior and widely used method for removal of various pollutants from water because of its convenience when applied in water current treatment processes (Bahamon et al. 2017; Domínguez et al. 2011; Suárez-Iglesias et al. 2017). In other terms, the adsorption method is comparatively low cost, easily operable, and fewer harmful secondary products as byproducts. Also it is an appropriate method which can be effectively applied in drinking water production (Cermakova et al. 2017).

To date, several adsorbents have been developed to remove different toxic chemicals from water with high efficiency as well as high adsorption capacity (Akhtar et al. 2016; Essandoh et al. 2015; Hasan and Jhung 2015; Krajišnik et al. 2011; Ortiz-Martínez et al. 2016; Suriyanon et al. 2013). Carbonaceous materials are the most widely used/studied adsorbents to remove organic micro pollutants from water because of their better performance in aqueous phase due to their high porosity, hydrophobicity, and surface functionality (Bhadra et al. 2016, 2017; Guedidi et al. 2013; Jauris et al. 2016). Carbonaceous materials such as activated carbon (Baccar et al. 2012; Bautista-Toledo et al. 2005; Guedidi et al. 2013; Kumar et al. 2016b; Liu et al. 2009; Melillo et al. 2004; Mestre et al. 2007, 2009), porous carbon (Abbasi et al. 2016; Asada et al. 2004; Bhadra et al. 2017; Koubaissy et al. 2011; Mohammadi and Mirghaffari 2015; Park et al. 2017; Wang et al. 2017c) have evolved attention for adsorption of toxic chemical constituent from polluted water. It is important to seek new novel carbonaceous adsorbents with high adsorption capacity, fast adsorption rate, and specific surface reactivity.

Graphene, 2D allotrope of carbon, a single atomic layer of sp^2-hybridized carbon arranged in a honeycomb structure (Ambrosi et al. 2016; Geim and Novoselov 2007; Kumar et al. 2013a; Singh et al. 2011) and carbon nanotubes (CNTs) (categorized as single-walled CNTs (SWCNTs) and multi-walled CNTs (MWCNTs)) as the 1D allotrope of carbon (Awasthi et al. 2009, 2010; Iijima 1991; Kumar et al. 2011a, b, 2013b; Kuo 2009; Lin et al. 2004) always attracted attention due to their chemical stability, large specific surface area, abundant pore size distribution, and feasibility of mass production. The derivative of graphene as graphene oxide (GO) (Gao et al. 2012b; Li et al. 2015e; Zhang et al. 2011) contains various oxygen functional groups on its surfaces is the promising material for the adsorption of polluted toxic chemicals present in water with the help of these functional groups. In recent times, it has been reported that graphene shows excellent adsorption capacity for the removal of heavy metal ions

(Pb^{2+}, Cd^{2+}, Cr^{6+}, etc.) (Deng et al. 2010; Zhu et al. 2012) fluoride ions (Li et al. 2011) and dyes (methyl orange, methyl blue, rhodamine B, etc.) (Liu et al. 2012; Ramesha et al. 2011) from aqueous solutions. Compared to other carbonaceous materials, the advantage of graphene is the selective adsorption ability of aromatic compounds with benzene rings through strong π–π interaction (Cai et al. 2011). Therefore, graphene and CNTs are highly expected to be a promising adsorbent for the removal of aromatic compounds in water treatment and removal of bisphenol A (Andra et al. 2015; Chowdhury and Balasubramanian 2014; Li et al. 2015a; Ragavan et al. 2013; Wang et al. 2013b).

2 Endocrine-Disrupting Chemicals

Endocrine-disrupting chemicals (EDCs) are synthetic toxic chemicals that can interfere with endocrine (or hormone) systems at certain doses and these disruptions can cause cancerous tumors, birth defects, obesity, diabetes, female reproduction, male reproduction, hormone-sensitive cancers in females, thyroid (Diamanti-Kandarakis et al. 2009; Gore et al. 2015). The endocrine system in our body is the collection of glands of an organism that secrete hormones directly into the circulatory system to be carried towards distant target organs. The phenomenon of biochemical processes serving to regulate distant tissues by means of secretions directly into the circulatory system is called endocrine signaling. The major endocrine glands include the pineal gland, pituitary gland, pancreas, thyroid gland, parathyroid gland, and adrenal glands etc. Any system in our body controlled by hormones can be derailed by hormone disruptors. Endocrine disruptors may be found in many everyday products including plastic bottles, metal food cans, detergents, flame retardants, food, toys, cosmetics, and pesticides etc. Health threat posed by EDCs, which are substances in our environment, food, and consumer products that interfere with hormonal biosynthesis, metabolism, or action resulting in a deviation from normal homeostatic control or reproduction are growing. EDCs can imitate the biological activity of natural hormones, occupy the hormone receptors, or interfere with the transport and metabolic processes of natural hormones, finally pose a risk to animals and humans (Chang et al. 2009; Ternes et al. 1999; Wang and Deng 2015; Xu et al. 2012). Thus, EDCs have drawn considerable social and scientific concern in recent years. EDCs are present in the form of bisphenols and its removal from wastewater is considered to be important for biosphere protection.

3 Bisphenols

The bisphenols are a group of chemical compounds containing two hydroxyl/ phenyl (–OH) functional groups and most of them are based on diphenylmethane ($(C_6H_5)_2CH_2$) (but exceptions are bisphenol S, P, and M). The diphenylmethane is an organic compound consisting methane wherein two hydrogen atoms are replaced by two phenyl groups. There are several kind of bisphenols such as bisphenol A ($C_{15}H_{16}O_2$), bisphenol AP, bisphenol AF, bisphenol B, bisphenol BP, bisphenol C, bisphenol C2, bisphenol E, bisphenol F, bisphenol G, bisphenol M, bisphenol S, bisphenol P, bisphenol PH, bisphenol TMG and bisphenol Z etc.

Bisphenol A (BPA) is the most popular representative of this group, often simply called "bisphenol". BPA, one of EDCs, is used as an intermediate in the industrial manufacturing of polycarbonate, epoxy, resins and polysulfone. Polycarbonate plastics are often used in containers that store food and beverages, such as water bottles etc. Epoxy resins are used to coat the inside of metal products, such as food cans, bottle tops and water supply lines. BPA can leach into food from the protective internal epoxy resin coatings of canned foods and from consumer

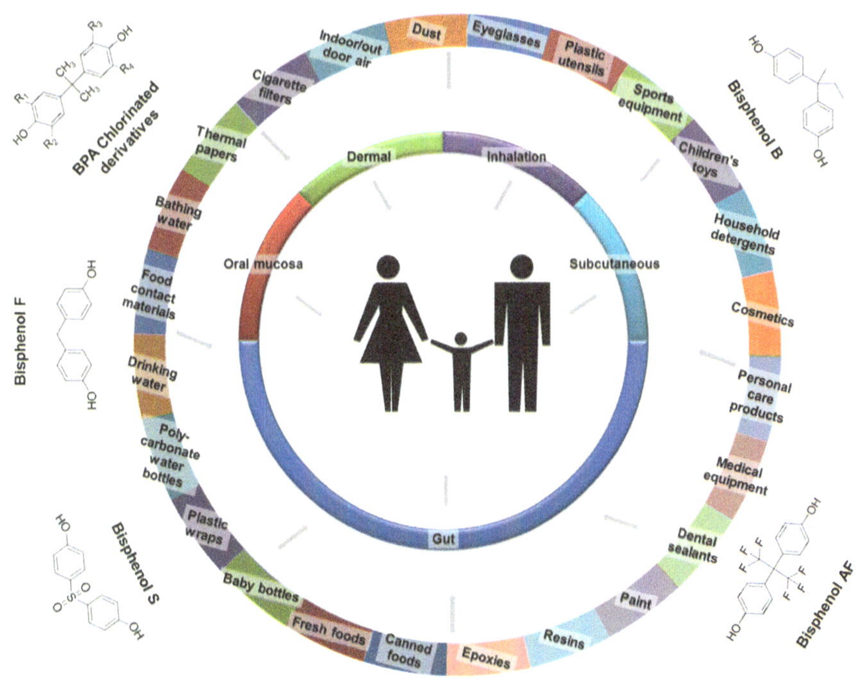

Fig. 1 Proposed exposure sources for BPA alternatives and derivatives, and possible exposure routes in humans. Reprinted (adapted) with permission from Andra et al. (2015). Copyright © 2015, All rights reserved: ATLAS of SCIENCE (https://atlasofscience.org/human-exposures-to-bisphenol-a-alternatives-and-derivatives-with-equal-or-more-harmful-effects/)

products such as polycarbonate tableware, food storage containers, compact discs, water bottles, dental sealants, canned foods and drinks, baby bottles, sippy cups, hair care products and many other products. The degree to which BPA leaches from polycarbonate bottles into liquid may depend more on the temperature of the liquid or bottle, than the age of the container. Moreover, BPA has been reportedly observed in waste landfill leachates, rivers, seas and soils (Staples et al. 1998). Although rapid biodegradation of BPA can be observed in river water (Kang and Kondo 2002, 2005), still BPA presents a risk to humans and animals.

The widespread utilization of BPA through polycarbonate plastics and epoxy resins there is increasing interest in the effective remediation technologies for its destruction in contaminated waters (Bautista-Toledo et al. 2014; Bhatnagar and Anastopoulos 2017; Vandenberg et al. 2007). BPA has been detected in all types of environmental water at concentrations ranging from 17.2 mg L^{-1} in hazardous waste landfill leachate (Yamamoto et al. 2001), to 12 μg L^{-1} in stream water (Liu et al. 2009) and 3.5–59.8 ng L^{-1} in drinking water (Santhi et al. 2012). Figure 1 proposed exposure sources for BPA alternatives and derivatives, and possible exposure routes in humans (Andra et al. 2015).

4 Adsorption of Bisphenols to Carbon Nanomaterials

Significant breakthroughs related to antimicrobial and adsorption properties of carbon-based nanomaterials have led to new avenues for the removal of various biological and organic/inorganic contaminants, bacterial pathogens, natural organic matter, and cyanobacterial toxins (Jariwala et al. 2013; Pan et al. 2008). The different classes of biologically toxic materials using different forms of carbon nanomaterials: carbon nanoparticles, CNTs, graphene, and GO, has been used for adsorption and removal of toxic contaminations (Ray and Jana 2017). Also, carbon material with porosity ranging from micro-to meso- and even macropores could allow the adsorption of a whole range of pollutants and provides good accessibility to the different adsorption sites. The tunability towards pore size, specific surface area, pore volume and surface chemistry provides a toolbox for the synthesis of a new generation of adsorbents for large organic pollutants (Libbrecht et al. 2017). Nitrogen doped nanocarbon materials also shows promising metal-free catalysis towards peroxymonosulfate (PMS) activation for environmental remediation (Wang et al. 2017a). Adsorption of bisphenols on carbon nanomaterials is one of the best and oldest water treatments (Dias et al. 2007; Moreno-Castilla and Rivera-Utrilla 2012; Rivera-Utrilla et al. 2011), available for removing organic and inorganic compounds from water intended for human consumption (Si et al. 2012). The value of carbons nanomaterials in this respect resides in the chemical and textural properties of their surface, including: large surface area, well-developed internal microporosity structure, and a wide spectrum of surface attached functional groups (Moreno-Castilla 2004).

5 Adsorption and Removal of Bisphenols Using Activated/ Commercial/Porous Carbon

Commercial carbon, activated carbon and porous carbon has been frequently used for different applications including adsorption and removal of BPA using various techniques (Sun et al. 2016; Wang et al. 2017a, c; Xu et al. 2017; Yeo et al. 2017; Zhang et al. 2017a, b). Liu et al. (2009) investigated BPA adsorption on two type of commercial carbons (W20 and F20) that had been selectively modified with nitric acid and thermally treated. Thermally modified sample (W20 N) with the largest surface area had the greatest BPA adsorption capacity, reaching up to 432 mg g^{-1}. The amount of BPA adsorbed was lower with an increase in temperature from 288 to 318 K and was slightly reduced with an increase in the pH from 5 to 9. Bautista-Toledo et al. (2005) studied the behavior of different activated carbons in the adsorption and removal of BPA from aqueous solutions. Two commercial activated carbons and one prepared from almond shells were used. The adsorption of BPA on activated carbon fundamentally depends on the chemical nature of the carbon surface and the pH of the solution. The most favorable experimental conditions for the process were those in which the net charge density of the carbon is zero and the BPA is in molecular form. The presence of mineral matter in carbons reduces their adsorption capacity because of the hydrophilic nature of the matter. The presence of electrolytes in the solution favor the adsorption process because of the screening effect produced between the positively charged carbon surface and the BPA molecules, with a resulting increase in adsorbent-adsorbate interactions.

The porous carbon has attracted extensive attention due to its unique properties such as high specific surface area and hierarchically porous structure (micro-, meso-, or even macropores) (Liu et al. 2008; Park et al. 2006; Srinivas et al. 2014; Wang et al. 2016c, 2017a). Such properties endow the porous carbon with abundant exposed active sites and shortened diffusion pathways for reactant (Sun and Xu 2014; Zhang et al. 2016). Asada et al. (2004) studied the adsorption properties of BPA and β-estradiol (E2). The adsorption amount of BPA gets increased as the carbonization temperature increases. The adsorption effect of a porous carbon carbonized at 400 °C was minimal, while the adsorption effect of a porous carbon carbonized at 1000 °C was significant. Although the adsorption effect of the activated carbon was also significant, but was lower than that of the 1000 °C porous carbon in the low concentration range, while was higher than that of the 1000 °C porous carbon in the high concentration range. The adsorption amount of E2 increases as the carbonization temperature was increased. The adsorption effect of the activated carbon was midway between 700 and 1000 °C. Porous carbon carbonized at 1000 °C with a low surface polarity was more effective for the removal of EDCs from environmental water than activated carbon with a large pore volume.

Carbon dots, a kind of nanomaterial, have been synthesized from various kinds of carbon sources including fullerene, graphite, GO, CNTs and carbohydrates have been used for the determination of BPA (Baker and Baker 2010; Li et al. 2015d, 2017a; Lim et al. 2015; Yuan et al. 2016; Zuo et al. 2016). Li et al. (2015c)

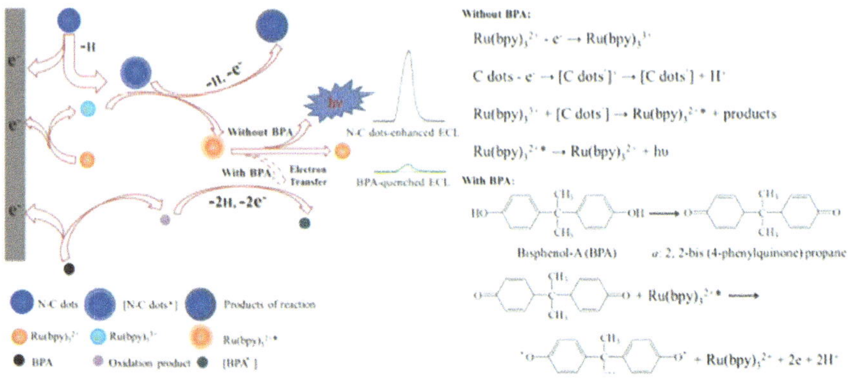

Fig. 2 The possible enhancement mechanism by N-C dots and the quench mechanism by BPA of the system (Li et al. 2015c). Reprinted (adapted) with permission from Li et al. (2015c). Copyright © 2015 Elsevier B. V. All rights reserved

demonstrated a novel $Ru(bpy)_3^{2+}$-based electro-chemiluminescence (ECL) sensing platform, using N-doped carbon nanodots (N-C dots) as co-reactant, for sensitive and selective detection of BPA. The enhancement mechanism by N-C dots and quenching mechanism are shown in Fig. 2. N-C dots enhance the ECL signal and significantly improve the reproducibility and stability of $Ru(bpy)_3^{2+}$ ECL system. The N-C dots-enhances the ECL mechanism and BPA quenches the ECL mechanism and under optimum conditions, the inhibited ECL intensity shows good linear relationship with the concentration of BPA in the range of 0.03–1.0 μM with the detection limit of 10 nM. The relative standard deviation (RSD) for 25 successive measurements with one sensor was 1.1% and for eight different sensors was 2.5%, respectively.

6 Adsorption and Removal of Bisphenols Using 1D Carbon Nanotubes

The unique morphology and structure of 1D CNTs keep attracting the researchers to explore the novel properties owing from their high aspect ratio, large specific surface area, high-temperature resistance and good electrical/thermal conductivities (Kaiser and Skakalova 2011; Karousis et al. 2010; Kumar et al. 2014; Peng et al. 2009; Wang et al. 2013a). These abnormal properties enhances various aspects which are utilized as catalyst carriers, electrochemical probes, energy storage electrodes, supercapacitors, high-temperature filters and many other applications (Andrews et al. 2002; Dhibar et al. 2014; Lawal 2016; Lee and Moon 2014; Li and Lee 2017; Simmons et al. 2012). The CNTs with different sizes/diameters provides unique surface structures, functionalities, and array of intriguing applications (Miners et al. 2016; Singh et al. 2009; Yan et al. 2015). Also, CNTs have aroused widespread attention as a new type of adsorbents for environmental pollutants and green engineering due to their outstanding ability for the removal of various inorganic and organic pollutants, and radio-nuclides from large volumes of wastewater (Ong et al. 2010; Ren et al. 2011). Also, CNTs contributes in terms of sustainable environment and green technologies perspective, such as waste water treatment, air pollution monitoring and biotechnologies. Recent developments show that the CNTs have potential and can be used for the detection and removal of BPA from various kinds of materials.

6.1 Removal by Pristine Carbon Nanotubes

Based on the literature studied, pristine CNTs pose a great prospective as a promising material as sensor in removal of BPA from solutions. Electrochemical (EC) sensing approaches have exploited the use of CNTs as electrode materials owing to their unique structures and properties, which provide strong electrocatalytic activity with minimal surface fouling. The novel device integration technologies have emerged along with advances in the synthesis, purification, conjugation and bio-functionalization of CNTs and combined efforts have contributed towards the rapid development of CNT-based sensors (Vashist et al. 2011). Several studies show that surface modified CNTs helps for the adsorption and removal of BPA. Vega et al. (2007) investigated the electrochemical detection of phenolic estrogenic compounds at CNTs-modified glassy carbon electrodes (GCE) (CNT-GCE). The analysis of underground well water and tap water samples spiked with some phenolic estrogenic compounds at a 14 nmol L^{-1} concentration level. Also, the detection limit for BPA with amperometric detection at the CNT-GCE is 0.6 ng mL^{-1}. Later in 2009, Kuo (2009) examined the adsorption potential of as-grown and modified CNTs (by $SOCl_2/NH_4OH$) using microwave irradiation for removing BPA. The adsorption capacity of BPA on the surface of CNTs fluctuates very little over the pH range 3–9. The maximum adsorption capacity of as-grown CNTs at 7, 27 and 47 °C was 60.98, 59.17 and 53.48 mg/g, respectively, and that of modified CNTs at 7 °C was 69.93 mg g^{-1} as shown in Fig. 3.

Tu et al. (2009) developed a sensitive electrochemical method for the determination of BPA at a GCE modified MWCNTs-gold (Au) nanoparticles (GNPs) (MWCNT-GNPs/GCE) hybrid film. The film was prepared based on the electrostatic interaction between positively charged cetyltrimethylammonium bromide (CTAB) and negatively charged MWCNTs and GNPs. Under the optimal conditions, the differential pulse voltammetric anodic peak current of BPA was linear with the BPA concentration from 2.0×10^{-8} to 2×10^{-5} mol L^{-1}, with a limit of detection of 7.5 nmol L^{-1}. In 2011, Huang et al. (2011) fabricated a novel electrochemical imprinted sensor for convenient determination of BPA. MWCNTs and gold nanoparticles (GNPs) were introduced for the enhancement of electronic transmission and sensitivity. Thin film of molecularly imprinted sol-gel polymers

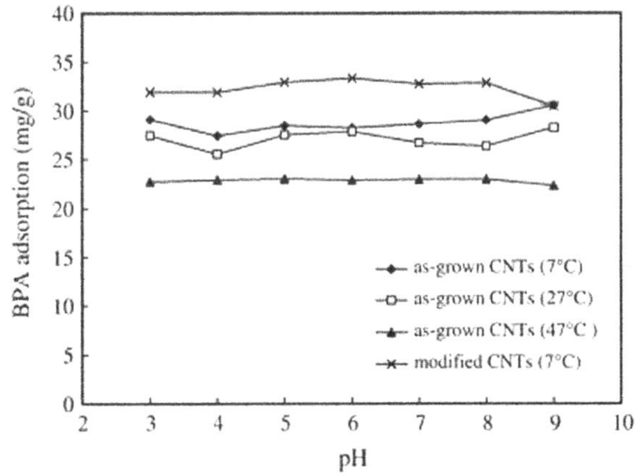

Fig. 3 Effects of pH for the adsorption of BPA by as-grown and modified CNTs (BPA = 10 mg L^{-1}, CNTs dose = 0.125 g L^{-1} and reaction time = 24 h) (Kuo 2009). Reprinted (adapted) with permission from Kuo (2009). Copyright © 2009 Elsevier B. V. All rights reserved

with specific binding sites for BPA was cast on gold electrode by electrochemical deposition as shown in Fig. 4. The linear dependence range was from 1.13×10^{-7} to 8.21×10^{-3} mol L^{-1}, with the limit of detection (LOD) of 3.6×10^{-9} mol L^{-1} ($S/N = 3$).

Gao et al. (2012a) also demonstrated a highly sensitive electrochemical sensor for the determination of BPA in aqueous solution using SWCNTs/β-cyclodextrin (β-CD) conjugate (SWCNT-CD) modified GCE. The detection limit of BPA was 1.0 nM ($S/N = 3$) and the sensor was applied to determine BPA leached from real plastic samples with good recovery, ranging from 95% to 103%. Derikvandi et al. (2016) reported sensitive electrochemical aptasensor based on a novel signal amplification strategy for the determination of BPA. Construction of the aptasensor began with the deposition of highly dispersed platinum nanoparticles (PtNPs)/acid-oxidized CNTs–COOH functionalized with polyethyleneimine (PEI) at the surface of glassy carbon (PtNPs/PEI/CNTs–COOH/GC) electrode. By adding BPA as a target, the aptamer specifically bound to BPA and its end folded into a BPA-binding junction. Sensitive quantitative detection of BPA was carried out by monitoring the decrease of differential pulse voltammetric responses of $[Fe(CN)_6]^{3-/4-}$ peak current with increasing BPA concentrations.

Moraes et al. (2013) compared two sensors based on vertically aligned and dispersed SWCNTs on the electrochemical determination of an endocrine disruptor. The vertical alignment of CNTs enabled an electrocatalytic effect and enhanced the sensitivity of BPA oxidation compared with an electrode modified with dispersed CNTs. Regarding the use of an Au electrode modified with a dispersed film of SWCNT in differential pulse voltammetry (DPV) measurements, the Au/ssDNA/SWCNT system improved the analytical sensitivity almost 2-fold. In 2014, Lin et al. (2014) studied the application of polyglutamate acid (PGA) and amino-functionalized MWCNTs ($MWCNTs\text{-}NH_2$) nanocomposite modified GCE for the electrochemical determination of BPA. The PGA/$MWCNTs\text{-}NH_2$ nanocomposite exhibits excellent electrocatalytic activity for the oxidation of BPA

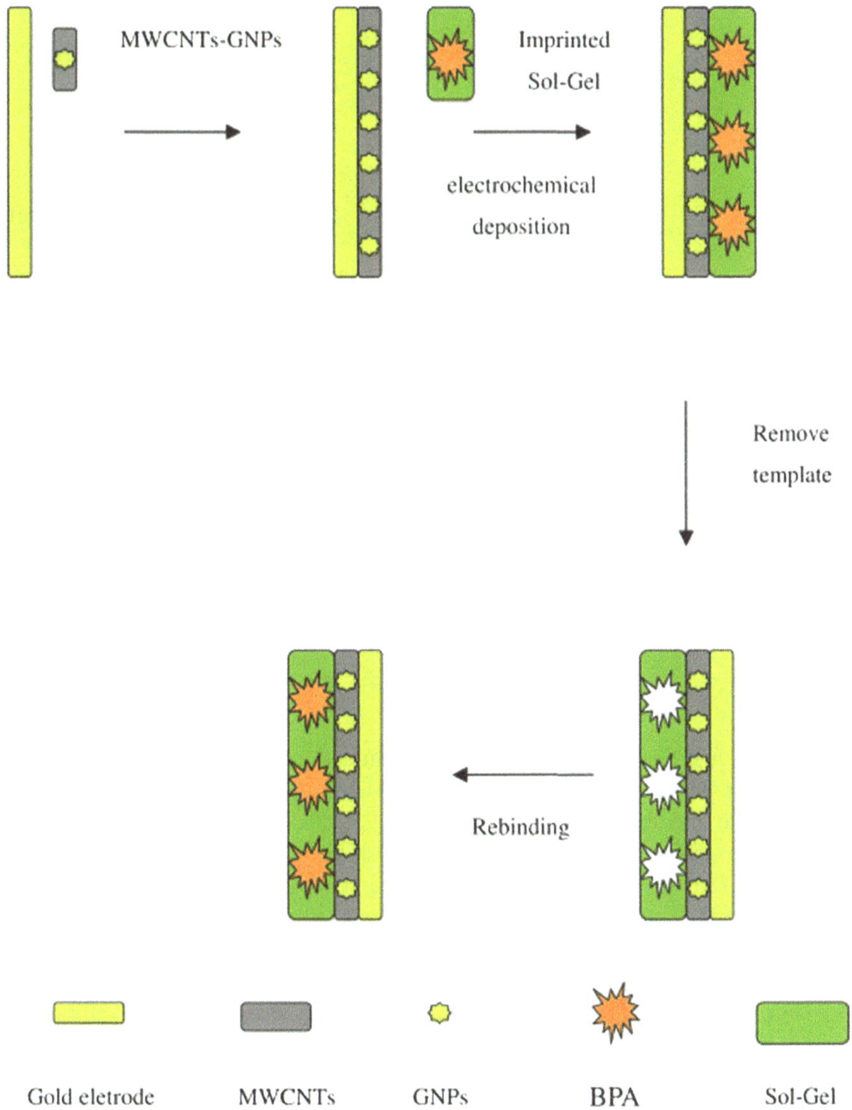

Fig. 4 The simplified sketch for the fabrication process of the MIPs/MWCNTs-GNPs electrode (Huang et al. 2011). Reprinted (adapted) with permission from Huang et al. (2011). Copyright © 2011 Elsevier B. V. All rights reserved

by substantially enhancing the current response and decreasing the BPA oxidation over potential. The composite modified GCE exhibited good performance for detecting BPA due to the enhanced electron transfer kinetics and large active surface area. Under the optimized conditions and according to the results from DPV, the BPA oxidation current is linear in a concentration range of 0.1 to 10 μM

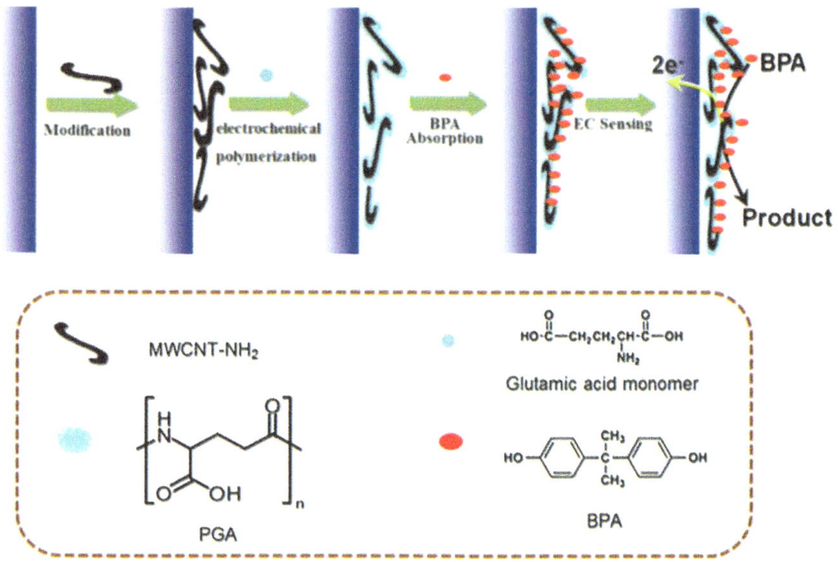

Fig. 5 Schematic illustration showing the electrode modification process and electrochemical reactions occurring at the PGA/MWCNT-NH_2 nanocomposite modified GCE to generate the analytical signal for detecting BPA (Lin et al. 2014). Reprinted (adapted) with permission from Lin et al. (2014). Copyright © 2014 Elsevier B. V. All rights reserved

(R = 0.998), and the detection limit was determined to be 0.02 μM ($S/N = 3$). The proposed sensors were successfully employed to determine BPA in real plastic products, and the recoveries were between 95 and 108%. Figure 5 shows the determination of BPA by electrochemical deposition of the glutamic acid monomer.

In 2012, Heo et al. (2012) studied the retention and adsorption of BPA and 17b-estradiol (E2) using three commercially available ultra filtration membranes based on CNTs. The results suggested that BPA and E2 transport was influenced by natural organic matter (NOM), which fouls the membrane through pore blockage and cake/gel formation. A strong linear correlation between the retention and adsorption of BPA and E2 was observed, indicating that retention by the ultrafiltration membranes was mainly due to the adsorption of BPA and E2 onto the membrane, the SWCNTs, and/or the NOM. Size-exclusion is unlikely to be a key factor in the retention of E2, however, BPA retention showed a slight dependence on the membrane pore size. Joseph et al. (2013) studied the combined coagulation–adsorption treatment using SWCNTs, MWCNTs, and powdered activated carbon (PAC) for the removal of EDCs (BPA and 17a-ethinyl estradiol (EE2)) from various water sources. Natural surface water was obtained from Broad River (BRW) in Columbia, South Carolina. Synthetic seawater (SW) and brackish water (BW) were produced to reflect the typical composition observed in previously published literature. Synthetic landfill leachates were created to replicate the characteristics of leachate produced in young landfills (YLs) and old landfills (OLs). The adsorption capacity of the SWCNTs was greater for EE2 than for BPA for the majority of the aqueous solutions with the exceptions of BW and BRW. The adsorption capacity for BPA: BRW (log K_f = 3.18) > SW (log K_f = 3.01) > YL (log K_f = 2.32) > BW (log K_f = 1.96) > OL (log K_f = 1.25), while the adsorption capacity of EE2 followed a different trend: SW (log Kf = 3.41) > YL (log K_f = 3.38) > BRW (log K_f = 3.14) > OL (log K_f = 2.09) > BW (log K_f = 1.38). The sole use of coagulation yielded small percentages of BPA removal (<10%) in each aqueous solution but higher percentages of EE2 removal. In these processes, the SWCNTs outperformed the MWCNTs for each combination, while PAC out performed both the SWCNTs and the MWCNTs, with more than 90% removal of both BPA and EE2.

Joseph et al. (2011) studied the adsorption of BPA and 17a-ethinyl estradiol (EE2) from landfill leachate onto SWCNTs. Different leachate solutions were prepared by altering the pH, ionic strength, and concentration of dissolved organic carbon (DOC) in the solutions to mimic the varying water conditions that occur in leachate during the various stages of waste decomposition. The youngest and oldest leachate solutions were represented by leachate Type A (pH = 5.0; DOC = 2500 mg L^{-1}; conductivity = 12,500 mS/cm; $[Ca^{2+}]$ = 1200 mg L^{-1}; $[Mg^{2+}]$ = 470 mg L^{-1}) and Type E (pH = 7.5; DOC = 250 mg L^{-1}; conductivity = 3250 mS cm^{-1}; $[Ca^{2+}]$ = 60 mg L^{-1}; $[Mg^{2+}]$ = 180 mg/L). The removal of BPA and EE2 decreases for older leachate solutions, with the adsorptive capacity of SWCNTs decreases in the order of leachate Type A > Type B > Type C > Type D > Type E. An increment in the pH from 3.5 to 11 decreases the adsorption of BPA by 22% in young leachate and by 10% in old leachate. In 2014, Zehani et al. (2015) fabricated a sensitive electrochemical biosensor for the detection of BPA in water by immobilizing tyrosinase onto a diazonium-functionalized boron doped diamond electrode (BDD) modified with MWCNTs. The fabricated biosensor exhibits excellent electroactivity towards o-quinone, a product of enzymatic reaction of BPA oxidation catalyzed by tyrosinase. The developed BPA biosensor displays a large linear range from 0.01 to 100 nM, with a detection limit (LOD) of 10 pM. The feasibility of the proposed biosensor was demonstrated on BPA spiked water river samples. Figure 6 show the schematic illustration of the functionalizing procedure and enzyme attachment on the BDD surface.

Lerman et al. (2013) used SWCNTs as sorbent, carbamazepine as primary adsorbate, and BPA and phenanthrene as competitors. Strong competition with BPA and no effect of phenanthrene on adsorption of carbamazepine was obtained. The hydrophobic neutral fraction of the DOM exhibited the strongest reductive effect on carbamazepine adsorption, most probably due to interactions in solution. The hydrophobic acid fraction decreased the carbamazepine adsorption mainly via direct competition. When DOM and BPA were co-introduced, the adsorption of carbamazepine was significantly reduced. In 2014, Zhang et al. (2014c) reported a novel magnetic molecularly imprinted polymers based on MWCNTs (MWCNTs@MMIPs) with specific selectivity toward BPA. It was synthesized using BPA as the template molecule, methacrylic acid, and β-cyclodextrin as binary functional monomers and ethylene glycol dimethacrylate as the cross-linker. Figure 7 illustrates the overall procedure for the fabrication of MWCNTs@MMIPs.

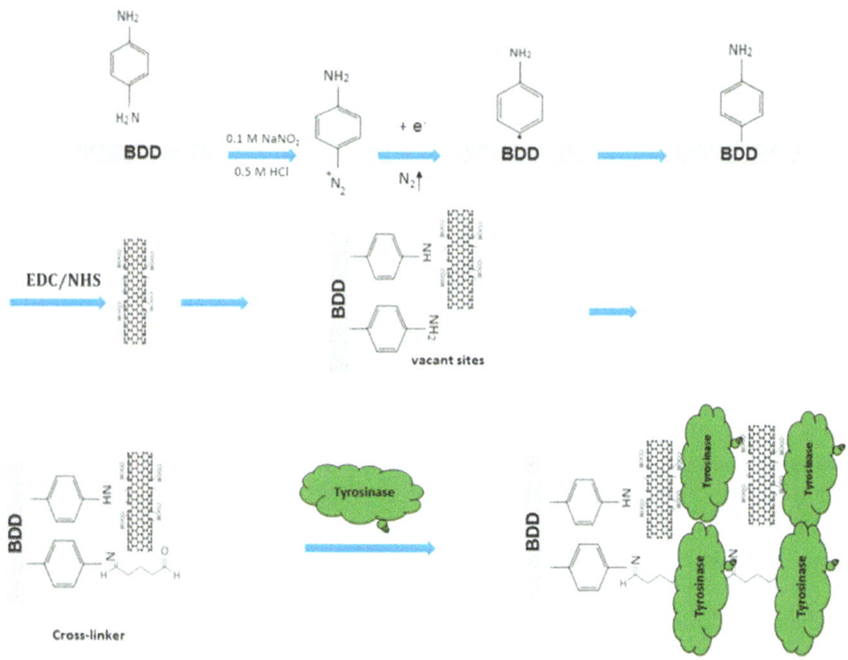

Fig. 6 Schematic illustration of the functionalizing procedure and enzyme attachment on the BDD surface (Zehani et al. 2015). Reprinted (adapted) with permission from Zehani et al. (2015). Copyright © 2015 Elsevier B. V. All rights reserved

The MWCNTs@MMIPs exhibited good affinity with a maximum adsorption capacity of 49.26 μmol g^{-1} and excellent selectivity toward BPA. Combined with high-performance liquid chromatography analysis, the MWCNTs@MMIPs were employed to extract BPA in tap water, rain water and lake water successfully with the recoveries of 89.8–95.4, 89.9–93.4, and 87.3–94.1%, respectively.

Recently in 2016, Goulart et al. (2016) analysed the influence of fictionalization and MWCNTs sizes on the detection of BPA. Samples with diameters of 20 to 170 nm were functionalized in HNO_3 5.0 mol L^{-1} and a concentrated sulphonitric solution. The size and acid treatment affected the oxidation of BPA. The MWCNTs with a 20–40 nm diameter improves the method sensitivity and achieved a detection limit of 84.0 nmol L^{-1}. Other kind of bisphenol such as bisphenol F (BPF) was reported to be absorbed on surface of MWCNTs and was removed and recovered from the solution efficiently (Zhang et al. 2013b). The results indicated that the removal percentages of BPF could reach up to 96.2% within 4 min under pH 4–10. Also the BPF adsorption kinetics was found to be very fast and the equilibrium was attained within 4.0 min with observed rate constants (k) 0.707–3.175 g mg^{-1} min^{-1} (at varied temperatures). The adsorption mechanisms involves the non-electrostatic dispersion interaction, hydrogen bonds and hydrophobic

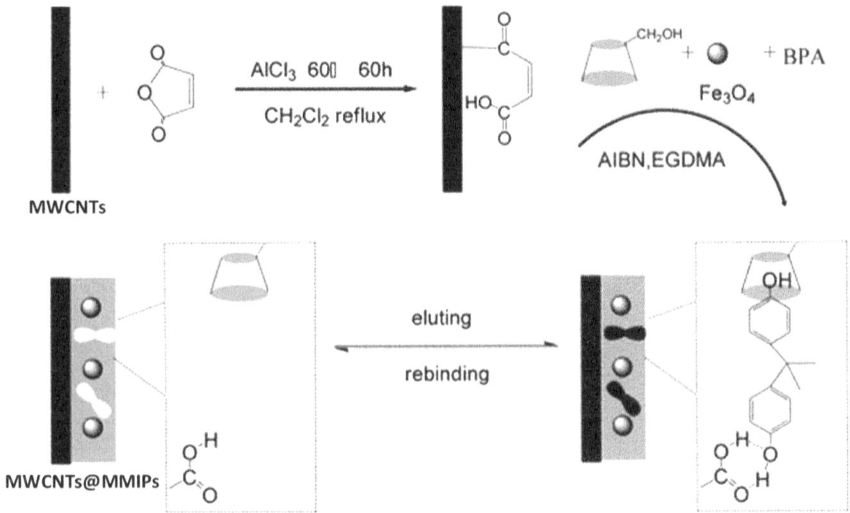

Fig. 7 The protocol for synthesis of MWCNTs@MMIPs (Zhang et al. 2014c). Reprinted (adapted) with permission from Zhang et al. (2014c). Copyright © 2014 Elsevier B. V. All rights reserved

interaction between BPF and MWCNTs and the overall rate process is mainly governed by the external mass transfer.

6.2 *Removal by Metal Oxide Nanoparticle Decorated Carbon Nanotubes*

The combination of CNTs and metal oxides/metal nanoparticles provides unique properties which are different corresponding to their individual components (Jiménez-Marín et al. 2017; Kumar et al. 2017d, e; Mallakpour and Khadem 2016; Rajarao et al. 2014; Sapurina et al. 2016; Ye and Guo 2013). The CNTs decorated with metal oxide exhibit strength, stiffness and toughness (Li et al. 2009) and also gives higher conductivity by a few orders of magnitude (Farbod and Ahangarpour 2016; Subramaniam et al. 2013). It is also attractive to consider the preparation of MWCNTs decorated with metal oxide/metal nanoparticles for developing sensors applications (Devadoss et al. 2015; Hrapovic et al. 2004). In 2010, Yin et al. (2010) reported an amperometric BPA biosensor fabricated by immobilizing tyrosinase on MWCNTs-cobalt phthalocyanine (CoPc)-silk fibroin (SF) composite modified GCE and the mechanism for electrocatalytic oxidation of BPA on the fabricated biosensor is expressed in Fig. 8. In MWCNTs-CoPc-SF composite film, SF provides a biocompatible microenvironment for the tyrosinase to retain its bioactivity, MWCNTs gives excellent inherent conductivity to enhance the electron transfer

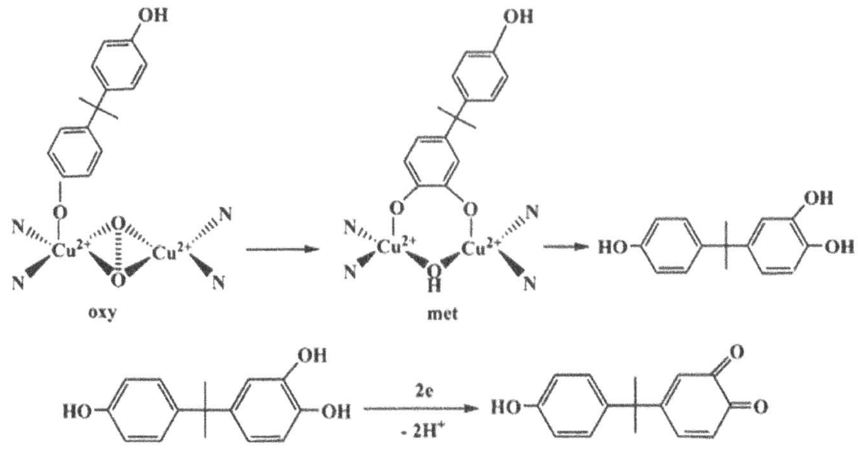

Fig. 8 The possible mechanism for BPA oxidation at Tyr-SF-MWCNTs-CoPc/GCE biosensor (Yin et al. 2010). Reprinted (adapted) with permission from Yin et al. (2010). Copyright © 2010 Elsevier B. V. All rights reserved

rate and CoPc presents good electrocatalytic activity to electrooxidation of BPA. The cyclic voltammogram of BPA at this biosensor exhibited a well defined anodic peak at 0.625 V. Compared with bare GCE, the oxidation signal of BPA significantly increases and hence, this oxidation signal was used to determine BPA. Under optimum conditions, the oxidation current was proportional to BPA concentration in the range from 5.0×10^{-8} to 3.0×10^{-6} M with correlation coefficient of 0.9979 and detection limit of 3.0×10^{-8} M ($S/N = 3$). The proposed method was successfully applied in determination of BPA in plastic products and the recovery was in the range from 95.36 to 104.39%.

Cleveland et al. (2014) studied the heterogeneous Fenton oxidation using Fe_3O_4 amended MWCNT (Fe_3O_4/MWCNT) which shows effective degradation of aqueous BPA. The Fe_3O_4/MWCNT catalyst-driven Fenton oxidation achieved a high removal percentage of BPA (97% removal in 6 h) at operating conditions (pH of 3, 0.5 g catalyst/L, [H_2O_2]:[BPA] of 4 (mol/mol), 50 °C). The Fenton oxidation of BPA demonstrated significant removal of BPA at 0.5–1 g catalyst/L and at the initial pH of 3. The [H_2O_2]:[BPA] of 4 in this study was found to be the cost-effective condition to achieve higher removal of BPA at low concentration of H_2O_2. Recently in 2017, Santana et al. (2017) reported GCE modified with a film of ferro ferric oxide nanoparticles (Fe_3O_4NPs) over a film of Au NPs, both stabilized in a polymer solution of 3-n-propyl-4-picolinium silsesquioxane chloride ($Si_4Pic^+Cl^-$), for the electroanalytical determination of BPA. The modified electrode, cyclic voltammograms of BPA in B-R buffer (pH 9.0) showed a diffusion-controlled irreversible oxidation peak at around +0.470 V. Under optimized experimental conditions for differential pulse voltammetry experiments, the peak current was found to vary linearly with the concentration of BPA in the range of 20–1400 nmol L^{-1} (R2 = 0.996) with a detection limit of 7.0 nmol L^{-1}.

The modified electrode was employed for the determination of BPA in different commercial plastic samples. The recoveries were obtained in the range of 90–120%, in utilizing the comparative technique (UV–vis spectroscopy).

In 2016 Li et al. (2016) reported the disposable electrochemical sensor for the determination of BPA. The working electrodes were fabricated by sputtering Au NPs on commercial art paper modified with MWCNTs. The determination of BPA was investigated by linear sweep voltammetry and shows wide linearity in the range from 0.2 to 20 mg L^{-1} with a detection limit of 0.03 mg L^{-1} ($S/N = 3$). The between-sensor reproducibility was 5.7% (n = 8) for 0.5 mg/L BPA. In same year, Goulart and Mascaro (2016) reported the electrochemical determination of BPA, hydroquinone (HQ) and catechol (CC) using GCE modified with MWCNT and NiO NPs. MWCNT were functionalized with sulfonitric solution ($3H_2SO_4$:$1HNO_3$)

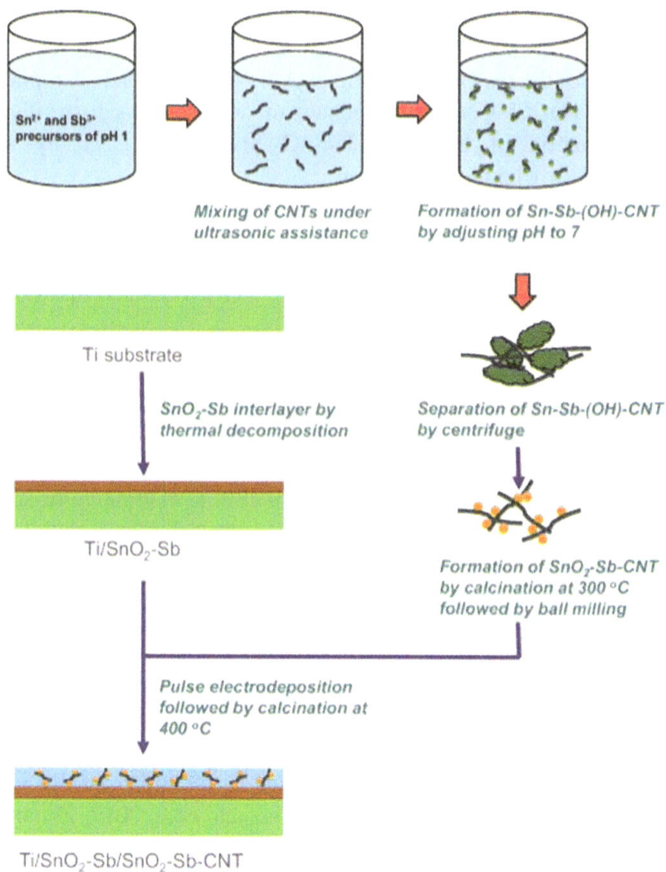

Fig. 9 Schematic illustration of SnO_2-Sb-CNT composite and Ti/SnO_2-Sb/SnO_2-Sb-CNT electrode preparation (Wu et al. 2017). Reprinted (adapted) with permission from Wu et al. (2017).

and dispersed in dimethylformamide for the MWCNT/GCE manufacturing. The NiO/MWCNT/GCE formed presented the lowest charge transfer resistance. The electrochemical detection of BPA, HQ and CC was performed using differential pulse voltammetry. The analytical curves showed an excellent linear response with the detection limits for the simultaneous determination of BPA, HQ and CC as 2.8×10^{-8} mol L^{-1}, 2.70×10^{-8} mol L^{-1} and 5.9×10^{-8} mol L^{-1}, respectively.

Recently in 2017, Wu et al. (2017) studied, a duplex-structured SnO_2-Sb-CNT composite anode (Ti/SnO_2-Sb/SnO_2-Sb-CNT) composed of 4–5 nm SnO_2-Sb particles. Figure 9 presents the schematic diagram for the preparation of SnO_2-Sb-CNT composites and duplex-structured Ti/SnO_2-Sb/SnO_2-Sb-CNT electrodes. Bulk electrolysis of BPA was conducted to evaluate the electrocatalytic performance of the electrodes at various current densities (10–40 mA cm^{-2}) and initial pH (3–11) of solution by determining BPA degradation, total organic carbon (TOC) removal and mineralization current efficiency (MCE). Ti/SnO_2-Sb/SnO_2-Sb-CNT intended to adsorb more BPA molecules, which may contribute to its high current efficiency for electrochemical oxidation of BPA. The presence of humic acid (HA) decreases the performance efficiency by inhibiting the adsorption of BPA by Ti/SnO_2-Sb/SnO_2-Sb-CNT. A mechanism was proposed to illustrate the enhanced degradation of BPA by Ti/SnO_2-Sb/SnO_2-Sb-CNT by synergistic adsorption and electrochemical oxidation.

7 Removal of Bisphenols Using 2D Graphene/Graphene Oxide

One of the derivatives of graphene, GO has become an attractive and novel material and can be easily produced in bulk quantity via chemical exfoliation process of natural graphite (Hummers and Offeman 1958; Poh et al. 2012; Singh et al. 2016; Staudenmaier 1898). The GO have been explored in a wide range of applications, like energy generation/storage, optical devices, electronic and photonic devices, drug delivery and clean energy (Aleksandrzak et al. 2017; Aliabadi et al. 2017; Bi et al. 2017; Ding et al. 2017; Dong et al. 2017; Kumar et al. 2016e, 2017c; Li et al. 2015f; Luo et al. 2017; Oribayo et al. 2017; Paek et al. 2017; Singh et al. 2016; Song et al. 2017; Wang et al. 2017b). Graphene-based material has recently been evaluated to remove toxic contaminant from water. Several studies confirmed GO as a promising adsorbent, due to its large surface area with abundant surface oxygen-containing groups supportive for the adsorption. (Aboubaraka et al. 2017; Chandra et al. 2010; Chen et al. 2017; Paul et al. 2015; Sheng et al. 2012; Varaprasad et al. 2017; Yoon et al. 2017; Zambianchi et al. 2017; Zhu et al. 2017). Graphene-based sensors have also been developed for the detection of persistent hydrocarbons, insecticides, hormone disruptors, and pharmaceutical contaminants (Choi et al. 2010; Wu et al. 2012; Yang et al. 2013a). The hydrogen bonding is mainly responsible in the adsorption of polar hydrocarbons by graphene/GO materials, including several toxics compounds as anthracenemethanol, naphthol, and 1-naphthylamine (Pei et al. 2013; Perreault et al. 2015; Wang et al. 2014; Yang et al. 2013b; Zhang et al. 2013a). The formation of hydrogen bonds helps in the adsorption of BPA with the interactions between hydroxyl groups on BPA molecules and the remaining oxygenated groups on the GO on graphene sheets obtained by the chemical reduction of GO (Xu et al. 2012). Because BPA shows aromatic nature, the hydrogen bonding likely to be coexists with π–π stacking interaction during the adsorption process (Perez and Martin 2015; Xu et al. 2012). The excellent adsorption capacity of graphene is due to its unique sp^2-hybridized single-atom-layer structure. On the above facts, graphene could be regarded as a promising adsorbent for BPA removal in water treatment.

7.1 Removal by Pristine Graphene/Graphene Oxide

Extensive research interest in GO and reduced GO (rGO) owes to relatively easy and cheap production of the material, highly attractive for sensing application, r removal and detection of BPA from water. Xu et al. (2012) investigated the decontamination of BPA from aqueous solution by adsorption on graphene. The maximum adsorption capacity (qm) of graphene calculated from Langmuir isotherm was 182 mg g^{-1} at 302.15 K. Figure 10 shows th π–π interactions and hydrogen bonds responsible for the adsorption of BPA on graphene. The high adsorption capacity of graphene was due to its unique sp^2-hybridized single-atom-layer structure. The presence of NaCl in the solution facilitates the adsorption process, whereas the alkaline pH range and higher temperature of the solution were unfavorable for BPA adsorption.

Ndlovu et al. (2012) used exfoliated graphite (EG) electrode as electrochemical sensors for detection of BPA in water. The oxidation peak of BPA was observed at

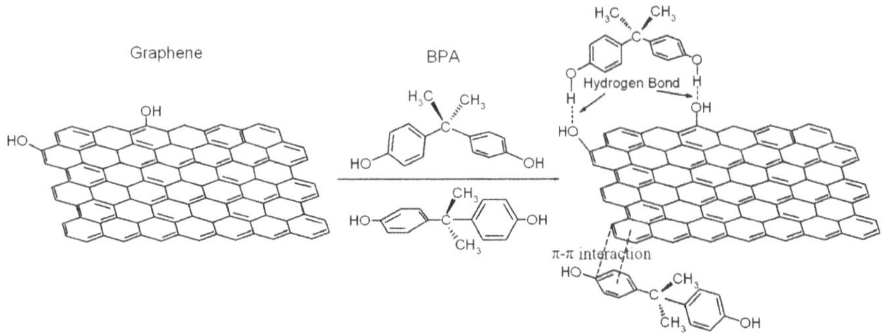

Fig. 10 Schematic of π–π interaction and hydrogen bonding between BPA and graphene (Xu et al. 2012). Reprinted (adapted) with permission from Xu et al. (2012). Copyright © 2012 Elsevier B. V. All rights reserved

about 0.45 V in phosphate buffer solution at pH 10. The current response exhibits a linear relationship with concentration over a range from 1.56 μM–50 μM with the detection limit 0.76 μM. Ntsendwana et al. (2012) reported the electrochemical detection of BPA using graphene-modified GCE. The GCE was modified with graphene by a simple drop and dry method and cyclic voltammetry was used to study the electrochemical properties of the prepared graphene-modified GCE using potassium ferricyanide as a redox probe. Under the optimum conditions, the oxidation peak current of BPA varied linearly with concentration over a wide range of 5×10^{-8} mol L^{-1} to 1×10^{-6} mol L^{-1} and the detection limit was as low as 4.689×10^{-8} M. This method was also employed to determine BPA in a real sample. Deng et al. (2013) presented a strategy to electrochemically synthesize rGO modified acetylene black paste electrode (GR/ABPE) for the determination of BPA. This modified electrode was obtained by one-step potentiostatic reduction of exfoliated GO sheets on the surface of ABPE. Under the optimized conditions, the linear range for the detection of BPA was from 0.8 nM to 10.0 nM, 10.0 nM to 8.0 lM and 8.0 lM to 0.1 mM, with a detection limit ($S/N = 3$) of 0.6 nM. The practical analytical performance of the GR/ABPE was examined by evaluating the detection of BPA in plastic products with satisfied recovery.

Bele et al. (2016) reported the effect of the reduction degree of graphite oxide on the adsorption of BPA. Graphite oxide was reduced to different reduction degrees by using hydrazine hydrate and finally to graphene using NH_4OH. The adsorption capacities for graphite oxide, rGO and graphene were found to be 17.27, 80.81 and 94.06 mg g^{-1}, respectively at 25 °C. The adsorption capacities also increases with increase in the degree of reduction of graphite oxide with the maximum adsorption capacity (Qmax = 94.06 mg g^{-1}) shown by graphene resulted by optimum reduction. The adsorption followed pseudo-second order kinetics and the thermodynamic analysis indicated that it was spontaneous and endothermic. The increase in the degree of graphite oxide reduction reduced the amount of oxygen-containing functional groups on the surface of reduced samples, resulting to the increase of the π–π interaction between sorbent-adsorbate and hence shows linear increase in adsorption capacity. Figure 11a presents the effect of the solution pH on BPA adsorption by graphite oxide, rGO and graphene, for initial pH range 3.0–10.0. Graphite oxide was negatively charged over the whole pH range, as presented in the relative proton-binding curve from the potentiometric titration results (Fig. 11b).

7.2 Removal by N-Doped Graphene/Graphene Oxide

There are so far a large number of reports on nitrogen (N) doped graphene since N have similar structural properties. The electronic properties of N doped graphene modifies with doping concentration, and therefore applied to developments of sensor applications (Bai and Zhou 2007; Fujimoto and Saito 2016; Huang et al. 2017; Jain and Chauhan 2017; Jiang et al. 2016; Kaur et al. 2017; Li et al. 2017b; Travlou et al. 2016; Yang et al. 2017). It is also reported that N-doped graphene

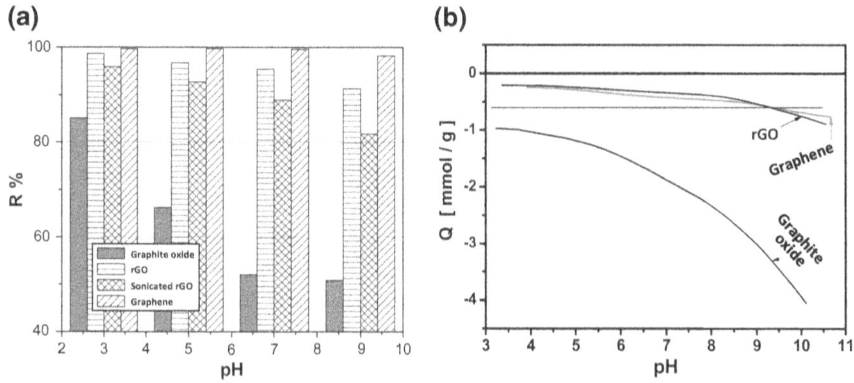

Fig. 11 **a** Effect of pH on BPA adsorption onto graphite oxide, rGO, sonicated rGO and graphene. **b** Potentiometric titration results for graphite oxide, rGO and graphene (Bele et al. 2016). Reprinted (adapted) with permission from Bele et al. (2016).

exhibits unique properties different than perfect graphene (Lotfi and Neek-Amal 2017). Fan et al. (2012) reported the electrochemical sensing of BPA using N-doped graphene sheets and chitosan. The sensor exhibits a sensitive response to BPA in the range of 1.0×10^{-8} to 1.3×10^{-6} mol L^{-1} with a low detection limit of 5.0×10^{-9} mol L^{-1} under the optimal conditions. Jiao et al. (2014) used tannic acid functionalized N-doped graphene (TA/N-G) modified GCE for the determination of BPA in food package. Under the optimal conditions, the oxidation current increases linearly with increasing concentration of BPA in the range of 0.05–13 μM with the detection limit of 4.0 nM and the fabricated electrode showed good reproducibility, stability and anti-interference.

Wang et al. (2015) investigated the synthesis of nitrogen modified rGO (N-rGO) by a hydrothermal method and found that the nitrogen modification enhances its adsorption and catalysis ability. For an initial bisphenol concentration of 0.385 mmol L^{-1}, the adsorption capacity of N-rGO was evaluated as 1.56 and 1.43 mmol g^{-1} for BPA and 1.43 mmol g^{-1} for bisphenol F (BPF), respectively, both of which were about 1.75 times that (0.90 and 0.84 mmol g^{-1}) on N-free rGO. The apparent degradation rate constant of BPA on N-rGO was 0.71 min^{-1}, being about 700 times that (0.001 min^{-1}) on N-free rGO. In mixtures of various phenols, the degradation rate constant of each phenol linearly increases with its adsorption capacity. Fan et al. (2012) reported the fabrication of electrochemical BPA sensor using nitrogen-doped graphene sheets (N-GS) and chitosan (CS). Compared with graphene, N-GS has favorable electron transfer ability and electrocatalytic property, which could enhance the response signal towards BPA. CS also exhibits excellent film forming ability and improves the electrochemical behavior of N-GS modified electrode. The sensor exhibits a sensitive response to BPA in the range of 1.0×10^{-8} to 1.3×10^{-6} mol L^{-1} with a low detection limit of 5.0×10^{-9} mol L^{-1} under the optimal conditions. Xin et al. (2015) reported that the GO nanoribbons (GONRs) synthesized from unzipping of MWCNTs with the help of microwave energy and modified with chitosan (CS) forms a sensitive electrochemical BPA sensor. Compared with graphene, MWCNTs/GONRs shows favorable adsorption capacity, electron transfer ability and electrocatalytic property, which could enhance the response signal toward BPA. The sensor displays a sensitive response to BPA within a wide concentration range (0.005–150 μg L^{-1}). The proposed electrochemical sensor shows low detection limit (1 ng L^{-1}), good reproducibility (RSD = 5.2%), selectivity, and acceptable stability.

Shen et al. (2015) developed a sensitive and reliable electrochemical sensor for the determination of BPA based on rGO/melamine (rGO/M)-modified GCE. Due to the combination of the electrocatalytic property of rGO and the electrostatic attraction of protonated M, the modified electrode exhibited the enhanced performance on the oxidation of BPA. Under the optimal conditions, good linearity was observed between the differential pulse voltammetric peak current and the concentration of BPA in the range of 1.0×10^{-8} to 2.0×10^{-4} M in PBS at pH 7.0 with the detection limit of 4.0×10^{-9} M ($S/N = 3$). The fabricated sensor was applied in determining BPA in real plastic samples with good recoveries (97.00–100.96%). The schematic diagrams for the fabrication of rGO/M-GCE and the electrochemical detection of BPA are shown in Fig. 12.

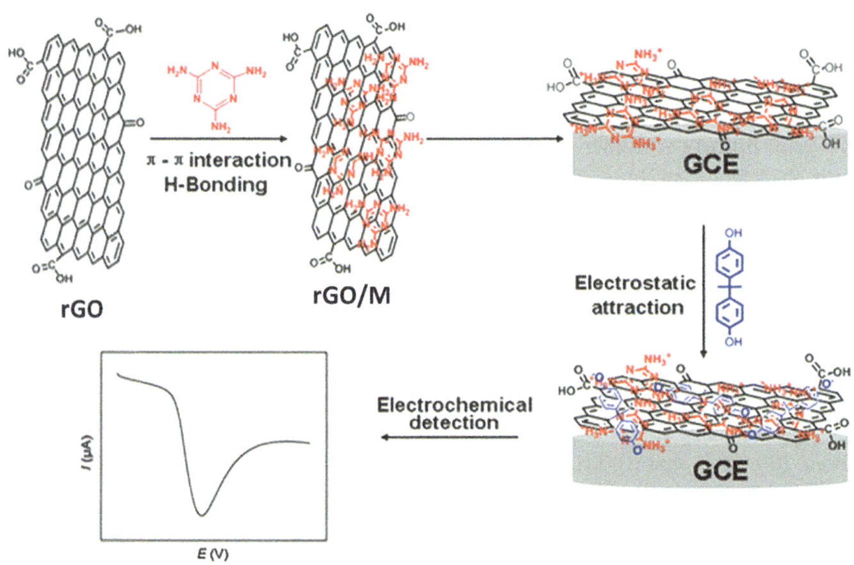

Fig. 12 Schematic diagrams for the fabrication of rGO/M-GCE and the electrochemical detection of BPA (Shen et al. 2015). Reprinted (adapted) with permission from Shen et al. (2015). Copyright © 2015, Springer-Verlag Gmbh and Co. KG

7.3 *Removal by Metal Oxide Decorated Graphene/Graphene Oxide Composites*

The adsorption of external molecular species on the surface of graphene may influence the local carrier concentration and result in variation in conductivity (Qi et al. 2003; Sakhaee-Pour et al. 2008; Tang et al. 2016; Wehling et al. 2008). However, the lack of chemically active defect sites on the surface leads to the low sensitivity and poor selectivity of pristine graphene (Schedin et al. 2007). Graphene decorated with metal oxide represents a new type of hybrid nanostructure, which could display good physical and chemical properties, improving the performance for various applications including sensing (Cui et al. 2013; Kumar et al. 2015a, b, 2016a, d, f, 2017a, b, f). Owing to the high reaction activity, the metal oxides are used to decorate the low-dimensional carbon nanomaterials for gas sensing. Researchers have explored graphene based sensors modified with a number of metal oxides to fabricate metal oxide–graphene hybrid sensors (Gupta Chatterjee et al. 2015). Zhang et al. (2013c) developed an electrochemical sensor for the detection of BPA using magnetic nanoparticles (MNPs) (Fe_3O_4 nanoparticles) decorated on reduced GO. Electrochemical studies show that MNPs-rGO composites can lower the oxidation over potential and enhance electrochemical response of BPA due to the synergetic effects of MNPs and rGO. The oxidation peak current was proportional to the concentration of BPA over the range of 6.0×10^{-8} to 1.1×10^{-5} mol L^{-1} with a detection limit of 1.7×10^{-8} mol L^{-1}. The electrochemical sensor was applied to the determination of BPA in real samples with satisfactory results.

Zhang et al. (2014b) synthesized the rGO nanosheets decorated with tunable MNPs (Fe_3O_4 NPs) by a simple co-precipitation method and employed it for recyclable removal of BPA from aqueous solution. The adsorbents could be recovered conveniently by magnetic separation and recyclably used because of the easy desorption of BPA. In 2016, Wang et al. (2016d) studied, a novel and quick method for selective removal of BPA from aqueous solutions, which used magnetic GO containing Fe_3O_4 particles-based molecularly imprinted polymers as the adsorbent. Adsorption experiments were carried out to examine the effect of pH, initial concentration of BPA, isotherms and sorption kinetics on the adsorption of BPA by magnetic molecularly imprinted polymers (MMIPs). Results revealed the maximum adsorption capacity of BPA by MMIPs was 106.38 mg g^{-1} at 298 K and the equilibrium data of MMIPs were described well by a Langmuir isotherm model. The preparation route of MMIPs and their application for removal of BPA with the help of an external magnetic field is presented in Fig. 13.

Du et al. (2016) reported the synthesis of heterogeneous manganese/magnetite/GO (Mn-MGO) hybrid catalyst through the reduction of $KMnO_4$ by ethylene glycol in the presence of magnetite/GO (MGO) particles for the removal of BPA. The Mn-MGO catalyst exhibited high efficacy and long-term stability in activating peroxymonosulfate (PMS) to generate sulfate radicals for the removal of BPA from water. Recently, Hu et al. (2017) reported a simple, sensitive and selective detection

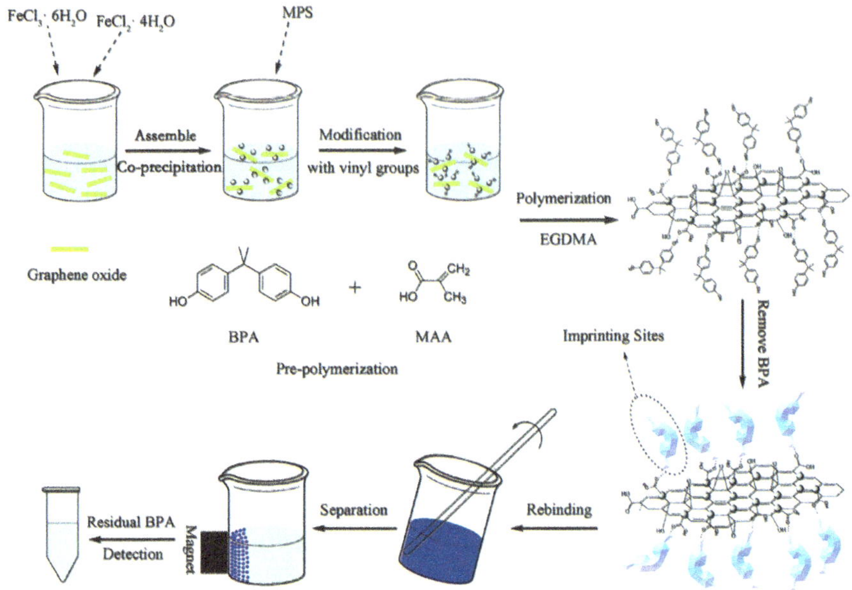

Fig. 13 Synthesis route of MMIPs and their application for the removal of BPA with the help of an applied magnetic field (Wang et al. 2016b). Reprinted (adapted) with permission from Wang et al. (2016b). Copyright (2016) The Royal Society of Chemistry

method to detect BPA in environmental water based on magnetic graphene oxidation (MGO) and the specific binding of aptamer to PBA. The proposed method has high selectivity and anti-interference capability, the recovery experiments shown that this method could be applied to the detection of real environmental samples. The detection limit of 0.071 ng mL^{-1} was obtained with the linear range of 0.2–10 ng mL^{-1}. The biosensor exhibited excellent anti-interference ability and selectivity in actual water samples.

Ragavan and Rastogi (2016) reported a microwave assisted method for the synthesis of graphene–copper(II) oxide (G–CuO) nanocomposite and utilized it for the sensing of BPA. Nanocatalytic activity of G–CuO nanocomposite was examined by its efficiency to improve the signals of luminol–H_2O_2 chemiluminescence (CL) reaction. Analytical application of luminol–H_2O_2 chemiluminescence reaction was applied for detection of BPA in feeding bottles and polycarbonate water bottles. The chemiluminescence method has linearity in the range 10,000–1 ng mL^{-1} with limit of detection of 0.55 ng mL^{-1} and limit of quantification of 8.3 ng mL^{-1}. In 2017, Shi et al. (2017) also reported a novel BPA sensor based on a facile one step electrochemical reduction of GO and cuprous oxide (Cu_2O) nanocomposite modified GCE. The details of the preparation of the Cu_2O-rGO composite are given as Fig. 14a. The prepared Cu_2O-rGO electrode gives a fast response, high sensitivity and low background current. The response of BPA on prepared electrode was 2.15 times higher than reduced graphene modified electrode. Under the optimized

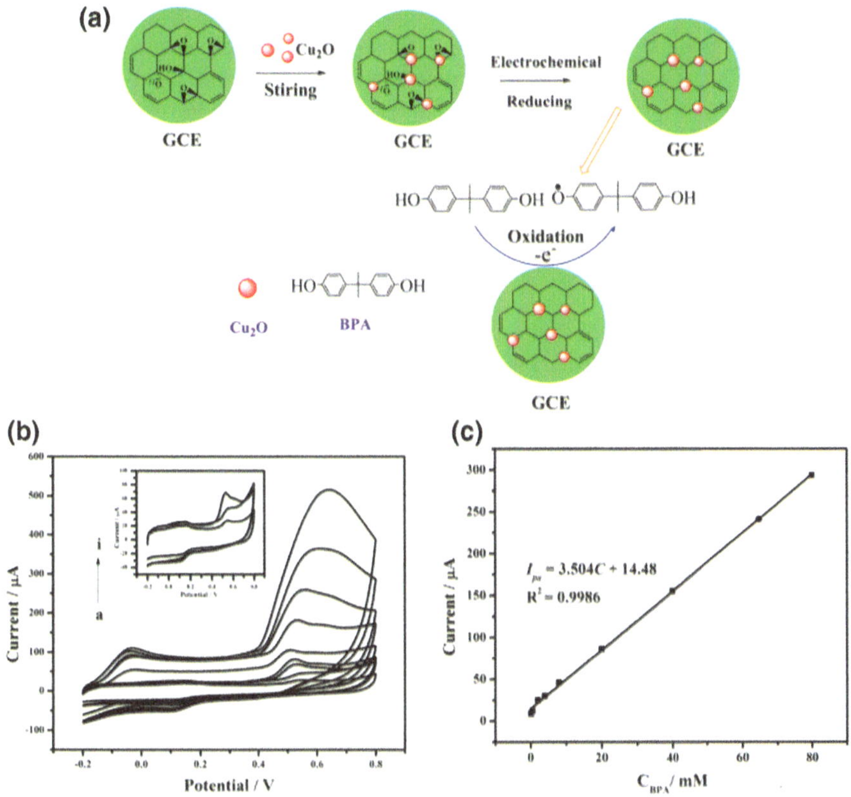

Fig. 14 **a** Schematic diagram for one step electrochemical synthesis of Cu_2O-rGO electrode, and electrochemical oxidation process of BPA at Cu_2O-rGO/GCE. **b** The CVs of the different concentrations of BPA (a–i: the concentrations of BPA are 0.1, 0.5, 2, 4, 8, 20, 40, and 80 μM). **c** The linear relationship between the peak current (Ipa) and the concentrations of BPA (Shi et al. 2017). Reprinted (adapted) with permission from Shi et al. (2017).

experimental parameters, the detection range of the modified electrode was from 1×10^{-7} to 8×10^{-5} M and the limit of detection was 5.3×10^{-8} M ($S/N = 3$) as shown in Fig. 14b, c.

Niu et al. (2013) reported an electrochemical sensor for the determination of BPA based on stacked graphene nanofibers (SGNF) and Au nanoparticles (AuNPs) composite modified GCE. The AuNPs/SGNF modified electrode shows an efficient electrocatalytic role for the oxidation of BPA. The oxidation overpotentials of BPA decreases and the peak current increases significantly compared with bare GCE and other modified electrode. The transfer electron number (n) and the charge transfer coefficient (α) were 0.52 for BPA. The effective surface areas of AuNPs/SGNF/GCE increased for about 1.7-fold larger than that of the bare GCE. Linear sweep voltammetry was applied as a sensitive analytical method for the determination of

BPA and a good linear relationship between the peak current and BPA concentration was obtained in the range from 0.08 to 250 μM with a detection limit of 3.5×10^{-8} M. Zhou et al. (2014a) developed a simple and label-free electrochemical aptasensor for BPA determination based on AuNPs dotted graphene (GNPs/GR) nanocomposite film on GCE. The electrochemical probe of ferricyanide was used to investigate the interactions between aptamer and BPA. The peak current change (ΔI) of ferricyanide varies linearly with the concentration of BPA in the range from 0.01 to 10 μM with the detection limit of 5 nM. The aptasensor was applied successfully to determine BPA in milk products, and the average recovery was 105%.

Pan et al. (2015) fabricated an electrochemical biosensor based on functional graphene-Au nano-composite for direct and rapid detection of BPA as shown via synthesis protocol (Fig. 15). The proposed biosensor exhibited excellent performance for BPA determination with a wide linear range (2.5×10^{-3} to 3.0 mM), a highly reproducible response (RSD of 2.7%), low interferences and long-term stability. The calculated detection limit of the proposed biosensor was as low as 1 nM. Recently in 2017, Su et al. (2017) reported a sensitive BPA electrochemical sensor, based on a surfactant-free AuPd nanoparticles-loaded graphene nanosheets (AuPdNPs/GNs) modified electrode. The AuPdNPs monodispersed on GNs were prepared by spontaneous redox reaction between bimetallic precursors and GNs. Due to the synergetic effect of Au and Pd, AuPdNPs/GNs displayed high electrochemical activity with voltammetric peaks of BPA oxidation and lower over potential compared with monometallic PdNPs and AuNPs supported GNs. In the results of DPV, under optimized conditions, linear response was observed for the concentration of BPA in the range of 0.05–10 μM with a detection limit of 8 nM. The developed electrochemical sensor was also applied to determine BPA in food package.

Li et al. (2015b) reported a 3D sea urchin-like α-MnO_2 nanoarchitectures and rGO for the elimination of BPA from water in the presence of ozone.

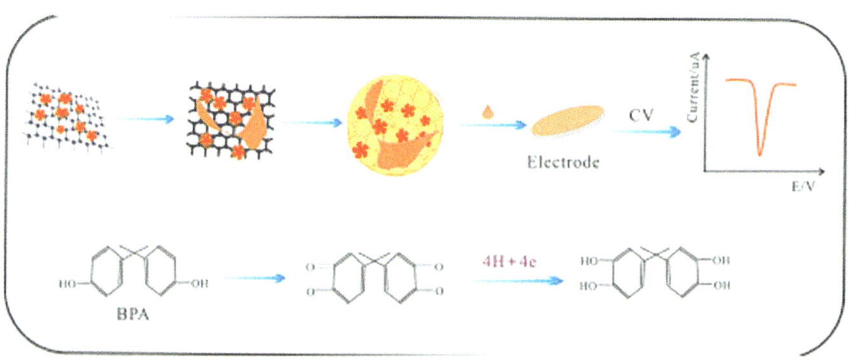

Fig. 15 The schematic diagram of biosensor preparation and the BPA sensing mechanism (Pan et al. 2015). Reprinted (adapted) with permission from Pan et al. (2015). Copyright © 2015 Elsevier B. V. All rights reserved

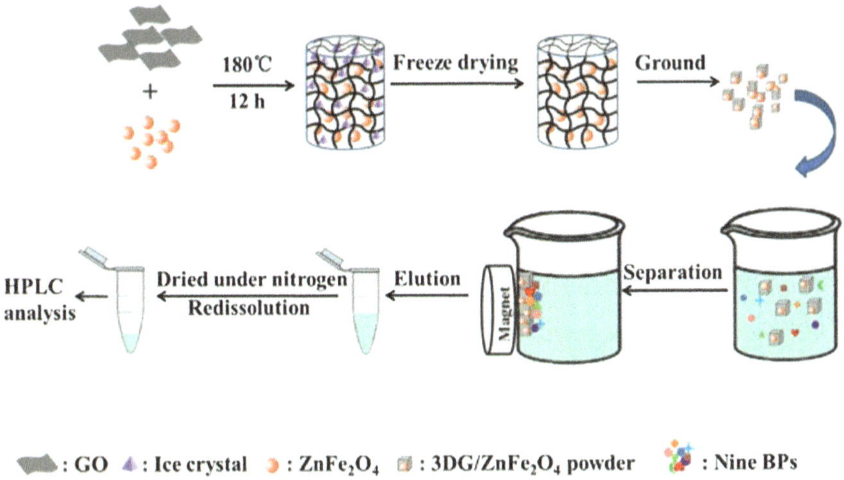

Fig. 16 Schematic procedure for MSPE of target compounds from water samples (Wang et al. 2016a). Reprinted (adapted) with permission from Wang et al. (2016a). Copyright © 2016 Elsevier B. V. All rights reserved

The as-synthesized α-MnO_2/rGO nanocomposite is proved to be highly efficient catalyst to eliminate BPA from water in the presence of ozone. Wang et al. Magnetic solid-phase extraction using nanoporous 3D graphene hybrid materials for high-capacity enrichment and simultaneous detection of nine bisphenol analogs from water sample has been reported (Wang et al. 2016b). The synthesis of a magnetic nanoporous 3D graphene (3DG)/$ZnFe_2O_4$composite is shown in Fig. 16. Under optimized condition, the lower method detection limits (0.05–0.18 ng mL^{-1}), the higher enrichment factors (800 fold) and good recoveries (95.1–103.8%) with relative standard deviation (RSD) values less than 6.2% were achieved.

Cheng et al. (2013) investigated a facile method using biopolymers mediated synthesis of 3D GO based gels, which efficiently remove cationic dyes and heavy metal ions from wastewater. The GO-biopolymer gels displayed the adsorption capacity as high as 1100 mg g^{-1} for MB dye and 1350 mg g^{-1} for MV dye, respectively. The 3D GO gels were also used for the removal of BPA from aqueous solution. Zinc phthalocyanine tetrasulfonic acid (ZnTsPc)-functionalized graphene nanocomposites (ZnTsPc/f-GN) was synthesized using efficient electrostatic self-assembly method (Fig. 17) (Hou et al. 2015). The synthesized ZnTsPc/f-GN nanocomposite modified with GCE act as a sensitive sensor for the determination of BPA. Under the optimal conditions, the oxidation peak current increased linearly with the concentration of BPA in the range of 5.0×10^{-8} to 4.0×10^{-6} M with correlation coefficient 0.998 and limits of detection 2.0×10^{-8} M.

Zhou et al. (2014b) reported electrochemical approach for the in situ determination of BPA in the milk and mineralised water using graphene-hypercross linked

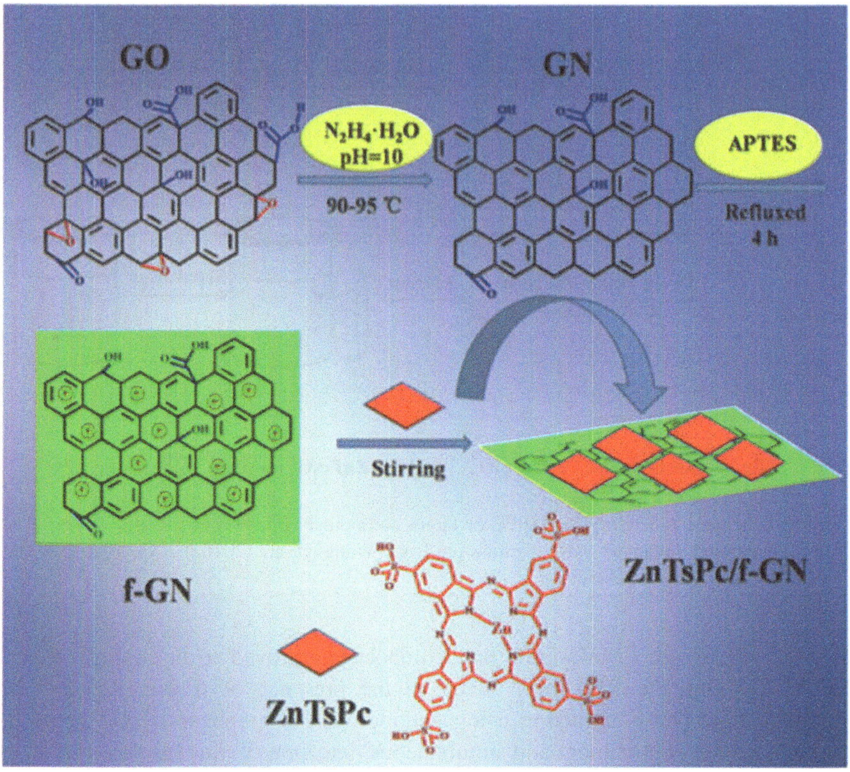

Fig. 17 Schematic illustration for electrostatic self-assembly of ZnTsPc/f-GN nanocomposites (Hou et al. 2015). Reprinted (adapted) with permission from Hou et al. (2015). Copyright © 2015 Elsevier B. V. All rights reserved

resin MN202 composite (MN202) modified electrode. The oxidation peak current of BPA enhances remarkably while the oxidation over potential decreases significantly. Under the optimized conditions, the oxidation peak current was proportional to BPA concentration in a wide range between 0.005 and 20.0 μmol/L, and the detection limit was 1.02 nmol L^{-1} ($S/N = 3$). Zhang et al. (2015) reported a ratiometric electrochemical sensor based on poly-β-cyclodextrin/electroreduced graphene modified GCE as highly sensitive and selective detection of BPA. The assay strategy was based on the competitive host–guest interaction between poly-β-cyclodextrin/electroreduced graphene (Pb-CD/EG) and Rhodamine B (RhB) probe or BPA target molecules. In the presence of BPA, the RhB molecules are displaced by BPA due to the stronger host–guest interaction between β-CD and BPA than β-CD and RhB. As a result, the oxidation peak current of RhB (I_{RhB}) decreases and the oxidation peak current of BPA (I_{BPA}) increases correspondingly. The logarithmic value of I_{BPA}/I_{RhB} is linear with the logarithm of BPA concentration in the range of 1–6000 nM and the detection limit is 52 pM ($S/N = 3$).

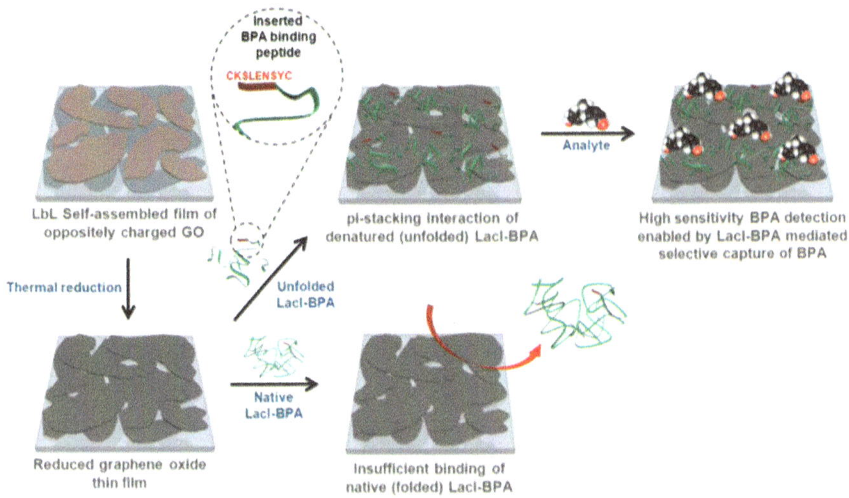

Fig. 18 Schematics of fabricating a BPA sensing platform based on pi-stacking interaction between LacI-BPA recombinant protein and the rGO surface (Kim et al. 2015). Reprinted (adapted) with permission from Kim et al. (2015).

This strategy provides a new approach for sensitive detection of BPA, and has promising applications in the detection of organic pollutants in real environmental samples. Kim et al. (2015) reported a novel BPA sensor comprising of layer-by-layer assembled rGO electrode and a designer probe specifically for recognizing of BPA as shown in Fig. 18. The BPA detection probe, a recombinant protein (LacI-BPA), was constructed by fusing a disulfide-constrained high affinity BPA binding peptide (CKSLENSYC) to the C-terminus of Lac repressor (LacI). The fabricated BPA sensor (LacI-BPA/rGO) shows a wide linear dynamic range of BPA detection spanning from 100 fM to 10 nM.

Electrochemical determination of BPA in plastic bottled drinking water and canned beverages using a molecularly imprinted chitosan–graphene composite film modified electrode was reported (Deng et al. 2014). The electrochemical sensor based on an acetylene black paste electrode modified with molecularly imprinted chitosan–graphene composite film for sensitive and selective detection of BPA was fabricated. Under the optimal conditions, a linear range from 8.0 nM to 1.0 μM and 1.0 to 20 μM for the detection of BPA was observed with the detection limit of 6.0 nM ($S/N = 3$). Zheng et al. (2013) suggested a facile and green method to synthesize the graphene-CNTs nanocomposite with a sandwich lamination structure loaded with Pt to prepare an electrochemical sensor for determining BPA on thermal printing paper as shown in Fig. 19. The electrochemical behavior of BPA on the Pt/graphene-CNTs nanocomposite has been investigated by cyclic voltammetry and chronocoulometry. The oxidation peak current was proportional to the

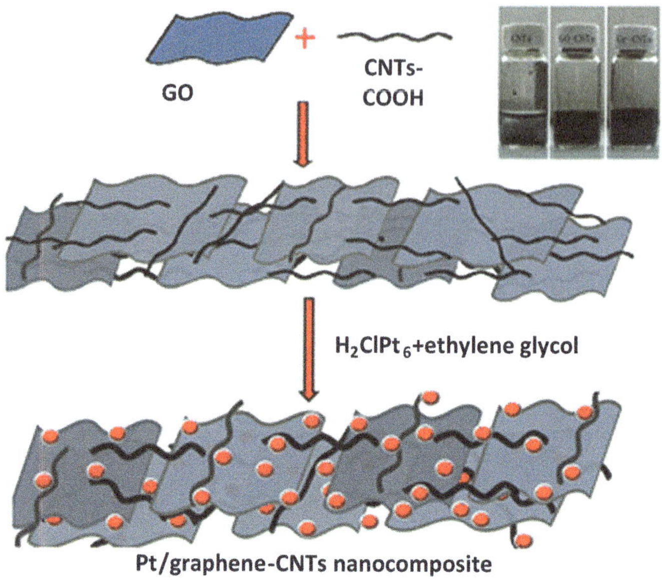

Fig. 19 Schematic diagram of the structure formation of Pt/graphene-CNTs nanocomposite, inset are the digital pictures of CNTs, GO-CNTs and graphene-CNTs solution, respectively (Zheng et al. 2013). Reprinted (adapted) with permission from Zheng et al. (2013). Copyright (2013) The Royal Society of Chemistry

BPA concentration in the range from 6.0×10^{-8} to 1.0×10^{-5} M and 1.0×10^{-5} to 8.0×10^{-5} M with a correlation coefficient of 0.987 and 0.998, respectively. The detection limit was 4.2×10^{-8} M ($S/N = 3$).

8 Conclusions, Challenges and Perspectives

Among carbon materials, GO and CNTs are the superstar carbon nano materials, and it has been widely studied by researchers throughout the world and has drawn significant curiosity in various research fields. These two materials have been applied widely in sensing applications because of its unique structure and exceptional chemical and physical properties. This chapter reports the selectively review of recent researches and developments in the field of GO and CNTs based sensors used for monitoring to detection and removal of BPA from waters and wastewater by adsorption. The GO and CNTs based carbon materials are mainly used in sensors following electrochemical principles due to their electrical conductivity, high electron transfer rate and large surface area. The recent development in this field corresponds to the synthesis of hybrids/composites combining GO and CNTs carbon materials with other metal nanoparticles/metal oxide, allowing the fabrication of electrochemical sensors with improved analytical performance since the combination of electrical characteristics of hybrids/composite constituents make advantageous its integration on such sensing platforms. GO and CNTs based electrochemical sensors have recently received increasing attention in the field of electro analysis. In addition, these have exhibited good sensitivity and selectivity toward the detection of BPA and heavy metal ions. In addition, the use of GO and CNTs is advantageous due to easy preparation and low cost of manufacturing. The GO and CNTs enables more sensitive detection when compared with the other type of nanomaterials used for sensing application. We believe that all these unique features of GO and CNTs would prove advantageous in manufacturing low-cost, portable, and implantable devices. It is also evident that GO/CNTs with metal nanoparticles/metal oxide are promising materials for fabricating of advanced sensors for BPA detection thereby improving the sensitivity of the assay. Nanotechnology is one of the likely areas to show encouraging prospects for developing sensors for detection of BPA by overcoming the shortcomings of currently-available analytical procedures. Also, justification of sensors and sensing systems is a critical step to ensure accurateness and exactness in quantification of BPA for extensive acceptance and use.

Acknowledgements RK and SAM would like to acknowledge CNPq and FAPESP (Brazil) for financial support.

References

Abbasi Z, Shamsaei E, Leong SK, Ladewig B, Zhang X, Wang H (2016) Effect of carbonization temperature on adsorption property of ZIF-8 derived nanoporous carbon for water treatment. Microporous Mesoporous Mater 236:28–37

Aboubaraka AE, Aboelfetoh EF, Ebeid E-ZM (2017) Coagulation effectiveness of graphene oxide for the removal of turbidity from raw surface water. Chemosphere 181:738–746

Ahamad T, Naushad M, Al-Maswari BM et al (2017) Synthesis of a recyclable mesoporous nanocomposite for efficient removal of toxic Hg^{2+} from aqueous medium. J Ind Eng Chem 53:268–275. https://doi.org/10.1016/j.jiec.2017.04.035

Akhtar J, Amin NAS, Shahzad K (2016) A review on removal of pharmaceuticals from water by adsorption. Desalin Water Treat 57(27):12842–12860

Aleksandrzak M, Kukulka W, Mijowska E (2017) Graphitic carbon nitride/graphene oxide/reduced graphene oxide nanocomposites for photoluminescence and photocatalysis. Appl Surf Sci 398:56–62

Aliabadi M, Shagholani H, Yunessnia lehi A (2017) Synthesis of a novel biocompatible nanocomposite of graphene oxide and magnetic nanoparticles for drug delivery. Int J Biol Macromol 98:287–291

Ambrosi A, Chua CK, Latiff NM, Loo AH, Wong CHA, Eng AYS, Bonanni A, Pumera M (2016) Graphene and its electrochemistry—an update. Chem Soc Rev 45(9):2458–2493

Andra SS, Charisiadis P, Arora M, van Vliet-Ostaptchouk JV, Makris KC (2015) Biomonitoring of human exposures to chlorinated derivatives and structural analogs of bisphenol A. Environ Int 85:352–379

Andrews R, Jacques D, Qian D, Rantell T (2002) Multiwall carbon nanotubes: synthesis and application. Acc Chem Res 35(12):1008–1017

Asada T, Oikawa K, Kawata K, Ishihara S, Iyobe T, Yamada A (2004) Study of removal effect of bisphenol A and β-estradiol by porous carbon. J Health Sci 50(6):588–593
Awasthi S, Awasthi K, Kumar R, Srivastava ON (2009) Functionalization effects on the electrical properties of multi-walled carbon nanotube-polyacrylamide composites. J Nanosci Nanotechnol 9(9):5455–5460
Awasthi K, Kumar R, Tiwari RS, Srivastava ON (2010) Large scale synthesis of bundles of aligned carbon nanotubes using a natural precursor: turpentine oil. J Exp Nanosci 5(6):498–508
Baccar R, Sarrà M, Bouzid J, Feki M, Blánquez P (2012) Removal of pharmaceutical compounds by activated carbon prepared from agricultural by-product. Chem Eng J 211:310–317
Bahamon D, Carro L, Guri S, Vega LF (2017) Computational study of ibuprofen removal from water by adsorption in realistic activated carbons. J Colloid Interface Sci 498:323–334
Bai L, Zhou Z (2007) Computational study of B- or N-doped single-walled carbon nanotubes as NH_3 and NO_2 sensors. Carbon 45(10):2105–2110
Baker SN, Baker GA (2010) Luminescent carbon nanodots: emergent nanolights. Angew Chem Int Ed 49(38):6726–6744
Bautista-Toledo I, Ferro-García MA, Rivera-Utrilla J, Moreno-Castilla C, Vegas Fernández FJ (2005) Bisphenol A removal from water by activated carbon. Effects of carbon characteristics and solution chemistry. Environ Sci Technol 39(16):6246–6250
Bautista-Toledo MI, Rivera-Utrilla J, Ocampo-Pérez R, Carrasco-Marín F, Sánchez-Polo M (2014) Cooperative adsorption of bisphenol-A and chromium(III) ions from water on activated carbons prepared from olive-mill waste. Carbon 73:338–350
Bele S, Samanidou V, Deliyanni E (2016) Effect of the reduction degree of graphene oxide on the adsorption of Bisphenol A. Chem Eng Res Des 109:573–585
Bhadra BN, Seo PW, Jhung SH (2016) Adsorption of diclofenac sodium from water using oxidized activated carbon. Chem Eng J 301:27–34
Bhadra BN, Ahmed I, Kim S, Jhung SH (2017) Adsorptive removal of ibuprofen and diclofenac from water using metal-organic framework-derived porous carbon. Chem Eng J 314:50–58
Bhatnagar A, Anastopoulos I (2017) Adsorptive removal of bisphenol A (BPA) from aqueous solution: a review. Chemosphere 168:885–902
Bi J, Wang Z, Cui Y, Chang J, Lu C (2017) A new type of graphene oxide and its application in laser devices. Opt Mater 66:277–280
Bora T, Sathe P, Laxman K, Dobretsov S, Dutta J (2017) Defect engineered visible light active ZnO nanorods for photocatalytic treatment of water. Catal Today 284:11–18
Cai X, Tan S, Lin M, Xie A, Mai W, Zhang X, Lin Z, Wu T, Liu Y (2011) Synergistic antibacterial brilliant blue/reduced graphene oxide/quaternary phosphonium salt composite with excellent water solubility and specific targeting capability. Langmuir 27(12):7828–7835
Cermakova L, Kopecka I, Pivokonsky M, Pivokonska L, Janda V (2017) Removal of cyanobacterial amino acids in water treatment by activated carbon adsorption. Sep Purif Technol 173:330–338
Chandra V, Park J, Chun Y, Lee JW, Hwang I-C, Kim KS (2010) Water-dispersible magnetite-reduced graphene oxide composites for arsenic removal. ACS Nano 4(7):3979–3986
Chang H-S, Choo K-H, Lee B, Choi S-J (2009) The methods of identification, analysis, and removal of endocrine disrupting compounds (EDCs) in water. J Hazard Mater 172(1):1–12
Chaplin BP (2014) Critical review of electrochemical advanced oxidation processes for water treatment applications. Environ Sci Process Impact 16(6):1182–1203
Chen L, Li Y, Du Q, Wang Z, Xia Y, Yedinak E, Lou J, Ci L (2017) High performance agar/graphene oxide composite aerogel for methylene blue removal. Carbohyd Polym 155:345–353
Cheng C, Deng J, Lei B, He A, Zhang X, Ma L, Li S, Zhao C (2013) Toward 3D graphene oxide gels based adsorbents for high-efficient water treatment via the promotion of biopolymers. J Hazard Mater 263(Part 2):467–478
Choi BG, Park H, Park TJ, Yang MH, Kim JS, Jang S-Y, Heo NS, Lee SY, Kong J, Hong WH (2010) Solution chemistry of self-assembled graphene nanohybrids for high-performance flexible biosensors. ACS Nano 4(5):2910–2918

Chowdhury S, Balasubramanian R (2014) Recent advances in the use of graphene-family nanoadsorbents for removal of toxic pollutants from wastewater. Adv Coll Interface Sci 204:35–56

Cleveland V, Bingham J-P, Kan E (2014) Heterogeneous Fenton degradation of bisphenol A by carbon nanotube-supported Fe_3O_4. Sep Purif Technol 133:388–395

Cui S, Mao S, Lu G, Chen J (2013) Graphene coupled with nanocrystals: opportunities and challenges for energy and sensing applications. J Phys Chem Lett 4(15):2441–2454

Deng X, Lü L, Li H, Luo F (2010) The adsorption properties of Pb(II) and Cd(II) on functionalized graphene prepared by electrolysis method. J Hazard Mater 183(1):923–930

Deng P, Xu Z, Kuang Y (2013) Electrochemically reduced graphene oxide modified acetylene black paste electrode for the sensitive determination of bisphenol A. J Electroanal Chem 707:7–14

Deng P, Xu Z, Kuang Y (2014) Electrochemical determination of bisphenol A in plastic bottled drinking water and canned beverages using a molecularly imprinted chitosan–graphene composite film modified electrode. Food Chem 157:490–497

Derikvandi Z, Abbasi AR, Roushani M, Derikvand Z, Azadbakht A (2016) Design of ultrasensitive bisphenol A–aptamer based on platinum nanoparticles loading to polyethyleneimine-functionalized carbon nanotubes. Anal Biochem 512:47–57

Devadoss A, Lee JW, Terashima C, Fujishima A, Kim Y-P, Kang JK, Paik U (2015) Synergistic oxidation of NADH on bimetallic CoPt nanoparticles decorated carbon nitride nanotubes. Sens Actuators B: Chem 208:204–211

Dhibar S, Bhattacharya P, Ghosh D, Hatui G, Das CK (2014) Graphene–single-walled carbon nanotubes–poly(3-methylthiophene) ternary nanocomposite for supercapacitor electrode materials. Ind Eng Chem Res 53(33):13030–13045

Diamanti-Kandarakis E, Bourguignon J-P, Giudice LC, Hauser R, Prins GS, Soto AM, Zoeller RT, Gore AC (2009) Endocrine-disrupting chemicals: an endocrine society scientific statement. Endocr Rev 30(4):293–342

Dias JM, Alvim-Ferraz MCM, Almeida MF, Rivera-Utrilla J, Sánchez-Polo M (2007) Waste materials for activated carbon preparation and its use in aqueous-phase treatment: a review. J Environ Manage 85(4):833–846

Ding F, Gao Y, He X (2017) Recent progresses in biomedical applications of aptamer-functionalized systems. Bioorg Med Chem Lett

Domínguez JR, González T, Palo P, Cuerda-Correa EM (2011) Removal of common pharmaceuticals present in surface waters by Amberlite XAD-7 acrylic-ester-resin: influence of pH and presence of other drugs. Desalination 269(1–3):231–238

Dong Y, Wu D, Chen X, Lin Y (2010) Adsorption of bisphenol A from water by surfactant-modified zeolite. J Colloid Interface Sci 348(2):585–590

Dong Y, Wu Z-S, Ren W, Cheng H-M, Bao X (2017) Graphene: a promising 2D material for electrochemical energy storage. Sci Bull 62(10):724–740

Du J, Bao J, Liu Y, Ling H, Zheng H, Kim SH, Dionysiou DD (2016) Efficient activation of peroxymonosulfate by magnetic Mn-MGO for degradation of bisphenol A. J Hazard Mater 320:150–159

El-Naas MH, Al-Muhtaseb SA, Makhlouf S (2009) Biodegradation of phenol by Pseudomonas putida immobilized in polyvinyl alcohol (PVA) gel. J Hazard Mater 164(2):720–725

Essandoh M, Kunwar B, Pittman CU, Mohan D, Mlsna T (2015) Sorptive removal of salicylic acid and ibuprofen from aqueous solutions using pine wood fast pyrolysis biochar. Chem Eng J 265:219–227

Fan H, Li Y, Wu D, Ma H, Mao K, Fan D, Du B, Li H, Wei Q (2012) Electrochemical bisphenol A sensor based on N-doped graphene sheets. Anal Chim Acta 711:24–28

Farbod M, Ahangarpour A (2016) Improved thermal conductivity of Ag decorated carbon nanotubes water based nanofluids. Phys Lett A 380(48):4044–4048

Fujimoto Y, Saito S (2016) Gas adsorption, energetics and electronic properties of boron- and nitrogen-doped bilayer graphenes. Chem Phys 478:55–61

Gao Y, Cao Y, Yang D, Luo X, Tang Y, Li H (2012a) Sensitivity and selectivity determination of bisphenol A using SWCNT–CD conjugate modified glassy carbon electrode. J Hazard Mater 199–200:111–118

Gao Y, Li Y, Zhang L, Huang H, Hu J, Shah SM, Su X (2012b) Adsorption and removal of tetracycline antibiotics from aqueous solution by graphene oxide. J Colloid Interface Sci 368 (1):540–546

Geim AK, Novoselov KS (2007) The rise of graphene. Nat Mater 6(3):183–191

Gore AC, Chappell VA, Fenton SE, Flaws JA, Nadal A, Prins GS, Toppari J, Zoeller RT (2015) Executive summary to EDC-2: the endocrine society's second scientific statement on endocrine-disrupting chemicals. Endocr Rev 36(6):593–602

Goulart LA, Mascaro LH (2016) GC electrode modified with carbon nanotubes and NiO for the simultaneous determination of bisphenol A, hydroquinone and catechol. Electrochim Acta 196:48–55

Goulart LA, de Moraes FC, Mascaro LH (2016) Influence of the different carbon nanotubes on the development of electrochemical sensors for bisphenol A. Mater Sci Eng, C 58:768–773

Guedidi H, Reinert L, Lévêque J-M, Soneda Y, Bellakhal N, Duclaux L (2013) The effects of the surface oxidation of activated carbon, the solution pH and the temperature on adsorption of ibuprofen. Carbon 54:432–443

Gupta Chatterjee S, Chatterjee S, Ray AK, Chakraborty AK (2015) Graphene–metal oxide nanohybrids for toxic gas sensor: a review. Sens Actuators B: Chem 221:1170–1181

Hasan Z, Jhung SH (2015) Removal of hazardous organics from water using metal-organic frameworks (MOFs): plausible mechanisms for selective adsorptions. J Hazard Mater 283:329–339

Heo J, Flora JRV, Her N, Park Y-G, Cho J, Son A, Yoon Y (2012) Removal of bisphenol A and 17β-estradiol in single walled carbon nanotubes–ultrafiltration (SWNTs–UF) membrane systems. Sep Purif Technol 90:39–52

Hou K, Huang L, Qi Y, Huang C, Pan H, Du M (2015) A bisphenol A sensor based on novel self-assembly of zinc phthalocyanine tetrasulfonic acid-functionalized graphene nanocomposites. Mater Sci Eng, C 49:640–647

Hrapovic S, Liu Y, Male KB, Luong JHT (2004) Electrochemical biosensing platforms using platinum nanoparticles and carbon nanotubes. Anal Chem 76(4):1083–1088

Hu L-Y, Niu C-G, X-y Wang, Huang D-W, Zhang L, Zeng G-M (2017) Magnetic separate "turn-on" fluorescent biosensor for Bisphenol A based on magnetic oxidation graphene. Talanta 168:196–202

Huang J, Zhang X, Lin Q, He X, Xing X, Huai H, Lian W, Zhu H (2011) Electrochemical sensor based on imprinted sol–gel and nanomaterials for sensitive determination of bisphenol A. Food Control 22(5):786–791

Huang B, Xiao L, Dong H, Zhang X, Gan W, Mahboob S, Al-Ghanim KA, Yuan Q, Li Y (2017) Electrochemical sensing platform based on molecularly imprinted polymer decorated N, S co-doped activated graphene for ultrasensitive and selective determination of cyclophosphamide. Talanta 164:601–607

Hummers WS, Offeman RE (1958) Preparation of graphitic oxide. J Am Chem Soc 80(6):1339

Iijima S (1991) Helical microtubules of graphitic carbon. Nature 354(6348):56–58

Jain U, Chauhan N (2017) Glycated hemoglobin detection with electrochemical sensing amplified by gold nanoparticles embedded N-doped graphene nanosheet. Biosens Bioelectron 89(Part 1):578–584

Jariwala D, Sangwan VK, Lauhon LJ, Marks TJ, Hersam MC (2013) Carbon nanomaterials for electronics, optoelectronics, photovoltaics, and sensing. Chem Soc Rev 42(7):2824–2860

Jauris IM, Matos CF, Saucier C, Lima EC, Zarbin AJG, Fagan SB, Machado FM, Zanella I (2016) Adsorption of sodium diclofenac on graphene: a combined experimental and theoretical study. Phys Chem Chem Phys 18(3):1526–1536

Jiang D, Du X, Chen D, Li Y, Hao N, Qian J, Zhong H, You T, Wang K (2016) Facile wet chemical method for fabricating p-type BiOBr/n-type nitrogen doped graphene composites:

Efficient visible-excited charge separation, and high-performance photoelectrochemical sensing. Carbon 102:10–17

Jiao S, Jin J, Wang L (2014) Tannic acid functionalized N-doped graphene modified glassy carbon electrode for the determination of bisphenol A in food package. Talanta 122:140–144

Jiménez-Marín E, Villalpando I, Trejo-Valdez M, Cervantes-Sodi F, Vargas-García JR, Torres-Torres C (2017) Coexistence of positive and negative photoconductivity in nickel oxide decorated multiwall carbon nanotubes. Mater Sci Eng, B 220:22–29

Joseph L, Zaib Q, Khan IA, Berge ND, Park Y-G, Saleh NB, Yoon Y (2011) Removal of bisphenol A and 17α-ethinyl estradiol from landfill leachate using single-walled carbon nanotubes. Water Res 45(13):4056–4068

Joseph L, Boateng LK, Flora JRV, Park Y-G, Son A, Badawy M, Yoon Y (2013) Removal of bisphenol A and 17α-ethinyl estradiol by combined coagulation and adsorption using carbon nanomaterials and powdered activated carbon. Sep Purif Technol 107:37–47

Kaiser AB, Skakalova V (2011) Electronic conduction in polymers, carbon nanotubes and graphene. Chem Soc Rev 40(7):3786–3801

Kang J-H, Kondo F (2002) Effects of bacterial counts and temperature on the biodegradation of bisphenol A in river water. Chemosphere 49(5):493–498

Kang J-H, Kondo F (2005) Bisphenol A degradation in seawater is different from that in river water. Chemosphere 60(9):1288–1292

Karousis N, Tagmatarchis N, Tasis D (2010) Current progress on the chemical modification of carbon nanotubes. Chem Rev 110(9):5366–5397

Kaur M, Mehta SK, Kansal SK (2017) A fluorescent probe based on nitrogen doped graphene quantum dots for turn off sensing of explosive and detrimental water pollutant, TNP in aqueous medium. Spectrochim Acta Part A Mol Biomol Spectrosc 180:37–43

Kim Y-H, Lee B, Choo K-H, Choi S-J (2011) Selective adsorption of bisphenol A by organic–inorganic hybrid mesoporous silicas. Microporous Mesoporous Mater 138(1):184–190

Kim KS, J-r Jang, Choe W-S, Yoo PJ (2015) Electrochemical detection of Bisphenol A with high sensitivity and selectivity using recombinant protein-immobilized graphene electrodes. Biosens Bioelectron 71:214–221

Koubaissy B, Toufaily J, Kafrouny L, Joly G, Magnoux P, Hamieh T (2011) Industrial water treatment, by adsorption, using organized mesoporous materials. Phys Procedia 21:228–233

Krajišnik D, Daković A, Milojević M, Malenović A, Kragović M, Bogdanović DB, Dondur V, Milić J (2011) Properties of diclofenac sodium sorption onto natural zeolite modified with cetylpyridinium chloride. Colloids Surf, B 83(1):165–172

Kulkarni S, Junghare S (2013) Robot based indoor autonomous trash detection algorithm using ultrasonic sensors, 16–18 Dec 2013, pp 1–5

Kumar R, Tiwari RS, Srivastava ON (2011a) Scalable synthesis of aligned carbon nanotubes bundles using green natural precursor: neem oil. Nanoscale Res Lett 6(1):92

Kumar R, Yadav RM, Awasthi K, Tiwari RS, Srivastava ON (2011b) Effect of nitrogen variation on the synthesis of vertically aligned bamboo-shaped C-N nanotubes using sunflower oil. Int J Nanosci 10(04n05):809–813

Kumar R, Singh RK, Dubey PK, Kumar P, Tiwari RS, Oh I-K (2013a) Pressure-dependent synthesis of high-quality few-layer graphene by plasma-enhanced arc discharge and their thermal stability. J Nanopart Res 15(9):1847

Kumar R, Yadav RM, Awasthi K, Shripathi T, Sinha ASK, Tiwari RS, Srivastava ON (2013b) Synthesis of carbon and carbon–nitrogen nanotubes using green precursor: jatropha-derived biodiesel. J Exp Nanosci 8(4):606–620

Kumar R, Singh RK, Kumar P, Dubey PK, Tiwari RS, Srivastava ON (2014) Clean and efficient synthesis of graphene nanosheets and rectangular aligned-carbon nanotubes bundles using green botanical hydrocarbon precursor: sesame oil. Sci Adv Mater 6(1):76–83

Kumar R, Singh RK, Dubey PK, Singh DP, Yadav RM (2015a) Self-assembled hierarchical formation of conjugated 3D cobalt oxide nanobead–CNT–graphene nanostructure using microwaves for high-performance supercapacitor electrode. ACS Appl Mater Interfaces 7 (27):15042–15051

Kumar R, Singh RK, Vaz AR, Moshkalev SA (2015b) Microwave-assisted synthesis and deposition of a thin ZnO layer on microwave-exfoliated graphene: optical and electrochemical evaluations. RSC Adv 5(83):67988–67995

Kumar R, Dubey PK, Singh RK, Vaz AR, Moshkalev SA (2016a) Catalyst-free synthesis of a three-dimensional nanoworm-like gallium oxide-graphene nanosheet hybrid structure with enhanced optical properties. RSC Adv 6(21):17669–17677

Kumar R, Kim H-J, Oh I-K (2016b) Bio-inspired engineering of 3D carbon nanostructures. In: Zhang M, Naik RR, Dai L (eds) Carbon nanomaterials for biomedical applications. Springer International Publishing, Cham, pp 365–420

Kumar A, Guo C, Sharma G et al (2016c) Magnetically recoverable ZrO2/Fe3O4/chitosan nanomaterials for enhanced sunlight driven photoreduction of carcinogenic Cr(VI) and dechlorination & mineralization of 4-chlorophenol from simulated waste water. RSC Adv 6:13251–13263. https://doi.org/10.1039/C5RA23372K

Kumar R, Singh RK, Savu R, Dubey PK, Kumar P, Moshkalev SA (2016d) Microwave-assisted synthesis of void-induced graphene-wrapped nickel oxide hybrids for supercapacitor applications. RSC Adv 6(32):26612–26620

Kumar R, Singh RK, Singh DP (2016e) Natural and waste hydrocarbon precursors for the synthesis of carbon based nanomaterials: Graphene and CNTs. Renew Sustain Energy Rev 58:976–1006

Kumar R, Singh RK, Singh DP, Savu R, Moshkalev SA (2016f) Microwave heating time dependent synthesis of various dimensional graphene oxide supported hierarchical ZnO nanostructures and its photoluminescence studies. Mater Des 111:291–300

Kumar R, Savu R, Singh RK, Joanni E, Singh DP, Tiwari VS, Vaz AR, da Silva ETSG, Maluta JR, Kubota LT, Moshkalev SA (2017a) Controlled density of defects assisted perforated structure in reduced graphene oxide nanosheets-palladium hybrids for enhanced ethanol electro-oxidation. Carbon 117:137–146

Kumar R, Singh RK, Singh AK, Vaz AR, Rout CS, Moshkalev SA (2017b) Facile and single step synthesis of three dimensional reduced graphene oxide-$NiCoO_2$ composite using microwave for enhanced electron field emission properties. Appl Surf Sci 416:259–265

Kumar R, Singh RK, Singh DP, Joanni E, Yadav RM, Moshkalev SA (2017c) Laser-assisted synthesis, reduction and micro-patterning of graphene: recent progress and applications. Coord Chem Rev 342:34–79

Kumar R, Singh RK, Singh DP, Vaz AR, Yadav RR, Rout CS, Moshkalev SA (2017d) Synthesis of self-assembled and hierarchical palladium-CNTs-reduced graphene oxide composites for enhanced field emission properties. Mater Des 122:110–117

Kumar R, Singh RK, Tiwari VS, Yadav A, Savu R, Vaz AR, Moshkalev SA (2017e) Enhanced magnetic performance of iron oxide nanoparticles anchored pristine/N-doped multi-walled carbon nanotubes by microwave-assisted approach. J Alloy Compd 695:1793–1801

Kumar R, Singh RK, Vaz AR, Savu R, Moshkalev SA (2017f) Self-assembled and one-step synthesis of interconnected 3D network of Fe_3O_4/reduced graphene oxide nanosheets hybrid for high-performance supercapacitor electrode. ACS Appl Mater Interfaces 9(10):8880–8890

Kuo C-Y (2009) Comparison with as-grown and microwave modified carbon nanotubes to removal aqueous bisphenol A. Desalination 249(3):976–982

Lawal AT (2016) Synthesis and utilization of carbon nanotubes for fabrication of electrochemical biosensors. Mater Res Bull 73:308–350

Lee W-H, Moon JH (2014) Monodispersed N-doped carbon nanospheres for supercapacitor application. ACS Appl Mater Interfaces 6(16):13968–13976

Lerman I, Chen Y, Xing B, Chefetz B (2013) Adsorption of carbamazepine by carbon nanotubes: effects of DOM introduction and competition with phenanthrene and bisphenol A. Environ Pollut 182:169–176

Li J, Lee E-C (2017) Functionalized multi-wall carbon nanotubes as an efficient additive for electrochemical DNA sensor. Sens Actuators B: Chem 239:652–659

Li H, Misra A, Horita Z, Koch CC, Mara NA, Dickerson PO, Zhu Y (2009) Strong and ductile nanostructured Cu-carbon nanotube composite. Appl Phys Lett 95(7):071907

Li Y, Zhang P, Du Q, Peng X, Liu T, Wang Z, Xia Y, Zhang W, Wang K, Zhu H, Wu D (2011) Adsorption of fluoride from aqueous solution by graphene. J Colloid Interface Sci 363(1): 348–354

Li F, Jiang X, Zhao J, Zhang S (2015a) Graphene oxide: a promising nanomaterial for energy and environmental applications. Nano Energy 16:488–515

Li G, Lu Y, Lu C, Zhu M, Zhai C, Du Y, Yang P (2015b) Efficient catalytic ozonation of bisphenol-A over reduced graphene oxide modified sea urchin-like α-MnO_2 architectures. J Hazard Mater 294:201–208

Li L, Yu B, Zhang X, You T (2015c) A novel electrochemiluminescence sensor based on Ru $(bpy)_3^{2+}$/N-doped carbon nanodots system for the detection of bisphenol A. Anal Chim Acta 895:104–111

Li X, Rui M, Song J, Shen Z, Zeng H (2015d) Carbon and graphene quantum dots for optoelectronic and energy devices: a review. Adv Func Mater 25(31):4929–4947

Li Z, Liu Z, Sun H, Gao C (2015e) Superstructured assembly of nanocarbons: fullerenes, nanotubes, and graphene. Chem Rev 115(15):7046–7117

Li Z, Zhang W, Guo J, Yang B, Yuan J (2015f) Improved synthesis of fluffy and wrinkled reduced graphene oxide for energy storage application. Vacuum 117:35–39

Li H, Wang W, Lv Q, Xi G, Bai H, Zhang Q (2016) Disposable paper-based electrochemical sensor based on stacked gold nanoparticles supported carbon nanotubes for the determination of bisphenol A. Electrochem Commun 68:104–107

Li M, Yu C, Hu C, Yang W, Zhao C, Wang S, Zhang M, Zhao J, Wang X, Qiu J (2017a) Solvothermal conversion of coal into nitrogen-doped carbon dots with singlet oxygen generation and high quantum yield. Chem Eng J 320:570–575

Li X, Zhao H, Shi L, Zhu X, Lan M, Zhang Q, Hugh Fan Z (2017b) Electrochemical sensing of nicotine using screen-printed carbon electrodes modified with nitrogen-doped graphene sheets. J Electroanal Chem 784:77–84

Libbrecht W, Verberckmoes A, Thybaut JW, Van Der Voort P, De Clercq J (2017) Soft templated mesoporous carbons: tuning the porosity for the adsorption of large organic pollutants. Carbon 116:528–546

Lim SY, Shen W, Gao Z (2015) Carbon quantum dots and their applications. Chem Soc Rev 44 (1):362–381

Lin Y, Taylor S, Li H, Fernando KAS, Qu L, Wang W, Gu L, Zhou B, Sun Y-P (2004) Advances toward bioapplications of carbon nanotubes. J Mater Chem 14(4):527–541

Lin Y, Liu K, Liu C, Yin L, Kang Q, Li L, Li B (2014) Electrochemical sensing of bisphenol A based on polyglutamic acid/amino-functionalised carbon nanotubes nanocomposite. Electrochim Acta 133:492–500

Liu B, Shioyama H, Akita T, Xu Q (2008) Metal-organic framework as a template for porous carbon synthesis. J Am Chem Soc 130(16):5390–5391

Liu G, Ma J, Li X, Qin Q (2009) Adsorption of bisphenol A from aqueous solution onto activated carbons with different modification treatments. J Hazard Mater 164(2):1275–1280

Liu T, Li Y, Du Q, Sun J, Jiao Y, Yang G, Wang Z, Xia Y, Zhang W, Wang K, Zhu H, Wu D (2012) Adsorption of methylene blue from aqueous solution by graphene. Colloids Surf, B 90:197–203

Lotfi E, Neek-Amal M (2017) Temperature distribution in graphene doped with nitrogen and graphene with grain boundary. J Mol Graph Model 74:100–104

Luo H, Li C, Wu C, Zheng W, Dong X (2015) Electrochemical degradation of phenol by in situ electro-generated and electro-activated hydrogen peroxide using an improved gas diffusion cathode. Electrochim Acta 186:486–493

Luo H, Ao H, Li G, Li W, Xiong G, Zhu Y, Wan Y (2017) Bacterial cellulose/graphene oxide nanocomposite as a novel drug delivery system. Curr Appl Phys 17(2):249–254

Malato S, Fernández-Ibáñez P, Maldonado MI, Blanco J, Gernjak W (2009) Decontamination and disinfection of water by solar photocatalysis: recent overview and trends. Catal Today 147 (1):1–59

Mallakpour S, Khadem E (2016) Carbon nanotube–metal oxide nanocomposites: Fabrication, properties and applications. Chem Eng J 302:344–367
Mantzavinos D, Kalogerakis N (2005) Treatment of olive mill effluents. Environ Int 31(2): 289–295
Marrot B, Barrios-Martinez A, Moulin P, Roche N (2006) Biodegradation of high phenol concentration by activated sludge in an immersed membrane bioreactor. Biochem Eng J 30 (2):174–183
Melillo M, Gun'ko VM, Tennison SR, Mikhalovska LI, Phillips GJ, Davies JG, Lloyd AW, Kozynchenko OP, Malik DJ, Streat M, Mikhalovsky SV (2004) Structural characteristics of activated carbons and ibuprofen adsorption affected by bovine serum albumin. Langmuir 20 (7):2837–2851
Mestre AS, Pires J, Nogueira JMF, Carvalho AP (2007) Activated carbons for the adsorption of ibuprofen. Carbon 45(10):1979–1988
Mestre AS, Pires J, Nogueira JMF, Parra JB, Carvalho AP, Ania CO (2009) Waste-derived activated carbons for removal of ibuprofen from solution: role of surface chemistry and pore structure. Biores Technol 100(5):1720–1726
Miners SA, Rance GA, Khlobystov AN (2016) Chemical reactions confined within carbon nanotubes. Chem Soc Rev 45(17):4727–4746
Mohammadi S, Mirghaffari N (2015) A preliminary study of the preparation of porous carbon from oil sludge for water treatment by simple pyrolysis or KOH activation. New Carbon Mater 30(4):310–318
Moraes FC, Silva TA, Cesarino I, Machado SAS (2013) Effect of the surface organization with carbon nanotubes on the electrochemical detection of bisphenol A. Sens Actuators B: Chem 177:14–18
Moreno-Castilla C (2004) Adsorption of organic molecules from aqueous solutions on carbon materials. Carbon 42(1):83–94
Moreno-Castilla C, Rivera-Utrilla J (2012) Carbon materials as adsorbents for the removal of pollutants from the aqueous phase. MRS Bull 26(11):890–894
Mousset E, Frunzo L, Esposito G, Hullebusch EDV, Oturan N, Oturan MA (2016) A complete phenol oxidation pathway obtained during electro-Fenton treatment and validated by a kinetic model study. Appl Catal B: Environ 180:189–198
Ndlovu T, Arotiba OA, Sampath S, Krause RW, Mamba BB (2012) An exfoliated graphite-based bisphenol A electrochemical sensor. Sensors 12(9):11601
Nerger BA, Peiris RH, Moresoli C (2015) Fluorescence analysis of NOM degradation by photocatalytic oxidation and its potential to mitigate membrane fouling in drinking water treatment. Chemosphere 136:140–144
Niu X, Yang W, Wang G, Ren J, Guo H, Gao J (2013) A novel electrochemical sensor of bisphenol A based on stacked graphene nanofibers/gold nanoparticles composite modified glassy carbon electrode. Electrochim Acta 98:167–175
Nowakowska M, Szczubiałka K (2017) Photoactive polymeric and hybrid systems for photocatalytic degradation of water pollutants. Polym Degrad Stab
Ntsendwana B, Mamba BB, Sampath S, Arotiba OA (2012) Electrochemical detection of bisphenol a using graphene-modified glassy carbon electrode. Int J Electrochem Sci 7(4): 3501–3512
Odukkathil G, Vasudevan N (2013) Toxicity and bioremediation of pesticides in agricultural soil. Rev Environ Sci Bio/Technol 12(4):421–444
Ong YT, Ahmad AL, Zein SHS, Tan SH (2010) A review on carbon nanotubes in an environmental protection and green engineering perspective. Braz J Chem Eng 27:227–242
Oribayo O, Feng X, Rempel GL, Pan Q (2017) Synthesis of lignin-based polyurethane/graphene oxide foam and its application as an absorbent for oil spill clean-ups and recovery. Chem Eng J 323:191–202
Ortiz-Martínez K, Reddy P, Cabrera-Lafaurie WA, Román FR, Hernández-Maldonado AJ (2016) Single and multi-component adsorptive removal of bisphenol A and 2,4-dichlorophenol from

aqueous solutions with transition metal modified inorganic–organic pillared clay composites: effect of pH and presence of humic acid. J Hazard Mater 312:262–271

Paek J, Kim J, Wan An B, Park J, Ji S, Kim S-Y, Jang J, Lee Y, Park Y-G, Cho E, Jo S, Ju S, Hyung Cheong W, Park J-U (2017) Stretchable electronic devices using graphene and its hybrid nanostructures. FlatChem

Pan B, Lin D, Mashayekhi H, Xing B (2008) Adsorption and hysteresis of bisphenol A and 17α-ethinyl estradiol on carbon nanomaterials. Environ Sci Technol 42(15):5480–5485

Pan D, Gu Y, Lan H, Sun Y, Gao H (2015) Functional graphene-gold nano-composite fabricated electrochemical biosensor for direct and rapid detection of bisphenol A. Anal Chim Acta 853:297–302

Park KS, Ni Z, Côté AP, Choi JY, Huang R, Uribe-Romo FJ, Chae HK, O'Keeffe M, Yaghi OM (2006) Exceptional chemical and thermal stability of zeolitic imidazolate frameworks. Proc Natl Acad Sci 103(27):10186–10191

Park J-A, Jung S-M, Yi I-G, Choi J-W, Kim S-B, Lee S-H (2017) Adsorption of microcystin-LR on mesoporous carbons and its potential use in drinking water source. Chemosphere 177:15–23

Paul B, Parashar V, Mishra A (2015) Graphene in the Fe_3O_4 nano-composite switching the negative influence of humic acid coating into an enhancing effect in the removal of arsenic from water. Environ Sci: Water Res Technol 1(1):77–83

Pei Z, Li L, Sun L, Zhang S, X-q Shan, Yang S, Wen B (2013) Adsorption characteristics of 1,2,4-trichlorobenzene, 2,4,6-trichlorophenol, 2-naphthol and naphthalene on graphene and graphene oxide. Carbon 51:156–163

Peiris BRH, Hallé C, Haberkamp J, Legge RL, Peldszus S, Moresoli C, Budman H, Amy G, Jekel M, Huck PM (2008) Assessing nanofiltration fouling in drinking water treatment using fluorescence fingerprinting and LC-OCD analyses. Water Sci Technol: Water Supply 8(4):459

Peng X, Chen J, Misewich JA, Wong SS (2009) Carbon nanotube-nanocrystal heterostructures. Chem Soc Rev 38(4):1076–1098

Perez EM, Martin N (2015) [small pi]-[small pi] interactions in carbon nanostructures. Chem Soc Rev 44(18):6425–6433

Perreault F, Fonseca de Faria A, Elimelech M (2015) Environmental applications of graphene-based nanomaterials. Chem Soc Rev 44(16):5861–5896

Pintor AMA, Vilar VJP, Boaventura RAR (2011) Decontamination of cork wastewaters by solar-photo-Fenton process using cork bleaching wastewater as H_2O_2 source. Sol Energy 85 (3):579–587

Poh HL, Sanek F, Ambrosi A, Zhao G, Sofer Z, Pumera M (2012) Graphenes prepared by Staudenmaier, Hofmann and Hummers methods with consequent thermal exfoliation exhibit very different electrochemical properties. Nanoscale 4(11):3515–3522

Qi P, Vermesh O, Grecu M, Javey A, Wang Q, Dai H, Peng S, Cho KJ (2003) Toward large arrays of multiplex functionalized carbon nanotube sensors for highly sensitive and selective molecular detection. Nano Lett 3(3):347–351

Ragavan KV, Rastogi NK (2016) Graphene–copper oxide nanocomposite with intrinsic peroxidase activity for enhancement of chemiluminescence signals and its application for detection of Bisphenol-A. Sens Actuators B: Chem 229:570–580

Ragavan KV, Rastogi NK, Thakur MS (2013) Sensors and biosensors for analysis of bisphenol-A. TrAC Trends Anal Chem 52:248–260

Rajarao R, Jayanna RP, Sahajwalla V, Bhat BR (2014) GreenApproach to decorate multi-walled carbon nanotubes by metal/metal oxide nanoparticles. Procedia Mater Sci 5:69–75

Ramesha GK, Vijaya Kumara A, Muralidhara HB, Sampath S (2011) Graphene and graphene oxide as effective adsorbents toward anionic and cationic dyes. J Colloid Interface Sci 361 (1):270–277

Ray SC, Jana NR (2017) Application of carbon-based nanomaterials for removal of biologically toxic materials (Chap. 2). In: Carbon nanomaterials for biological and medical applications. Elsevier, pp 43–86

Ren X, Chen C, Nagatsu M, Wang X (2011) Carbon nanotubes as adsorbents in environmental pollution management: a review. Chem Eng J 170(2):395–410

Rivera-Utrilla J, Sánchez-Polo M, Gómez-Serrano V, Álvarez PM, Alvim-Ferraz MCM, Dias JM (2011) Activated carbon modifications to enhance its water treatment applications. An overview. J Hazard Mater 187(1):1–23
Roostaei N, Tezel FH (2004) Removal of phenol from aqueous solutions by adsorption. J Environ Manag 70(2):157–164
Saien J, Nejati H (2007) Enhanced photocatalytic degradation of pollutants in petroleum refinery wastewater under mild conditions. J Hazard Mater 148(1):491–495
Sakhaee-Pour A, Ahmadian MT, Vafai A (2008) Potential application of single-layered graphene sheet as strain sensor. Solid State Commun 147(7):336–340
Sampaio MJ, Lima MJ, Baptista DL, Silva AMT, Silva CG, Faria JL (2017) Ag-loaded ZnO materials for photocatalytic water treatment. Chem Eng J 318:95–102
Santana ER, de Lima CA, Piovesan JV, Spinelli A (2017) An original ferroferric oxide and gold nanoparticles-modified glassy carbon electrode for the determination of bisphenol A. Sens Actuators B: Chem 240:487–496
Santhi VA, Sakai N, Ahmad ED, Mustafa AM (2012) Occurrence of bisphenol A in surface water, drinking water and plasma from Malaysia with exposure assessment from consumption of drinking water. Sci Total Environ 427:332–338
Sapurina I, Stejskal J, Šeděnková I, Trchová M, Kovářová J, Hromádková J, Kopecká J, Cieslar M, Abu El-Nasr A, Ayad MM (2016) Catalytic activity of polypyrrole nanotubes decorated with noble-metal nanoparticles and their conversion to carbonized analogues. Synth Met 214:14–22
Schedin F, Geim AK, Morozov SV, Hill EW, Blake P, Katsnelson MI, Novoselov KS (2007) Detection of individual gas molecules adsorbed on graphene. Nat Mater 6(9):652–655
Sharma G, Pathania D, Naushad M, Kothiyal NC (2014) Fabrication, characterization and antimicrobial activity of polyaniline Th(IV) tungstomolybdophosphate nanocomposite material: Efficient removal of toxic metal ions from water. Chem Eng J (Amsterdam, Netherlands) 251:413–421. https://doi.org/10.1016/j.cej.2014.04.074
Shen R, Zhang W, Yuan Y, He G, Chen H (2015) Electrochemical detection of bisphenol A at graphene/melamine nanoparticle-modified glassy carbon electrode. J Appl Electrochem 45 (4):343–352
Sheng G, Li Y, Yang X, Ren X, Yang S, Hu J, Wang X (2012) Efficient removal of arsenate by versatile magnetic graphene oxide composites. RSC Adv 2(32):12400–12407
Shi R, Liang J, Zhao Z, Liu A, Tian Y (2017) An electrochemical bisphenol A sensor based on one step electrochemical reduction of cuprous oxide wrapped graphene oxide nanoparticles modified electrode. Talanta 169:37–43
Si Y, Ren T, Li Y, Ding B, Yu J (2012) Fabrication of magnetic polybenzoxazine-based carbon nanofibers with Fe_3O_4 inclusions with a hierarchical porous structure for water treatment. Carbon 50(14):5176–5185
Simmons TJ, Rivet CJ, Singh G, Beaudet J, Sterner E, Guzman D, Hashim DP, Lee S-H, Qian G, Lewis KM, Karande P, Ajayan PM, Gilbert RJ, Dordick JS, Linhardt RJ (2012) Application of carbon nanotubes to wound healing biotechnology. In: Nanomaterials for biomedicine. American Chemical Society, pp 155–174
Singh P, Campidelli S, Giordani S, Bonifazi D, Bianco A, Prato M (2009) Organic functionalisation and characterisation of single-walled carbon nanotubes. Chem Soc Rev 38 (8):2214–2230
Singh V, Joung D, Zhai L, Das S, Khondaker SI, Seal S (2011) Graphene based materials: past, present and future. Prog Mater Sci 56(8):1178–1271
Singh RK, Kumar R, Singh DP (2016) Graphene oxide: strategies for synthesis, reduction and frontier applications. RSC Adv 6(69):64993–65011
Song M-M, Xu H-L, Liang J-X, Xiang H-H, Liu R, Shen Y-X (2017) Lactoferrin modified graphene oxide iron oxide nanocomposite for glioma-targeted drug delivery. Mater Sci Eng, C 77:904–911

Srinivas G, Krungleviciute V, Guo Z-X, Yildirim T (2014) Exceptional CO_2 capture in a hierarchically porous carbon with simultaneous high surface area and pore volume. Energy Environ Sci 7(1):335–342
Staples CA, Dome PB, Klecka GM, Oblock ST, Harris LR (1998) A review of the environmental fate, effects, and exposures of bisphenol A. Chemosphere 36(10):2149–2173
Staudenmaier L (1898) Verfahren zur Darstellung der Graphitsäure. Ber Dtsch Chem Ges 31 (2):1481–1487
Su B, Shao H, Li N, Chen X, Cai Z, Chen X (2017) A sensitive bisphenol A voltammetric sensor relying on AuPd nanoparticles/graphene composites modified glassy carbon electrode. Talanta 166:126–132
Suárez-Iglesias O, Collado S, Oulego P, Díaz M (2017) Graphene-family nanomaterials in wastewater treatment plants. Chem Eng J 313:121–135
Subramaniam C, Yamada T, Kobashi K, Sekiguchi A, Futaba DN, Yumura M, Hata K (2013) One hundred fold increase in current carrying capacity in a carbon nanotube–copper composite. Nat Commun 4:2202
Sun JK, Xu Q (2014) Functional materials derived from open framework templates/precursors: synthesis and applications. Energy Environ Sci 7(7):2071–2100
Sun W, Lipka SM, Swartz C, Williams D, Yang F (2016) Hemp-derived activated carbons for supercapacitors. Carbon 103:181–192
Suriyanon N, Punyapalakul P, Ngamcharussrivichai C (2013) Mechanistic study of diclofenac and carbamazepine adsorption on functionalized silica-based porous materials. Chem Eng J 214:208–218
Tang S, Wu W, Yu J (2016) Interfacial interaction of Ag nanoparticles with graphene oxide supports for improving NH_3 and NO adsorption: a first-principles study. Phys Chem Chem Phys 18(11):7797–7807
Ternes TA, Stumpf M, Mueller J, Haberer K, Wilken RD, Servos M (1999) Behavior and occurrence of estrogens in municipal sewage treatment plants—I. Investigations in Germany, Canada and Brazil. Sci Total Environ 225(1):81–90
Travlou NA, Seredych M, Rodríguez-Castellón E, Bandosz TJ (2016) Insight into ammonia sensing on heterogeneous S- and N- co-doped nanoporous carbons. Carbon 96:1014–1021
Tsoumachidou S, Velegraki T, Antoniadis A, Poulios I (2017) Greywater as a sustainable water source: A photocatalytic treatment technology under artificial and solar illumination. J Environ Manag 195(Part 2):232–241
Tu X, Yan L, Luo X, Luo S, Xie Q (2009) Electroanalysis of bisphenol a at a multiwalled carbon nanotubes-gold nanoparticles modified glassy carbon electrode. Electroanalysis 21(22):2491–2494
Vandenberg LN, Hauser R, Marcus M, Olea N, Welshons WV (2007) Human exposure to bisphenol A (BPA). Reprod Toxicol 24(2):139–177
Varaprasad K, Jayaramudu T, Sadiku ER (2017) Removal of dye by carboxymethyl cellulose, acrylamide and graphene oxide via a free radical polymerization process. Carbohyd Polym 164:186–194
Vashist SK, Zheng D, Al-Rubeaan K, Luong JHT, Sheu F-S (2011) Advances in carbon nanotube based electrochemical sensors for bioanalytical applications. Biotechnol Adv 29(2):169–188
Vega D, Agüí L, González-Cortés A, Yáñez-Sedeño P, Pingarrón JM (2007) Electrochemical detection of phenolic estrogenic compounds at carbon nanotube-modified electrodes. Talanta 71(3):1031–1038
Wang X, Deng C (2015) Preparation of magnetic graphene @polydopamine @Zr-MOF material for the extraction and analysis of bisphenols in water samples. Talanta 144:1329–1335
Wang C, Takei K, Takahashi T, Javey A (2013a) Carbon nanotube electronics—moving forward. Chem Soc Rev 42(7):2592–2609
Wang H, Yuan X, Wu Y, Huang H, Peng X, Zeng G, Zhong H, Liang J, Ren M (2013b) Graphene-based materials: Fabrication, characterization and application for the decontamination of wastewater and wastegas and hydrogen storage/generation. Adv Coll Interface Sci 195:19–40

Wang F, Haftka JJH, Sinnige TL, Hermens JLM, Chen W (2014) Adsorption of polar, nonpolar, and substituted aromatics to colloidal graphene oxide nanoparticles. Environ Pollut 186:226–233

Wang X, Qin Y, Zhu L, Tang H (2015) Nitrogen-doped reduced graphene oxide as a bifunctional material for removing bisphenols: synergistic effect between adsorption and catalysis. Environ Sci Technol 49(11):6855–6864

Wang DK, Elma M, Motuzas J, Hou W-C, Schmeda-Lopez DR, Zhang T, Zhang X (2016a) Physicochemical and photocatalytic properties of carbonaceous char and titania composite hollow fibers for wastewater treatment. Carbon 109:182–191

Wang L, Zhang Z, Zhang J, Zhang L (2016b) Magnetic solid-phase extraction using nanoporous three dimensional graphene hybrid materials for high-capacity enrichment and simultaneous detection of nine bisphenol analogs from water sample. J Chromatogr A 1463:1–10

Wang M, Xu X, Liu Y, Li Y, Lu T, Pan L (2016c) From metal-organic frameworks to porous carbons: a promising strategy to prepare high-performance electrode materials for capacitive deionization. Carbon 108:433–439

Wang R-Z, Huang D-L, Liu Y-G, Peng Z-W, Zeng G-M, Lai C, Xu P, Huang C, Zhang C, Gong X-M (2016d) Selective removal of BPA from aqueous solution using molecularly imprinted polymers based on magnetic graphene oxide. RSC Adv 6(108):106201–106210

Wang G, Chen S, Quan X, Yu H, Zhang Y (2017a) Enhanced activation of peroxymonosulfate by nitrogen doped porous carbon for effective removal of organic pollutants. Carbon 115:730–739

Wang H, Feng B, Ye Y, Guo J, Fang H-T (2017b) Tuning inner-layer oxygen functional groups of reduced graphene oxide by potentiostatic oxidation for high performance electrochemical energy storage devices. Electrochim Acta 240:122–128

Wang Y, Zhu M, Li Y, Zhang M, Xue X, Shi Y, Dai B, Guo X, Yu F (2017) Heteroatom-doped porous carbon from methyl orange dye wastewater for oxygen reduction. Green Energy Environ

Wehling TO, Novoselov KS, Morozov SV, Vdovin EE, Katsnelson MI, Geim AK, Lichtenstein AI (2008) Molecular doping of graphene. Nano Lett 8(1):173–177

Wu C, Sun D, Li Q, Wu K (2012) Electrochemical sensor for toxic ractopamine and clenbuterol based on the enhancement effect of graphene oxide. Sens Actuators B: Chem 168:178–184

Wu W, Huang Z-H, Hu Z-T, He C, Lim T-T (2017) High performance duplex-structured SnO_2-Sb-CNT composite anode for bisphenol A removal. Sep Purif Technol 179:25–35

Xin X, Sun S, Li H, Wang M, Jia R (2015) Electrochemical bisphenol A sensor based on core–shell multiwalled carbon nanotubes/graphene oxide nanoribbons. Sens Actuators B: Chem 209:275–280

Xu J, Wang L, Zhu Y (2012) Decontamination of bisphenol A from aqueous solution by graphene adsorption. Langmuir 28(22):8418–8425

Xu X, Gao J, Tian Q, Zhai X, Liu Y (2017) Walnut shell derived porous carbon for a symmetric all-solid-state supercapacitor. Appl Surf Sci 411:170–176

Yamamoto T, Yasuhara A, Shiraishi H, Nakasugi O (2001) Bisphenol A in hazardous waste landfill leachates. Chemosphere 42(4):415–418

Yan Y, Miao J, Yang Z, Xiao F-X, Yang HB, Liu B, Yang Y (2015) Carbon nanotube catalysts: recent advances in synthesis, characterization and applications. Chem Soc Rev 44(10):3295–3346

Yang S, Liang J, Luo S, Liu C, Tang Y (2013a) Supersensitive detection of chlorinated phenols by multiple amplification electrochemiluminescence sensing based on carbon quantum dots/graphene. Anal Chem 85(16):7720–7725

Yang X, Li J, Wen T, Ren X, Huang Y, Wang X (2013b) Adsorption of naphthalene and its derivatives on magnetic graphene composites and the mechanism investigation. Colloids Surf, A 422:118–125

Yang S, Li G, Liu L, Wang G, Wang D, Qu L (2017) Preparation of nickel oxide nanoparticles on N-doped reduced graphene oxide: a two-dimensional hybrid for electrocatalytic sensing of l-cysteine. J Alloy Compd 691:834–840

Ye Y, Guo T (2013) Improvement of the field emission of carbon nanotubes-metal nanocomposite. J Mater Sci: Mater Electron 24(6):1775–1781

Yeo H, Jung J, Song HJ, Choi Y-M, Wee J-H, You N-H, Joh H-I, Yang C-M, Goh M (2017) Preparation and formation mechanism of porous carbon cryogel. Microporous Mesoporous Mater 245:138–146

Yin H, Zhou Y, Xu J, Ai S, Cui L, Zhu L (2010) Amperometric biosensor based on tyrosinase immobilized onto multiwalled carbon nanotubes-cobalt phthalocyanine-silk fibroin film and its application to determine bisphenol A. Anal Chim Acta 659(1–2):144–150

Yoon Y, Zheng M, Ahn Y-T, Park WK, Yang WS, Kang J-W (2017) Synthesis of magnetite/non-oxidative graphene composites and their application for arsenic removal. Sep Purif Technol 178:40–48

Yuan F, Li S, Fan Z, Meng X, Fan L, Yang S (2016) Shining carbon dots: synthesis and biomedical and optoelectronic applications. Nano Today 11(5):565–586

Zambianchi M, Durso M, Liscio A, Treossi E, Bettini C, Capobianco ML, Aluigi A, Kovtun A, Ruani G, Corticelli F, Brucale M, Palermo V, Navacchia ML, Melucci M (2017) Graphene oxide doped polysulfone membrane adsorbers for the removal of organic contaminants from water. Chem Eng J 326:130–140

Zehani N, Fortgang P, Saddek Lachgar M, Baraket A, Arab M, Dzyadevych SV, Kherrat R, Jaffrezic-Renault N (2015) Highly sensitive electrochemical biosensor for bisphenol A detection based on a diazonium-functionalized boron-doped diamond electrode modified with a multi-walled carbon nanotube-tyrosinase hybrid film. Biosens Bioelectron 74:830–835

Zhang Y, Causserand C, Aimar P, Cravedi JP (2006) Removal of bisphenol A by a nanofiltration membrane in view of drinking water production. Water Res 40(20):3793–3799

Zhang W, Zhou C, Zhou W, Lei A, Zhang Q, Wan Q, Zou B (2011) Fast and considerable adsorption of methylene blue dye onto graphene oxide. Bull Environ Contam Toxicol 87(1):86

Zhang C, Wu L, Cai D, Zhang C, Wang N, Zhang J, Wu Z (2013a) Adsorption of polycyclic aromatic hydrocarbons (fluoranthene and anthracenemethanol) by functional graphene oxide and removal by pH and temperature-sensitive coagulation. ACS Appl Mater Interfaces 5 (11):4783–4790

Zhang L, Pan F, Liu X, Yang L, Jiang X, Yang J, Shi W (2013b) Multi-walled carbon nanotubes as sorbent for recovery of endocrine disrupting compound-bisphenol F from wastewater. Chem Eng J 218:238–246

Zhang Y, Cheng Y, Zhou Y, Li B, Gu W, Shi X, Xian Y (2013c) Electrochemical sensor for bisphenol A based on magnetic nanoparticles decorated reduced graphene oxide. Talanta 107:211–218

Zhang T, Su J, Wang K, Zhu T, Li X (2014a) Ursolic acid reduces oxidative stress to alleviate early brain injury following experimental subarachnoid hemorrhage. Neurosci Lett 579:12–17

Zhang Y, Cheng Y, Chen N, Zhou Y, Li B, Gu W, Shi X, Xian Y (2014b) Recyclable removal of bisphenol A from aqueous solution by reduced graphene oxide–magnetic nanoparticles: adsorption and desorption. J Colloid Interface Sci 421:85–92

Zhang Z, Chen X, Rao W, Chen H, Cai R (2014c) Synthesis and properties of magnetic molecularly imprinted polymers based on multiwalled carbon nanotubes for magnetic extraction of bisphenol A from water. J Chromatogr B 965:190–196

Zhang X, Wu L, Zhou J, Zhang X, Chen J (2015) A new ratiometric electrochemical sensor for sensitive detection of bisphenol A based on poly-β-cyclodextrin/electroreduced graphene modified glassy carbon electrode. J Electroanal Chem 742:97–103

Zhang H, Ma Z, Duan J, Liu H, Liu G, Wang T, Chang K, Li M, Shi L, Meng X, Wu K, Ye J (2016) Active sites implanted carbon cages in core-shell architecture: highly active and durable electrocatalyst for hydrogen evolution reaction. ACS Nano 10(1):684–694

Zhang C, Kong R, Wang X, Xu Y, Wang F, Ren W, Wang Y, Su F, Jiang J-X (2017a) Porous carbons derived from hypercrosslinked porous polymers for gas adsorption and energy storage. Carbon 114:608–618

Zhang W, Cui T, Yang L, Zhang C, Cai M, Sun S, Yao Y, Zhuang X, Zhang F (2017b) Hollow-structured conjugated porous polymer derived Iron/Nitrogen-codoped hierarchical porous carbons as highly efficient electrocatalysts. J Colloid Interface Sci 497:108–116

Zheng Z, Du Y, Wang Z, Feng Q, Wang C (2013) Pt/graphene-CNTs nanocomposite based electrochemical sensors for the determination of endocrine disruptor bisphenol A in thermal printing papers. Analyst 138(2):693–701

Zhou L, Wang J, Li D, Li Y (2014a) An electrochemical aptasensor based on gold nanoparticles dotted graphene modified glassy carbon electrode for label-free detection of bisphenol A in milk samples. Food Chem 162:34–40

Zhou W, Sun C, Zhou Y, Yang X, Yang W (2014b) A facial electrochemical approach to determinate bisphenol A based on graphene-hypercrosslinked resin MN202 composite. Food Chem 158:81–87

Zhu J, Wei S, Gu H, Rapole SB, Wang Q, Luo Z, Haldolaarachchige N, Young DP, Guo Z (2012) One-pot synthesis of magnetic graphene nanocomposites decorated with core@double-shell nanoparticles for fast chromium removal. Environ Sci Technol 46(2):977–985

Zhu S, Y-G Liu, S-b Liu, G-m Zeng, L-h Jiang, X-f Tan, Zhou L, Zeng W, T-t Li, C-p Yang (2017) Adsorption of emerging contaminant metformin using graphene oxide. Chemosphere 179:20–28

Zuo P, Lu X, Sun Z, Guo Y, He H (2016) A review on syntheses, properties, characterization and bioanalytical applications of fluorescent carbon dots. Microchim Acta 183(2):519–542

Chapter 15
Toxic Metal Ions in Drinking Water and Effective Removal Using Graphene Oxide Nanocomposite

Marija Nujić and Mirna Habuda-Stanić

Abstract The discharge of heavy metal ions into the environment, due to extensive industrialization and inadequate waste disposal, has become a worldwide issue. Since heavy metals are not biodegradable, they accumulate in living organisms. The most common toxic metal ions in water are arsenic, lead, copper, cadmium, chromium, nickel, zinc, cobalt and manganese. Different treatment technologies such as coagulation, chemical precipitation, ion-exchange and filtration have been employed for pollutant removal from water, but adsorption is still one of the most suitable technologies for heavy metal ions removal. Recent studies show that graphene oxide, functionalized graphene oxide and their composites can efficiently remove heavy metal ions from water. This chapter reviews the application of graphene oxide nanocomposites in the removal of toxic heavy metal ions from aqueous solution.

Keywords Heavy metals · Toxicity · Adsorption · Removal · Graphene oxide

1 Introduction

The presence of heavy metals in drinking water can cause serious health hazards to humans. Even though there are naturally occurring elements, their multiple use in industries, household, agriculture and medicine have contributed to their wide distribution in the environment (Gambhir et al. 2012), thus increasing recognition of their potential harmful effects on human life and the environment (Tchounwou et al. 2012). Metal ions such as arsenic, copper, chromium, nickel, lead and mercury cannot be degraded into less detrimental constituents wherefore they are classified

M. Nujić (✉) · M. Habuda-Stanić
Faculty of Food Technology Osijek, Josip Juraj Strossmayer
University of Osijek, Franje Kuhača 20, 31000 Osijek, Croatia
e-mail: marija.nujic@ptfos.hr

M. Habuda-Stanić
e-mail: mirna.habuda-stanic@ptfos.hr

M. Naushad (ed.), *A New Generation Material Graphene: Applications in Water Technology*, https://doi.org/10.1007/978-3-319-75484-0_15

as non-biodegradable metal ions with high toxicity (Awad et al. 2017). The contamination of drinking water with some toxic metals is becoming significantly an increasing issue everywhere. Long-term exposure to toxic metal ions can cause bladder, lung, kidney, and skin cancer (Zhu et al. 2013). Therefore, the removal of toxic metals from water, as well as wastewater, is a serious problem (Alam et al. 2013; Naushad et al. 2016; Shahat et al. 2015).

There are several physical and chemical processes available for metal ions removal, such as adsorption, ion exchange, chemical precipitation, coagulation, membrane filtration etc. (Wang et al. 2012). Due to its simplicity, high efficiency and relatively low cost, adsorption has been recognized as the most suitable technology for toxic metal removal. Different materials, such as activated carbons, oxide minerals, polymers, resins and biosorbents have been successfully used for metal ions removal. The efficiency of metal ions removal depends on the functional groups on the adsorbent surface as well as on the physical and chemical characteristics of metal ions in water (Youssef and Malhat 2014; Habuda-Stanić and Nujić 2015). However, most of these materials possess low efficiency or adsorption capacity. In addition, it is crucial to find cost-effective and efficient adsorbents for toxic metal removal from aqueous media (Awad et al. 2017). Over the past decade, nanotechnology becomes a rapidly growing research area, especially for preparation of new materials for water remediation (Dargahi et al. 2016). Some of the new investigated materials (carbon nanotubes, fullerens, nanoparticles etc.) often showed better removal efficiencies due to their large surface area, high specificity, adsorption capacity and thermal stability (Ren et al. 2011; Liu et al. 2012; Turhanen et al. 2015).

Between the carbon based nanomaterials, graphene oxide (GO) and reduced graphene oxide (rGO) has tempted great attention as an efficient and low-cost adsorbent for toxic metal and dye removal from water (Ahamad et al. 2015; Gao et al. 2011; Hoan et al. 2016; Pathania et al. 2016). However, reduced graphene oxide is hydrophobic and has limited access to water molecules (Gao et al. 2011). Graphene and its composites can be used in various applications due to its unique two-dimensional structure (Jastrzębska et al. 2012). The structure of graphene is a carbon nanomaterial with sp^2-hybridized single atom layer structure (Nupearachchi et al. 2017). Graphene oxide, a product of oxidation and exfoliation of graphite (Gi), offers a large surface specific area, including hydroxyl, epoxide, carboxyl and carbonyl functional groups. Metal ions can likely bind with the oxygen-containing functional groups (Tien et al. 2017). Heavy metals (cations) can be relatively easily removed because of the negative surface charge of the GO and rGO (Nupearachchi et al. 2017). Furthermore, the idea of synergism by combining different materials is the key for successful development of new materials with improved properties and effectiveness. Regarding this, in order to prepare graphene nanocomposite material with polymers or metallic oxides is one of the most successful approach of material technology (Salavagione et al. 2011).

2 Heavy Metal Toxicity

Heavy metals are the most common contaminants in urban dust and runoff. Due to their toxicity and detrimental effects to human health and the nature, emissions of heavy metal are controlled by various Regulations throughout the world.

Human exposure to toxic heavy metals can occurs by contact or consumption of contaminated water or food and inhalation of contaminated. Diseases related to water is attributed with the exposure to increased heavy metal concentrations of organic and inorganic contaminants. Many of these components occur naturally, but in most cases, their elevated concentrations are result of human activities (Cobbina et al. 2015).

Some of metallic ions such as cobalt, copper, chromium, iron, magnesium, manganese, molybdenum, nickel, selenium and zinc are essential for various physiological functions (Tchounwou et al. 2012). In trace concentrations, heavy metals are considered as important compounds of several enzymes significant for chemical reactions like oxidation and reduction (WHO 1996). However, cationic contaminants cannot be decomposed by chemical reactions or bioprocesses, which can lead to some serious health problems consuming contaminated water (Table 1).

Table 1 Effect of some heavy metals in water on human health

Toxic metal	Occurrence/application	Effects	References
Arsenic	Former extensive use in pesticides, industry, wood preservation, pyrotechnics. light-emitting diodes, lasers	Hypertension, cardiovascular system, hepatic damage, hyperkeratosis, hyper and hypo pigmentation, spontaneous abortion, damage of nervous systems	Thürmer et al. (2002), Chouhan et al. (2016)
Cadmium	Anthropogenic source, industrial air emissions, fertilizers, protective coatings (electroplating), preparation of Cd–Ni batteries, stabilizers for PVC	Kidney and bone damage, lung damage	Chouhan et al. (2016)
Chromium	Occurs naturally by the burning of oil and coal, petroleum from ferro cromate refractory material, pigment oxidants, catalyst, chromium steel, fertilizers, oil well drilling and metal plating tanneries	Oxidative stress in cell causing damage to DNA, mutagenic properties	Jaishankar et al. (2014)
Lead	Industrial processes, food, smoking, drinking water, gasoline, house paint, lead bullets, plumbing pipes, batteries, toys, faucets	Learning disabilities (decreased IQ), attention deficit disorder, behavior issues, nervous system damage, decreased muscle and bone growth, kidney damage	Thürmer et al. (2002), Payne (2008)
Mercury	Fossil fuels, dental amalgams, old latex paint, incinerators, thermometers	Tremors, emotional liability, insomnia, memory loss, neuromuscular changes, headaches, effects on kidney and thyroid, vomiting, abdominal pain	Jaishankar et al. (2014)

Essential heavy metals (Fe, Mn or Cu) are important for biochemical and physiological functions in living organisms (Vaishaly et al. 2015). However, serious problems occurs when they become carcinogenic. In this case, it is important to remove them, especially from drinking water by various technologies like adsorption, ion exchange, reverse osmosis etc.

Arsenic, the most abundant element on earth, and its inorganic forms As(III) and As(V) are very toxic to humans and animals. Drinking water may be contaminated by pesticides containing arsenic or inadequate disposal of chemicals (Jaishankar et al. 2014). Cadmium is distributed in the earth's crust and the highest level of cadmium is accumulated in sedimentary rocks. This element is used for the production of batteries, pigments and alloys. The main exposure to cadmium is via inhalation or cigarette smoke, even tough, cadmium is present in certain foods like liver, mushrooms, shellfish etc. In water bodies, cadmium occur mainly through industrial activities and/or effluents. The increase in chromium concentrations is linked to the air and wastewater discharged from metallurgical and chemical industries. Chromium in hexavalent form is linked with anthropogenic activity. Lead is used in different industrial, agricultural and domestic applications, but it is important to mention that in recent years, the industrial use of lead is significantly reduced. Exposure to this element occurs via inhalation of dust particles contain lead, or it can be ingested by food and water (WHO 1996; Tchounwou et al. 2012). Mercury is one of the most toxic heavy metal ions in the environment. Its presence is due to volcanic activity, erosion or mercury-containing sediments, anthropogenic activities such as mining operations, tanneries etc. (Kyzas et al. 2014).

As it is evident, with the increment of industrial activity, toxic metal pollution of the environment becomes a significant issue. The pollution of aquifers with heavy metals shows not immediately negative effects on human health, but has great negative long-term impact (Chouhan et al. 2016). Therefore, monitoring the pollution of water with heavy metals in the environment and prevention/removal of such pollution is the only key to prevent their toxicity and negative impact on the environment (Jaishankar et al. 2014).

3 Graphene Oxide Nanocomposite for Toxic Metal Removal

Various adsorbents have been studied for toxic metal removal including activated carbon, zeolites, clays and other inorganic materials, but nanomaterials showed promising results due to their higher specific area and enhanced reactivity. Carbon-based nanomaterials are a good example of superior adsorbents for environmental applications (Madadrang et al. 2012). Since GO is a functional form of graphene containing oxygenated groups, it makes GO a hydrophilic material improving the dispersion in water (Pérez-Ramírez et al. 2016). Graphene oxide can be used as competent electron acceptor to intensify the photoinduced charge transfer for straighten photocatalytic activity (Hashim et al. 2016). Therefore, for the removal of heavy metal ions, various composites of graphene oxide have been investigated. The studied materials are chitosan, ethylenediaminetetraaceticacid (EDTA), metal oxides, different polymers, etc.

Graphene oxide is often used for metal ions removal due to its hydrophilic properties and the presence of functional groups containing oxygen atoms. For example, the adsorption capacities toward copper, zink, cadmium and lead ions were 294, 345, 530, 1119 mg g^{-1} according to Sitko et al. (2013) who used GO nanosheets as adsorbents. The kinetic study revealed that the adsorption of metal ions was chemisorption, and according to Langmuir isotherm model, the sorption on GO nanosheets was monolayer. Wu et al. (2013) reported on copper removal from aqueous solution using GO prepared by hummers method. The maximum adsorption capacity was 117.5 mg g^{-1} at pH 5.3 which is lower than that obtained in the research by Sitko et al. (2013).

To improve the characteristics of adsorbent, different methods for synthesis are employed. There are four common methods for synthesis of graphene oxide. These are Brodie, Staudenmaier, Hummers and Hofmann method. Then, GO can be reduced to rGO via chemical/thermal/solvothermal/hydrothermal, microwave and photo catalyst reduction method. There are differences between GO and rGO, which can be observed in terms of functional groups on the surface, crystallinity, optical property and defective structure, and both are considered as ideal material for water treatment (Nupearachchi et al. 2017). Besides, emphasize is given to new, cost effective, environmental friendly modification methods of GO and rGO. Therefore, introducing more functional groups on the GO, increase in pollutant adsorption can be expected. This chapter attempted to summarize recent work on graphene oxide based adsorbents for water remediation.

3.1 Chitosan/GO Nanocomposites

The removal efficiency of graphene oxide can be improved by functionalization with chemical groups on the GO. In theory, any molecule can be bonded trough interactions with oxygen domains, such as –OH, –COOH and R–O groups, and hence, extend the physico-chemical properties of the adsorbent (Pendolino and Armata 2017). Table 2 summarizes some chitosan-based adsorbents and their adsorption capacities on some heavy metal ions.

The porous graphene oxide/chitosan (PGOC) was developed by He et al. (2011) using a green method—the unidirectional freeze-drying method. The authors concluded that the PGOC materials in wet state show high mechanical strength. In addition, the PGOC demonstrate strong physical adsorption of Pb^{2+}, i.e. stronger attraction between Pb^{2+} and the functional groups on the GO than that of Cu^{2+}. The unidirectional freeze-drying method was also used for preparation of chitosan-gelatin/graphene oxide (CGGO) monoliths as adsorbent for removal of Cu^{2+} and Pb^{2+} from aqueous solutions. The adsorption ability of the CGGO monoliths for Cu^{2+} and Pb^{2+} was greatly influenced by the pH of solution, which can be explained

Table 2 Adsorption capacity of some chitosan/GO-based adsorbents used for heavy metal ions removal from aqueous solution

Ion	Adsorbent	Maximum adsorption capacity (mg g^{-1})	References
Cd^{2+}	Chitosan/iron oxide (CH–FeO)	201.84 (Langmuir)	Keshvardoostchokami et al. (2017)
Cr^{6+}	MCGO-IL	145.35 (Langmuir)	Li et al. (2014)
	TGOCS	219.5 (Langmuir)	Ge and Ma (2015)
	Chitosan/GO	310.4 (Langmuir)	Najafabadi et al. (2015)
Cu^{2+}	Porous spongy CSGO monolith	53.69	Wang et al. (2014)
	PVA/CS/GO	163.97 (Langmuir)	Li et al. (2015)
	Chitosan/GO	423.8 (Langmuir)	Najafabadi et al. (2015)
Pb^{2+}	PGOC	99.9 (Langmuir)	He et al. (2011)
	GO/chitosan/ FeOOH	111.11 (Langmuir)	Sheshmani et al. (2015)
	GO/chitosan/ FeOOH	2.073 (Freundlich)	Sheshmani et al. (2015)
	Chitosan/GO	461.3 (Langmuir)	Najafabadi et al. (2015)
	MCGO	76.94 (Langmuir)	Fan et al. (2013)
	MHCGO composite	66.27 (Langmuir)	Wang et al. (2016)
	Chitosan/iron oxide (CH–FeO)	11.69 (Langmuir)	Keshvardoostchokami et al. (2017)
Ni^{2+}	Chitosan/iron oxide (CH–FeO)	57.86 (Langmuir)	Keshvardoostchokami et al. (2017)

by the fact that amino groups of the chitosan macromolecules become protonated, and therefore can inhibit the adsorption of Cu^{2+} on CGGO monoliths. With increasing pH, amino groups become deprotonated and, thus, can form complexes with metal ions (Zhang et al. 2011). In the research performed by Wang et al. (2014) microporous spongy chitosan monoliths doped with graphene oxide (CSGO monoliths) was used as adsorbent for methyl orange and Cu^{2+} removal. The adsorption capacity, as a function of pH, contact time, amount of doped GO and initial Cu^{2+} and methyl orange concentrations, for Cu^{2+} is 53.69 mg g^{-1}. Zhang et al. (2014) prepared porous graphene oxide/carboxymethyl cellulose (GO/CMC) monoliths by a green unidirectional freeze-drying method. This adsorbent showed good adsorption properties with capacities of 82.93, 76.7, 72.04, 59.99 and 46.13 mg g^{-1} for Cu^{2+}, Pb^{2+}, Ni^{2+}, Co^{2+} and Cd^{2+}, respectively.

Another chitosan/graphene oxide composite developed by electrospinning process for the removal of Cu^{2+}, Pb^{2+} and Cr^{6+} was investigated by (Najafabadi et al. 2015). The effects of several parameters (GO concentrations, pH, contact time and temperature) were investigated. The maximum adsorption capacities were 423.8, 461.3 and 310.4 mg g^{-1} for Cu^{2+}, Pb^{2+}, and Cr^{6+}, respectively. This type of composite followed external and internal diffusion adsorption kinetics at equilibrium time of 30 min. Additionally, the chitosan/graphene oxide could be used up to the fifth cycle of regeneration.

Sheshmani et al. (2015) prepared a graphene oxide/chitosan/FeOOH (GO/Ch/FeOOH) nanostructured composite for the removal of Pb^{2+} from aqueous solution. The results showed that the amino and hydroxyl functional groups of chitosan are suitable for Pb^{2+} removal, which was also proved by Langmuir and Freundlich model with adsorption capacities of 111.11 and 2.073 mg g^{-1}, respectively. In addition, the adsorption increased with increasing temperature (25–80 °C) with and optimum pH of 5.5. The authors proposed the mechanism of lead onto GO/Ch/FeOOH to be due to ionic interactions of the metal with amino groups of the nanocomposite.

A composite hydrogel, consists of polyvinyl alcohol (PVA), chitosan (CS) and graphene oxide (GO), was synthesized by a gelation method for the removal of Cu^{2+} from aqueous solutions by Li et al. (2015). According to Langmuir isotherm parameters, achieved maximum adsorption capacities were 159.26–163.97 mg g^{-1} at temperatures 20–40 °C. The pseudo-second order fitted well with the experimental values what assumes that the rate-determining step could be a chemical sorption (Qiu et al. 2009).

Adsorption of Cr^{6+} on triethylentetramine modified graphene oxide/chitosan composite (TGOCS), synthesized by microwave irradiation method, was studied by Ge and Ma (2015). The results showed that the highest adsorption capacity was at acidic pH (pH 2) and that adsorption followed the Langmuir isotherm (219.5 mg g^{-1}) and the pseudo second order kinetic model. The authors observed that the adsorption was completed within 20 min and reached 85% of maximum uptake. In addition, the adsorbent was effectively regenerated with NaOH. Magnetic ionic liquid/chitosan/graphene oxide composite (MCGO–IL) for Cr^{6+} removal from wastewater was proposed by (Li et al. 2014). Maximum adsorption capacities of 145.35 mg g^{-1} were determined at lower pH value of solution. The desorption was carried out with HCl. The conclusion of their study was that the sorption properties of the prepared material exhibited remarkable removal capacity in short time because of the ample amine and hydroxyl groups, which are available for linkage of chromium.

For the removal of lead ions from aqueous solution, magnetic hydroxypropyl chitosan/graphene oxide (MHCGO) composites have been used. The results revealed that the adsorption of lead ions on MHCGO was pH dependent and that the adsorption was quick, while the thermodynamic study proved that the adsorption process was spontaneous (Wang et al. 2016). Due to the known excellent adsorption properties, GO has been widely employed as biosorbents. Thus, Fan et al. (2013) prepared magnetic chitosan/graphene oxide (MCGO) and tested its sorption properties for Pb^{2+} removal from aqueous solution. The MCGO showed an increased surface area with adsorption capacity of 76.94 mg g^{-1}, with an optimum time for adsorption of Pb^{2+} of 60 min, and remained stable even after five regeneration cycles (75% adsorption efficiency). However, the pH value has been found to be the most important parameter for metal sorption. Usually, metal adsorption do not take place at low pH. However, the effect of pH on the adsorption of lead onto MCGO was tested from 1 to 7, and the results showed that maximum adsorption capacity was 50.23 mg g^{-1} at pH 5, because the functional groups of inserted chitosan on the surface of the biomass offers more metal adsorption sites. Decreasement of lead uptake in acidic pH can be ascribed to the protonation of the long pair of nitrogen that inhibits the formation of a complex (Fan et al. 2013).

3.2 *Metal Oxide/GO*

Iron minerals have good adsorbent capacities for pollutant removal from water. Modification of graphene oxide with metal oxides gives various nanocomposites with improved adsorption capacity and removal efficiency (Wang et al. 2013a).

Jiang et al. (2011) reported on graphene oxide/TiO_2 composite. This nanoc omposite showed great photocatalytic ability for the oxidation of organic pollutants, but also exhibited great reductive conversion of toxic Cr^{6+} to Cr^{3+} by increment of irradiation time, which the composite makes a promising material for environmental photocatalysis.

In the research by Luo et al. (2013) hydrated zirconium oxide nanoparticles modified with graphene oxide (GO–$ZrO(OH)_2$) was employed for arsenic removal. Results showed that the composite worked in a wide pH range and the maximum adsorption capacities were 95.15 and 84.89 mg g^{-1} for As(III) and As(V), respectively. The good adsorption capacity can be explained by the good dispersion of zirconium oxide nanoparticles on the GO.

The simultaneous adsorption of heavy metal ions and humic acids on poly-ethyleneimine (PEI)-modified magnetic mesoporous silica and graphene oxide (MMSP–GO) were investigated by Wang et al. (2013b). In batch adsorption experiments, adsorption capacities of Pb^{2+} and Cd^{2+} increased as the pH increased. For Pb^{2+} the maximum adsorption capacity was 220 mg g^{-1}, and for Cd^{2+} 65 mg g^{-1}. All composites containing GO were more efficient in adsorbing metal ions. In the presence of humic acid (HA), the adsorption capacity of assed metal ions was greater than in the absence of HA, which can be attributed to the joint effects of the PEI modified magnetic mesoporous silica and GO sheet. This means that adsorption of HA can neutralize the positive surface charge of the MMSP and thus enhance adsorption of heavy metals.

Jiang et al. (2015) prepared a graphene oxide (GO)/β-FeOOH composite by the liquid insert method. Under optimal conditions, the removal of copper ions was 93.8% (pH 6, 60 min, 60 mg of adsorbent). The increment of adsorbent dose (to 80 mg) had little impact on Cu^{2+} removal, because the adsorption equilibrium was reached at adsorbent dose of 60 mg. The (GO)/β-FeOOH composite exhibited higher removal efficiencies under weak acidic conditions, while at lower pH value. low removal rate was observed. The removal efficiency was compared in single metal system (Cu^{2+}, Pd^{2+} and Cd^{2+}). The maximum uptake on the (GO)/β-FeOOH composite followed the order of Cu^{2+}, Pd^{2+} and Cd^{2+}, corresponding to the sequence of electronegativity.

Pourbeyram (2016) synthesized a graphene oxide-zirconium phosphate (GO–Zr–P) nanocomposite and tested it on lead, cadmium, copper and zinc removal. The effect of different process parameters were investigated (initial metal ion concentration, pH and contact time). The results showed that the adsorption occurred in two steps. In the first 10 min the adsorption was very rapid, and after that gradually increased. The equilibrium was reached after 20 min. Agglomeration and precipitation of the nanocomposite was evident after adsorption process. The amount of heavy metals adsorbed on the nanocomposite increased with higher initial metal concentration (10–200 mg L^{-1}). Maximum adsorption capacities for Pb^{2+}, Cd^{2+}, Cu^{2+} and Zn^{2+} at pH 6 were 363.42, 232.36, 328.56 and 251.58 mg g^{-1}, respectively. The removal efficiency increased by increasing the dosage of nanocomposite too. Adsorption isotherm and kinetic studies proposed that the adsorption was monolayer and controlled by chemical adsorption including surface complexation of heavy metals with phosphate groups on the surface of GO–Zr–P.

In the study conducted by Samadi and Abbaszadeh (2017) magnesium oxide/ polyethylene glycol/GO (MgO/PEG/GO) nanocomposite was prepared by a sol-gel and solvothermal method. By changing process parameters, pH, adsorbent dosage, contact time and solution volume, the best condition for maximum copper removal efficiency was evaluated. The removal efficiency increases (about 99%) when the pH increases and the maximum removal rate was obtained at pH 6 and 7. Interestingly, with the increment of nanocomposite dosages, the removal efficacy decreased. The interference of other ions in solution was also tested and the conclusion is that other ions had no significant impact on copper removal from aqueous solution.

Tien et al. (2017) synthesized silica from rice husk ash by precipitation method (RHA-SiO_2), while GO was prepared by hummers method. The RHA-SiO_2/GO nanocomposite was prepared by in situ one-step method with 3-aminopro pyltriethoxysilane (APS) as coupling agent to functionalize the surface by binding monolayer –NH_2. The adsorption experiments were performed at room temperature, pH 7, initial Ni^{2+} concentration varied from 5 to 260 mg L^{-1}, in different time intervals. Results showed that most of the nickel ions were removed within 180 min and the equilibrium adsorption time was 360 min. The adsorption of nickel ions onto SiO_2/GO increased in pH range from 2 to 5. However, the adsorption capacity increased in pH range from 5 to 7, which can be explained by the competition between H^+ and Ni^{2+} ions for the same adsorption sites in acidic media and inversely. The calculated isotherm parameters revealed that the adsorption of nickel ions onto the nanocomposite followed the Langmuir isotherm model in other words, the adsorption was a monolayer process. The maximum adsorption capacity of RHA-SiO_2/GO was 256.4 mg g^{-1}, which is higher than that of pure GO (243.9 mg g^{-1}) and SiO_2 (212.77 mg g^{-1}). The presence of SiO_2 inhibited the aggregation of GO sheets and had higher surface area.

Novel researches evaluate metal organic frameworks (MOF) and graphene oxide nanocomposites for different kinds of organic and inorganic pollutants from aqueous solutions according to their high capacity and selectivity (Rahimi and Mohaghegh 2017). Therefore, copper terephthalate MOF and few-layered GO

(Cu(tpa).GO) were prepared by solvothermal and hummers method. This adsorbent possess a great number of active sites and the metal ions are chemically sorbed via inclosing with some functional groups like epoxides, hydroxides, carboxylic and carboxylate which are on the surface of the adsorbent, and the porous structure enables metal ions to diffuse into the MOF. The proposed mechanism for metal ions removal is in agreement with Langmuir and Freundlich adsorption isotherms. The maximum achieved adsorption capacities were 88, 243 and 131 mg g^{-1} for Pb^{2+}, Cu^{2+} and Zn^{2+}, respectively. In addition, data calculated from the pseudo-second order kinetic model is coherent with the presumption that the determining rate step is chemical sorption. This adsorbent was tested on a real wastewater sample as well and showed great potential to be applied as an adsorbent for simultaneous heavy metal removal (Rahimi and Mohaghegh 2017).

If graphene oxide is modified with magnetic materials, it is possible to obtain new materials with large surface area and excellent mechanical properties (GO) and simultaneously, magnetic properties for easy separation (Al-Nafiey 2016). For example, magnetic graphene nanocomposites (MGNC), fabricated by one-pot thermodecomposition method, was used for fast chromium removal. For comparison, MGNCs and pure graphene were used for Cr^{6+} removal. Results showed that removal efficiencies with pure graphene (44.6% with MGNcs concentration of 3 g L^{-1}) were much lower than that of the MGNcs (52.6% with MGNcs concentration of 0.25 g L^{-1}). In addition, the removal efficiency increased by about 100% when adding 10% of nanoparticles in the MGNCs. The chromium was completely removed in 5 min using 3 g L^{-1} MGNCs. Authors concluded that the adsorption mechanism of Cr^{6+} on pure graphene and MGNCs is different: the adsorption on pure graphene is a single layer adsorption, while on MGNCs is a combined process of surface complexation between sulfur and chromium on the outside, and single layer adsorption on the surface of graphene (Zhu et al. 2013).

The preparation of iron oxide magnetic nanoparticles (MGO), performed by coprecipitation method of iron oxide nanoparticles on the surface of GO, was performed by Deng et al. (2013). Results showed that MGO simultaneously removed Cd^{2+} and ionic dyes (orange G (OG) and methylene blue (MB)). The adsorbent was tested in mono and binary systems in ultrapure and tap water. With increasing the pH, sorption capacities for Cd^{2+} and MB increased, while for OG decreased. The maximum sorption capacities in ultrapure water for Cd^{2+}, MB and OG were 91.29, 64.23 and 20.85 mg g^{-1}, respectively, in mono system. But, in binary system, the sorption capacity of Cd^{2+} decreased as with increasing MB concentration, while MB was not affected with increasing Cd^{2+} concentration. However, the sorption capacity of Cd^{2+} was not affected by OG concentration, but dependent on Cd^{2+} initial concentration in Cd^{2+}–OG system. Real water samples were spiked with Cd^{2+} or dyes, and the maximum adsorption capacities were 58.08, 20.48 and 59.69 mg g^{-1}, respectively, for MB, OG and Cd^{2+} which are lower than that achieved in ultrapure water. The decrease of Cd^{2+} uptake is attributed by the increase in ionic strength in tap water (Deng et al. 2013). Hu et al. (2015) synthesized magnetic graphene oxide-supported β-cyclodextrin (MGO/β-CD) and tested it for Cu^{2+} removal. The effect of pH was studied in batch experiments at different $NaNO_3$ concentrations (0, 0.001, 0.01 and 0.1 mol L^{-1}). The Cu^{2+} adsorption on MGO/β-CD was pH dependent, and in the system without $NaNO_3$, the removal of Cu^{2+} slowly increased from pH 2.5 to 6, then quickly increased at pH 6–8, and maintained high at pH >8. Results also showed that the presence of citric acid enhances Cu^{2+} adsorption at various pH. Graphene oxide functionalization with cyclodextrin was also performed by Chauke et al. (2015). First, the GO was functionalized with α-cyclodextrin (α-CD) to form GO-α-CD nanocomposites (NC). After that, the GO-α-CD nanocomposite was modified with polypyrrole (PPY) (GO-α-CD-PPY NC) to make the adsorbent selective to Cr^{6+} ions. From the results it can be concluded that the adsorption of Cr^{6+} onto GO-α-CD-PPY NC is pH and temperature dependent, where optimum adsorption was achieved at pH 2 and increased with increasing temperature. The maximum adsorption capacities ranged from 606.06 to 666.67 mg g^{-1} according to Langmuir isotherm. Other cations and anions had little or no impact on the adsorption of toxic hexavalent chromium (Chauke et al. 2015).

Khazaei et al. (2016) prepared magnetic graphene oxide nanocomposite (Fe_3O_4@SiO_2–GO) and tested it on Pb^{2+} removal. The removal process was rapid and the adsorption process was almost completed in 10 min. The adsorption capacity of Pb^{2+} after 16 min was 505.81 mg g^{-1} at initial lead concentration of 90 mg L^{-1}. The authors concluded that the mechanism of adsorption was chemisorption due to the mean free energy (E = 9.901 kJ/mol). The adsorption was sensitive to pH and contact time (Khazaei et al. 2016). For efficient dye removal from aqueous solution, Namvari and Namazi (2014) employed Fe_3O_4-clicked GO.

Table 3 Adsorption capacity of some metal oxide/GO-based adsorbents used for heavy metal ions removal from aqueous solution

Ion	Adsorbent	Maximum adsorption capacity (mg g^{-1})	References
As^{3+}	GO-$ZrO(OH)_2$	95.15	Luo et al. (2013)
As^{5+}	GO-$ZrO(OH)_2$	84.89	Luo et al. (2013)
Cd^{2+}	MMSP-GO	65	Wang et al. (2013b)
	MGO	91.29	Deng et al. (2013)
	GO-Zr-P	232.36	Pourbeyram (2016)
Cu^{2+}	GO-Zr-P	328.56	Pourbeyram (2016)
	Cu(tpa).GO	243	Rahimi and Mohaghegh (2017)
Cr^{6+}	GO-α-CD-PPY NC	666.67	Chauke et al. (2015)
Pb^{2+}	MMSP-GO	220	Wang et al. (2013b)
	GO-Zr-P	363.42	Pourbeyram (2016)
	Fe_3O_4@SiO_2-GO	505.81	Khazaei et al. (2016)
	Cu(tpa).GO	88	Rahimi and Mohaghegh (2017)
Ni^{2+}	RHA-SiO_2/GO	256.4	Tien et al. (2017)
Zn^{2+}	GO-Zr-P	251.58	Pourbeyram (2016)
	Cu(tpa).GO	131	Rahimi and Mohaghegh (2017)

High adsorption capacities were achieved (109.5 mg g^{-1} for methylene blue and 98.8 mg g^{-1} for Congo red). The adsorbent was for the first time prepared by covalent attachment of nanoparticles on the GO surface using powerful click reaction.

Table 3 summarizes the adsorption capacities of some metal oxide/GO composites for some heavy metal removal.

3.3 Graphene-Based Polymer Oxide Nanocomposite

Graphene-based polymer nanocomposites have also demonstrated great potential to be applied in water treatment. These nanohybrid materials showed improved properties in comparison conventional composites or pure polymers (Salavagione et al. 2011). Excellent potential application on Cr^{6+} ion removal showed polypyrrole/graphene oxide composite nanosheets (PPy/GO) synthesized by sacrificial-template polymerization method. The removal capacity of PPy/GO was two times higher than that of conventional PPy. The equilibrium was reached after 180 min at initial Cr^{6+} concentration of 0.38 mmol L^{-1}, and with increasing the concentration of Cr^{6+} to 1.97 mmol L^{-1}, 24 h needed for reaching equilibrium. The adsorption efficiency decreased (from 100 to 70.8%) with increasing initial Cr^{6+} concentration (from 0.38 to 1.97 mmol L^{-1}) (Li et al. 2012).

Musico et al. (2013) reported on poly(*N*-vinylcarbazole)-graphene oxide (PVK-GO) nanocomposite as adsorbent for Pb^{2+} removal. This research studied different PVK and GO concentrations, pH of solutions and contact time. The results showed that the highest adsorption capacity of 887.98 mg g^{-1} (according Langmuir adsorption isotherm) was achieved using 10:90 wt% ratio of PVK and GO, respectively at pH 7 and 90 min of contact time. The adsorption capacity of lead increased with increasing the GO concentration, the available oxygen-containing functional groups were increased.

As it is already known, nitrogen-based ligands, such as amino, imidiazole and hydrazine groups, on adsorbents surfaces, are efficient in forming complexation with metal ions. Therefore, Xing et al. (2015) synthesized NH_2-rich polymer/graphene oxide composite for the removal of Cu^{2+} ions from aqueous solutions (PAH-GO). Results showed that the adsorption mechanism of Cu^{2+} on GO is a single-layer adsorption trough ion-exchange process, while on PAH-GO it is a process of surface complexation and single layer adsorption. The maximum achieved adsorption capacity of Cu^{2+} on PAH-Go was 349.03 mg g^{-1}. In the experiment with coexisting ions (Mg^{2+}, Ca^{2+}, Cr^{6+}, Ni^{2+} and Cd^{2+}), results showed that the adsorption of copper ions is slightly decreased, but PAH-GO composite showed much higher affinity for Cu^{2+} ions in mixed heavy metal solution. In addition, the adsorption-desorption experiments show that the reusability of PAH-GO is possible (even after adsorption 5 cycles).

Kumar et al. (2013) reported on TOA (long chain tertiary amine) in conjunction with EGO (exfoliated graphene oxide) for chromium adsorption. This adsorbent performed high adsorption capacity (232.55 mg g^{-1}), and the adsorption of Cr^{6+} could involve pore, surface or intraparticle diffusion. Sui et al. (2015) synthesized Fe_3O_4 nanoparticles on graphene oxide and then mixed polyethyleneimine to get GO/Fe_3O_4/PEI nanocomposites. This nanocomposite showed high affinity toward Cu^{2+} ions. The maximum removal efficiency was as high as 157 mg g^{-1}, and even after five adsorption/desorption cycles, the removal efficiency remained 84%.

3.4 Hydrogels

Hydrogels are natural or synthesized aggregations of hydrophilic polymer networks and water (Yu et al. 2015). For copper removal, hyperbranched polymers or/and graphene based nanosheets were used to synthesize poly(acrylic acid)-based hybrid hydrogels (PAA and PAA hybrid composites), which have good mechanical properties. The adsorption and desorption studies of Cu^{2+} showed that the optimum pH for Cu ions uptake was at pH 5, at lower pH the Cu^{2+} adsorption capacity decreased, while at higher pH values, the copper ions hydrolyzed and formed $Cu(OH)_2$ colloids, and precipitated from solution. The PAA showed highest adsorption capacity (230.8 mg g^{-1}) (Yu et al. 2015).

Zhou et al. (2015) prepared a low cost, highly adsorptive and reusable adsorbent for heavy metal removal. The sponge-like polysiloxane-graphene oxide (PS-GO) gel adsorbent was prepared by an easy one-step sol-gel method. As expected, the removal efficiency was influenced by the pH of solution and the sorption of Pb^{2+} and Cd^{2+} increased with increasing the pH. The sorption of metal ions on PS–GO increased rapidly in 30 min and the equilibrium was reached at 60 min. The rapid adsorption is attributed to amino groups and porous structure of the adsorbent. The calculated isotherm parameters showed a better fit to Langmuir model. The maximum adsorption capacities were 256.41 and 136.98 mg g^{-1} at pH 6 and 30 °C for Pb^{2+} and Cd^{2+}, respectively.

Liu et al. (2016) synthesized three GO hyperbranched polyethylenimine (GO-HPEI) gels by a cross-linking reaction. The HPEIs had different molecular weight. The adsorption capacities of the prepared GO-HPEI gels for Pb^{2+} are pH dependant and the maximum adsorption capacity was 438.6 mg g^{-1}. The adsorbents exhibited good stability in aqueous media, and showed only slight decrease of adsorption capacity after eight cycles of adsorption and desorption.

The adsorption of Pb^{2+}, Cd^{2+} and Ag^{+} onto composite hydrogels based on gum tragacanth (GT) carbohydrate and graphene oxide was investigated by Sahraei and Ghaemy (2017). The adsorption kinetic followed the pseudo first-order model. According Langmuir adsorption isotherm, the maximum adsorption capacity was 142.5, 112.5 and 132.12 mg g^{-1} for lead, cadmium and silver ions, respectively. Results showed that maximum removal efficiencies at pH 6 were achieved. The prepared hydrogels could be reused after regeneration in HNO_3. This indicates that even after four desorption cycles the chelating properties of the hydrogel's functional groups were not decreased.

Recently, Dong et al. (2018) combined a functional nanocomposite hydrogel with a highly photocatalytic Fenton reaction for the removal of heavy metals ions and degradation of organic dyes (Rhodamine B) from industrial wastewater. The hydrogel consisted of Fe_3O_4 nanoparticles, reduced graphene oxide and polyacrylamide (PAM). Results showed that the removal efficiencies were 66.3% for Ce^{3+}, 34.8% for Cu^{2+}, 52.4% for Ag^+ and 52.9% for Cd^{2+}, respectively.

3.5 EDTA/GO Nanocomposites

Introducing ethylenediaminetetraaceticacid (EDTA) groups to the GO surface can significantly increase the adsorption capacity of GO for toxic metal removal. The hydroxyl and carboxyl groups together with EDTA groups make EDTA-GO an effective adsorbents for the removal of lead, copper, nickel and cadmium in water (Madadrang et al. 2012). For example, the adsorption capacity of EDTA-GO for Pb^{2+} is 479 $\pm$ 46 mg g^{-1} according to Langmuir isotherm, which is higher than that of GO (Madadrang et al. 2012).

Cui et al. (2015) used EDTA functionalized magnetic graphene oxide (EDTA-mGO) for Pb^{2+}, Hg^{2+} and Cu^{2+} adsorption. This adsorbent is composed of three units (EDTA, Fe_3O_4 and GO) and every unit has a different role in the adsorption process. The removal efficiency of mono-components (EDTA, Fe_3O_4 and GO) for Pb^{2+}, Hg^{2+} and Cu^{2+} were less than 45%, for bi-components (EDTA-GO, EDTA-Fe_3O_4 and m-GO)were less than 80%, while EDTA-mGO exhibited the highest removal efficiencies of 96.2, 95.1 and 96.5% for Pb^{2+}, Hg^{2+} and Cu^{2+}. It has been concluded that the adsorption of metal ions depend on the metal chelation of EDTA and electrostatic attractions of functional groups on the GO surface. The kinetic data fitted well with the pseudo-second order model and the equilibrium data followed the Freundlich and Temkin isotherm model. According to Langmuir isotherm model, the maximum adsorption capacities were 508.4, 268.4 and 301.2 mg g^{-1} for Pb^{2+}, Hg^{2+} and Cu^{2+}, respectively. Additionally, the good magnetic performance of the adsorbent contributes to easy operation.

Recently, EDTA-functionalized magnetic chitosan graphene oxide nanocomposite (EDTA-MCS/GO) was synthesized by a reduction-precipitation method and used for the removal of lead, copper and arsenic from aqueous solution. This adsorbent showed excellent adsorption capacities of 206.52 (Pb^{2+}), 207.26 (Cu^{2+}) and 42.75 (As^{3+}) mg g^{-1}. The EDTA-MCS/GO was tested on a real domestic wastewater spiked with Pb^{2+}, Cu^{2+} and As^{3+}. The results showed a slightly decrease in removal efficiency which might occur due to the increase in ionic strength in wastewater. The reusability of the adsorbent was tested in four adsorption and desorption cycles. Although a small decrease in adsorption efficiency is evident, this adsorbent shows a great potential in simultaneous removal of divalent and trivalent metal ions (Shahzad et al. 2017).

The development of multi-functional nanomaterials for heavy metals removal coupled with anti-microbial properties, is a very important scientific achievement for various applications. For this purpose, antimicrobial activity of graphene oxide silanized *N*-(trimethoxysilypropyl) ethylenediamine triacetic acid (GO-EDTA) against Cu^{2+} and Pb^{2+} and Gram-negative, *Cupriavidus metallidurans* CH_4 was evaluated. FTIR analysis showed that carboxyl and carbonyl functional groups were responsible for Cu^{2+} and Pb^{2+} uptake. The maximum adsorption capacity was 454.6 and 108.7 mg g^{-1} for Pb^{2+} and Cu^{2+} which were higher than that of some activated carbons. The GO-EDTA showed not only antimicrobial capability, but also good adsorption properties, thus making this adsorbent a multi-functional material for water remediation (Mejias Carpio et al. 2014).

3.6 Other GO Nanocomposites

Graphene oxide-porphyrin nanocomposites were used for mercury removal from aqueous solutions. The Hg^{2+} removal efficiency was higher with the nanocomposite than with graphene oxide. The advantages of this removal process is simple operation, low cost, rapidity and higher adsorption capacities. Besides, the graphene oxide-porphyrin composite could be regenerated with EDTA and showed excellent selectivity towards mercury ions (Rahimi et al. 2014).

Sitko et al. (2014) synthesized aminosilanized graphene oxide (GO–NH_2) for selective Pb^{2+} adsorption. The batch adsorption experiments confirmed the high selectivity toward Pb^{2+} ions and the calculated Langmuir adsorption isotherm suggest that adsorption of lead ions on the adsorbent is monolayer, while the whole process is controlled by surface complexation of lead ions with the nitrogen containing groups on the surface of GO–NH_2. The maximum adsorption capacity is lower than with pure GO (96 mg g^{-1}) and was achieved at pH 6. In addition, the GO–NH_2 was used for the sensitive determination of Pb^{2+} ions by electrochemical atomic absorption spectrometry (ET-AAS): after the separation of solid phase, the suspension of GO–NH_2 with adsorbed lead ions can directly be injected into the graphite tube and analyzed by ET-AAS. The adsorptive properties of GO–NH_2 has potential for application as an adsorbent in analytical chemistry with an extremely low limit of detection (9.4 ng L^{-1}). The adsorption capacity of Pb^{2+} on graphene oxide and thiol functionalized graphene oxide was investigated by Yari et al. (2015). Different concentrations of cysteamine (60, 50 and 100 mg) were used as functionalizing agents for conversion of GO to thiol functionalized GO (GO-SH). It has been observed that ideal conditions for optimum Pb^{2+} adsorption was 60 min, 25 mg L^{-1} initial Pb^{2+} concentration, pH 6 and 20 mg of added adsorbent to solution at 25 °C (Yari et al. 2015). Enhanced metal removal efficiency by hydrophilic GO nanosheets containing thiol groups was also reported by (Gao et al. 2011).

Another efficient adsorbent for Pb^{2+} removal was synthesized by crosslinking reaction between GO sheets and poly3-aminopropyltriethoxysilane (PAS) oligomers.

The PAS-GO exhibited high adsorption capacity (312.5 mg g^{-1} at 30 °C). The adsorbent is effective in a broad pH range (4–7). This adsorbent showed a great tendency to remove Pb, Cu and Fe in mixed (real) metal ion solution (Luo et al. 2014).

3.7 Reduced GO Nanocomposites

For adsorbents applicable in environmental remediation, it is crucial that they show features such large surface area and presence of various functional groups. But, the possibility to make GO via chemical methods and its reduction to reduced graphene oxide (rGO) revealed that there are different ways to produce graphene oxide, and additionally, various possibilities to enhance the properties of GO and rGO trough chemical modifications (Sreeprasad et al. 2011). For example, Chandra et al. (2010) employed magnetite-reduced graphene oxide (M-RGO) composites for As(III) and As(V) removal. Due to the increased adsorption sites, these composites showed high binding capacity for arsenic. Sreeprasad et al. (2011) synthesized various RGO-composites (RGO-MnO_2 and RGO-Ag) and supported these composites on river sand (RS). The unsupported and supported composites were assessed for Hg^{2+} removal from water. In this research the kinetic of Hg^{2+} adsorption was performed with the aim to perceive the time needed for mercury removal with the prepared adsorbents. Results showed that tested adsorbents could remove Hg^{2+} ions with no significant changes in the equilibrium uptake capacity. The authors concluded that a specific type of functional groups (most probably the carboxylic acid groups) is present in both GO and RGO which are responsible for mercury uptake. In order to assess the Hg^{2+} removal ability of the prepared unsupported and supported composites, experiments were performed in distilled and real groundwater sample spiked with 1 mg L^{-1} Hg^{2+}. The results showed complete removal of Hg^{2+} in both systems and the co-ions did not affect the removal efficiency. The reported method is simple and eco-friendly, and the composites can be used in water purification as well as in other scientific fields.

Rodriguez (2015) synthesized cerium oxide (CeO_2) nanoparticles anchored to GO for the adsorption of lead, chromium and cadmium from water. The CeO_2 nanoparticles were mixed with GO, and after that a reduction process of the GO to rGO was conducted. To obtain negatively charged GO nanosheets, polyvinilypyrrolidine (PVP) was dispersed in the solution (self-assembly strategy). After that positively charged CeO_2 suspension was added, and at the end ascorbic acid. The obtained product was designated as CeO_2/RGO-AHA. To obtain CeO_2/GO-AM, acrylamide (AM) was dissolved in *N,N*.dimethylformamide (DMF), while ascorbic acid was added for the reduction of GO to rGO. In third modification, hexamethylenetetramine (HMT) was used based on an *in situ* growth. It was evident that Pb^{2+} was effectively removed in greater extent than Cd^{2+} ions. The adsorption of Cr^{6+} was rather inefficient. For example, maximum adsorption capacities for lead (initial concentration 250 mg L^{-1}) were 46.782, 62.798 and

95.756 for CeO_2/rGO/-AHA, CeO_2/rGO/-AM, and CeO_2/rGO/-HMT, respectively, which can be explained by the fact that the CeO_2 nanoparticle concentration is higher in the in situ synthesis.

The removal efficiency of lead, arsenate and cadmium ions using amine cross-linked reduced graphene oxide investigated Babu et al. (2015). The graphene oxide was synthesized by hummers method and reduced with hydrazine hydrate. (3-aminopropyl)trimethoxysilane was used as a precursor for amine functionalized rGO. FTIR, TGA, Raman, N_2 sorption, SEM and TEM techniques were used to characterize these adsorbent. The surface area was found to be 346 $m^2\ g^{-1}$. The adsorption of toxic metal ions were high for the rGO nanocomposite in comparison to bare amine functionalized nanoparticles. The adsorption equilibrium occurred within 15 min (Babu et al. 2015).

Hoan et al. (2016) synthesized Fe_3O_4/rGO nanocomposite and investigated it for arsenic, nickel and lead removal. The adsorption isotherm model suggested that monolayer takes place on the surface of Fe_3O_4/rGO. The kinetic study of metal ions adsorption shows that the pseudo-second order model provides excellent correlation of the sorption data.

Recently, Du et al. (2017) prepared molybdenum disulfide/reduced graphene oxide (MoS_2/rGO) composites for Pb^{2+} removal from aqueous solutions. The results showed that the adsorption was pH dependent and followed the pseudo-second order model. Maximum adsorption capacities were 384.16 mg g^{-1} at pH 5. The pH influenced both, the species and ionization degree of Pb^{2+}, and the surface charge and functional groups of the sorbent. In general, the adsorption efficiency increased with the increment of pH. The authors concluded that the adsorption process was dominated by surface complexation of Pb^{2+} and electrostatic interaction. In addition, the influence of humic acid on metal ion adsorption was studied, because humic acid have influence on the metal ion mobility and dispersion in the environment. Results showed that due to the negatively charged functional groups of humic acid (0.1 g L^{-1}) the adsorption of positively charged Pb^{2+} on MoS_2/rGO was improved.

Awad et al. (2017) presented a novel adsorbent 2-Imino-4-Thiobiuret partially reduced graphene oxide (IT-PRGO) for heavy metal ion removal from water with high selectivity for Hg^{2+}. The novelty of this adsorbent was that a partial modification of GO was never done before. The selectivity for Hg^{2+} was achieved by introducing amidoxime (AO) functional group within the surface of GO. Since the authors have predicted that the pH of solution will affect the adsorption capacity of metal ions, a series of batch equilibrium measurements were carried out to determine the effect of pH on the adsorption of Hg^{2+}, Pb^{2+}, Cu^{2+}, Cr^{6+} and As^{5+} on IT-PRGO. The pH was adjusted from 1 to 8. Since the predominant species of arsenic and chromium ion are $H_2AsO_4^-$ and $HCrO_4^-$, protonation of the NH_2 groups on the surface of IT-PRGO will result in electrostatic attraction between the amino groups and anions (As and Cr). However, the adsorption capacities of mercury, lead and copper increases with increasing the pH, and finally start to precipitate as hydroxides at pH > 6, so all experiments are carried out at pH 5–5.5. Briefly, IT-PRGO showed excellent capacity and selectivity, 100% removal efficiency for Hg^{2+}, rapid adsorption in short time (at initial Hg^{2+} concentration of 50 mg L^{-1}). The removal efficiencies of other examined metal ions were 85, 92 and 98% respectively for Cu^{2+}, Cr^{6+} and Pb^{2+}. At the end, the desorption study showed that the IT-PRGO could be regenerated performing still excellent removal efficiencies, suggesting this adsorbent as a top performing adsorbent for toxic metal ions contaminated water.

4 Conclusion

Recent studies on graphene oxide nanocomposites as adsorbents for toxic metal removal from water indicate that these materials can effectively be applied in water remediation technologies. The efficiency of these adsorbents mostly depend on the solutions pH, temperature, initial pollutant concentration, but also, on the surface charge and content of functional groups. Therefore, this review provides information about recently developed/modified graphene oxide nanocomposites, their efficiency in heavy metal removal from water, adsorption kinetics and thermodynamic models, as well as their regeneration potential. However, there is always a need to develop more efficient and cost effective adsorbents for remediation processes and their further development to make them applicable in real water and wastewater treatment processes.

References

Ahamad T, Naushad M, Inamuddin (2015) Heavy metal ion-exchange kinetic studies over cellulose acetate Zr(IV) molybdophosphate composite cation-exchanger. Desalin Water Treat 53:1675–1682. https://doi.org/10.1080/19443994.2013.855676

Alam MM, ALOthman ZA, Naushad M (2013) Analytical and environmental applications of polyaniline Sn(IV) tungstoarsenate and polypyrrole polyantimonic acid composite cation-exchangers. J Ind Eng Chem 19:1973–1980. https://doi.org/10.1016/j.jiec.2013.03.006

Al-Nafiey AKH (2016) Reduced graphene oxide-based nanocomposites: synthesis, characterization and applications. Dissertation, University de Lille 1, France

Awad FS, AbouZeid KM, El-Maaty WMA, El-Wakil AM, El-Shall MS (2017) Efficient Removal of Heavy Metals from Polluted Water with High Selectivity for Mercury(II) by 2-Imino-4-thiobiuret–Partially Reduced Graphene Oxide (IT-PRGO). ACS Appl Mater Interfaces acsami.7b10021. https://doi.org/10.1021/acsami.7b10021

Babu CM, Vinodh R, Abidov A, Ravikumar R, Peng MM, Cha WS, Jang H-T (2015) Removal of heavy metals using amine crosslinked reduced graphene oxide. Adv Sci Technol Lett 120:430–433. https://doi.org/10.14257/astl.2015.120.83

Chandra V, Park J, Chun Y, Lee JW, Hwang IC, Kim KS (2010) Water-dispersible magnetite-reduced graphene oxide composites for arsenic removal. ACS Nano 4:3979–3986. https://doi.org/10.1021/nn1008897

Chauke VP, Maity A, Chetty A (2015) High-performance towards removal of toxic hexavalent chromium from aqueous solution using graphene oxide-alpha cyclodextrin-polypyrrole nanocomposites. J Mol Liq 211:71–77. https://doi.org/10.1016/j.molliq.2015.06.044

Chouhan B, Meena P, Poonar N (2016) Effect of heavy metal ions in water on human health. Int J Sci Eng Res 4:2015–2017

Cobbina SJ, Duwiejuah AB, Quansah R, Obiri S, Bakobie N (2015) Comparative assessment of heavy metals in drinking water sources in two small-scale mining communities in Northern Ghana. Int J Environ Res Public Health 12:10620–10634. https://doi.org/10.3390/ijerph120910620

Cui L, Wang Y, Gao L, Hu L, Yan L, Wei Q, Du B (2015) EDTA functionalized magnetic graphene oxide for removal of Pb(II), Hg(II) and Cu(II) in water treatment: adsorption mechanism and separation property. Chem Eng J 281:1–10. https://doi.org/10.1016/j.cej.2015.06.043

Dargahi A, Golestanifar H, Darvishi P, Karami A, Hasan SH, Poormohammadi A, Behzadnia A (2016) An investigation and comparison of removing heavy metals (lead and chromium) from aqueous solutions using magnesium oxide nanoparticles. Pol. J. Environ. Stud 25:557–562. https://doi.org/10.15244/pjoes/60281

Deng JH, Zhang XR, Zeng GM, Gong JL, Niu QY, Liang J (2013) Simultaneous removal of Cd (II) and ionic dyes from aqueous solution using magnetic graphene oxide nanocomposite as an adsorbent. Chem Eng J 226:189–200. https://doi.org/10.1016/j.cej.2013.04.045

Dong C, Lu J, Qiu B, Shen B, Xing M, Zhang J (2018) Developing stretchable and graphene-oxide-based hydrogel for the removal of organic pollutants and heavy metal ions. Appl Catal B Environ 222:146–156. https://doi.org/10.1016/j.apcatb.2017.10.011

Du Y, Wang J, Zou Y, Yao W, Hou J, Xia L, Peng A, Alsaedi A, Hayat T, Wang X (2017) Synthesis of molybdenum disulfide/reduced graphene oxide composites for effective removal of Pb(II) from aqueous solutions. Sci Bull 62:913–922. https://doi.org/10.1016/j.scib.2017.05.025

Fan L, Luo C, Sun M, Li X, Qiu H (2013) Highly selective adsorption of lead ions by water-dispersible magnetic chitosan/graphene oxide composites. Colloids Surf B Biointerfaces 103:523–529. https://doi.org/10.1016/j.colsurfb.2012.11.006

Gambhir RS, Kapoor V, Nirola A, Sohi R, Bansal V (2012) Water pollution: impact of pollutants and new promising techniques in purification process. J Hum Ecol 37:103–110

Gao W, Majumder M, Alemany LB, Narayanan TN, Ibarra MA, Pradhan BK, Ajayan PM (2011) Engineered graphite oxide materials for application in water purification. ACS Appl Mater Interfaces 3:1821–1826

Ge H, Ma Z (2015) Microwave preparation of triethylenetetramine modified graphene oxide/chitosan composite for adsorption of Cr(VI). Carbohydr Polym 131:280–287. https://doi.org/10.1016/j.carbpol.2015.06.025

Habuda-Stanić M, Nujić M (2015) Arsenic removal by nanoparticles: a review. Environ Sci Pollut Res 22:8094–8123. https://doi.org/10.1007/s11356-015-4307-z

Hashim N, Muda Z, Hussein MZ, Isa IM, Mohamed A, Kamari A, Bakar SA, Mamat M, Jaafar AM (2016) A brief review on recent graphene oxide-based material nanocomposites: synthesis and applications. J Mater Environ Sci 7:3225–3243

He YQ, Zhang NN, Wang XD (2011) Adsorption of graphene oxide/chitosan porous materials for metal ions. Chinese Chem Lett 22:859–862. https://doi.org/10.1016/j.cclet.2010.12.049

Hoan NTV, Thu NTA, Van Duc H, Cuong ND, Khieu DQ, Vo V (2016) Fe_3O_4/reduced graphene oxide nanocomposite: synthesis and its application for toxic metal ion removal. J Chem 2016:1–10. https://doi.org/10.1155/2016/2418172

Hu XJ, Liu YG, Wang H, Zeng GM, Hu X, Guo YM, Li TT, Chen AW, Jiang LH, Guo FY (2015) Adsorption of copper by magnetic graphene oxide-supported β-cyclodextrin: effects of pH, ionic strength, background electrolytes, and citric acid. Chem Eng Res Des 93:675–683. https://doi.org/10.1016/j.cherd.2014.06.002

Jaishankar M, Tseten T, Anbalagan N, Mathew BB, Beeregowda KN (2014) Toxicity, mechanism and health effects of some heavy metals. Interdiscip Toxicol 7:60–72. https://doi.org/10.2478/intox-2014-0009

Jastrzębska AM, Kurtycz P, Olszyna AR (2012) Recent advances in graphene family materials toxicity investigations. J Nanoparticle Res 14:1–21. https://doi.org/10.1007/s11051-012-1320-8

Jiang G, Lin Z, Chen C, Zhu L, Chang Q, Wang N, Wei W, Tang H (2011) TiO_2 nanoparticles assembled on graphene oxide nanosheets with high photocatalytic activity for removal of pollutants. Carbon N Y 49:2693–2701. https://doi.org/10.1016/j.carbon.2011.02.059

Jiang T, Yan L, Zhang L, Li Y, Zhao Q, Yin H (2015) Fabrication of a novel graphene oxide/ β-FeOOH composite and its adsorption behavior for copper ions from aqueous solution. Dalt Trans 44:10448–10456. https://doi.org/10.1039/c5dt01030f

Keshvardoostchokami M, Babaei L, Zamani AA, Parizanganeh AH, Piri F (2017) Synthesized chitosan/iron oxide nanocomposite and shrimp shell in removal of nickel, cadmium and lead from aqueous solution. Glob J Environ Sci Manag 3:267–278. https://doi.org/10.22034/gjesm.2017.03.03.004

Khazaei M, Nasseri S, Ganjali MR, Khoobi M, Nabizadeh R, Mahvi AH, Nazmara S, Gholibegloo E (2016) Response surface modeling of lead (II) removal by graphene oxide-Fe_3O_4 nanocomposite using central composite design. J Environ Heal Sci Eng 14:1–14. https://doi.org/10.1186/s40201-016-0243-1

Kumar ASK, Kakan SS, Rajesh N (2013) A novel amine impregnated graphene oxide adsorbent for the removal of hexavalent chromium. Chem Eng J 230:328–337. https://doi.org/10.1016/j.cej.2013.06.089

Kyzas GZ, Deliyanni EA, Matis KA (2014) Graphene oxide and its application as an adsorbent for wastewater treatment. J Chem Technol Biotechnol 89:196–205. https://doi.org/10.1002/jctb.4220

Li S, Lu X, Xue Y, Lei J, Zheng T, Wang C (2012) Fabrication of polypyrrole/graphene oxide composite nanosheets and their applications for Cr(VI) removal in aqueous solution. PLoS ONE 7:1–7. https://doi.org/10.1371/journal.pone.0043328

Li L, Luo C, Li X, Duan H, Wang X (2014) Preparation of magnetic ionic liquid/chitosan/ graphene oxide composite and application for water treatment. Int J Biol Macromol 66:172–178. https://doi.org/10.1016/j.ijbiomac.2014.02.031

Li L, Wang Z, Ma P, Bai H, Dong W, Chen M (2015) Preparation of polyvinyl alcohol/chitosan hydrogel compounded with graphene oxide to enhance the adsorption properties for Cu(II) in aqueous solution. J Polym Res 22:1–10. https://doi.org/10.1007/s10965-015-0794-3

Liu L, Li C, Bao C, Jia Q, Xiao P, Liu X, Zhang Q (2012) Preparation and characterization of chitosan/graphene oxide composites for the adsorption of Au(III) and Pd(II). Talanta 93:350–357. https://doi.org/10.1016/j.talanta.2012.02.051

Liu Y, Xu L, Liu J, Liu X, Chen C, Li G, Meng Y (2016) Graphene oxides cross-linked with hyperbranched polyethylenimines: preparation, characterization and their potential as recyclable and highly efficient adsorption materials for lead(II) ions. Chem Eng J 285:698–708. https://doi.org/10.1016/j.cej.2015.10.047

Luo X, Wang C, Wang L, Deng F, Luo S, Tu X, Au C (2013) Nanocomposites of graphene oxide-hydrated zirconium oxide for simultaneous removal of As(III) and As(V) from water. Chem Eng J 220:98–106. https://doi.org/10.1016/j.cej.2013.01.017

Luo S, Xu X, Zhou G, Liu C, Tang Y, Liu Y (2014) Amino siloxane oligomer-linked graphene oxide as an efficient adsorbent for removal of Pb(II) from wastewater. J Hazard Mater 274:145–155. https://doi.org/10.1016/j.jhazmat.2014.03.062

Madadrang CJ, Kim HY, Gao G, Wang N, Zhu J, Feng H, Gorring M, Kasner ML, Hou S (2012) Adsorption behavior of EDTA-graphene oxide for Pb (II) removal. ACS Appl Mater Interfaces 4:1186–1193. https://doi.org/10.1021/am201645g

Mejias Carpio IE, Mangadlao JD, Nguyen HN, Advincula RC, Rodrigues DF (2014) Graphene oxide functionalized with ethylenediamine triacetic acid for heavy metal adsorption and anti-microbial applications. Carbon N Y 77:289–301. https://doi.org/10.1016/j.carbon.2014.05.032

Musico YLF, Santos CM, Dalida MLP, Rodrigues DF (2013) Improved removal of lead (II) from water using a polymer-based graphene oxide nanocomposite. J Mater Chem A 1:3789–3796. https://doi.org/10.1039/c3ta01616a

Najafabadi HH, Irani M, Roshanfekr Rad L, Heydari Haratameh A, Haririan I (2015) Removal of Cu 2+, Pb 2+ and Cr 6+ from aqueous solutions using a chitosan/graphene oxide composite nanofibrous adsorbent. RSC Adv 5:16532–16539. https://doi.org/10.1039/C5RA01500F

Namvari M, Namazi H (2014) Clicking graphene oxide and Fe_3O_4 nanoparticles together: an efficient adsorbent to remove dyes from aqueous solutions. Int J Environ Sci Technol 11:1527–1536. https://doi.org/10.1007/s13762-014-0595-y

Naushad M, Ansari AA, ALOthman ZA, Mittal J (2016) Synthesis and characterization of YVO 4: Eu 3+ nanoparticles: kinetics and isotherm studies for the removal of Cd 2+ metal ion. Desalin Water Treat 57:2081–2088. https://doi.org/10.1080/19443994.2014.986205

Nupearachchi CN, Mahatantila K, Vithanage M (2017) Application of graphene for decontamination of water; implications for sorptive removal. Groundw Sustain Dev 5:206–215. https://doi.org/10.1016/j.gsd.2017.06.006

Pathania D, Gupta D, Al-Muhtaseb AH et al (2016) Photocatalytic degradation of highly toxic dyes using chitosan-g-poly(acrylamide)/ZnS in presence of solar irradiation. J Photochem Photobiol A Chem 329:61–68. https://doi.org/10.1016/j.jphotochem.2016.06.019

Payne M (2008) Lead in drinking water. Can Med Assoc J 179:253–254. https://doi.org/10.1503/cmaj.071483

Pendolino F, Armata N (2017) Graphene oxide in environmental remediation process. Springer, Switzerland

Pérez-Ramírez EE, De Luz-Asunción M, Martínez-Hernández AL, Velasco-Santos C (2016) Graphene materials to remove organic pollutants and heavy metals from water: photocatalysis and adsorption. In: Cao W (ed) Semiconductor photocatalysis—materials, mechanisms and applications, 1st edn. InTech, Rijeka, Croatia, pp 493–522

Pourbeyram S (2016) Effective removal of heavy metals from aqueous solutions by graphene oxide-zirconium phosphate (GO–Zr–P) nanocomposite. Ind Eng Chem Res 55:5608–5617. https://doi.org/10.1021/acs.iecr.6b00728

Qiu H, Lv L, Pan B, Zhang QQ, Zhang W, Zhang QQ (2009) Critical review in adsorption kinetic models. J Zhejiang Univ Sci A 10:716–724. https://doi.org/10.1631/jzus.A0820524

Rahimi E, Mohaghegh N (2017) New hybrid nanocomposite of copper terephthalate MOF-graphene oxide: synthesis, characterization and application as adsorbents for toxic metal ion removal from Sungun acid mine drainage. Environ Sci Pollut Res 24:22353–22360. https://doi.org/10.1007/s11356-017-9823-6

Rahimi R, Dorabei RZ, Koohi A, Zargari S (2014) Synthesis of graphene oxide-porphyrin nanocomposite and its application in removal of toxic metals. https://doi.org/10.3390/ecsoc-18-a031

Ren X, Chen C, Nagatsu M, Wang X (2011) Carbon nanotubes as adsorbents in environmental pollution management: a review. Chem Eng J 170:395–410. https://doi.org/10.1016/j.cej.2010.08.045

Rodriguez ARC (2015) Removal of cadmium (II), lead (II) and chromium (VI) in water with nanomaterials. Dissertation, University Autonoma de Barcelona

Sahraei R, Ghaemy M (2017) Synthesis of modified gum tragacanth/graphene oxide composite hydrogel for heavy metal ions removal and preparation of silver nanocomposite for antibacterial activity. Carbohydr Polym 157:823–833. https://doi.org/10.1016/j.carbpol.2016.10.059

Salavagione HJ, Martínez G, Ellis G (2011) Recent advances in the covalent modification of graphene with polymers. Macromol Rapid Commun 32:1771–1789. https://doi.org/10.1002/marc.201100527

Samadi S, Abbaszadeh M (2017) Synthesis and characterization of MgO/PEG/GO nanocomposite and its application for removal of copper (II) from aquatic media. Bull la Soc R des Sci Liege 86:271–280

Shahat A, Awual MR, Khaleque MA et al (2015) Large-pore diameter nano-adsorbent and its application for rapid lead (II) detection and removal from aqueous media. Chem Eng J 273:286–295. https://doi.org/10.1016/j.cej.2015.03.073

Shahzad A, Miran W, Rasool K, Nawaz M, Jang J, Lim S-R, Lee DS (2017) Heavy metals removal by EDTA-functionalized chitosan graphene oxide nanocomposites. RSC Adv 7:9764–9771. https://doi.org/10.1039/C6RA28406J

Sheshmani S, Akhundi Nematzadeh M, Shokrollahzadeh S, Ashori A (2015) Preparation of graphene oxide/chitosan/FeOOH nanocomposite for the removal of Pb(II) from aqueous solution. Int J Biol Macromol 80:475–480. https://doi.org/10.1016/j.ijbiomac.2015.07.009

Sitko R, Turek E, Zawisza B, Malicka E, Talik E, Heimann J, Gagor A, Feist B, Wrzalik R (2013) Adsorption of divalent metal ions from aqueous solutions using graphene oxide. Dalt Trans 42:5682. https://doi.org/10.1039/c3dt33097d

Sitko R, Janik P, Feist B, Talik E, Gagor A (2014) Suspended aminosilanized graphene oxide nanosheets for selective preconcentration of lead ions and ultrasensitive determination by electrothermal atomic absorption spectrometry. ACS Appl Mater Interfaces 6:20144–20153. https://doi.org/10.1021/am505740d

Sreeprasad TS, Maliyekkal SM, Lisha KP, Pradeep T (2011) Reduced graphene oxide-metal/metal oxide composites: facile synthesis and application in water purification. J Hazard Mater 186:921–931. https://doi.org/10.1016/j.jhazmat.2010.11.100

Sui N, Wang L, Wu X, Li X, Sui J, Xiao H, Liu M, Wan J, Yu WW (2015) Polyethylenimine modified magnetic graphene oxide nanocomposites for Cu 2+ removal. RSC Adv 5:746–752. https://doi.org/10.1039/C4RA11669K

Tchounwou PB, Yedjou CG, Patlolla AK, Sutton DJ (2012) Heavy metals toxicity and the environment. In: Luch A (ed) Molecular, clinical and environmental toxicology, vol 3. Springer, Berlin

Thürmer K, Williams E, Reutt-Robey J (2002) Autocatalytic oxidation of lead crystallite surfaces. Science 297(80):2033–2035. https://doi.org/10.1126/science.297.5589.2033

Tien TTT, Tu TH, Thao HNP, Hieu NH (2017) Preparation of RHA-silica/graphene oxide nanocomposite for removal of nickel ions from water. In: International conference on chemical engineering, food and biotechnology (ICCFB2017), p 1878

Turhanen PA, Vepsäläinen JJ, Peräniemi S (2015) Advanced material and approach for metal ions removal from aqueous solutions. Sci Rep 5:1–8. https://doi.org/10.1038/srep08992

Vaishaly A, Blessy MB, Krishnamurthy N (2015) Health effects caused by metal contaminated ground water. Int J Adv Sci Res 1:60–64. https://doi.org/10.7439/ijasr

Wang X, Guo Y, Yang L, Han M, Zhao J, Cheng X (2012) Nanomaterials as sorbents to remove heavy metal ions in wastewater treatment. J Environ Anal Toxicol 2:154. https://doi.org/10.4172/2161-0525.1000154

Wang S, Sun H, Ang HM, Tade MO (2013a) Adsorptive remediation of environmental pollutants using novel graphene-based nanomaterials. Chem Eng J 226:336–347. https://doi.org/10.1016/j.cej.2013.04.070

Wang Y, Liang S, Chen B, Guo F, Yu S, Tang Y (2013b) Synergistic removal of Pb (II), Cd (II) and humic acid by Fe_3O_4@ mesoporous silica-graphene oxide composites. PLoS ONE 8:2–9. https://doi.org/10.1371/journal.pone.0065634

Wang Y, Liu X, Wang H, Xia G, Huang W, Song R (2014) Microporous spongy chitosan monoliths doped with graphene oxide as highly effective adsorbent for methyl orange and copper nitrate ($Cu(NO_3)_2$) ions. J Colloid Interface Sci 416:243–251. https://doi.org/10.1016/j.jcis.2013.11.012

Wang Y, Yan T, Gao L, Cui L, Hu L, Yan L, Du B, Wei Q (2016) Magnetic hydroxypropyl chitosan functionalized graphene oxide as adsorbent for the removal of lead ions from aqueous solution. Desalin Water Treat 57:3975–3984. https://doi.org/10.1080/19443994.2014.989273

WHO (1996) Trace elements in human nutrition and health, World Health Organization. World Heal Organ Geneva, pp 1–360

Wu W, Yang Y, Zhou H, Ye T, Huang Z, Liu R, Kuang Y (2013) Highly efficient removal of Cu (II) from aqueous solution by using graphene oxide. Water Air Soil Pollut 224:1372. https://doi.org/10.1007/s11270-012-1372-5

Xing HT, Chen JH, Sun X, Huang YH, Su ZB, Hu SR, Weng W, Li SX, Guo HX, Wu WB, He YS, Li FM, Huang Y (2015) NH_2-rich polymer/graphene oxide use as a novel adsorbent for removal of Cu(II) from aqueous solution. Chem Eng J 263:280–289. https://doi.org/10.1016/j.cej.2014.10.111

Yari M, Rajabi M, Moradi O, Yari A, Asif M, Agarwal S, Gupta VK (2015) Kinetics of the adsorption of Pb(II) ions from aqueous solutions by graphene oxide and thiol functionalized graphene oxide. J Mol Liq 209:50–57. https://doi.org/10.1016/j.molliq.2015.05.022

Youssef AM, Malhat FM (2014) Selective removal of heavy metals from drinking water using titanium dioxide nanowire. Macromol Symp 337:96–101. https://doi.org/10.1002/masy.201450311

Yu Y, De Andrade LCX, Fang L, Ma J, Zhang W, Tang Y (2015) Graphene oxide and hyperbranched polymer-toughened hydrogels with improved absorption properties and durability. J Mater Sci 50:3457–3466. https://doi.org/10.1007/s10853-015-8905-4

Zhang N, Qiu H, Si Y, Wang W, Gao J (2011) Fabrication of highly porous biodegradable monoliths strengthened by graphene oxide and their adsorption of metal ions. Carbon N Y 49:827–837. https://doi.org/10.1016/j.carbon.2010.10.024

Zhang Y, Liu Y, Wang X, Sun Z, Ma J, Wu T, Xing F, Gao J (2014) Porous graphene oxide/carboxymethyl cellulose monoliths, with high metal ion adsorption. Carbohydr Polym 101:392–400. https://doi.org/10.1016/j.carbpol.2013.09.066

Zhou G, Liu C, Tang Y, Luo S, Zeng Z, Liu Y, Xu R, Chu L (2015) Sponge-like polysiloxane-graphene oxide gel as a highly efficient and renewable adsorbent for lead and cadmium metals removal from wastewater. Chem Eng J 280:275–282. https://doi.org/10.1016/j.cej.2015.06.041

Zhu J, Wei S, Chen M, Gu H, Rapole SB, Pallavkar S, Ho TC, Hopper J, Guo Z (2013) Magnetic nanocomposites for environmental remediation. Adv Powder Technol 24:459–467. https://doi.org/10.1016/j.apt.2012.10.012

Chapter 16
New Analytical Approaches for Pharmaceutical Wastewater Treatment Using Graphene Based Materials

P. Senthil Kumar and A. Saravanan

Abstract Recently, water pollution is the serious environmental threat throughout the world. Water environment can be polluted by several ways. Amongst that, pharmaceutical industry wastewater assumes a noteworthy part, which varies colossally in flow and composition, contingent upon variables, for example, production rate, the particular preparation being completed, which activities are producing the waste water, etc. Every one of these factors imply that the contamination of the last emanating can be exceptionally various and variable after some time. In perspective of the shortage of water assets, it is important to comprehend and create methodologies for treatment of pharmaceutical wastewater as a feature of water administration. The arrival of graphene and grapheme based materials in water appears an inescapable outcome of the monstrous future utilization of these carbonaceous allotropes. From a natural designing perspective, it ought to be noticed that piece of the watery streams containing these nanomaterials will wind up in wastewater treatment plants, and there will be cooperation between the nanomaterials, other toxins in the sewage, which could influence the viability of the depuration procedure.

Keywords Pharmaceutical effluent · Graphene · Water pollution
Nanomaterials · Wastewater reuse

P. Senthil Kumar (✉)
Department of Chemical Engineering, SSN College of Engineering,
Chennai 603110, India
e-mail: psk8582@gmail.com; senthilkumarp@ssn.edu.in

A. Saravanan
Department of Biotechnology, Vel Tech High Tech Dr. Rangarajan
Dr. Sakunthala Engineering College, Chennai 600062, India
e-mail: sara.biotech7@gmail.com

M. Naushad (ed.), *A New Generation Material Graphene: Applications in Water Technology*, https://doi.org/10.1007/978-3-319-75484-0_16

1 Introduction

The worldwide interest for quality water, regardless of whether for reasons for drinking, sanitation, water system, and industrial purposes, has been on a persistent ascent, and there has been overpowering worry in late years about water treatment and reuse (Gadipelly et al. 2014). Rising concerns related with water shortage are winding up even more a critical issue over a more noteworthy number of locales, and particularly in territories helpless to dry spell and water deficiencies. Reducing water supply is especially troublesome to enterprises that are dependent on brilliant process water for use in assembling. To relieve potential water deficiency dangers and increase more prominent water security, mechanical organizations are progressively executing techniques that encourage higher water effectiveness. Industrial manufacturing forms in the pharmaceutical business create wastewater that is for the most part portrayed as high quality organic profluent—waste streams that can be trying to make do with traditional wastewater treatment. In the course of the most recent years an extensive variety of pharmaceutically dynamic compound deposits have been found in a few natural grids because of their broad utilization and pseudo-ingenuity in the earth (Verlicchi et al. 2012; Yang et al. 2011). Pharmaceuticals, recently perceived classes of natural toxins, are ending up progressively risky contaminants of either surface water or ground water around mechanical and private groups (Kyzas et al. 2015). Pharmaceutical products have been broadly utilized as a part of many fields, for example, pharmaceutical, industry, domesticated animals cultivating, aquaculture and people's day by day life. They are getting to be pervasive in the conditions because of their broad applications also; poor expulsion by the regular natural wastewater treatment plants (Wang and Wang 2016). The pharmaceutical items can be brought into the earth by two pathways—direct and indirect way.

Elaborately, pharmaceutical items can enter the surface water by means of straightforwardly releasing into the surface water by enterprises, healing facilities, family units and wastewater treatment plants furthermore, through land overflow in the event of bio solids spread on rural arrive, which can come to the groundwater by draining. The contents of pharmaceutical wastewater have shown in Fig. 1. Inside the surface water compartment, silt can absorb the pharmaceutical items since it has an assortment of restricting destinations (Kastner et al. 2014).

1.1 Wellsprings of Dangers in Pharmaceutical Businesses

- Assembling and definition establishments.
- Handling and capacity of dangerous chemicals including distribution centres, good possesses; tank shapes in ports/fuel warehouses/docks.
- Transportation (street, rail, air, water, pipelines).

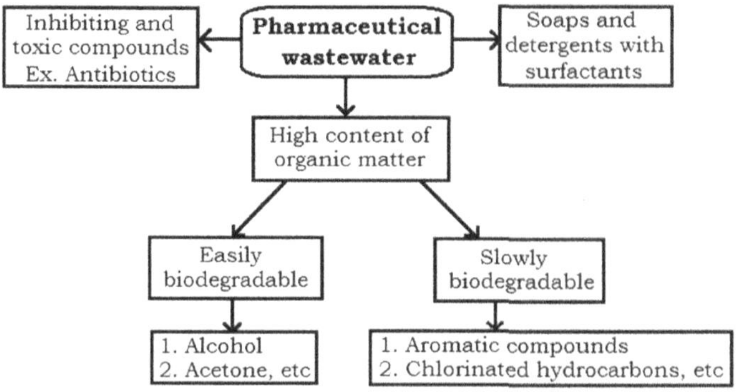

Fig. 1 Contents of pharmaceutical wastewater

- Emission of pollutants–the air toxins incorporate carbon monoxide (CO), Nitrogen dioxide (NO_2), particulate matter of 10 μm or less (PM10), Total suspended particulate issue (SPM), sulphur dioxide (SO_2), and Volatile natural mixes (VOCs). The most basic VOCs incorporate methanol, dichloromethane, toluene, ethylene glycol, *N*, *N* dimethyl formamide, and acetonitrile.
- Effluents, particularly those that are not effectively biodegradable and harmful in nature. The profluent discharges could go straightforwardly to streams, waterways, lakes, seas, or different waterways. The discharges due to overflow, including storm water spill over, could likewise be a potential risk.

2 Pharmaceutical Manufacturing Process and Wastewater Arrangements

Pharmaceuticals producing process for the most part includes forms, to be specific, aging, extraction, refinement, concoction blend, plan, and bundling. Each of these procedures creates strong, fluid, and vaporous squanders.

The water utilization in the pharmaceutical business relies upon the procedure, the nature of the crude materials utilized, and items produced. The purging procedure likewise devours immense amounts of water. Fluid effluents coming about because of cleaning operations after cluster generation may likewise contain poisonous natural deposits. Notwithstanding the assembling waste, medicates that are not completely used in the body might be discharged into the sewer framework. The classification of pharmaceutical industries was shown in Fig. 2.

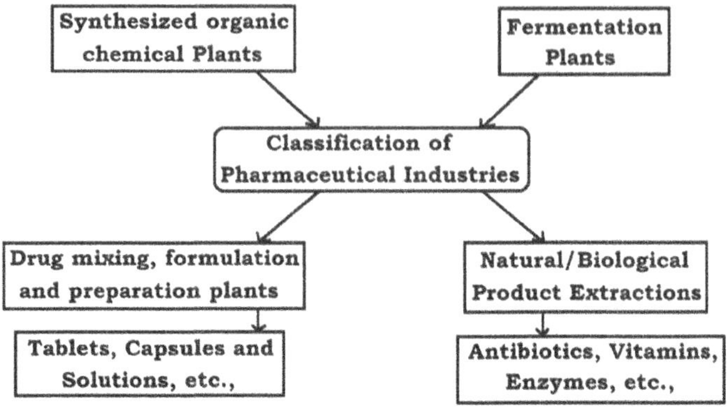

Fig. 2 Classification of pharmaceutical industry

2.1 *Synthetic Blend Process*

Compound union procedures utilize natural and inorganic chemicals in clump operations to create drugs with characterized pharmacological activity or intermediates. Wastewaters from substance combination operations are various because of numerous operations and responses occurring in the reactor and in addition at various stages. Practically every stage produces mother alcohol that contains unreacted reactants, items, coproduces/side-effects, and leftover items in the natural dissolvable base. Acids, bases, halides, nitrates, sulfates, cyanides, and metals may likewise be produced. As a rule, the spent dissolvable recuperation prompts dissolvable wastewater at the scrubber stage after dissipation. Wastewater is created at the filtration steps including solvents, completed items, cleaning water, and spills. This sewage has a high poisonous quality level; therefore, it requires prompt treatment instead of discharge into WWTP. Wastewaters from amalgamation forms ordinarily have high organic oxygen request (BOD), synthetic oxygen request (COD), and aggregate suspended solids (TSS) levels and pH.

2.2 *Fermentation Process*

The aging procedure creates a lot of waste, for example, spent fluid maturation juices and dead cell squander. As in the greater part of the watery stage maturations the microbes don't get by at higher convergences of the item in light of hindrance of the microscopic organisms because of gathering of the item. The waste stream has an expansive amount of unconsumed crude materials, for example, the supplement stock, metal salts, starch, nitrates, and phosphates with high COD, BOD, and TSS with pH esteems extending from 4 to 8. Steam and little measures of modern chemicals (phenols, cleansers, and disinfectants) keep up the sterility in the process plant and in this manner their scraps likewise add to the fluid waste stream. A significant amount of metal and halogen pollutions is additionally found because of utilization for the precipitation of the item from the mother alcohol (Awual et al. 2015, 2016; Nabi et al. 2011). A lot of solvents are additionally utilized for the filtration of the coveted item, and amid the reusing of the solvents watery waste having miscible natural solvents is produced.

Various pharmaceutical mixes have been appeared to go through sewage treatment plants and sully the oceanic condition. Wastewaters in a pharmaceutical assembling industry more often than not begin from the blend and plan of the medications.

Since the reactors and separators utilized as a part of a multiproduct pharmaceutical industry are not planned per the limit but rather regularly curiously large or utilized wastefully, the amount of wastewaters produced is expanded. There are various sub forms happening in a pharmaceutical industry, and it is a troublesome undertaking to describe every single item squander. A more expounded arrangement in light of crude materials, last items, and uniqueness of plants has been endeavoured. The characterization is done on the premise of the likenesses of substance procedures and medications and additionally certain classes of items. In light of the procedures engaged with assembling, pharmaceutical enterprises can be subdivided into the accompanying five noteworthy subcategories.

2.3 *Characteristic/Biological Extraction Process*

A lot of common materials are prepared to extricate the dynamic pharmaceutical fixing from the source. In each progression, a huge volume of water input is required and the item recuperation diminishes until the point when the last item is come to. Ordinarily hexane is utilized as dissolvable for characteristic item or natural extraction, which is discharged into the air and furthermore the water.

Nowadays forms in light of supercritical carbon dioxide are created to contain natural polluting influences in the last item and in addition to decrease profluent. Spent crude material and solvents, wash water, and spills are the essential wellsprings of wastewater. Natural and inorganic chemicals might be available as build-ups in these waste streams. Additionally, the use of an assortment of low-bubbling natural solvents creates wastewater with solvents. More often than not, wastewaters have low BOD, COD, and TSS, with moderately impartial pH esteem. Schematic flow diagram for Natural/Biological extraction process was shown in Fig. 3.

A wide assortment of items is made in the compound and pharmaceutical assembling ventures, regularly requiring extensive volumes of chemicals, materials, and substances that are utilized all through procedure operations. The blends of pharmaceuticals, hormones, and other wastewater contaminants can happen at low focuses in streams that are vulnerable to different wastewater sources, and the volumes will change from industry to industry or site to site for a similar

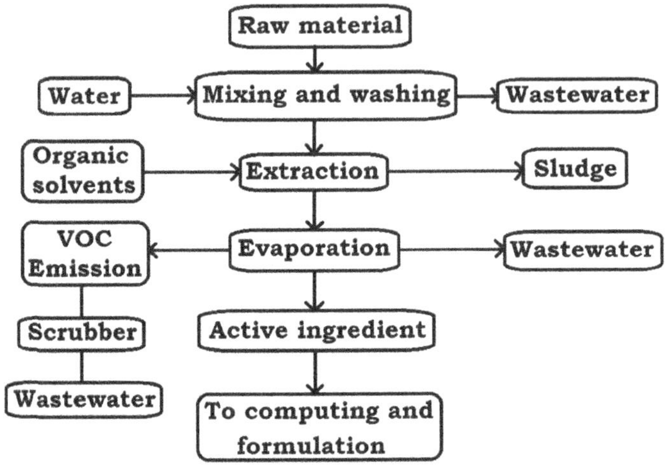

Fig. 3 Schematic flow diagram for natural/biological extraction process

compound. Squander streams produced in these enterprises can be vigorously weighed down with contaminants, poisons, supplements, and organics, displaying one of some kind difficulties as far as treatment in perspective of stringent directions. It is critical that for reuse in both approved and non-approved frameworks the treated wastewater quality must surpass the nourish water quality for high operational proficiency, water quality, and item wellbeing. In this way, it might be conceivable to extend generation limit without surpassing water release limits, definitely decrease crude water necessities and waste transfer cost of operation, and diminish particular organics while leaving other inorganic species in place.

The wastewater leaving pharmaceutical units shifts in substance and focus and in this manner a novel treatment is not endeavoured since the volumes are little and distinctive items are produced from a similar battery of reactors and separators. Water reuse gives investment funds through the lessening of waste transfer expenses and sustains water necessities, balancing operational expenses related with the waste reuse process.

3 Environmental Impacts by Pharmaceutical Industries

Pharmaceutical medications being utilized for human and in addition veterinary prescriptions are developing as ecological toxins. Diverse pharmaceuticals have been named Analgesics, Antibiotics, Antiepileptic, Antiseptics, Beta-blockers, Antihypertensive, Hormones, Contraceptives, Psychotherapeutics and Anti Virals (Halling-Sorensen et al. 1998).

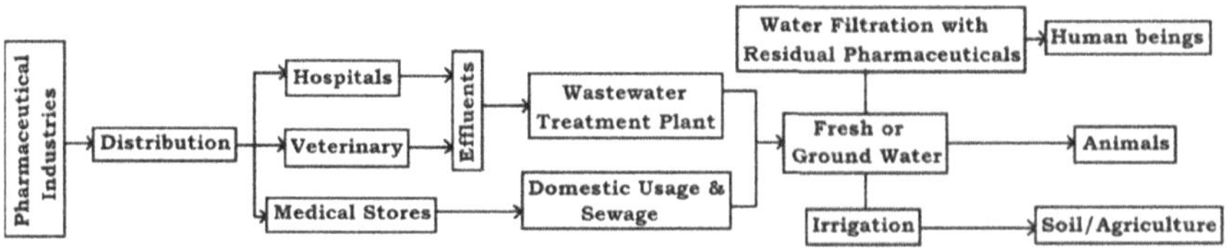

Fig. 4 Overview of pharmaceutical industry

The risks from the pharmaceuticals could be sorted as:

- Eco toxic-harm is caused to nature.
- Carcinogenic-add to the causation of disease.
- Persistent-stay perilous for quite a while.
- Bio-accumulative–accumulates as it advances up the sustenance chain.
- Disastrous because of a fiasco, disaster, catastrophe or grave event in any territory.

The ecological presentation courses of pharmaceuticals into nature are producing units and doctor's facility effluents, arrive applications (e.g., bio solids and water reuse) and so forth. The sewage treatment administrations are not generally effective in expelling the dynamic chemicals from squander water. Therefore, pharmaceuticals discover their way into the amphibian condition, where they straightforwardly influence oceanic life forms and can be fused into natural ways of life (Daughton and Ternes 1999). The overview of pharmaceutical industries was shown in Fig. 4.

Concentrates on antimicrobials have demonstrated that up to 95% of anti-microbial mixes can be discharged unaltered into the sewage framework. Besides, higher centralizations of anti-infection agents can prompt change in microbial group structure and eventually influence evolved ways of life. Non-steroidal mitigating drugs, like ibuprofen, naproxen and diclofenac are broadly being utilized and thus are every now and again identified in sewage, surface water and might be found in ground water framework. Ibuprofen, ketoprofen, naproxen, indomethacin, diclofenac, acetyl salicylic corrosive and phenazone have been found in surface water framework (Tixer et al. 2003).

3.1 *Health Risk of Pharmaceutical Effluents*

The long haul presentation of lower grouping of complex pharmaceutical blends on stream biota may result in intense and unending harms behavioural changes aggregation in tissues conceptive harm and restraint of cell proliferation. Several think about have shown that fish presented to wastewater effluents can display regenerative irregularities. Additionally, angle presented to follow levels of contraception pharmaceuticals in the scope of fixations found in the earth demonstrate sensational diminishes in regenerative achievement, recommending populace level effects are conceivable (Gaworecki and Klaine 2008; Quinn et al. 2008).

The greater part of the fabricating pharmaceutical plants and scholarly/inquire about establishments are situated inside the city where they chip away at the union of new nanomaterial, chemicals and pharmaceuticals. Then again, the untreated or mostly treated wastewater that contains distinctive chemicals and substantial metals may discover its way to some neighbourhood drinkable water wells. Wastewater conveys three noteworthy concoction perils classes with toxicological ramifications that incorporates intense and unending poisonous quality, cancer-causing nature, and regenerative, formative, and neurotoxicity. It is hypothesized that cancer-causing and neurotoxic impacts are definitely not bound to limits.

4 Graphene and Graphene Based Materials

As of late, graphene, a solitary nuclear layer of hybridized carbon particles covalently reinforced in a honeycomb grid has risen as a 'ponder material' with various potential applications. Graphene additionally shows uncommon properties, for example, incredible mechanical, electrical, warm, optical properties and high particular surface territory.

Furthermore, graphene has additionally been utilized as a fantastic adsorbent for various toxins because of its extensive, delocalized π-electron framework, which can shape solid associations with different toxins. Graphene based materials was shown in Fig. 5.

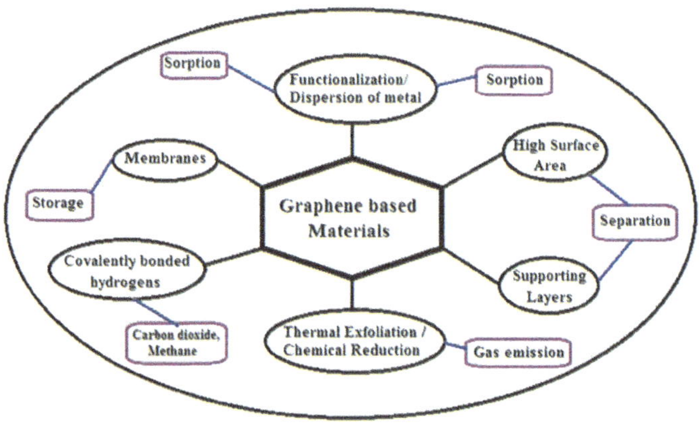

Fig. 5 Graphene based materials

4.1 Properties

Graphene has a few properties that make it appealing for natural applications. Graphene additionally has great optical, warm and mechanical properties. Single sheet graphene is a very straightforward material yet each layer in thickness assimilates up to 2.3% of white light, with under 0.1% reflectance. The most considered part of graphene is presumably its electronic properties.

4.1.1 Electronic Properties

A standout amongst the most valuable properties of graphene is that it is a zero-cover semimetal (with the two gaps and electrons as charge transporters) with high electrical conductivity. Carbon particles have a sum of six electrons; two in the internal shell and four in the external shell. The four external shell electrons in an individual carbon iota are accessible for substance holding, however in graphene, every particle is associated with 3 other carbon iotas on the two dimensional plane, departing 1 electron unreservedly accessible in the third measurement for electronic conduction. These profoundly portable electrons are called pi (π) electrons and are situated above and beneath the graphene sheet. These pi orbitals cover and help to upgrade the carbon to carbon bonds in graphene. In a general sense, the electronic properties of graphene are directed by the holding and hostile to holding (the valance and conduction groups) of these pi orbitals.

4.1.2 Mechanical Strength

Another of graphene's emerge properties is its characteristic quality. Because of the quality of its 0.142 Nm-long carbon bonds, graphene is the most grounded material at any point found, with an extreme elasticity of 130,000,000,000 Pa (or 130 GPa), contrasted with 400,000,000 for A36 basic steel, or 375,700,000 for Aramid (Kevlar). Not exclusively is graphene exceptionally solid, it is likewise light at 0.77 mg for every square meter (for correlation purposes, 1 m^2 of paper is around 1000 times heavier). It is frequently said that a solitary sheet of graphene (being just 1 iota thick), adequate in measure enough to cover an entire football field, would weigh under 1 single gram.

4.1.3 Optical Properties

Graphene's capacity to retain a somewhat extensive 2.3% of white light is likewise an extraordinary and fascinating property, particularly considering that it is just 1 particle thick. This is because of its previously mentioned electronic properties; the electrons acting like mass less accuse bearers of high portability. Because of these great qualities, it has been watched that once optical force achieves a specific edge storable ingestion happens (high power light causes a diminishment in assimilation). This is a vital trademark with respect to the mode-locking of fibre lasers. Because of graphene's properties of wavelength-cold hearted ultrafast saturable ingestion, full-band mode locking has been accomplished utilizing an erbium-doped dissipative solution fibre laser equipped for acquiring wavelength tuning as huge as 30 nm.

4.2 Concepts

Graphene, similar to all nanoscale materials, likewise has a high particular surface territory. Truth be told, graphene speaks to the most extraordinary instance of high-surface materials, since each particle of a solitary layer graphene sheet is uncovered from the two sides to its condition (Sanchez et al. 2012). Graphene has the most noteworthy particular surface territory of all materials, with a hypothetical estimation of 2630 m^2 g^{-1}. The high surface range of graphene makes it a perfect possibility for forms including adsorption or surface responses. In addition to that, graphene speaks to a superb support to grapple substance functionalities or nanomaterials and, in this manner, graphene-based nanocomposites destinations have been a dynamic territory of explore for novel materials (Compton and Nguyen 2010).

4.3 Synthesis

Properties and utilizations of graphene are connected to its shape, size, and morphology. Subsequently, it is vital to have amalgamation systems to control shape, size, and morphology. While considering business applications, researchers need to create synthetic courses which can give high return of graphene with great control over morphology. Genuine Graphene is just a single nuclear layer thick and it ordinarily exists as a film yet it can be drifted off the substrate and can be redepositing onto another substrate or utilized as a part of its separated shape. There are, in any case, a few sorts of graphene containing powder frame materials, for example, graphene oxide, graphene nanoplatelets, graphene nanoribbons, and graphene quantum dabs and also graphene empowered items, for example, graphene ink or graphene master batches. Graphene, because of its one of kind physicochemical properties, was the most proficient, which could be combined by various techniques.

There are 3 primary approaches to blend graphene, they are:

- Concoction Vapour Deposition
- Concoction or Plasma Exfoliation from common Graphite
- Mechanical cleavage from characteristic Graphite.

4.4 *Evacuation of Pharmaceutical Contaminants by Graphene and Graphene-Based Materials*

In the course of the most recent decade, investigate on graphene has expanded many overlap because of its extraordinary properties for ecological remediation and recovery. They can be utilized to diminish contamination stack by adsorption, disintegrate natural poisons, and diligent natural contaminations. Quickly developing pharmaceutical businesses and different exercises have been releasing plentiful measures of natural, inorganic, biodegradable, and non-biodegradable waste into the earth. As of late, pharmaceuticals have been recognized as "rising poisons" extraordinarily contaminating the water streams, making noteworthy danger sea-going life frameworks, and people. These contaminants, which are extremely differing in nature, speak to a major ecological and general wellbeing concern. As a result, a worldwide exertion exists to create hearty advancements to successfully expel contaminants from both air and water. Among these innovations, adsorption is a quick, economical, and compelling strategy for expulsion of contaminants from amphibian environments. Adsorption is a procedure where the contamination (adsorbate) is caught by the nanomaterial (adsorbent) by means of physicochemical associations. Thus, the researcher depicts the utilization of graphene-based materials as adsorbents for the expulsion of inorganic, natural, and vaporous contaminants.

4.4.1 Adsorption of Conceivably Lethal Contaminants

A large portion of the pharmaceutical toxins are natural in nature and have unfriendly effect on condition and human beings. The natural poisons normally have a high oxygen request and low biodegradability, and furthermore have a high bio-aggregation rate along the evolved way of life because of their lipophilicity.

The adsorption component of natural toxin on graphene is subject to π-electron arrangement of natural atoms and π electron of the fragrant ring of graphene. When all is said in done, five conceivable cooperation including hydrophobic impact, π–π bonds, hydrogen bonds, and covalent and electrostatic collaborations have been seen in carbon materials and accepted to be in charge of the adsorption of natural chemicals on the surface of carbon nanosized particles. Several research studies stated that adsorption of pharmaceutical waste on to graphene or other graphene materials happen due to π–π cooperation. The vast majority of the pharmaceutical details, for example, antibiotic medication, ibuprofen, diclofenac, paracetamol, headache medicine, and so on, comprises of at least one fragrant rings and consequently π electrons. The above examinations recommend that the principle between sub-atomic drive between the adsorbent and adsorbate could be π–π association. Other than π–π association, the above research articles likewise propose electrostatic powers and H–bonding to have a viable part in the adsorption instrument for the pharmaceutical mixes. Thus, these are the significant components that can happen amid adsorption procedure to encourage poison expulsion utilizing graphene and graphene-based material.

An assortment of studies has portrayed the use of graphene-based materials as adsorbents for the expulsion of pharmaceutical effluents from water condition. The vast majority of these examinations have utilized graphene as a model adsorbent for remediation of pharmaceutical poisons in water. Graphene is desirable over immaculate graphene for natural compound adsorption because of graphene's high substance of oxygen gatherings accessible to connect with natural compound. The significance of these oxygen-containing utilitarian gatherings was exhibited by looking at the adsorption execution of perfect and oxidized graphene sheets. Various elements, for example, ionic quality, pH, number of oxygen-containing gatherings of graphene, and nearness of characteristic natural issue were found to impact the adsorption limit of graphene. For example, the impact of ionic quality on the adsorption limit might be because of rivalry between electrolytes (NaCl, KCl, and $NaClO_4$). Indeed, the presentation of electrolytes may influence the electrical twofold layer of hydrated particles, in this manner changing the way organic compounds tie to the graphene sheets.

Notwithstanding expanding the adsorption limit, functionalization of graphene materials with natural atoms can likewise be utilized to improve the material recuperation process. Thermo-responsive properties were conferred to graphene-based adsorbent materials utilizing a non-covalent get together of graphene-based adsorbent material with poly (*N*-isopropyl acrylamide).

Graphene-based materials have likewise been investigated for the expulsion of anionic toxins from watery arrangements, for example, phosphate, perchlorate, and fluoride. Not at all like the immobilization of cationic metal species, was the instrument of anion adsorption beforehand credited to anion–π connections. This anion–π affiliation depends on the collaboration between the adversely charged anion (and solitary electron match) with an electron-insufficient fragrant ring on graphene layer.

4.4.2 Graphene-Based Photo Catalytic Materials for Water Disinfecting

Despite the fact that adsorption can expel contaminants from water, this strategy does not corrupt the mixes, which require assist transfer. Finish mineralization or pulverization of contaminants conceivably can be accomplished utilizing photo catalytic treatment. In this attempt, photo catalysis has emerged as an alluring methodology for water remediation and wastewater treatment, since it is low in cost and powerful. In this segment, we portray the diverse techniques to plan graphene-based photo catalysts and their part in the corruption of natural and natural contaminants.

(i) **Arrangement of Graphene-Based Photocatalysts**

A standout amongst the most imperative attributes of graphene for photocatalysis is its capacity to tune the band crevice vitality of semiconductors. Moreover, the nearness of graphene, due to its high electron versatility, adds to the concealment of the quick recombination of electron–hole sets, accordingly prompting an improvement in photocatalytic action. Graphene-based photocatalysts are set up by mooring photoactive nanostructures on graphene. Earlier surveys on this subject portrayed in detail the methods used to plan graphene nanocomposites for photocatalysis purposes. Graphene-based photo catalytic nanocomposites are combined utilizing three fundamental systems. The primary approach includes the development of nanoparticles specifically on graphene surface utilizing the oxygen-containing gatherings of graphene as nucleation destinations for the nanoparticle development.

4.4.3 Graphene in Film and Desalination Innovations

Graphene, in spite of being just a single molecule in thickness, is an impermeable material in its immaculate frame. The delocalized electron billows of the orbitals hinder the hole that would be found in the fragrant rings in graphene, adequately obstructing the entry of even the littlest atomic species. The impermeable idea of graphene has permitted its application as an obstruction for gas and fluid permeation, or to ensure metallic surfaces against consumption. In the territory of water treatment, this novel property of graphene has activated broad endeavours to utilize graphene for the outline of ultrathin graphene-based water-detachment films. Two methodologies have been investigated to utilize graphene nanomaterials in film forms: nanoporous graphene sheets and stacked graphene boundaries.

Until the point when the specialized and temperate confinements of graphene-based films can be beaten, polymeric layers will remain the cutting edge for layer based detachments. While the vitality utilization of a few weight driven film forms is high, their penetrability, selectivity, and reasonableness stay unchallenged by unadulterated graphene-based layers. Be that as it may, by incorporating graphene nano-materials in the plan of polymeric layers, it is conceivable to enhance the execution of polymeric layers by expanding their mechanical properties or diminishing their natural and organic fouling penchant. The phenomenal mechanical properties of graphene nano materials can be utilized to enhance the mechanical quality of polymeric films. Solid film materials are alluring to keep away from film disappointment and to lessen the effect of layer compaction underweight, particularly for film forms subject to high water powered weights like weight impeded osmosis.

Graphene nanomaterials can possibly essentially enhance layer based water treatment. Although a few specialized difficulties stay with a specific end goal to outline graphene-based films for substantial scale applications, huge advances have been made towards accomplishing high selectivity from either nanoporous or stacked graphene films. The principle impediments may stay monetary. Contrasted with the settled polymeric films innovation, graphene-based layer creation will presumably stay costly and constrained to little scale gadgets, for example, microfluidic frameworks, where the superior of graphene-based layers might be required.

5 Future Standpoint

Amid the previous decade, critical advance has been made in seeing how graphene and graphene-based materials can be used to address natural difficulties. The one of kind properties of graphene have opened new promises to enhance the execution of various natural procedures. Be that as it may, in different cases, the change brought by the utilization of graphene was just like what was accomplished with other carbon-based nanomaterials, or even with customary carbonaceous materials like actuated carbon. The decision of whether to utilize graphene as a carbon-based nanocomposite will be dictated by the cost, process ability, and natural ramifications of every material.

In this respect, natural applications in light of graphene offer more sensible conceivable outcomes thought about to perfect graphene because of graphene's bring down creation costs. Notwithstanding monetary contemplations, ecological ramifications of graphene-based materials will speak to a vital factor in the improvement of graphene-based advancements. The significance of precisely assessing the natural ramifications of graphene- based materials must be underscored. Thorough Eco toxicological appraisals and life-cycle investigations still should be performed, in request to recognize the types of graphene-based nano materials that will enable us to use the properties of graphene, while limiting the related wellbeing and natural effects. Graphene remains a one of a kind material with properties that could lead possibly to critical advancement in the expulsion of pharmaceutical effluents. From molecularly thin layers to ultra-high surface range materials is giving new answers for the worldwide natural difficulties about the pharmaceutical wastewater.

References

Awual MR, Hasan MM, Naushad M et al (2015) Preparation of new class composite adsorbent for enhanced palladium(II) detection and recovery. Sens Actuators, B Chem 209:790–797. https://doi.org/10.1016/j.snb.2014.12.053

Awual MR, Hasan MM, Eldesoky GE et al (2016) Facile mercury detection and removal from aqueous media involving ligand impregnated conjugate nanomaterials. Chem Eng J 290:243–251. https://doi.org/10.1016/j.cej.2016.01.038

Compton OC, Nguyen ST (2010) Graphene oxide, highly reduced graphene oxide, and graphene: versatile building blocks for carbon-based materials. Small 6:711–723

Daughton CG, Ternes TA (1999) Pharmaceuticals and personal care products in the environment: agents of subtle change? Environ Health Perspect 107:907–938

Gadipelly C, Perez-Gonzalez A, Yadav GD et al (2014) Pharmaceutical industry wastewater: review of the technologies for water treatment and reuse. Ind Eng Chem Res 53:11571–11592

Gaworecki KM, Klaine SJ (2008) Behavioural and biochemical responses of hybrid striped bass during and after fluoxetine exposure. Aquat Toxicol 88:207–213

Halling-Sorensen B, Nielsen SN, Lanzky PF et al (1998) Occurrence, fate and effects of pharmaceutical substances in the environment—a review. Chemosphere 36:357–394

Kastner M, Nowak KM, Miltner A et al (2014) Classification and modelling of nonextractable residue (NER) formation of xenobiotics in soil—a synthesis. Crit Rev Environ Sci Technol 44:2107–2171

Kyzas GZ, Fu J, Lazaridis NK et al (2015) New approaches on the removal of pharmaceuticals from wastewaters with adsorbent materials. J Mol Liq 209:87–93

Nabi SA, Bushra R, Al-Othman ZA, Naushad M (2011) Synthesis, characterization, and analytical applications of a new composite cation exchange material acetonitrile stannic(IV) selenite: adsorption behavior of toxic metal ions in nonionic surfactant medium. Sep Sci Technol 46:847–857. https://doi.org/10.1080/01496395.2010.534759

Quinn B, Gagne F, Blaise C (2008) An investigation into the acute and chronic toxicity of eleven pharmaceuticals (and their solvents) found in wastewater effluent on the cnidarian, Hydra attenuata. Sci Total Environ 389:306–314

Sanchez VC, Jachak A, Hurt Rh et al (2012) Biological interactions of graphene-family nanomaterials: an interdisciplinary review. Chem Res Toxicol 25:15–34

Tixer C, Singer HP, Oellers S et al (2003) Occurrence and fate of carbamazepine, clofibric acid, diclofenac, ibuprofen, ketoprofen, and naproxen in surface waters. Environ Sci Technol 37:1061–1068

Verlicchi P, Al Aukidy M, Zambello E (2012) Occurrence of pharmaceutical compounds in urban wastewater: removal, mass load and environmental risk after a secondary treatment—a review. Sci Total Environ 429:123–155

Wang J, Wang S (2016) Removal of pharmaceuticals and personal care products (PPCPs) from wastewater: a review. J Environ Manag 182:620–640

Yang Y, Fu J, Peng H et al (2011) Occurrence and phase distribution of selected pharmaceuticals in the Yangtze Estuary and its coastal zone. J Hazard Mater 190:588–596

Chapter 17
Photocatalytic Degradation of Organic Pollutants in Water Using Graphene Oxide Composite

Suneel Kumar, Chiaki Terashima, Akira Fujishima, Venkata Krishnan and Sudhagar Pitchaimuthu

Abstract Developing sustainable and less-expensive technique is always challenging task in water treatment process. This chapter explores the recent development of photocatalysis technique in organic pollutant removal from the water. Particularly, advantages of graphene oxide in promoting the catalytic performance of semiconductor, metal nanoparticle and polymer based photocatalyst materials. Owing to high internal surface area and rapid electron conducting property of graphene oxide fostering as backbone scaffold for effective hetero-photocatalyst loading, and rapid photo-charge separation enables effective degradation of pollutant. This chapter summaries the recent development of graphene oxide composite (metal oxide, metal nanoparticle, metal chalcogenides, and polymers) in semiconductor photocatalysis process towards environmental remediation application.

Keywords Photocatalyst · Graphene composite · Organic pollutant degradation Advanced oxidation process

S. Kumar · V. Krishnan (✉)
School of Basic Sciences and Advanced Materials Research Center,
Indian Institute of Technology Mandi, Kamand, Mandi 175005,
Himachal Pradesh, India
e-mail: vkn@iitmandi.ac.in

S. Kumar
e-mail: sssuneel44@gmail.com

C. Terashima · A. Fujishima
Photocatalysis International Research Center, Tokyo University of Science,
2641 Yamazaki, Noda, Chiba 278-8510, Japan
e-mail: terashima@rs.tus.ac.jp

A. Fujishima
e-mail: fujishima_akira@admin.tus.ac.jp

S. Pitchaimuthu (✉)
Multi-functional Photocatalyst & Coatings Group, SPECIFIC, College of Engineering,
Swansea University (Bay Campus), Swansea SA1 8EN, Wales, UK
e-mail: S.Pitchaimuthu@swansea.ac.uk

M. Naushad (ed.), *A New Generation Material Graphene: Applications in Water Technology*, https://doi.org/10.1007/978-3-319-75484-0_17

1 Introduction

Clean and non-polluted water is one of the basic requirements for all living organisms including human beings. But its availability is a major issue nowadays. In the future, this issue will further increase due to global industrialization and population growth. Natural water is being contaminated by the discharge of industrial, domestic, and agricultural wastes. Therefore, it is very important to remove the pollutants and pathogens from wastewater to fulfill the needs for irrigation as well as industrial and domestic use (Naushad et al. 2015, 2016; Pathania et al. 2014). In the past years, conventional biological and physical treatment methods (adsorption, ultrafiltration, coagulation, etc.) have been used to remove the organic pollutants (Albadarin et al. 2016; Awual et al. 2015). These methods are still adequate in removing lower concentration of organic pollutant from the water. Also, it requires sophisticated techniques to convert chemically toxic pollutants into environmental benign species. *Non-renewable energy* based water treatment technologies are currently formulated for organic pollutant removal from industrial waste, which are driven by oil, electricity and chemical processes. The depletion of fossil fuel resources and emission of undesired by-products (CO_2) into the environment challenges the sustainability of these technologies.

In this context, advanced oxidation processes (AOPs) are more efficient, cheap, and eco-friendly in the degradation of any kind of toxic pollutants (O'Shea and Dionysiou 2012). AOPs generate hydroxyl radical, a strong oxidant, which can completely degrade or mineralize the pollutants nonselectively into harmless products (Deng and Zhao 2015). In detail, hydroxyl radical (OH·) is the most reactive oxidizing agent in water treatment, and can be operated under broad range of applied potential window. For instance, the operating potential at pH 0 is about 2.8 V versus SCE (saturated calomel electrode) and for pH 14 is 1.95 V versus SCE. Furthermore OH· is nonselective and rapid (rate constant = 10^8–10^{10} M^{-1} s^{-1}) in oxidization of numerous organic species (Deng and Zhao 2015). Hydroxyl radicals can react with organic pollutants through following routes: (a) radical addition, (b) hydrogen abstraction, (c) electron transfer, and (d) radical combination. While, organic compounds interact with hydroxyl radicals produce carbon-centered radicals (R· or R·–OH), and organic peroxyl radicals (ROO·). These radicals further react with organic pollutant and forming more reactive species such as H_2O_2 and super oxide ($O_2^{\cdot-}$) lead chemical degradation and mineralization. Due to extremely short life time, hydroxyl radicals are only in situ produced during application through different process.

Recently 'photocatalysis' based environmental remediation using oxide semiconductors perceived great deal of attention compare to other AOP's techniques (Cates 2017; Chong et al. 2010; Dong et al. 2015). Because, it can be functioned without external applied potential. The light irradiation on semiconductor catalyst is a key source (UV light, solar light, visible light sources, etc.) to drive the catalysis

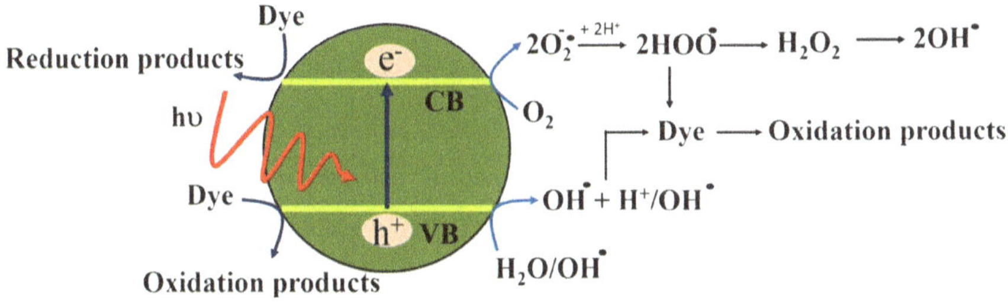

Fig. 1 Schematic illustration of photocatalytic dye degradation using light and semiconductor (this scheme is redrawn based on (Ajmal et al. 2014) work)

reaction. This technique had long before received considerable study from the perspective of the *electronic theory of chemical catalysis*. For the last one decade, a huge quantity of research has been progressed in this area, and historically it dates back at least to the 1960s (Rajeshwar et al. 2015). In the early studies, zinc oxide was a popular choice for the oxide semiconductor candidate. In 1979, Oliver et al. reported the influence of *sunlight and rocks on natural remediation processes* (Oliver et al. 1979). Followed this work few research work has been reported on photocatalysis based water treatment (Carey et al. 1976; Frank and Bard 1977; Oliver et al. 1979). Though these articles are pioneer in photocatalysis based environmental remediation, a ground-breaking phenomena *Fujishima–Honda* discovery (Fujishima and Honda 1972) at 1972 could offer an explanation on how the combination of sunlight and semiconductor involved in water oxidation process. Here, photoelectrocatalytic process, the generation of photocharge carriers' electron (e^-) and holes (h^+) at a light irradiated semiconductor drive the catalysis reaction that facilitates the conversion of light energy into stored chemical energy. In analogous to photoelectrocatalytic water oxidation (Fig. 1a), the photoholes generated from valance band of semiconductor drives the pollutant oxidation or degradation of ambiguous refractory organic species into biodegradable compounds, and eventually mineralizing them to carbon dioxide and water.

Though huge quantity of semiconductor photocatalyst materials are demonstrated in environmental remediation, yet photocatalysis reaction rate, selectivity and material stability were to be improved. In this view, graphene and graphene oxide derivatives can be applied in photocatalyst and it opens new avenue on composite based semiconductor photocatalyst. This chapter summarises the recent development of graphene oxide composite (metal oxide, metal nanoparticle, metal chalcogenides, and polymers) in semiconductor photocatalysis process towards environmental remediation application.

2 Theory

2.1 *Why Graphene Oxide Composite in Photocatalysis Reaction?*

The photocatalytic degradation of organic pollutants from water has gathered a great deal of interest as it utilizes renewable solar energy and produces non-toxic by-products during the reaction. On this note, several semiconductor materials including ZnO, SnO_2, WO_3, a-Fe_2O_3, and TiO_2 have been widely investigated (Fig. 2).

On the other hand, engineering the photocatalyst surface area also significantly influences the PC efficiency, as the degradation of organic pollutants takes place mostly at the semiconductor surface. Therefore, nanomaterials with very high surface area are considered to be effective for PC applications over micron-size materials. The nanostructured electrodes provide more adsorption sites for organic pollutants owing to their larger surface area (Al-othman and Naushad 2011). Among different nanostructures, two-dimensional (2D) nanostructures offer more effective channels for electron transport due to the reduced junctions and grain boundaries compared to spherical nanocrystals. Such fast electron transport decreases the rate of recombination and enhances PC degradation performance. In this line, 2-D structured graphene oxide is appropriate candidate to promote the semiconductor photocatalysis performance.

In recent years, the potential applications of graphene oxide (GO)-semiconductors nanocomposites have been extensively investigated as heterogeneous catalysts due to fascinating electronic and optical properties of GO (Kumar et al. 2017;

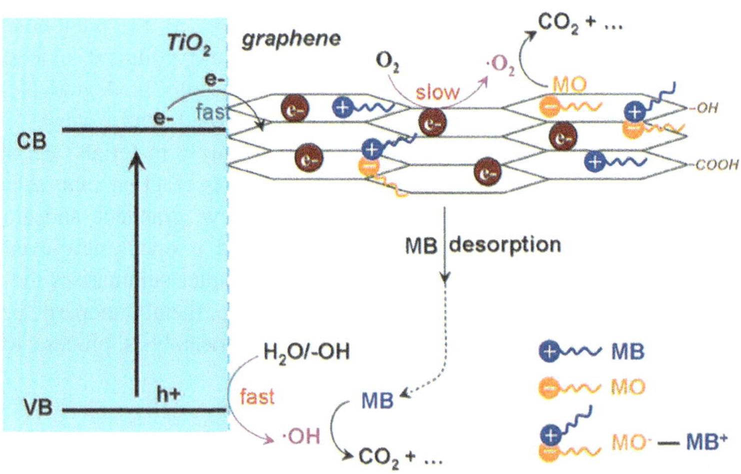

Fig. 2 Photocatalysis based organic dye pollutant removal using TiO_2-graphene oxide composite. Reprinted with permission from reference Kou et al. (2017), copyright 2013 American Chemical Society

Zhang et al. 2011). Currently, the role of GO based nanocomposites for mineralization of various kinds of organic pollutants is one of the hot research fields worldwide (Zhang et al. 2012b). As discussed above, owing to 2-D structure of GO, it possess high specific area with plenty of active sites, which facilitates the surface photocatalytic reactions (Zhang et al. 2012b). Therefore, GO acts as one of superior support material to various semiconductors and metals to form effective heterojunction with enhanced photocatalytic performance (Wang et al. 2014). From Fig. 2, the TiO_2 nanoparticle decorated graphene oxide (GO), the photoelectrons generated at conduction band of TiO_2 is rapidly transferred to graphene layer which promote the organic dye pollutant degradation rate. Moreover, the high surface area of GO provides more surface adsorption sites for pollutants during photocatalytic reactions to significantly facilitate the surface photocatalytic reactions, thereby enhancing the catalytic activity (Wang et al. 2014). It is noteworthy to mention that, GO gets always reduced to graphene or reduced GO during synthesis of nanocomposites to revive the conjugation and high conductivity (Marschall 2014). In photocatalytic reactions, graphene acts as an excellent electron sink to capture and shuttle the electrons due to its ultra-high electron conductivity (200,000 $cm^2\ V^{-1}\ S^{-1}$) (Han et al. 2012). Hence, graphene effectively retards the electron-hole recombination in semiconductors, which is one of the key steps in photocatalytic reactions (Kumar et al. 2016a). Furthermore, the Fermi level of graphene (0 V vs. NHE) is less negative than most of the semiconductor materials, which results in the favorable band gap alignment of semiconductors and graphene (Han et al. 2012). It is well known that, electrons transfer takes place 'down potential' while holes transfer lead 'up potential', hence in nanocomposites the electrons transfer occurs from conduction band (CB) of semiconductors to graphene and active species radical's formation take place, which results in the degradation of organic pollutants (Marschall 2014).

3 Semiconductor-Graphene Oxide Nanocomposites

3.1 Metal Oxide-Graphene Oxide Nanocomposites

Among semiconductor oxides, TiO_2 and ZnO are widely explored in photocatalytic pollutants removal due to their appropriate band gap energy, non-toxic nature and earth abundance (Konstantinou and Albanis 2004; Kumar and Rao 2015). Recently, Jiang et al. (2011) reported the graphene oxide-TiO_2 nanocomposite (GO-TiO_2) synthesis route using liquid phase deposition method. In this method, TiO_2 nanoparticles were grown in situ over 2-D GO nanosheets. The photocatalytic activity of this nanocomposite has been demonstrated for the degradation of methyl orange dye and reductive conversion of toxic, inorganic pollutant Cr(VI) to non-toxic Cr(III). The as prepared GO-TiO_2 nanocomposite exhibit several times higher activity than bare catalysts and P25 (commercial TiO_2) due to high surface area of nanocomposite (80 $m^2\ g^{-1}$), more pollutants adsorption ability and fast charge transfer. Moreover, due to the high work function of GO, the photoexcited electrons form CB of TiO_2 are easily transferred to GO, which suppresses their recombination. Thus, the transferred photo-induced electrons participate in the oxidative degradation of organic dyes and reductive conversion Cr(VI) to Cr(III), which leads to enhanced photocatalytic activity of nanocomposite. In addition to these two pollutants, the prepared graphene oxide-TiO_2 nanocomposite was also applied for degradation of other azo dyes such as methyl red, orange G, acid orange 7 and notably improved performance was observed.

The GO based nanocomposites have also shown their potential for the degradation of volatile aromatic pollutants in air along with their removal from water and is widely reported in literature. In this regard, Zhang et al. (2010) reported graphene-TiO_2 nanocomposite by hydrothermal reduction of GO. They varied the amount of graphene in nanocomposite and investigated the comparative photocatalytic activity of various compositions for removal of methylene blue from water and benzene, a volatile aromatic pollutant, in air. Interestingly, it was observed that the nanocomposites with higher amount of graphene exhibit decreased activity towards pollutants removal, which is analogous with liquid-phase degradation of organic pollutants. The graphene being a zero-band gap semiconductor exhibit absorption in entire solar spectrum. Thus, the improved photocatalytic performance of graphene-TiO_2 nanocomposite has been attributed to the extended solar energy utilization, fast charge transfer which results in the formation of highly reactive $^{*}OH$ which results in the mineralization of benzene/organic dyes into CO_2 and H_2O.

In addition to TiO_2, numerous studies have been carried out on other semiconductors oxides-GO nanocomposites for environmental remediation applications. In 2009, a very interesting study by Li et al. (2009) proved ZnO as better photocatalytic material than TiO_2 because of high efficiency of photoinduced charge generation and prolonged lifetime of charge carriers. This report has stimulated extensive research work on the fabrication of ZnO and GO based nanocomposite materials as efficient photocatalysts for the removal of various organic pollutants from water. Very recently, Bai et al. (2017) reported the hydrothermal synthesis of graphene-ZnO nanocomposites and demonstrated it for mycotoxin detoxification by degrading deoxynivalenol (DON) from water under UV light irradiation. DON is one of several mycotoxins produced by certain Fusarium species that frequently infect corn, wheat, oats, barley, rice, and other grains in the field or during storage which can possess a serious threat to water environment. ZnO is photo-excited by UV irradiation and once in hybridization with graphene, the photocatalytic activity of nanocomposite is substantially enhanced as compared to pure ZnO. Graphene accelerates photocatalytic process by suppressing photogenerated electron-hole recombination and adsorption of pollutants as presented in Fig. 3. Figure 3a describes the synergetic effect between graphene and ZnO wherein, photogenerated electrons transfer takes place form CB of ZnO to graphene to degrade the adsorbed pollutant by generating active species. The active species trapping experiments reveals that superoxide radicals (O_2^{-*}) and holes (h^+) were active species formed during the photocatalytic reaction. Figure 3b shows the adsorption effect more dominating with the increasing amount of graphene in nanocomposite but at same

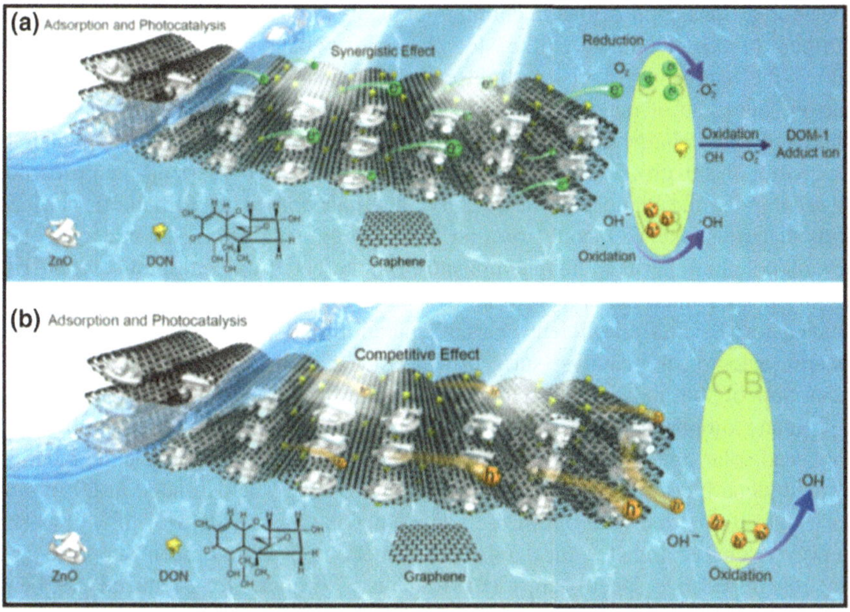

Fig. 3 Charge separation, transfer mechanism and adsorption-photocatalytic process over graphene-ZnO photocatalysts under UV light irradiation. Reprinted with permission from reference Bai et al. (2017), copyright 2016 Elsevier

time higher amount of graphene leads to the decrease in the photocatalytic activity due to the shading effect, which hinders ZnO to absorb light energy and hence there is lesser generation of charge carriers, which decreases photocatalytic performance.

It observed that nanocomposite with 0.3 wt% of graphene shows highest photocatalytic performance. Therefore, it implies that the trade-off between charge generation (synergetic effect) and light harvesting (competitive effect) leads to improved photocatalytic activity with the optimized amount of graphene, which possesses excellent adsorption. The optimized photocatalyst with 0.3 wt% of graphene degraded about 99% of DON (15 ppm) in 30 min under UV irradiation.

3.1.1 Metal Chalcogenide-Graphene Oxide Nanocomposites

For the past one decade, two dimensional materials has attracted immense attention of the scientists working in the field of nanomaterials due to their interesting electronic properties, which has lead research into device fabrication for diverse technological applications (Lu et al. 2016). In this regard, transition metal dichalcogenides which are semiconductors of type MX_2 where M is transition metal, such as Mo or W, and X is chalcogen atom (O, S, Se and Te) have been widely explored due to atomically thin nature of nanosheets with abundant reaction sites, for various applications, including pollutants removals (Tan and Zhang 2015). Furthermore, semiconductors like MoS_2 has narrow band gap of ~ 1.86 eV, which allows the harvesting of visible light energy (Kumar et al. 2016b). However, the narrow band gap of materials like MoS_2 results in high recombination rate of the photogenerated charge carriers and hence limits their use in potential applications (Kumar et al. 2016b). Therefore, a single material cannot possess all the properties to accomplish the photocatalytic activity. The nanocomposite formation of MoS_2 with materials like GO could be one of the promising strategies and can offer several potential merits to nanocomposites. Firstly, better utilization of solar energy spectrum, as semiconductors with narrow band gaps possess high light absorption efficiencies. Secondly, efficient charge separation and its transportation to active reaction sites on the catalyst surface to boost the photocatalytic activity. In consideration of the above facts, Ding et al. (2015) reported MoS_2-graphene oxide (GO) nanocomposite by one step hydrothermal hydrogel method with significantly enhanced solar light absorption and photocatalytic activity towards the degradation of methylene blue, a dye pollutant. About 99% of dye was degraded in 60 min of solar light irradiation with optimized nanocomposite having 10 wt% of MoS_2. The improved charge transfer mechanism of nanocomposite has been supported by electron impedance spectroscopy (EIS) analysis. Similar MoS_2-reduced graphene oxide (RGO) nanocomposite has been fabricated by Zhang et al. (2016) by simple hydrothermal synthesis method. This prepared nanocomposite exhibits effective separation of photogenerated charge carriers and has been supported by photoluminescence measurements, which reveals minimum recombination of photogenerated charge carriers. Upon visible light irradiation, electron-hole pair formation takes place and electrons form CB of MoS_2 gets transferred to RGO resulting in the degradation of RhB dye by the formation of superoxide radicals (O_2^{-*}) as active oxidizing species. Thus, the formation of hetero-interface between MoS_2 and RGO results in the separation of photogenerated charge carriers and suppresses their recombination, which is responsible for enhanced photocatalytic activity under visible light irradiation for pollutant removal.

Very recently, micron thin, flexible paper morphology of MoS_2-RGO hybrid layers has been synthesized and demonstrated for photocatalytic degradation of organic dyes by Jeffery et al. (2017). The potential of macrostructure, self-standing film has been shown in this work for environmental cleanup. For the preparation of such unique hybrid, a simple exfoliation method involving evaporation under ambient conditions has been demonstrated. Such kind of 2D-2D nanocomposites are formed by co-stacking of different layers with large face-to-face contact area, which allows more electronic interactions between the two 2D components. Therefore, 2D-2D composites exhibit improved activity due to more pollutant adsorption and excellent charge separation at interfacial regions. This nanocomposite was also proved to be active for the photocatalytic reduction reactions of nitro compounds to amino compounds, which has some medical significance.

In addition to dichalcogenides, metal sulfides such as cadmium sulfide (CdS) and zinc sulfide (ZnS) are of great research interest due to their potential applications for pollutant removals (2012). CdS is one of fascinating II-VI

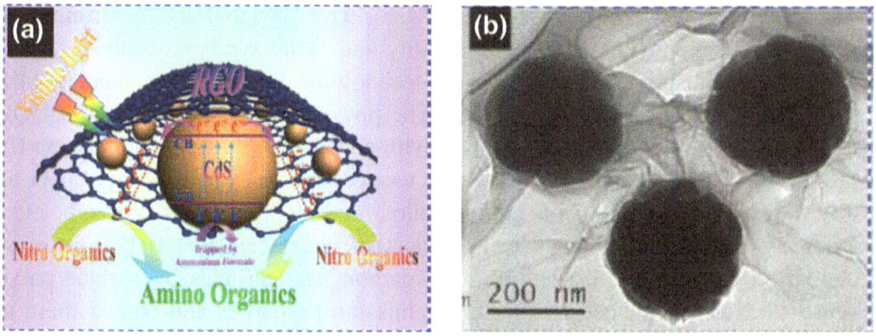

Fig. 4 **a** Schematic diagram of photocatalytic reduction of nitrocompounds over the surface of CdS-RGO catalyst, **b** TEM images of CdS nanospehers on RGO nanosheets. Reprinted with permission from reference Chen et al. (2013b), copyright 2013 American Chemical Society

semiconductor materials with direct band gap energy of ~2.4 eV, which makes it active under visible light (Ye et al. 2012). The CB potential and VB potential of CdS are −0.52 and 1.88 V, respectively indicates the strong reducing power of electrons, which is highly beneficial for photocatalytic reactions (Li et al. 2015). Therefore, CdS has been widely explored for various applications, including pollutants removal (Li et al. 2015). But pure CdS suffers from various limitations, such as fast charge recombination, poor pollutant adsorption capacity and low photostability, which restrict its practical applications. Thus, construction of nanocomposite of CdS with materials like GO can bring about the effective charge separation and transfer of photogenerated charge carriers because of less negative Fermi energy level. Chen et al. (2013) synthesized nanocomposite comprising of CdS nanospheres decorated over reduced graphene oxide (RGO) sheets by facile hydrothermal synthesis process. The interfacial contact between CdS nanospheres and RGO acts as bridge for fast transfer to photogenerated charge carriers, as presented schematically in Fig. 4a. Whereas, Fig. 4b presents the TEM images of CdS-RGO nanocomposite showing homogenous CdS nanospheres well dispersed over RGO nanosheeets to form the intimate interfacial contact between them. These nanocomposites with different loading of RGO were demonstrated to be very effective for selective reduction of nitro-aromatic compounds to amino compounds in aqueous phase. The nanocomposite with 5% of RGO exhibits the highest photocatalytic performance for 4-nitroaniline reduction. The excellent pollutant adsorption ability of RGO and improved photogenerated charge separation gives opportunity for the transferred electrons to reduce the adsorbed species.

3.1.2 Carbon Quantum Dot-Graphene Oxide Nanocomposites

The fluorescent carbon nanoparticles or carbon quantum dots (CQD) have gained much importance recently because of their fascinating electronic, optical and physicochemical properties (Lim et al. 2015). The low toxicity, environmental benignity, cost effectiveness, biocompatibility and facile synthesis routes of CQD makes them promising material for diverse technological applications, including the field of photocatalysis (Lim et al. 2015). The physicochemical properties of CQD can be tuned by surface functionalization with enhanced fluorescence emission (Li et al. 2010). These carbon nanomaterials were discovered in 2004 by Xu et al. (2004) during purification of single-walled carbon nanotubes through electrophoresis, provides a pathway to new class of fluorescent materials. Sun et al. (2006) in 2006 reported these fluorescent carbon nanoparticles by surface passivation with much enhanced fluorescence emission properties and named them as carbon quantum dots (CQD). CQD are usually prepared by laser ablation of graphite, electrochemical oxidation of graphite, electrochemical soaking of carbon nanotubes, thermal oxidation of suitable molecular precursors, vapor deposition of soot, proton-beam irradiation of nano diamonds, microwave synthesis and bottom-up methods (Lim et al. 2015). The CQD are advantageous over several of semiconductors based quantum dots, because of their facile surface passivation to enhance emission fluorescence, their benign chemical composition, easy surface functionalization and high resistance to photobleaching (Lim et al. 2015). Moreover, semiconductors based quantum dots involves the use of heavy metal ions in their preparation, which are highly toxic even in low concentration (Lim et al. 2015). CQD are conjugated materials with quasi spherical structure comprising of crystalline to amorphous sp^2 hybridized carbon atoms. It is noteworthy to mention here that, the photoluminescence of CQD can be quenched efficiently by either electron acceptor or electron donor molecules in solution, which reveals their excellent electron donor and electron acceptor behavior (Li et al. 2010). These desired properties of CQD can be exploited for technological applications such as photocatalysis, energy storage devices and sensing (Lim et al. 2015). In addition, CQD exhibit upconversion photoluminescence property, which makes them even more desirable material for photocatalysis applications, especially to make use of the larger infrared region of the solar spectrum. It has been reported that, CQD exhibit emission in visible region when excited by the femtosecond pulsed laser in near infrared range (NIR) (Li et al. 2010). The size dependent photoluminescence and excellent upconversion photoluminescence properties of CQD have been investigated in detail by excitation wavelength in range of 500–1000 nm with upconversion emission bands in 325–425 nm range. Therefore numerous studies have been devoted and significant advancement has been achieved to fabricate CQD-semiconductor based nanocomposite to exploit the maximum region of solar energy spectrum and investigate them for photocatalytic water purification by removing various kind of organic pollutants. In this regard, Li et al. (2010) designed photocatalysts of CQD-TiO_2 and CQD-SiO_2 by alkali assisted electrochemical synthesis route to harvest the full solar energy spectrum. On light irradiation to CQD-TiO_2 and CQD-SiO_2 photocatalyst, the CQD absorb visible light, and then emit light in UV region due to upconversion property. This upconverted light energy can excite TiO_2 or SiO_2 semiconductors to form electron-hole pairs (Fig. 5c). Furthermore, the nanocomposite formation between semiconductors and

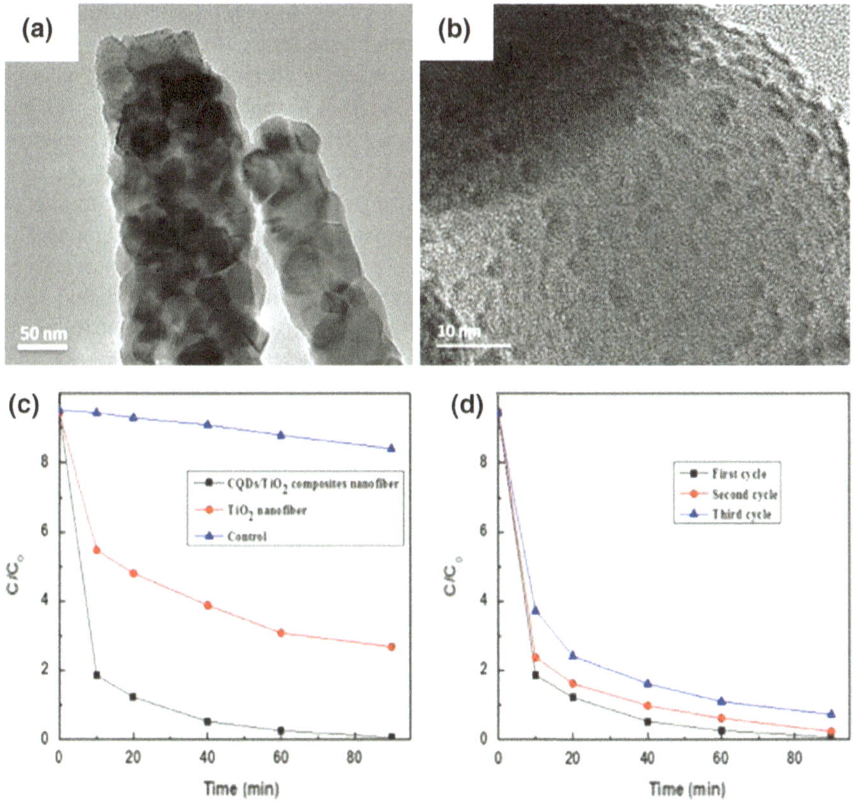

Fig. 5 **a**, **b** HRTEM images of CQD decorated TiO_2 fiber, **c** photocatalytic dye degradation experiments of pristine and CQD decorated TiO_2 fiber, **d** cycle test of CQD decorated TiO_2 fiber in dye degradation experiments. Reprinted with permission from reference Saud et al. (2015), copyright 2010 Wiley-VCH

CQD favors the electrons transfer from their CB to CQD as per suitable band gap potentials. Thus, photogenerated charge carriers are transferred rapidly through highly conducting CQD on catalyst surface to enhance the photocatalytic performance of the nanocomposites. These photogenerated charge carriers produce active oxygen species mainly O_2^{-*}, OH^* in aqueous photocatalytic reaction mixture with dissolved O_2, which are highly oxidizing in nature and causes the mineralization of pollutants into CO_2 and H_2O. In this line, Saud et al. (2015) demosntrated CQDs decorated TiO_2 nanofiber (Fig. 5a, b) in organic dye pollutant degradiation. Here, the anchored CQDs can both enhance the light absorption and suppress photogenerated electron–hole's recombination which results in the enhancement of photocatalytic dye degradation rate (Fig. 5c, d).

Subsequently, Zhang et al. (2012a) reported the facile fabrication of CQD-Ag_3PO_4 and CQD-Ag–Ag_3PO_4 nanocomposites and demonstrated them as superior photocatalysts for the degradation of organic pollutants by harvesting visible light. The dual functionality of CQD has been discussed in detail to enhance photocatalytic performance, by facilitated charge transfer and to protect the Ag_3PO_4 form photocorrosion. The CQD acts both as, electron acceptor and donor, wherein electrons can be transferred to the surface of Ag_3PO_4 for photocatalytic reaction and redundant electrons can be transferred back to CQD. Moreover, the CQD layer on the surface of Ag_3PO_4 and Ag–Ag_3PO_4 particles can effectively protect Ag_3PO_4 from dissolution in aqueous solution. The up-conversion photoluminescence behavior of CQD utilizes the wide range of solar energy spectrum. As compared to pure Ag_3PO_4 and Ag–Ag_3PO_4, the CQD-Ag–Ag_3PO_4 exhibit higher photocatalytic performance and degrades methyl orange (MO) dye in 10 min of visible light irradiation. The PL spectra of CQD reveal the presence of emission intensity peaks in 300–700 nm range when it was excited at higher wavelengths as 700–1000 nm. Hence this upconverted light excites Ag_3PO_4 which is one the narrow band gap (2.4 eV) semiconductor and results in the formation of photogenerated charge carriers. To confirm the role of upconverted light, the photocatalytic activity of Ag_3PO_4 and Ag–Ag_3PO_4 was checked by the researchers, under NIR light and almost negligible degradation was observed, while CQD-Ag_3PO_4 and CQD-Ag–Ag_3PO_4 exhibit good photocatalytic activity under NIR irradiation. Furthermore, in case of the CQD-Ag–Ag_3PO_4 photocatalyst, the surface plasmon resonance (SPR) of Ag particles (intense electric fields at the Ag particle surface) can further increase electron-hole pair's generation on the surface Ag_3PO_4 particles.

Furthermore, Yu and Kwak (2012) reported the nanocomposite of CQD with mesoporous hematite (α-Fe_2O_3) by solvothermal process as efficient, recyclable photocatalyst for the degradation of organic compounds. Pure α-Fe_2O_3 possesses low activity due to poor conductivity and fast recombination of photogenerated charge carriers. α-Fe_2O_3 is an n-type semiconductor with narrow band gap (2.1 eV), and is the most thermodynamically stable phase of iron oxide. It has been well explored in the field of photocatalysis due to its non-toxicity, facile synthesis and high photostability. The mesoporous α-Fe_2O_3 has high surface area to volume ratio and the well-defined nanoporous skeleton structure contributes for the its improved photocatalytic activity. The α-Fe_2O_3-CQD nanocomposites exhibit improved charge transfer from the photoexcited α-Fe_2O_3 to highly conducting CQD framework, which enhances the MB dye degradation by harvesting visible light. The improved photocatalytic activity of α-Fe_2O_3 was attributed first to the high surface area (187 $m^2\ g^{-1}$) of α-Fe_2O_3-CQD nanocomposite with abundant reaction sites for pollutant adsorption and photocatalytic reaction. Secondly, the excellent catalytic dispersion results in fast photogenerated charge transfer to generate reactive oxidative species (O_2^{-*}, OH^*), which degrades the pollutants.

3.1.3 Metal Nanoparticles-Graphene Oxide Nanocomposites

It is well known that, metal nanoparticles, such as Au, Ag, Pt, Pd, and Ru exhibit interesting electronic, optical, and magnetic properties (Subramanian et al. 2004). When light is incident on noble metals (Au, Ag, etc.), the oscillating electric field of light interacts with conduction electrons and resonance happens between the incident photon frequency and oscillating frequency of the conduction electrons, which results in resonance. This resonance is known as surface Plasmon resonance (SPR) and noble hence noble metals exhibit absorption in the visible region. Furthermore, the localized SPR of metal nanoparticles, mainly Au or Ag enables to tune their absorption bands in the visible light region (Haes et al. 2004). Hence, noble metal nanoparticles based nanocomposites can be exploited as very efficient visible light active photocatalyst materials for pollutants removal. The nanocomposite formation of noble metal nanoparticles with GO is one of the promising strategy to retard the surface recombination of photogenerated charge carriers, wherein the charge transfer takes place from one component to another in photocatalytic reaction. By varying some parameters, such as size, shape, chemical composition and dielectric environment, the absorption properties of noble metal based nanocomposites can be tuned. This has stimulated extensive morphology based research on noble metal nanostructures. Various kinds of morphology investigated includes nanowires, nanorods, nanoprisms, nanoplates, polyhedral structures, etc. (Lee et al. 2011; Qin et al. 2014; Schider et al. 2001; Zhou et al. 2012). Furthermore, the synthesis of nanoparticles with exposed high-energy or active facets has attracted considerable attention as well, because they usually exhibit fascinating interfacial behavior and have been applied in technological applications (Hayakawa et al. 2003; Jain et al. 2008). Most of the conventional methods of loading of metal nanoparticles over GO surface includes the use of some reducing agents such as surfactants and polymers, which have the limitation of impurity in crystal lattice of the final nanocomposite (Tang et al. 2010). Recently, Qin et al. (2014) have successfully synthesized size specific Au nanoparticles decorated over GO sheets via facile hydrothermal reduction and crystallization method. Hence this was one of the major breakthroughs in the development of reducing agent free pathway to grow Au nanocrystals over GO sheets. The morphology of Au nanocrystals can be tuned by varying the degree of oxidation on GO surface.

Ullah et al. (2014) have reported the one pot microwave assisted synthesis of Pt-graphene nanocomposite for photocatalytic degradation of organic dye pollutants RhB and MB by utilizing visible light energy. The uniform distribution of homogenous Pt nanoparticles over graphene sheets has been confirmed by transmission electron microscopy (TEM) images wherein intimate contact between both the constituents can be seen, which is highly advantageous to boost the photocatalytic activity of the nanocomposite (Fig. 6). On visible light irradiation, the electrons are photoexcited from ground state to graphene, which acts as excellent electrons acceptor and transporter. Thus, the photoelectrons are transferred to Pt due to ultra-high charge carrier mobility of graphene. Hence efficient charge transfer takes at metal-graphene interface, which further forms superoxide radicals (O_2^{-*}) by reacting with dissolved oxygen and hydroxyl radicals (OH^*) from water. Both of these radical species are highly oxidizing in nature and results in the mineralization of dyes into CO_2 and H_2O.

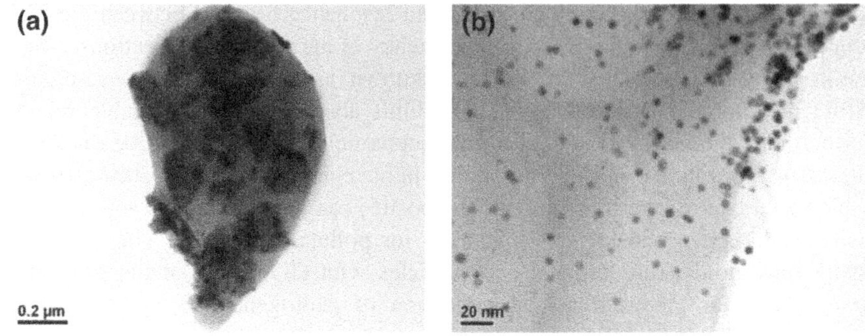

Fig. 6 **a** Low magnification TEM image of Pt-graphene composite, **b** high magnification TEM image presenting graphene sheet decorated with Pt nanoparticles. Reprinted with permission from reference Ullah et al. (2014), copyright 2013 Elsevier

Recently, Vilian et al. (2017) reported Pd nanoparticles decorated on reduced graphene oxide (RGO) by using gum arabic solution as reducing agent. The in situ synthesized Pd nanoparticles (5 nm) were found to be uniformly dispersed on RGO nanosheets and was employed as highly efficient catalyst for photocatalytic reduction of 4-nitrophenol (4-NP) pollutant to 4-amino phenol (4-AP) and its sensing. 4-NP and its derivatives have been found to be highly toxic and carcinogenic to human beings. Hence their reduction by Green, sustainable technology and sensing is of great significance. The RGO support facilitates the charge transfer process in nanocomposites and increases the photocatalytic efficiency of reaction.

In the recent years, tremendous research efforts have been devoted to explore the bimetallic alloys (Pt–Au, Au–Ag, Au–Pd) for pollutants degradation reactions because of extended light absorption range and increased charge transfer processes (Gallo et al. 2012; Tominaga et al. 2006; Zhang et al. 2014). Moreover, bimetallic systems offer more tunable parameters, synergistic effects and non-uniform charge distribution as compared to monometallic system, which can significantly enhance their photocatalytic activity for pollutants removal (Gallo et al. 2012). Moreover, coupling of such metallic alloys with GO can greatly enhance the charge transfer process in nanocomposite to boost their photocatalytic performance. Such bimetallic nanoalloys system composed of Au–Pd has been reported by Zhang et al. (2014) supported over 2D reduced graphene oxide (RGO) in aqueous phase by one pot synthesis method. Herein, they reported the formation of stabilizer free Au–Pd-RGO nanocomposite in which RGO plays the role of surfactant and support material to nanoalloys. The reduction of precursor materials to form bimetallic alloys and loading over RGO occurs simultaneously. 2D RGO acts as an excellent support platform and a unique macromolecular surfactant to promote the formation of nanocomposite of bimetallic alloys with RGO. This report gives new insight on the role GO as surfactant due to presence of large number oxygenated defects and its reduction to RGO makes it more hydrophobic in nature. The photocatalytic performance of Au–Pd-RGO nanocomposite was demonstrated for the degradation of model pollutant rhodamine B under visible light irradiation. It was very interesting to note that, the photocatalytic activity of bimetallic alloy was found to be higher than both the monometallic systems (Au-RGO and Pd-RGO), which indicate the prolonged lifetime of charge carriers and their fast transfer in bimetallic system leading to their enhanced photocatalytic activity as compared to their respective monometallic counterparts.

Furthermore, the physicochemical properties of noble metals have been exploited for catalysis applications. In order to enhance the catalytic performance, the carbon material mainly GO support plays crucial role by boosting the charge transfer process in nanocomposites. Therefore, owing to their advantages, Ye et al. (2016) recently reported the Green synthesis of Pt–Au dendrimer alloy supported over the surface of functionalized RGO. Thus Pt–Au exhibit dendrimer type morphology composed of nanoparticles and exhibit abundant edge sites and corner atoms, which can significantly enhance the catalytic performance. The surface of RGO was functionalized with polydopamine (PDA), which polymerizes by self-polymerization of dopamine and acts as reducing agent to graphene oxide (GO). PDA is biocompatible and stable polymer and hence, the use of toxic reducing agents such as hydrazine or sodium boron hydride could be avoided for reduction of GO to RGO and hence environmental benign route was adopted for synthesis. The various functional groups of PDA, such as amine and catechol, play a crucial role in the stabilization of metal nanoparticles by electrostatic interactions with their negatively charged precursor ions. The catalytic activity of Pt–Au alloy supported over PDA-RGO was tested for reduction of 4-NP to 4-AP. The loading effect of Pt to Au ratio has been systematically studied and the nanocomposite having 3:1 ratio was found to exhibit enhanced catalytic performance for reduction of 4-NP. Moreover, control experiments with PDA-RGO as catalyst reveal the strong adsorption of pollutant, which diminishes the absorption intensity of characteristic peak of 4-NP around 400 nm but no peak at 300 nm (4-AP) was detected. While in case of nanocomposites with Pt–Au alloy supported PDA-RGO surface, the decrease in absorption peak intensity at 400 nm occurs simultaneously with the appearance of peak at 300 nm, which corresponds to reduced product of 4-NP. The enhanced activity of optimized nanocomposite as compared to Pt-PDA-RGO, Au-PDA-RGO and Pt/C (commercial catalyst) has been attributed to the high electron density on the alloy due to PDA followed by electron transfer from Au to Pt, which facilitates the transfer to adsorbed 4-phenolate ions to reduce them.

3.1.4 Polymer-Graphene Oxide Nanocomposites

Polymer-GO nanocomposites have been identified as one of the promising materials for water purification by removal of pollutants. Therefore, research in this area has also been emphasized by scientific community due to their sustainability and environmental benignity, which has opened new paths for materials with improved physicochemical properties (Kuilla et al. 2010). The investigation on polymers-graphene nanocomposites for environmental remediation applications is one of the key additions in this field of materials science and technology (Kuilla et al. 2010). Polymer-graphene nanocomposites are fabricated by dispersing exfoliated graphene into polymer matrix, which leads to interesting properties different form bare polymers. There are several reports available in literature on polymer-graphene nanocomposite such as epoxy (Ganguli et al. 2008), PMMA (Jang et al. 2009), polypropylene (Kalaitzidou et al. 2007), polystyrene (Zheng et al. 2004), Nylon (Pan et al. 2000), polyaniline (Du et al. 2004), and silicone rubber (Cho et al. 2005). These reports have investigated that hybridization of graphene with polymer matrix results in the enhancement of mechanical properties such as tensile strength and storage modulus. This improvement in the mechanical properties occurs mainly due to the electrostatic interaction between the various functional groups in polymer and graphene. The parameters like molecular weight, hydrophobicity and polarity also affects the interaction in polymer-graphene nanocomposites.

Recently, polyaniline (PANI), an organic conducting polymer has attracted much attention as one of the promising materials for photocatalytic applications because of its high stability, ease of synthesis, electrical and photoelectrical properties (Ameen et al. 2012). PANI possesses a delocalized structure composed of benzenoid and quinonoid structural units, which indicate the high mobility of charge carriers upon visible light excitation (Ameen et al. 2012). Hence, many efforts have been devoted by researchers to design PANI based multicomponent nanocomposites with excellent performance for various technological applications (Deshpande et al. 2009; Li et al. 2008, 2013; Murugan et al. 2009). In this regard, Shin et al. (Ameen et al. 2012) prepared PANI-graphene nanocomposites by in situ polymerization of aniline using ammonium peroxydisulphate (APS) as initiator and utilized as efficient photocatalyst for removal of rose bengal (RB) dye form water. RB removal from water is of great significance because of its harmful effects on liver and stomach. It has been investigated with the help of Fourier transform infrared spectroscopy (FTIR) and X-ray photoelectron spectroscopy (XPS) that hydrogen bonding interactions exist between O=C–O$^-$ group of functionalized graphene and –NH group of PANI during polymerization process. The nanocomposite with 3 wt% of graphene in PANI offers well dispersion of the catalyst and high surface area, which results in better photocatalytic activity as compared to bare PANI, graphene and other nanocomposites with different contents of graphene. The high surface area of PANI-graphene nanocomposites possess high ability of RB dye adsorption due to existence of $\pi \rightarrow \pi^*$ stacking between RB molecules and aromatic groups of graphene. On light illumination, photogenerated charge carriers formation takes place in PANI and presence of graphene in nanocomposite facilitate electron-hole separation. Photogenerated electron-hole separation is followed by transfer of electrons form PANI to graphene, wherein they react with dissolved oxygen and results in the formation of active radical species in the form of O_2^{-*} and OH^* on the catalyst surface. These oxidizing species have potential to mineralize pollutants into CO_2 and H_2O. The schematic illustration of enhanced charge transfer over PANI-graphene nanocomposite to degrade RB dye is shown in Fig. 7.

In addition to this, polyvinyl alcohol (PVA) also demonstrated as promising biocompatible and non-toxic polymers (Wang et al. 2017). It is hydrophilic in nature

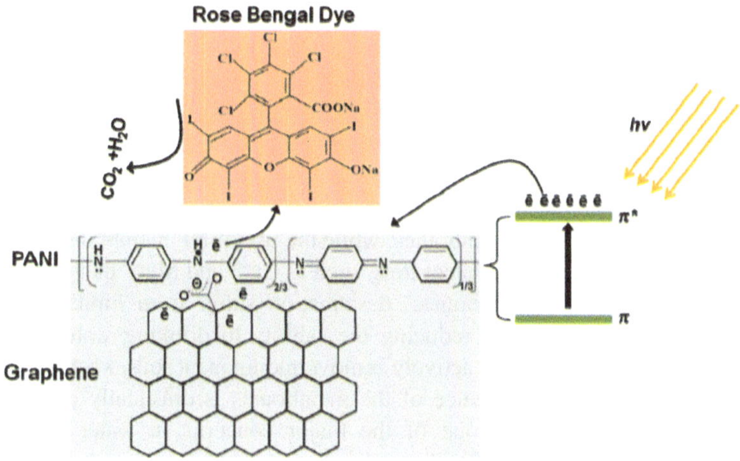

Fig. 7 Schematic illustrations of enhanced charge transfer to degrade RB dye over the surface of PANI–graphene nanocomposites. Reprinted with permission from reference Ameen et al. (2012), copyright 2012 Elsevier

with excellent thermal and chemical stability. PVA can react with various kinds of cross linking agents to form gels, which are highly useful material to be in medical, cosmetic and food industries (Wang et al. 2017). The interesting properties of PVA has also been utilized very recently in photocatalytic applications by Zhang et al. (Wang et al. 2017) who reported nanocomposites composed of 3D graphene oxide-PVA-TiO_2 microspheres for photocatalytic environmental remediation applications by removing methylene blue and methyl violet from water under simulated solar light. The PVA serves as excellent polymeric matrix for dye adsorption by hydrogen bonding interactions between them. Under simulated solar light irradiation for 120 min, the best graphene oxide-PVA-TiO_2 nanocomposites removes more than 90% of the dyes from water, which was many folds higher than bare samples. Thus the abundant reaction sites provided by graphene oxide-PVA interactions with high pollutant adsorption capacity results in excellent photocatalytic performance of the nanocomposites. The excellent pollutant adsorption ability of graphene oxide-polymer nanocomposites has been investigated by Han et al. (Sui et al. 2013), wherein they reported 3D porous graphene oxide-polyethylenimine (PEI) nanocomposite. This highly porous nanocomposite was found to exhibit high specific surface area of 476 m^2 g^{-1} and shows excellent adsorption capacity for amaranth and orange G (acidic dyes) adsorption. Therefore, such polymer nanocomposites with graphene oxide with high pollutant adsorption can be exploited as efficient photocatalysts for environmental remediation applications.

In the past decade, the fabrication of GO assisted free standing membranes has gained much importance and potentially applied for desalination and water purification (Roberts et al. 2010). Desalination is one of the promising methods of removing salt from water to obtain fresh water which contains less than 1000 mg L^{-1} of total dissolved solids (Hegab and Zou 2015). Thus with increasing population and industrialization, the demand for fresh water has increased globally. Desalination involves the removal of salt by thermal process or membrane process. Membrane based desalination involves reverse osmosis (RO), nanofiltration (NF) and electrodialysis (ED) (Greenlee et al. 2009; Van der Bruggen and Vandecasteele 2003). ED membranes operate under an electric current that causes ions movement through the membranes, while the use of NF membranes have been successfully used to remove divalent ions, such as Ca^{2+} and Mg^{2+} that contribute to water hardness. But NF membranes' desalination suffer from limitation of low efficiency and not effective in reducing the salinity to drinking water standards. However, RO membranes, can actively remove monovalent ions, such as Na^{+} and Cl^{-}. Furthermore, the performance of the membranes is drastically decreased by membrane fouling, which is one of the major concerns in water desalination technology (Roberts et al. 2010). Since the discovery of graphene in 2004, it has been explored extensively by scientific community for diverse technological applications due to its fascinating electrical and optical properties. Various research groups have successfully employed graphene in desalination application by introducing nanopores in graphene structure (Cohen-Tanugi and Grossman 2012; Surwade et al. 2015). First report on the use of GO incorporated membrane came in 2012 by Geim et al. (Nair et al. 2012), wherein they fabricated a sub-micrometer thick membrane of GO that allows the permeation of water molecules. Following this, various studies were successfully carried out based on GO incorporated membranes for nanofiltration and ultrafiltration (Qiu et al. 2009; Wang and Karnik 2012). Moreover, performance of GO based membranes was significantly improved by surface modification with polymers to form composites. In this regard, Kim et al. designed a membrane using layer-by-layer assembly method to form GO nanosheets deposited on the surface of amino polyether sulfone. This membrane was employed as RO membrane and showed much better resistance to chlorine and improved antifouling properties. In another report by Hung et al. (2014) polyacrylonitrile (PAN) was used and a membrane with GO was designed by pressure assisted self-assembly method. These membranes were exhibited remarkably improved performance with 99.5% water recovery, which was attributed to the perfect packing of GO over PAN surface to form composite film.

Recent photocatalyst developments on graphene oxide based composites were summarized and enlisted in Table 1.

4 Future Perspectives

As discussed above, the graphene and its nanocomposites with semiconductors oxides/sulfides, metal nanoparticles, carbon quantum dots and polymers have opened up new opportunities in photocatalysis domain. Such heterogeneous photocatalysts offers great advantages, such as cost effectiveness, high efficiency, good

Table 1 Photocatalytic degradation of various pollutants in aqueous solution by graphene based nanocomposites

Photocatalysts	Light source	Pollutant concentration	$t_{completion}$ (min)	Ref. (year)
P25-GR	100 W Hg lamp	MB (2.7×10^{-5} M)	55 min	Zhang et al. (2010)
P25-GO	20 W lamp	MB (2.5×10^{-5} M)	60 min	Nguyen-Phan et al. (2011)
TiO_2-GR	350 W Xe lamp	RhB (5.3×10^{-3} mM)	~200 min	Zhang et al. (2011)
TiO_2-GR	6 W UV lamp	MB (0.02 g L^{-1})	>80 min	Guo et al. (2011)
TiO_2-GO	1000 W Xe lamp	MO (12 mg L^{-1})	180 min (~40%)	Chen et al. (2010)
TiO_2-GO	20 W UV lamp	MO (10 mg L^{-1})	9 min	Jiang et al. (2011)
TiO_2-GO	11 W UV lamp	AO 7 (100 ppm) (acid orange)	30 min	Liu et al. (2011)
ZnO-GR	8 W UV lamp	MB (1.0×10^{-5} M)	40 min	Xu et al. (2011)
ZnO-GR	300 W Hg lamp	RhB (1.0×10^{-5} M)	90 min	Li and Cao (2011)
ZnO-GR	20 W UV lamp	MB (10 mg L^{-1})	56 min (~75%)	Fan et al. (2012)
ZnO-GR	Helogen lamp	MB (20 mg L^{-1})	90 min	Ahmad et al. (2013)
ZnS-GR	500 W Hg lamp	MB (15 mg L^{-1})	30 min	Hu et al. (2011)
ZnS-RGO	150 W Xe lamp	MB (6.25×10^{-5} M)	60 min (~80%)	Sookhakian et al. (2013)
CdS-RGO	300 W Xe lamp	MB (100 mg L^{-1})	150 min	Wang et al. (2012)
CuS-RGO	500 W Xe lamp	MB (4 mg L^{-1})	~ 120 min	Zhang et al. (2012c)
$BiVO_4$-GR	500 W Xe lamp	MB (20 mg L^{-1})	~300 min	Fu et al. (2011)
BiOBr-GR	500 W Xe lamp	MO (7.5 mg L^{-1})	140 min (~82%)	Song et al. (2012)
BiOI-GR	500 W Xe lamp	MO (7.5 mg L^{-1})	240 min	Li et al. (2013a)
WO_3-GR	150 W Xe lamp	MO (25 mg L^{-1})	120 min	Zhou et al. (2012a)
Ag_3PO_4-RGO	350 W Xe lamp	MO (10 mg L^{-1})	~210 min	Dong et al. (2013)
Bi_2WO_6-RGO	Helogen lamp	MB (15 mg L^{-1})	90 min	Xu et al. (2013)
Bi_2O_3-RGO	400 W helogen lamp	MB (5.0 mg L^{-1})	240 min	Li et al. (2013b)
MoS_2-RGO	5 W LED	MB (60 mg L^{-1})	60 min	Li et al. (2014)

(continued)

Table 1 (continued)

Photocatalysts	Light source	Pollutant concentration	$t_{completion}$ (min)	Ref. (year)
SnS_2-RGO	500 W Xe lamp	Rh B (10 mg L^{-1})	120 min	Chen et al. (2013a)
TiO_2–ZnO-RGO	300 W Xe lamp	MB (0.3 mg L^{-1})	120 min	Raghavan et al. (2015)
CdS–ZnO-RGO	Sunlight	MO (1.0×10^{-5} M)	60 min	Kumar et al. (2016c)
RGO-Ag–Bi_2MoO_6	300 W halogen lamp	Phenol (10 mg L^{-1})	300 min	Meng and Zhang (2016)
ZnO-MoS_2-RGO	Sunlight	MB (5.0×10^{-5} M)	60 min	Kumar et al. (2016c)
ZnO–Ag-RGO	300 W Xe lamp	Rh B (10 mg L^{-1})	~60 min	Surendran et al. (2017)
TiO_2–$BiVO_4$-GR	500 W helogen lamp	MB (10 ppm)	10 min	Nanakkal and Alexander (2017)
TiO_2–Zr-RGO	400 W Hg lamp	EB	90 min	Prabhakarrao et al. (2017)
TiO_2-RGO	160 W Hg lamp	IB, CM, SM (5.0 mg L^{-1})	~200 min	Lin et al. (2017)
TiO_2–Au-GR	125 W Hg lamp	AB 93 (20 ppm)	120 min (~60 min)	Ghasemi et al. (2017)
CeO_2-GR	400 W Hg lamp	MB	~210 min	Khan et al. (2017)
ZnO-RGO	75 W Xe lamp	RhB, MO, MB (10 mg L^{-1})	~ 300 min	Ranjith et al. (2017)

GR graphene; *GO* graphene oxide; *RGO* reduced graphene oxide; *MB* methylene blue; *RhB* rhodamine B; *MO* methyl orange; *EB* eosin blue; *IB* ibuprofen; *CM* carbamazepine; *CM* sulfamethoxazole; *AB* acid blue

thermal stability, tunable band gap structure and facile synthesis routes. Therefore, it has emerged as Green and sustainable approach to remove pollutants form water by utilizing solar energy. Despite the great achievements over the years with graphene based nanocomposites for pollutants removal, there are still many challenges in fabrication and application of these materials in catalysis. Although various method have been designed and developed for the fabrication of graphene based nanocomposites but still with the increasing demand for high efficiency and optimized parameters, some channelized efforts needs to be carried out in order to achieve large scale practical applications. Therefore, the industrialization of graphene is one of the highly anticipated tasks in coming years as low bulk density can severely hinder its applications. Hence, systematic efforts should be executed to design particular hybrid nanocomposite architectures with desired properties rather than random mixture. Furthermore, it is still challenging to fabricate a cost effective, thin, uniform, high quality graphene layer with controllable layer thickness in nanocomposites, which exhibit distinct electronic, optical and thermal properties.

It is noteworthy to mention here that hybridization of graphene with various polymers have drastically increased the electrical, thermal and mechanical properties of nanocomposites, which display excellent photocatalytic performance to remove the organic pollutants. The strong electronic interactions existing between graphene and conductive polymer matrix provides abundant reaction sites on large specific surface area. Fabrication of devices composed of nanocomposite membranes based on the conductive polymers, graphene and metal oxide/sulfides could be a promising approach to obtain materials with remarkably enhanced photocatalytic performance for pollutants removal form water. Going forward, the utilization of entire solar spectrum is another promising area for graphene based nanocomposites in photocatalysis by exploiting the up-conversion photoluminescence phenomenon of carbon quantum dots and other lanthanide based materials. The development of noble metal free, cost effective nanocomposites which can utilize the near infrared radiation should be further explored in future as it is still not very clear from mechanistic point of view about the up-conversion photoluminescence. Therefore, more studies are required for promoting the general understanding of optical and electronic properties of carbon quantum dots and other up conversion materials.

5 Conclusion

In summary, various graphene based nanocomposites have been designed and demonstrated as efficient heterogeneous photocatalysts for water purification. The introduction of graphene into such nanocomposite have made high impact such as, extended light absorption in visible region, excellent pollutants adsorption, retarded photogenerated electron-hole recombination to prolong their life time, which overall increases the photocatalytic performance as compared to conventional catalysts. In this regard, metal oxide semiconductors, metal nanoparticles and other carbon-based materials have been combined with 2D graphene and widely investigated as nanocomposites with improved charge separation and light absorption in which graphene acts as excellent support matrix to degrade pollutants from water. Hence collection of all useful properties of 2D graphene, abundance, environmental benignity and facile synthesis routes of its nanocomposites have made it a promising class of functional materials for environmental remediation applications. Unfortunately, despite of all the exciting results obtained so far for pollutant removal with graphene based nanocomposites, this field is still challenges and more efforts needs to be devoted to exploit the graphene based nanocomposites effectively in practical applications. Therefore, in order to maximize the advantages of graphene based nanocomposites, facile and robust synthesis routes should be further developed which can prepare material on large scale to fabricate devices and commercialize them. Moreover, the oxidation of graphite introduces impurities in

GO skeleton which leads to decrease in its conductance. The reduction of GO during synthesis to RGO is useful to revive its conductivity, hence the development of highly efficient nanocomposites is still a challenging domain.

Acknowledgements This work is supported by Sêr Cymru II-Rising Star Fellowship program through Welsh Government and European Regional Development Fund.

References

Ajmal A, Majeed I, Malik RN, Idriss H, Nadeem MA (2014) RSC Advances 4(70):37003–37026

Ahmad M, Ahmed E, Hong Z, Xu J, Khalid N, Elhissi A, Ahmed W (2013) Appl Surf Sci 274:273–281

Albadarin AB, Collins MN, Naushad M et al (2016) Activated lignin–chitosan extruded blends for efficient adsorption of methylene blue. Chem Eng J 307:264–272. https://doi.org/10.1016/j.cej.2016.08.089

Al-othman ZA, Naushad M (2011) Organic–inorganic type composite cation exchanger poly-o-toluidine Zr (IV) tungstate: preparation, physicochemical characterization and its analytical application in separation of heavy metals. Chem Eng J 172:369–375. https://doi.org/10.1016/j.cej.2011.06.018

Ameen S, Seo H-K, Akhtar MS, Shin HS (2012) Novel graphene/polyaniline nanocomposites and its photocatalytic activity toward the degradation of rose Bengal dye. Chem Eng J 210:220–228

Awual R, Eldesoky GE, Yaita T et al (2015) Schiff based ligand containing nano-composite adsorbent for optical copper (II) ions removal from aqueous solutions. Chem Eng J 279:639–647. https://doi.org/10.1016/j.cej.2015.05.049

Bai X et al (2017) Photocatalytic degradation of deoxynivalenol using graphene/ZnO hybrids in aqueous suspension. Appl Catal B 204:11–20

Carey JH, Lawrence J, Tosine HM (1976) Photodechlorination of PCB's in the presence of titanium dioxide in aqueous suspensions. Bull Environ Contam Toxicol 16:697–701. https://doi.org/10.1007/bf01685575

Cates EL (2017) Photocatalytic water treatment: so where are we going with this? Environ Sci Technol 51:757–758. https://doi.org/10.1021/acs.est.6b06035

Chen Z, Liu S, Yang M-Q, Xu Y-J (2013) Synthesis of uniform CdS nanospheres/graphene hybrid nanocomposites and their application as visible light photocatalyst for selective reduction of nitro organics in water. ACS Appl Mater Interface 5:4309–4319

Chen P, Su Y, Liu H, Wang Y (2013a) ACS Appl Mater Interfaces 5(22):12073–12082

Chen Z, Liu S, Yang M-Q, Xu Y-J (2013b) ACS Appl Mater Interfaces 5(10):4309–4319

Cho D, Lee S, Yang G, Fukushima H, Drzal LT (2005) Dynamic mechanical and thermal properties of phenylethynyl-terminated polyimide composites reinforced with expanded graphite nanoplatelets. Macromol Mater Eng 290:179–187

Chong MN, Jin B, Chow CWK, Saint C (2010) Recent developments in photocatalytic water treatment technology: a review. Water Res 44:2997–3027. https://doi.org/10.1016/j.watres.2010.02.039

Cohen-Tanugi D, Grossman JC (2012) Water desalination across nanoporous graphene. Nano Lett 12:3602–3608

Deng Y, Zhao R (2015) Advanced oxidation processes (AOPs) in Wastewater Treatment. Curr Pollut Rep 1:167–176. https://doi.org/10.1007/s40726-015-0015-z

Deshpande N, Gudage Y, Sharma R, Vyas J, Kim J, Lee Y (2009) Studies on tin oxide-intercalated polyaniline nanocomposite for ammonia gas sensing applications. Sens Actuator B: Chem 138:76–84

Ding Y, Zhou Y, Nie W, Chen P (2015) MoS 2–GO nanocomposites synthesized via a hydrothermal hydrogel method for solar light photocatalytic degradation of methylene blue. Appl Surf Sci 357:1606–1612
Dong P, Wang Y, Cao B, Xin S, Guo L, Zhang J, Li F (2013) Appl Catal B: Environ 132:45–53
Dong S et al (2015) Recent developments in heterogeneous photocatalytic water treatment using visible light-responsive photocatalysts: a review. RSC Adv 5:14610–14630. https://doi.org/10.1039/C4RA13734E
Du X, Xiao M, Meng Y (2004) Facile synthesis of highly conductive polyaniline/graphite nanocomposites. Eur Polym J 40:1489–1493
Fan H, Zhao X, Yang J, Shan X, Yang L, Zhang Y, Li X, Gao M (2012) Catal Commun 29:29–34
Frank SN, Bard AJ (1977) Heterogeneous photocatalytic oxidation of cyanide ion in aqueous solutions at titanium dioxide powder. J Am Chem Soc 99:303–304. https://doi.org/10.1021/ja00443a081
Fu Y, Sun X, Wang X (2011) Mater Chem Phys 131(1):325–330
Fujishima A, Honda K (1972) Electrochemical photolysis of water at a semiconductor electrode. Nature 238:37–38
Gallo A et al (2012) Bimetallic Au–Pt/TiO_2 photocatalysts active under UV-A and simulated sunlight for H_2 production from ethanol. Green Chem 14:330–333
Ganguli S, Roy AK, Anderson DP (2008) Improved thermal conductivity for chemically functionalized exfoliated graphite/epoxy composites. Carbon 46:806–817
Ghasemi S, Hashemian S, Alamolhoda A, Gocheva I, Setayesh SR (2017) Mater Res Bull 87:40–47
Greenlee LF, Lawler DF, Freeman BD, Marrot B, Moulin P (2009) Reverse osmosis desalination: water sources, technology, and today's challenges. Water Res 43:2317–2348
Guo J, Zhu S, Chen Z, Li Y, Yu Z, Liu Q, Li J, Feng C, Zhang D (2011) Ultrason Sonochem 18(5):1082–1090
Haes AJ, Zou S, Schatz GC, Van Duyne RP (2004) A nanoscale optical biosensor: the long range distance dependence of the localized surface plasmon resonance of noble metal nanoparticles. J Phys Chem B 108:109–116
Han L, Wang P, Dong S (2012) Progress in graphene-based photoactive nanocomposites as a promising class of photocatalyst. Nanoscale 4:5814–5825
Hayakawa K, Yoshimura T, Esumi K (2003) Preparation of gold–dendrimer nanocomposites by laser irradiation and their catalytic reduction of 4-nitrophenol. Langmuir 19:5517–5521
Hegab HM, Zou L (2015) Graphene oxide-assisted membranes: fabrication and potential applications in desalination and water purification. J Membr Sci 484:95–106
Hu H, Wang X, Liu F, Wang J, Xu C (2011) Synth Met 161(5):404–410
Hung W-S et al (2014) Pressure-assisted self-assembly technique for fabricating composite membranes consisting of highly ordered selective laminate layers of amphiphilic graphene oxide. Carbon 68:670–677
Jain PK, Huang X, El-Sayed IH, El-Sayed MA (2008) Noble metals on the nanoscale: optical and photothermal properties and some applications in imaging, sensing, biology, and medicine. Acc Chem Res 41:1578–1586
Jang JY, Kim MS, Jeong HM, Shin CM (2009) Graphite oxide/poly (methyl methacrylate) nanocomposites prepared by a novel method utilizing macroazoinitiator. Compos Sci Technol 69:186–191
Jeffery AA, Rao SR, Rajamathi M (2017) Preparation of MoS 2–reduced graphene oxide (rGO) hybrid paper for catalytic applications by simple exfoliation–costacking. Carbon 112:8–16
Jiang G et al (2011) TiO_2 nanoparticles assembled on graphene oxide nanosheets with high photocatalytic activity for removal of pollutants. Carbon 49:2693–2701
Kalaitzidou K, Fukushima H, Drzal LT (2007) Multifunctional polypropylene composites produced by incorporation of exfoliated graphite nanoplatelets. Carbon 45:1446–1452
Khan ME, Khan MM, Cho MH (2017) Sci Rep 7

Konstantinou IK, Albanis TA (2004) TiO_2-assisted photocatalytic degradation of azo dyes in aqueous solution: kinetic and mechanistic investigations: a review. Appl Catal B 49:1–14
Kou J, Lu C, Wang J, Chen Y, Xu Z, Varma R. S (2017) Chem Rev 117 (3):1445–1514
Kuilla T, Bhadra S, Yao D, Kim NH, Bose S, Lee JH (2010) Recent advances in graphene based polymer composites. Prog Polym Sci 35:1350–1375
Kumar SG, Rao KK (2015) Zinc oxide based photocatalysis: tailoring surface-bulk structure and related interfacial charge carrier dynamics for better environmental applications. RSC Adv 5:3306–3351
Kumar S, Sharma V, Bhattacharyya K, Krishnan V (2016c) New J Chem 40(6):5185–5197
Kumar S, Sharma R, Sharma V, Harith G, Sivakumar V, Krishnan V (2016a) Role of RGO support and irradiation source on the photocatalytic activity of CdS–ZnO semiconductor nanostructures. Beilstein J Nanotechnol 7:1684–1697
Kumar S, Sharma V, Bhattacharyya K, Krishnan V (2016b) Synergetic effect of MoS 2–RGO doping to enhance the photocatalytic performance of ZnO nanoparticles. New J Chem 40:5185–5197
Kumar S, Kumar A, Bahuguna A, Sharma V, Krishnan V (2017) Two-dimensional carbon-based nanocomposites for photocatalytic energy generation and environmental remediation applications Beilstein. J Nanotechnol 8:1571–1600
Lee SJ, Park G, Seo D, Ka D, Kim SY, Chung IS, Song H (2011) Coordination power adjustment of surface-regulating polymers for shaping gold polyhedral nanocrystals. Chem-A Eur J 17:8466–8471
Li X, Wang D, Cheng G, Luo Q, An J, Wang Y (2008) Preparation of polyaniline-modified TiO_2 nanoparticles and their photocatalytic activity under visible light illumination. Appl Catal B 81:267–273
Li Y et al (2009) Comparison of dye photodegradation and its coupling with light-to-electricity conversion over TiO_2 and ZnO. Langmuir 26:591–597
Li H et al (2010) Water-soluble fluorescent carbon quantum dots and photocatalyst design. Angew Chem Int Ed 49:4430–4434
Li Z-F, Zhang H, Liu Q, Sun L, Stanciu L, Xie J (2013a) Fabrication of high-surface-area graphene/polyaniline nanocomposites and their application in supercapacitors. ACS Appl Mater Interface 5:2685–2691
Li B, Cao H (2011) J Mater Chem 21(10):3346–3349
Li Z-F, Zhang H, Liu Q, Sun L, Stanciu L, Xie J (2013b) ACS Appl Mater Interfaces 5(7):2685–2691
Li J, Liu X, Pan L, Qin W, Chen T, Sun Z (2014) RSC Advances 4(19):9647–9651
Li Q, Li X, Wageh S, Al-Ghamdi A, Yu J (2015) CdS/graphene nanocomposite photocatalysts. Adv Energy Mater 5
Lin L, Wang H, Xu P (2017) Chem Eng J 310:389–398
Lim SY, Shen W, Gao Z (2015) Carbon quantum dots and their applications. Chem Soc Rev 44:362–381
Lu Q, Yu Y, Ma Q, Chen B, Zhang H (2016) 2D transition-metal-dichalcogenide-nanosheet-based composites for photocatalytic and electrocatalytic hydrogen evolution reactions. Adv Mater 28:1917–1933
Marschall R (2014) Semiconductor composites: strategies for enhancing charge carrier separation to improve photocatalytic activity. Adv Func Mater 24:2421–2440
Meng X, Zhang Z (2016) J Catal 344:616–630
Murugan AV, Muraliganth T, Manthiram A (2009) Rapid, facile microwave-solvothermal synthesis of graphene nanosheets and their polyaniline nanocomposites for energy strorage. Chem Mater 21:5004–5006
Nanakkal A, Alexander L (2017) J Mater Sci 52(13):7997–8006
Nair R, Wu H, Jayaram P, Grigorieva I, Geim A (2012) Unimpeded permeation of water through helium-leak–tight graphene-based membranes. Science 335:442–444
Naushad M, ALOthman ZA, Sharma G, Inamuddin (2015) Kinetics, isotherm and thermodynamic investigations for the adsorption of Co(II) ion onto crystal violet modified amberlite IR-120 resin. Ionics 21:1453–1459. https://doi.org/10.1007/s11581-014-1292-z

Naushad M, Ahamad T, Sharma G et al (2016) Synthesis and characterization of a new starch/ SnO_2 nanocomposite for efficient adsorption of toxic Hg^{2+} metal ion. Chem Eng J 300:306–316. https://doi.org/10.1016/j.cej.2016.04.084
O'Shea KE, Dionysiou DD (2012) Advanced oxidation processes for water treatment. J Phys Chem Lett 3:2112–2113. https://doi.org/10.1021/jz300929x
Oliver BG, Cosgrove EG, Carey JH (1979) Effect of suspended sediments on the photolysis of organics in water. Environ Sci Technol 13:1075–1077. https://doi.org/10.1021/es60157a011
Pan YX, Yu ZZ, Ou YC, Hu GH (2000) A new process of fabricating electrically conducting nylon 6/graphite nanocomposites via intercalation polymerization. J Polym Sci Part B: Polym Phys 38:1626–1633
Pathania D, Sharma G, Naushad M, Kumar A (2014) Synthesis and characterization of a new nanocomposite cation exchanger polyacrylamide Ce(IV) silicophosphate: photocatalytic and antimicrobial applications. J Ind Eng Chem 20:3596–3603
Prabhakarrao N, Chandra MR, Rao TS (2017) J Alloy Compd 694:596–606
Qin Y, Li J, Kong Y, Li X, Tao Y, Li S, Wang Y (2014) In situ growth of Au nanocrystals on graphene oxide sheets. Nanoscale 6:1281–1285
Qiu S, Wu L, Pan X, Zhang L, Chen H, Gao C (2009) Preparation and properties of functionalized carbon nanotube/PSF blend ultrafiltration membranes. J Membr Sci 342:165–172
Rajeshwar K, Thomas A, Janáky C (2015) Photocatalytic activity of inorganic semiconductor surfaces: myths, hype, and reality. J Phys Chem Lett 6:139–147. https://doi.org/10.1021/jz502586p
Raghavan N, Thangavel S, Venugopal G (2015) Mat Sci Semicon Proc 30:321–329
Ranjith KS, Manivel P, Rajendrakumar RT, Uyar T (2017) Chem Eng J 325:588–600
Roberts DA, Johnston EL, Knott NA (2010) Impacts of desalination plant discharges on the marine environment: a critical review of published studies. Water Res 44:5117–5128
Saud PS, Pant B, Alam A-M, Ghouri ZK, Park M, Kim H-Y (2015) Carbon quantum dots anchored TiO_2 nanofibers: effective photocatalyst for waste water treatment. Ceram Int 41:11953–11959. https://doi.org/10.1016/j.ceramint.2015.06.007
Schider G, Krenn J, Gotschy W, Lamprecht B, Ditlbacher H, Leitner A, Aussenegg F (2001) Optical properties of Ag and Au nanowire gratings. J Appl Phys 90:3825–3830
Sookhakian M, Amin Y, Basirun W (2013) Appl Surf Sci 283:668–677
Song S, Gao W, Wang X, Li X, Liu D, Xing Y, Zhang H (2012) Dalton Transactions 41 (34):10472–10476
Subramanian V, Wolf EE, Kamat PV (2004) Catalysis with TiO_2/gold nanocomposites. Effect of metal particle size on the Fermi level equilibration. J Am Chem Soc 126:4943–4950
Sui Z-Y, Cui Y, Zhu J-H, Han B-H (2013) Preparation of three-dimensional graphene oxide–polyethylenimine porous materials as dye and gas adsorbents. ACS Appl Mater Interface 5:9172–9179
Sun Y-P et al (2006) Quantum-sized carbon dots for bright and colorful photoluminescence. J Am Chem Soc 128:7756–7757
Surendran DK, Xavier MM, Viswanathan VP, Mathew S (2017) Environ Sci Pollut Res 1–9
Surwade SP, Smirnov SN, Vlassiouk IV, Unocic RR, Veith GM, Dai S, Mahurin SM (2015) Water desalination using nanoporous single-layer graphene. Nat Nanotechnol 10:459–464
Tan C, Zhang H (2015) Two-dimensional transition metal dichalcogenide nanosheet-based composites. Chem Soc Rev 44:2713–2731
Tang Z, Shen S, Zhuang J, Wang X (2010) Noble-metal-promoted three-dimensional macroassembly of single-layered graphene oxide. Angew Chem 122:4707–4711
Tominaga M, Shimazoe T, Nagashima M, Kusuda H, Kubo A, Kuwahara Y, Taniguchi I (2006) Electrocatalytic oxidation of glucose at gold–silver alloy, silver and gold nanoparticles in an alkaline solution. J Electroanal Chem 590:37–46
Ullah K, Ye S, Zhu L, Meng Z-D, Sarkar S, Oh W-C (2014) Microwave assisted synthesis of a noble metal-graphene hybrid photocatalyst for high efficient decomposition of organic dyes under visible light. Mater Sci Eng B 180:20–26

Van der Bruggen B, Vandecasteele C (2003) Removal of pollutants from surface water and groundwater by nanofiltration: overview of possible applications in the drinking water industry. Environ Pollut 122:435–445
Vilian AE, Choe SR, Giribabu K, Jang S-C, Roh C, Huh YS, Han Y-K (2017) Pd nanospheres decorated reduced graphene oxide with multi-functions: highly efficient catalytic reduction and ultrasensitive sensing of hazardous 4-nitrophenol pollutant. J Hazard Mater 333:54–62
Wang EN, Karnik R (2012) Nat Nanotechnol 7(9):552–554
Wang EN, Karnik R (2012) Water desalination: graphene cleans up water. Nat Nanotechnol 7:552–554
Wang H et al (2014) Semiconductor heterojunction photocatalysts: design, construction, and photocatalytic performances. Chem Soc Rev 43:5234–5244
Wang M et al (2017) One-pot composite synthesis of three-dimensional graphene oxide/poly (vinyl alcohol)/TiO_2 microspheres for organic dye removal. Sep Purif Technol 172:217–226
Xu X, Ray R, Gu Y, Ploehn HJ, Gearheart L, Raker K, Scrivens WA (2004) Electrophoretic analysis and purification of fluorescent single-walled carbon nanotube fragments. J Am Chem Soc 126:12736–12737
Xu J, Ao Y, Chen M (2013) Mater Lett 92:126–128
Ye A, Fan W, Zhang Q, Deng W, Wang Y (2012) CdS–graphene and CdS–CNT nanocomposites as visible-light photocatalysts for hydrogen evolution and organic dye degradation. Catal Sci Technol 2:969–978
Ye W, Yu J, Zhou Y, Gao D, Wang D, Wang C, Xue D (2016) Green synthesis of Pt–Au dendrimer-like nanoparticles supported on polydopamine-functionalized graphene and their high performance toward 4-nitrophenol reduction. Appl Catal B 181:371–378
Yu BY, Kwak S-Y (2012) Carbon quantum dots embedded with mesoporous hematite nanospheres as efficient visible light-active photocatalysts. J Mater Chem 22:8345–8353
Zhang Y, Tang Z-R, Fu X, Xu Y-J (2010) TiO_2–graphene nanocomposites for gas-phase photocatalytic degradation of volatile aromatic pollutant: is TiO_2–graphene truly different from other TiO_2–carbon composite materials? ACS Nano 4:7303–7314
Zhang J, Xiong Z, Zhao X (2011) Graphene–metal–oxide composites for the degradation of dyes under visible light irradiation. J Mater Chem 21:3634–3640
Zhang H, Huang H, Ming H, Li H, Zhang L, Liu Y, Kang Z (2012a) Carbon quantum dots/ Ag_3PO_4 complex photocatalysts with enhanced photocatalytic activity and stability under visible light. J Mater Chem 22:10501–10506
Zhang N, Zhang Y, Xu Y-J (2012b) Recent progress on graphene-based photocatalysts: current status and future perspectives. Nanoscale 4:5792–5813
Zhang N, Zhang Y, Xu Y-J (2012c) Nanoscale 4(19):5792–5813
Zhang Y, Zhang N, Tang Z-R, Xu Y-J (2014) Graphene oxide as a surfactant and support for in-situ synthesis of Au–Pd nanoalloys with improved visible light photocatalytic activity. J Phys Chem C 118:5299–5308
Zhang L, Sun L, Liu S, Huang Y, Xu K, Ma F (2016) Effective charge separation and enhanced photocatalytic activity by the heterointerface in MoS 2/reduced graphene oxide composites Rsc. Advances 6:60318–60326
Zheng W, Lu X, Wong SC (2004) Electrical and mechanical properties of expanded graphite-reinforced high-density polyethylene. J Appl Polym Sci 91:2781–2788
Zhou X, Liu G, Yu J, Fan W (2012) Surface plasmon resonance-mediated photocatalysis by noble metal-based composites under visible light. J Mater Chem 22:21337–21354
Zhou X, Liu G, Yu J, Fan W (2012a) J Mater Chem 22(40):21337–21354

Chapter 18
Recent Developments in Adsorption of Dyes Using Graphene Based Nanomaterials

A. Carmalin Sophia, Tanvir Arfin and Eder C. Lima

Abstract Dyes are frequently let out into the environment along with wastewater sans necessary treatment. Fast, cost-effective, scientific and suitable elimination of dyes from wastewaters has been an important problem for researchers. Adsorption technique is a robust, well studied, widely employed and promising water treatment method. In the last decade, nanocarbon based adsorbents have gained attention in water treatment. These adsorbents are usually produced from low cost substrate and are found to be highly efficient than other adsorbents. Recently, graphene based nanomaterials are widely used as adsorbents to sorb various toxic organic contaminants from aqueous solutions. It showed high efficiency due to its chemical stability, structure, surface area and surface functional groups. So graphene are called as 'miracle material'. Recently nanographene composites are proven to be a likely adsorbent for eliminating contaminants from the industrial effluents. In this chapter, we have presented briefly the synthesis of graphene and its other variants viz., GO, rGO and nano graphene composites. This chapter presents a small introduction to adsorption principles and adsorption isotherms. It explains the synthesis and use of nano graphene materials for the remediation of dyes. It also consolidates the recent literature available for dye adsorption using graphene materials and its mechanism.

A. Carmalin Sophia (✉)
National Environmental Engineering Research Institute (NEERI), Chennai Zonal Laboratory, CSIR Campus, Taramani, Chennai 600113, India
e-mail: ac_sophia@neeri.res.in

T. Arfin
Environmental Materials Division, CSIR-National Environmental Engineering Research Institute (CSIR-NEERI), Nagpur 440020, India
e-mail: t_arfin@neeri.res.in

E. C. Lima
Institute of Chemistry, Federal University of Rio Grande do Sul (UFRGS), Av. Bento Gonçalves 9500, P.O. Box 15003, Porto Alegre, RS 91501-970, Brazil
e-mail: eder.lima@ufrgs.br

M. Naushad (ed.), *A New Generation Material Graphene: Applications in Water Technology*, https://doi.org/10.1007/978-3-319-75484-0_18

Keywords Textile wastewater · Dye wastewater · Adsorption · Graphene
Graphene oxide · Graphene composites

1 Introduction

Scarcity of water is an obvious problem which prevails all over the world. Almost 70% of the total water resources are contaminated by various organic and inorganic pollutants (Ramphal 2013). The industrial and domestic sectors contribute to high incidence of water pollution in comparison to other sectors (Naushad et al. 2015a, b, c). The usage of synthetic chemical compounds in the industrial activities adds to pollutant load in water that exceeds the regeneration capacity of the natural water resources (Murty and Kumar 2011). Rapid industrialization in the recent past has rendered most of the water bodies in urban areas polluted with harmful pollutants due to which there is a growing demand for new scientific, effective, economical technologies for water treatment (Ramesha et al. 2011).

One of the major contaminating industries is dyes and dye intermediates manufacturing industries. Dyes are substances which has ability to show color when applied to different substrates like plastics, textiles etc. The process used for the applying of dyes to the substrate is decided in large extent by the nature of substrate (Rajaram and Das 2008; Sengupta 2004). Dyes are broadly categorized in the form of ionic charge which is shown in Fig. 1. Dyes, their structure, molecular formula, type and toxicity information are presented in Table 1 (Clarke and Anliker 1980; Adak et al. 2005; Doğan and Alkan 2003; Wu et al. 1998; Kazim et al. 2007; Iram et al. 2010; Baek et al. 2010; Mall et al. 2005; Lucas and Peres 2006; Namasivayam and Kavitha 2002; El Haddad et al. 2012).

Dyes are synthesised from different organic compounds using chemical process and they are usually hazardous all living beings (de Oliveira et al. 2014; Leme et al. 2014; Rocha et al. 2017). Dyes and their intermediates can be produced using several additives to synthesize desired colours.

Textile and dye industries generate huge amounts of wastewater (Shamik and Rajasekhar 2014). These effluents are toxic, carcinogenic, and xenobiotic and should be treated before it is released into the natural water resources (Swaminathan et al. 2003; Carneiro et al. 2010; de Lima et al. 2007). Most of the dyes employed for textile manufacturing are persistent and non-biodegradable in nature (Kurade et al. 2017; Pathania et al. 2016; Rajeswari et al. 2017). It is important to treat dye wastewaters before discharging them into the surrounding.

The enforcement of strict waste management and handling rules concerning the discharge of hazardous waste streams, leads to the development of a wide array of eco-friendly and efficient wastewater treatment technologies. Many approaches have been established for remediation of dyes, including coagulation/flocculation, adsorption, advanced oxidation methods etc. which are shown in Fig. 2. Adsorption is the most encouraging method for the removal of organic contaminants from the wastewater (Jauris et al. 2016; Shamik and Rajasekhar 2014;

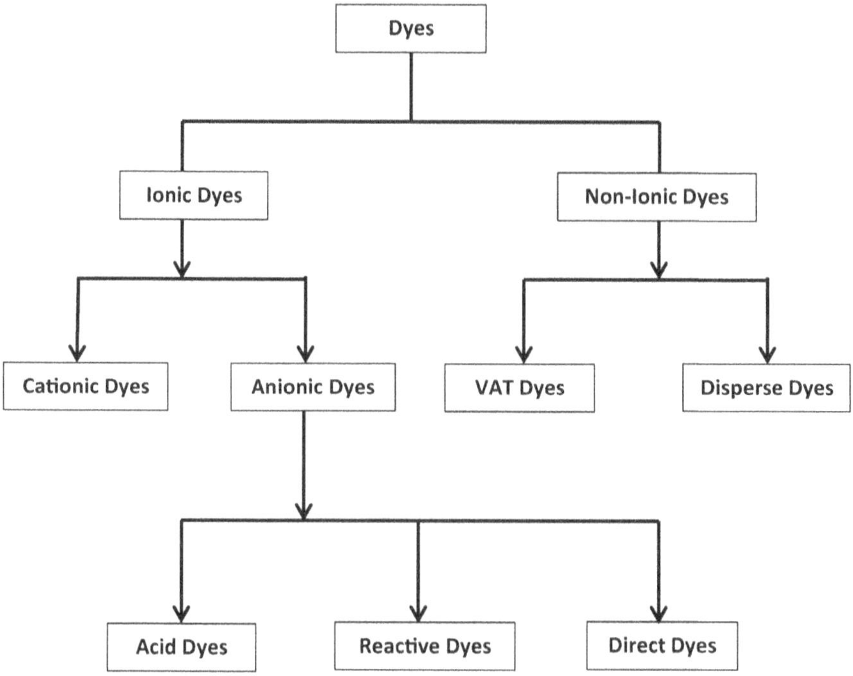

Fig. 1 Types of dyes on the basis of ionic charge

Niculescu et al. 2015) due of its versatility, applicability and economic feasibility (Puchana-Rosero et al. 2016).

Adsorption removes inorganic and organic pollutants from wastewater. When effluents come in contact with porous solid surface of the adsorbent, a solid-liquid intermolecular force of attraction is produced and it causes some of the contaminants to be deposited on the exterior of adsorbent (Machado et al. 2011). A representation of adsorption phenomenon is presented in Fig. 3. Various adsorbent materials have been used such as carbon (Cardoso et al. 2012), bio-sorbents (Cardoso et al. 2012), and agricultural wastes (Prola et al. 2013).

Owing to large surface area and porosity, usually adsorption using activated carbon is employed in wastewater treatment (Ribas et al. 2014). This process is simple and less expensive for removing pollutant. But, the disadvantages of activated carbon are as follows; the regeneration process is expensive and the efficiency is low (Machado et al. 2011) and they occasionally exhibit low adsorption capacity for some organic compounds and heavy metal ions (Huang et al. 2017; Kurniawan and Lo 2009).

Hence, there is a need for developing cost effective and reliable adsorbent to remove pollutants efficiently, so researchers have developed and used economical substrates such as cedar sawdust (Hamdaoui 2006), wood-shaving bottom ash (Leechart et al. 2009), brewery left overs (Tsai et al. 2008), sewage sludge

Table 1 Dyes, their structure, molecular formula, type and toxicity information

S. No	Name of dye	Structure	Molecular formula	Type	Toxicity	Reference
1.	Methylene blue	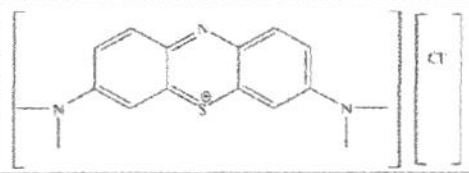	$C_{16}H_{18}ClN_3S$	Cationic (basic)	Carcinogenic	Clarke and Anliker (1980)
2.	Crystal violet	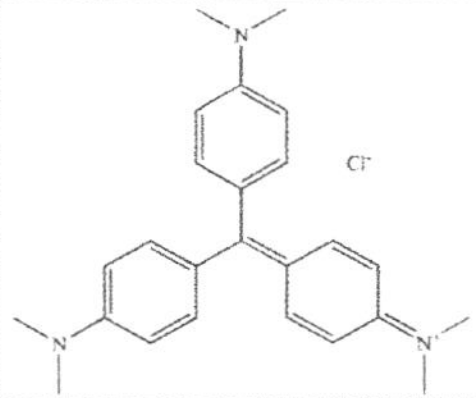	$C_{25}H_{30}ClN_3$	Cationic (basic)	Carcinogenic	Adak et al. (2005)
3.	Methyl violet	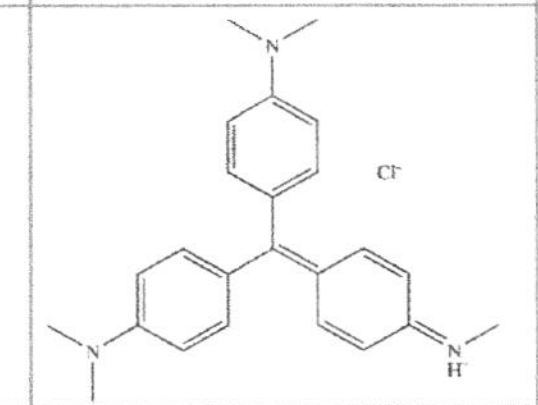	$C_{24}H_{28}N_3Cl$	Anionic (acid)	Mutagenic and mitotic	Doğan and Alkan (2003)
4.	Rhodamine-B	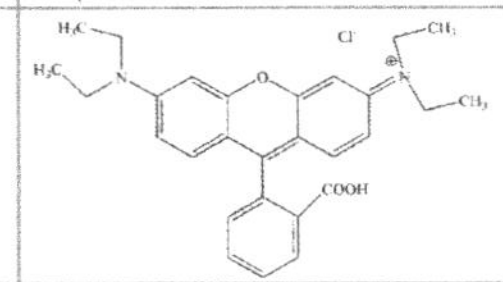	$C_{28}H_{31}ClN_2O_3$	Cationic (basic)	Carcinogenic	Wu et al. (1998)
5.	Acridine orange	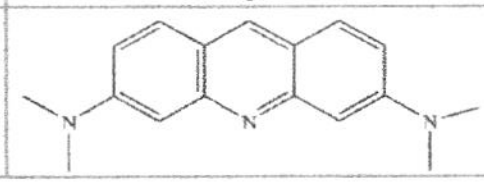	$C_{17}H_{19}N_3$	Cationic (basic)	Mutagenic	Kazim et al. (2007)

(continued)

Table 1 (continued)

S. No	Name of dye	Structure	Molecular formula	Type	Toxicity	Reference
6.	Neutral red	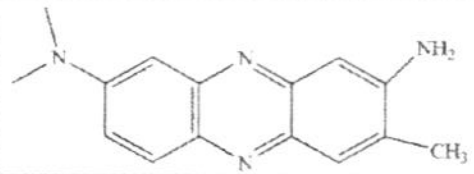	$C_{15}H_{17}ClN_4$	Basic		Iram et al. (2010)
7.	Malachite green	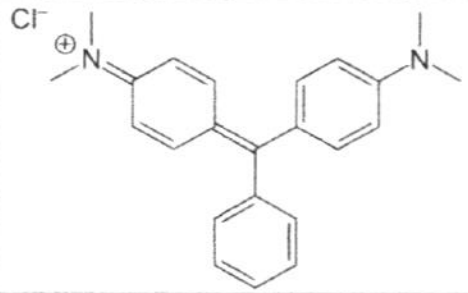	$C_{23}H_{25}ClN_2$ (chloride)	Acidic	Carcinogenic	Baek et al. (2010)
8.	Orange-G	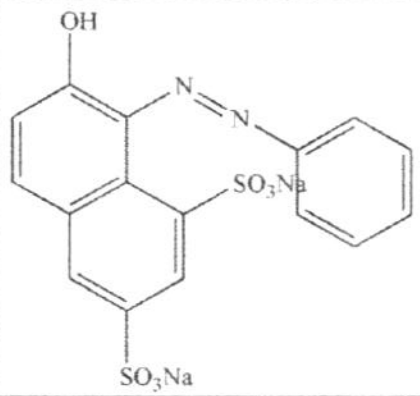	$C_{16}H_{10}N_2Na_2O_7S_2$	Acidic	Carcinogenic and neurotoxin	Mall et al. (2005)
9.	Reactive black-5	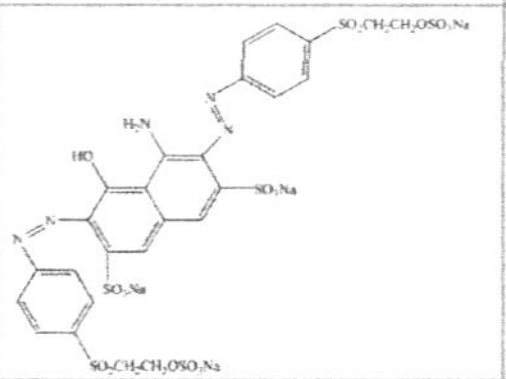	$C_{26}H_{21}N_5Na_4O_{19}S_6$	Acidic		Lucas and Peres (2006)
10.	Congo red	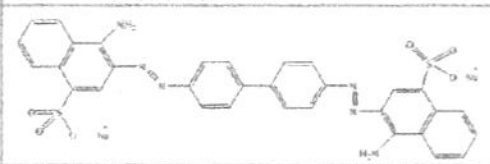	$C_{32}H_{22}N_6Na_2O_6S_2$	Acidic	Carcinogenic and mutagenic	Namasivayam and Kavitha (2002)

(continued)

Table 1 (continued)

S. No	Name of dye	Structure	Molecular formula	Type	Toxicity	Reference
11.	Basic red-12	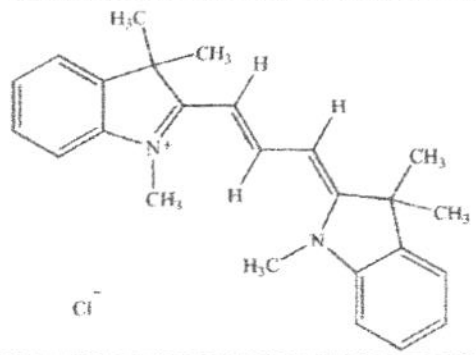	$C_{25}H_{29}ClN_2$	Basic	Mutagenic and Tumorigenic	El Haddad et al. (2012)

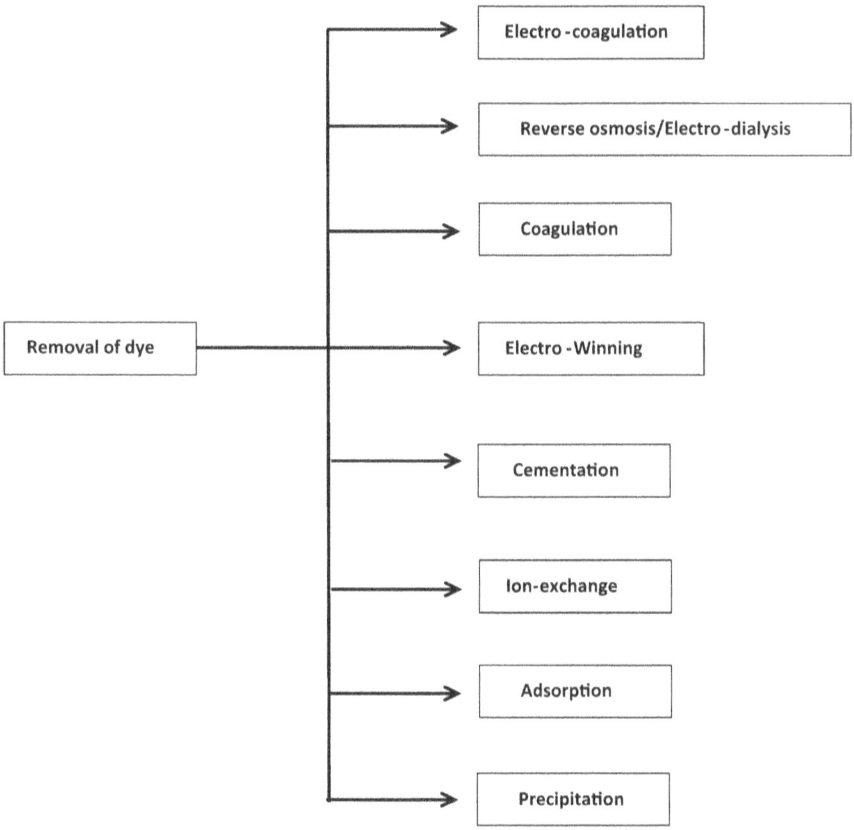

Fig. 2 Conventional process for remediation of dyes

(Rashed 2011) etc. for removing dyes. Among the above adsorbents, activated rice husk has proved to be a economical choice (Gupta et al. 2006).

The rapid development in nanotechnology has been considered effective in solving numerous water pollution problems (Anbia and Samira 2012). Nanomaterials have been examined for the sorption of dyes, antibiotics etc. Nano materials (e.g., carbon nano tubes (CNTs) and dendrimers) have contributed widely for effective water treatment (Machado et al. 2012; Shirmardi et al. 2013; Saito et al. 1998).

Graphene is a popular nanomaterial used as an efficient adsorbent. It is one-atom thick and made of graphite with distributed carbon atoms, tightly packed into 2-D honeycomb carbon lattice in the pattern of sp^2 hybridization (Yang et al. 2011). The structure of graphene is shown in Fig. 4. Graphene has high specific surface (2630 $m^2\ g^{-1}$) and plain geometrical structure is considered for usage in the wastewater treatment.

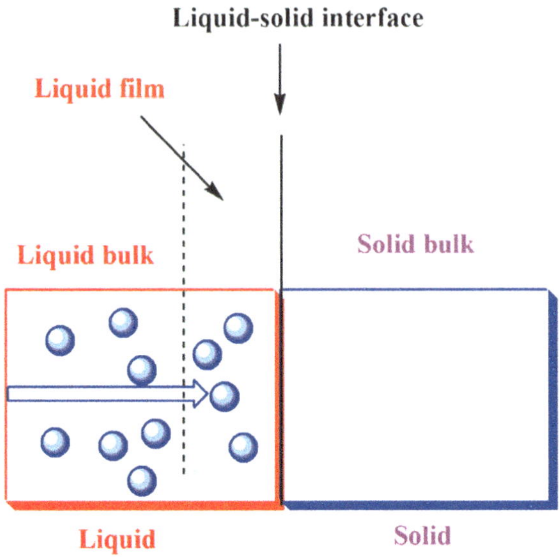

Fig. 3 Schematic representation of adsorption process

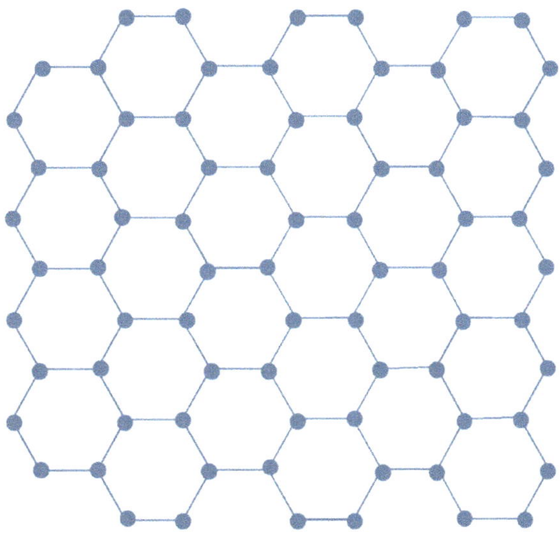

Fig. 4 Structure of graphene

These characteristics allow the manufacture of mixed adsorbents containing less quantity of graphene dose yet possessing maximum adsorption capacity (Amin et al. 2014). Hence, graphene and its family are widely used as nanoadsorbents (Ren et al. 2012).

2 Graphenes

The graphene is found to be an attractive carbonaceous adsorbent owing to its exclusive electrical, mechanical, chemical properties in addition to its low production costs compared to the other adsorbents (Shamik and Rajasekhar 2014). The present studies consolidate some basic information about the synthesis of graphene nanocomposites. It also provides the advantages of using graphene nanocomposites over conventional adsorbents, its potential benefits and challenges in usage.

2.1 *Pristine Graphene*

Scotch tape exfoliation method was the first method to produce pure graphene (Geim and Novoselov 2007). This method builds on continuous flaking of graphene layers from a section of graphite by applying adhesive tape and transferring it into a silicon substrate which yields good quality graphene sheets. Although this method is very simple and cost effective, the quantity produced is much lower and it is not suitable for using it as standard synthesis method for fundamental studies of its properties (Chen et al. 2012). Different techniques are also established by many researchers for manufacturing graphene of good quality (Park and Ruoff 2009). The methods of producing high quality graphene can be broadly categorized as two major classes, such as "top-down" approach as well as "bottom-up" approach which are shown in Fig. 5.

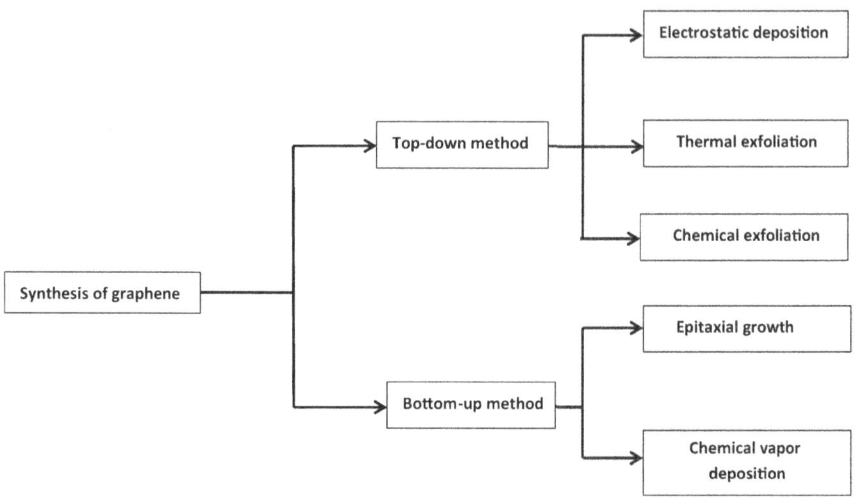

Fig. 5 Chemical synthesis of graphene. Adapted with permission from Park and Ruoff (2009)

The top-down approach is the most common method of producing large quantities of graphene at low-cost. It is synthesized by exfoliating natural or synthetic graphite using a strong oxidizing agent and followed by thermal exfoliation or chemical reduction. However, it is very difficult to obtain high quality graphene sheets, because during exfoliation defects get introduced (Aoki and Dresselhaus 2014). In contrast, the bottom-up approach yields a large quantity of defect-free materials with excellent physical properties and involves direct synthesis of graphene from organic starting materials such as methane and other hydrocarbon sources. Some of the bottom-up methods are epitaxial growth, and uses chemical vapour deposition (CVD) techniques. Bottom-up approaches, however, are reported to have a high manufacturing cost (Wolf 2004).

2.2 *Graphene Oxide (GO)*

GO is synthesized from graphene and the structure of GO is given in Fig. 6. Currently, most of the GO is manufactured via chemical oxidation process using agents such as $KClO_4$ and HNO_3 (Shaobin et al. 2013; Zhang et al. 2010). Hummers and Offeman (1958) modified the above method by employing H_2SO_4 and $KMnO_4$. The main step carried out in Hummers process is the contact of graphite powders to certain solvents, and later sonication.

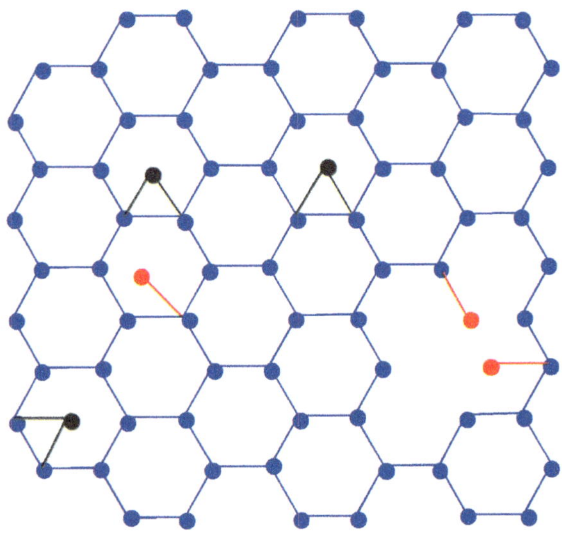

Fig. 6 Structure of GO

2.3 Reduced Graphene Oxide (rGO)

The reduction of GO through chemical, thermal or electrochemical process, produces rGO and its structure of it is given in Fig. 7. rGO is highly imperfect and it shows low conductivity in comparison to pristine graphene (Pei and Cheng 2012; Dreyer et al. 2010). P_{25}-reduced graphene oxide was synrthesized using a hydrothermal reaction by Wang et al. (2015a, b). This research group prepared GO through Hummer's method, and later treated the same with P_{25} ethanol suspension.

The graphene may be produced in large amount by using cheaper substances like cane sugar (disaccharides) was studied by Gupta et al. (2012). This is a fast and low cost process for preparing graphene to be used in water treatment.

2.4 Graphene-Based Nanocomposite

Graphene-based nanocomposites are developed as the interesting adsorbent materials owing to its outstanding features. The use of graphene as nanoadsorbents is dependent on its dispersion in liquid phase and it capability for removing various contaminants. Graphene in abundant amount is capable of agglomeration and restacking to yield graphite during the treatment of wastewater. Separation and collection of graphene and GO at the same time is difficult (Huang and Chen 2015). The chemical functionalization is used for preventing agglomeration during treatment process. Furthermore, it also enhances their interaction with different organic and inorganic pollutants (Han et al. 2012).

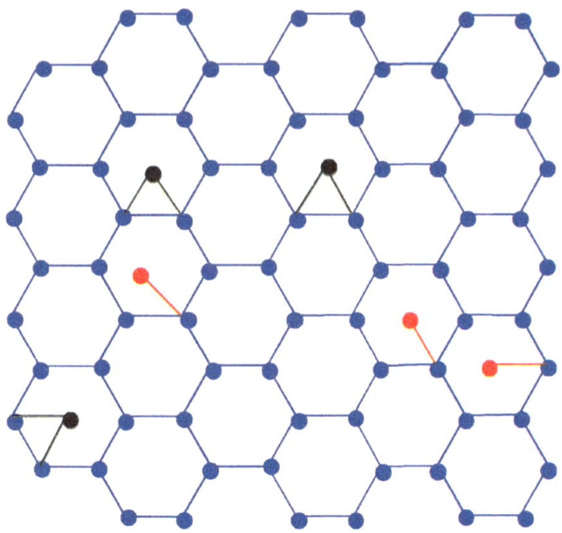

Fig. 7 Structure of rGO

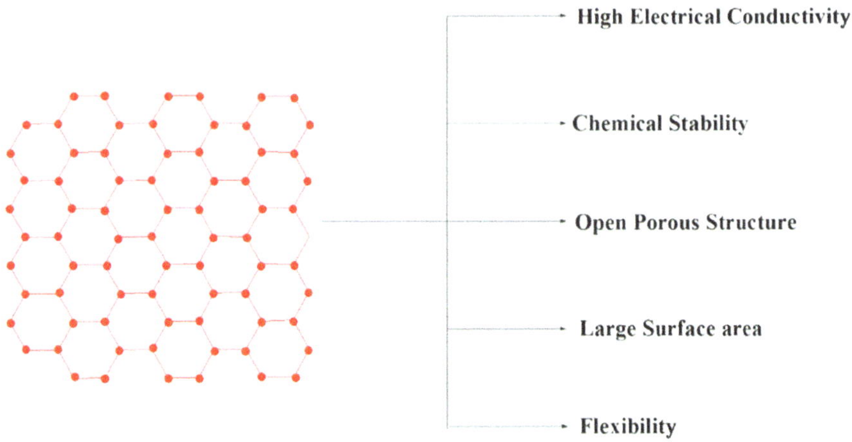

Fig. 8 Features of graphene

The surface modification of graphene materials with nanoparticles or other functional elements increases their sensitivity and selectivity (Shamik and Rajasekhar 2014; Kyzas et al. 2014).

3 Properties of Graphene Materials

Graphene has large surface area whereas GO has distinct functional group reflecting the capability for the process of adsorption (Xiang et al. 2012). The properties of graphene are shown in Fig. 8.

There are various features in graphene which play a vital role in adsorption. The graphene is hexagonal in shape with two C atoms at each unit and it has a surface area of 0.052 nm^2 and density 0.77 mg/m^2. It is transparent and it is able to absorb just 2.3% intensity. Graphene is 100 times stronger than steel. The electrical conductivity of graphene is higher than copper. Graphene shows a thermal conductivity of 5000 W/mK and it can conduct heat 10 times more than copper (Graphene 2010).

4 Adsorption Phenomenon

Sorption is the combination of absorption and adsorption. It is the accumulation of given specie (adsorbate) in a solid or liquid surface (adsorbent). When the adsorbate is released from the surface of the adsorbent, it is named as desorption (Lima et al. 2015; Waghmare and Arfin 2015).

$$\text{Adsorbent}_{(s)} + \text{Adsorbate}_{(aq)} \underset{\text{desorption}}{\overset{\text{adsorption}}{\rightleftharpoons}} \text{Adsorbent} - \text{Adsorbate}_{(s)} \tag{1}$$

Adsorption isotherm explains the relationship amongst the quantity of adsorbate (mg) adsorbed by the sorbent (g) defined as sorption capacity (q_e) and the adsorbate concentration remaining in solution after equilibrium is reached (C_e) at constant temperature

$$q_e = \frac{(C_o - C_e)}{m}.V \tag{2}$$

where q_e is sorption capacity at equilibrium (mg g^{-1}), C_o is the initial adsorbate concentration in contact with the adsorbent (mg L^{-1}), C_e is the adsorbate concentration (mg L^{-1}) remaining in the solution after the system attains equilibrium, V is the volume of adsorbate solutions (L) placed along with the adsorbent, and m is the mass (g) of adsorbent.

4.1 *Adsorption Isotherms*

There are several isotherm models for being applied for the equilibrium of adsorption of an adsorbate onto adsorbent. The most used is the Langmuir isotherm model (Langmuir 1918). Also, other adsorption equilibrium models such as Freundlich isotherm, (Freundlich 1906); Liu isotherm (Liu et al 2003) are also documented in the literature for adsorption of solid adsorbates dissolved in a solvent.

4.1.1 Langmuir Equilibrium Model

The Langmuir equilibrium model presents the following basic premises, adsorbates are taken up by a fixed number of dynamic sites accessible for the sorbent; a monolayer of the sorbate species covers the surface of the sorbent; each active site of the sorbent can hold just one unique sorbate species; all active zones of the adsorbent are energetically equivalent; the adsorbed species on the activate sites does not interacts themselves when they are sorbed onto the sorbent (Lima et al. 2015).

The Langmuir equilibrium model is represented by Eq. 3:

$$q_e = \frac{Q_{\max} \cdot K_L \cdot C_e}{1 + K_L \cdot C_e} \tag{3}$$

being q_e the sorption capacity of the adsorbent taken up at equilibrium (mg g^{-1}), C_e is the concentration of adsorbate left in the aqueous phase at equilibrium (mg L^{-1}),

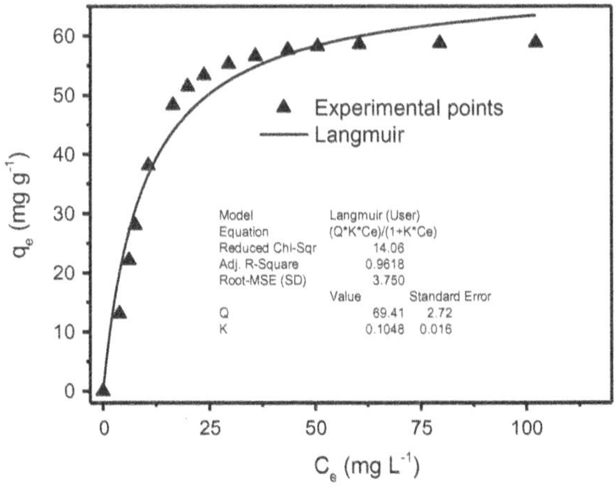

Fig. 9 Langmuir adsorption equilibrium isotherm of sodium diclofenac onto GO at 25 °C and pH 8.0. Adapted with permission from Jauris et al. (2016)

K_L is the Langmuir equilibrium constant (L mg^{-1}), $Q_{\max}$ is the maximum sorption capacity of the adsorbent (mg g^{-1}). $Q_{\max}$ is a theoretical value of a maximum sorption capacity of an adsorbent and this value does not corresponds to the q_e values of the saturation plateau of the isotherm.

In Fig. 9 is presented the Langmuir isotherm model of the sodium diclofenac adsorbed onto graphene oxide adsorbent (GO) (Jauris et al. 2016).

The one with the lowest value of Residual Standard Deviation (SD) is the best fitted model. It is also named as square root of mean square error (root MSE), and the one whose value of adjusted coefficient of determination (R^2_{adj}) is closer to 1.000. Equations 4–7 depicts the expressions of reduced chi-square, SD, coefficient of determination (R^2) and R^2_{adj}, respectively.

$$\text{Reduced Chi-squared} = \sum_i^n \frac{\left(q_{i,\exp} - q_{i,\text{model}}\right)^2}{n-p} \tag{4}$$

$$\text{SD} = \sqrt{\left(\frac{1}{n-p}\right) \cdot \sum_i^n \left(q_{i,\exp} - q_{i,\text{model}}\right)^2} \tag{5}$$

$$R^2 = \left[\frac{\sum_i^{np} \left(q_{i,\exp} - \bar{q}_{i,\exp}\right)^2 - \sum_i^{np} \left(q_{i,\exp} - q_{i,\text{model}}\right)^2}{\sum_i^n \left(q_{i,\exp} - \bar{q}_{i,\exp}\right)^2}\right] \tag{6}$$

$$R^2_{adj} = 1 - \left(1 - R^2\right) \cdot \left(\frac{n-1}{n-p-1}\right) \tag{7}$$

$q_{i,\text{ model}}$ value is q projected by the best fit model, $q_{i,\exp}$ is q quantified experimentally, $\bar{q}_{\exp}$ is the average of all q values quantified experimentally, n is the number of experimental points and p is the number of parameters of the model.

The reduced chi-squared is the residual sum of squares divided by n-p (degree of freedom) (Eq. 4). SD is also defined as square rood of mean square error (root-MSE). Both reduced Chi square as well as SD are useful for evaluation of adsorption data. Each experimental point ($q_{i,\exp}$) corresponds to a point that matches exactly to the point over the curve ($q,_{\text{ i model}}$). The lower the values of reduced chi-squared and the SD, the lower is the difference between the values of $q_{\text{experimental}}$ q and q_{model}, leading to the best fit model.

The R^2 and R^2_{adj} statistical parameters depicted on Eqs. 6 and 7, respectively are used to evaluate models. Their values are always limited between 0 and 1. If the values of R^2 and R^2_{adj} that are closer to 1, indicates that the model present the better fit for the experimental data. It should be highlighted that $\bar{q}_{i,\exp}$ of Eq. 6 is the average of all experimental data (q_e). If the interval of q_e values is too large, $\bar{q}_{i,\exp}$ could lead to a distortion in the interpretation of the best-fit. If there is an intermediate value for $q_{i,\exp}$, the values of R^2 will be close to 1 (Lima et al. 2015). The reader may think that there was a best fit, however this is a misinterpretation of the R^2 values. Direct comparison of R^2 values of 2-parameter models (Freundlich, Langmuir) with 3-parameter models (Sips 1948, Liu et al 2003) is not justified, because the equations with higher number of parameters have a mathematical tendency to exhibit higher R^2 values. Therefore the R^2_{adj} penalizes the model with a greater number of parameters to verify if the increase in the values of R^2 is due to the beneficial of the mathematical increases of parameters, or if indeed the models with more parameter explain better the physical behavior presented. For example if Langmuir isotherm model ($p = 2$) presented a R^2 of 0.9984 and the Sips isotherm model ($p = 3$) presented a R^2 of 0.9985, and the set of experimental points are 12 ($n_p = 12$). Using Eq. 7, the values of R^2_{adj} were 0.9980 and 0.9979 for Langmuir and Sips model, respectively, therefore the model of 2 parameters was better fitted than a model of 3 parameters.

In Fig. 9, the R^2_{adj} for Langmuir model was 0.9618, the Q_{max} was 69.41 mg g^{-1} and the K_L was 0.1047 L mg^{-1}. The SD was 3.750 mg g^{-1}.

4.1.2 Freundlich Equilibrium Model

The Freundlich isotherm model is the oldest model reported in the literature (Freundlich 1906) and it considers that the concentration of sorbate taken up by the sorbent will always increase exponentially as the adsorbate concentration get increased. This model would expect an infinite adsorption taking place. This isotherm model assumes that multiple layers of adsorption will appear instead of a monolayer. It wold be possible to form the second layer of the adsorbate, without

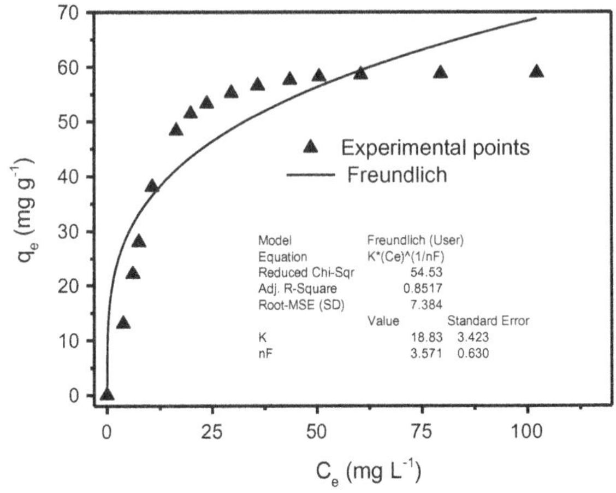

Fig. 10 Freundlich equilibrium model of sodium diclofenac onto GO at 25 °C and pH 8.0. Adapted with permission from Jauris et al. (2016)

the need of the first layer being completely fulfilled. The Freundlich model has being mainly applied for adsorbents presenting a heterogeneous surface.

$$q_e = K_F \cdot C_e^{1/n_F} \tag{8}$$

K_F the Freundlich equilibrium constant (mg g^{-1}(mg L^{-1})$^{-1/n}$), and n_F the Freundlich exponent (dimensionless).

Figure 10 shows the Freundlich equilibrium plot for sodium diclofenac adsorbed onto graphene oxide adsorbent (GO) (Jauris et al. 2016). The R^2_{adj} was 0.98517, and K_F was 18.83 (mg g^{-1}(mg L^{-1})$^{-1/nF}$), and n_F was 3.571. The total standard deviation of the fitting was 7.384 mg g^{-1}.

4.1.3 Liu Equilibrium Model

The Liu equilibrium model (Liu et al. 2003) is a junction of the Langmuir and Freundlich isotherm models. Here the premises of single layer of Langmuir model and the infinite uptake of the adsorbate of Freundlich are inconsiderate. The Liu model also disregards that all the dynamic zones of the sorbent is considered energetically equivalent, contradicting the Langmuir model. In the Liu model, it is predicted that a saturation of the adsorbent surface would take place at higher adsorbate concentration, in contrast with Freudlich isotherm model. Equation 9 describes the Liu equilibrium model:

$$q_e = \frac{Q_{\max} \cdot (K_g \cdot C_e)^{n_L}}{1 + (K_g \cdot C_e)^{n_L}} \tag{9}$$

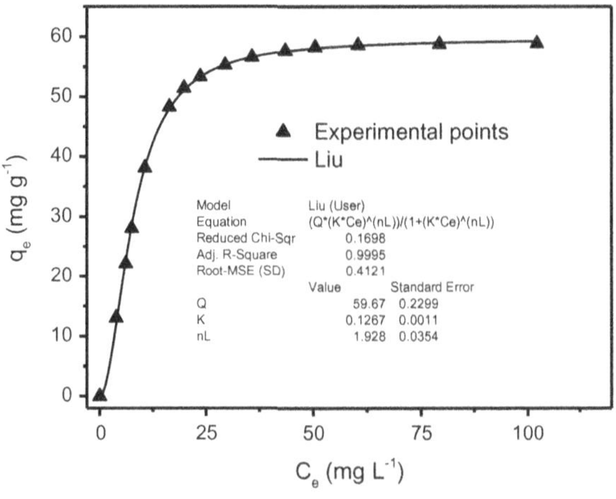

Fig. 11 Liu adsorption equilibrium isotherm of sodium diclofenac onto GO at 25 °C and pH 8.0. Adapted with permission from Jauris et al. (2016)

K_g is the Liu equilibrium constant (L mg^{-1}); Q_{max} is the maximum adsorption capacity of the adsorbent (mg g^{-1}), and n_L is the exponent in Liu model (dimensionless). It may be noted that Q_{max} is a theoretical parameter obtained by the fitting of the isotherms, and their values are not an average of maximum experimental q_e values obtained at the supposed plateau of the isotherm.

Figure 11 shows the Liu equilibrium plot for sodium diclofenac adsorbed onto graphene oxide adsorbent (GO) (Jauris et al. 2016).

The R^2_{adj} for Liu isotherm was 0.9995 while the Liu equilibrium constant (K_s) was 0.1267 L mg^{-1}, the value of nL exponent was 1.928. The total standard deviation of the fitting was 0.4121 mg g^{-1}.

5 Adsorption of Dyes Using Graphene Nanocomposites

Dyes usually have a complex and stable molecular structure. Dyes contamination in water is toxic to all form of living organisms. The efficiency of graphene and activated carbon are widely studied by many researchers but recent studies are mostly on the chemical modifications in graphene and its adsorption capacity on dyes (Radich et al. 2011).

Yang et al. (2011) explained the use of GO in methylene blue decolourisation. GO was able to absorb methylene blue to a large extent showing very good absorption capacity (714 mg g^{-1}). The removing competence was nearly 99% for dye concentrations <250 mg L^{-1}.

Li et al. (2014a, b) had investigated removal of methyl violet using graphite/ Fe_3O_4. At pH range 2 to 4, the adsorption capacity of methyl violet increased marginally whereas at pH range 4 to 10, it was higher. Equilibrium was reached at 10 min. The thermodynamic and kinetic parameters were determined. Langmuir model was the best fit. Thermodynamic study reveals that adsorption was rapid, effortless and endothermic. The material was highly stable and had significant regeneration capability.

Tannic acid-graphene nanocomposite adsorbent material was used for the removal of rhodamine blue (Liu et al. 2015a, b). This adsorbent exhibited good removal of rhodamine blue due to electrostatic attraction and π–π stacking present in the adsorbent. Langmuir isotherm was the best fit. The kinetic studies demonstrated that chemisorption followed the pseudo-second-order model.

Sun et al. (2012) had investigated adsorption of acid orange on graphene oxide. GO displayed very good adsorption capacity (2158 mg g^{-1}) which makes it suitable for the treatment process. The adsorption capacity decreased to 976 mg g^{-1} when the concentration of GO was 2 folds, however the removal efficacy rose to 95%. Langmuir model was found to best fit than Freundlich.

Graphene-SO_3H/Fe_3O_4 was used for removing red dye from solution by Sakuramoto et al. (2007). Graphene-SO_3H/Fe_3O_4 was found to remove 93% of the cationic dye in 10 min. The kinetic studies revealed that adsorption followed pseudo second order model. Both Langmuir and Freundlich isotherms were found to fit adsorption, however the sorption capability of the sorbent was estimated using Langmuir isotherm.

Magnetite/reduced graphene oxide nanocomposites were used for sorption of dyes (Sun et al. 2011). The adsorbent was synthesized using one step solvothermal synthesis. The material showed significant removal of malachite dyes (94%) in 2 h at ambient temperature. The highest sorption capacity of the sorbent was 22.0 mg g^{-1} for malachite green. The kinetics study fitted pseudo-second-order model. The adsorption study confirms to Langmuir and Freundlich isotherms.

Hosseinabadi-Farahani et al. (2015) synthesized surface modified graphene oxide nanosheet for the removal of Acid Blue 92. The material was characterised using Fourier transform infrared spectroscopy (FT-IR) and scanning electron microscopy (SEM). The adsorption kinetics confirmed to pseudo-second order model. Adsorption followed Langmuir isotherm model. The adsorption capacity of the nanosheet was 60 mg g^{-1}.

3D chitosan–graphene nanocomposites was synthesised to remove reactive black 5 (Cheng et al. 2012). The removal efficacy was observed to be 97.5%. Sorption was dependent on pH, temperature and initial concentration of the dye.

Li et al. (2014a, b) synthesized Fe_3O_4@mTiO_2@GO for sorption of congo red. The graphene oxide nanocomposite exhibited good sorption capability and fast remediation of congo red from solution. The adsorbent possessed vast surface area and variety of functional groups on the adsorbent surface. The kinetics followed pseudo-second-order model and thermodynamic studies confirmed to Freundlich model. The sorption capability of the graphene nanocomposite was 89.95 mg g^{-1}.

Moradi et al. (2015) synthesised graphene and graphene oxide for the sorption of basic red 12 from aqueous media. The kinetic studies confirmed to pseudo-second-order. The adsorbents displayed good adsorption capacity for the dye. The maximum sorption capacity was achieved at pH 9. Several other graphene based nanoadsorbents used for dye removal are presented in Table 2.

Table 2 Graphene based nanoadsorbents used for dye removal

Dye	Graphene based nano adsorbent	Details	Reference
Methylene blue	Reduced graphene oxide/iron oxide (rGO-IO) nanocomposite	Maximum adsorptive capacity—39 mg g^{-1}	Sharif et al. (2017)
	Graphene/alginate double network nanocomposite	Adsorbent displays good thermal stability and high specific surface area	Zhuang et al. (2016)
	Graphene oxide aerogel	Fits Langmuir isotherm, maximum adsorption capacity—416.6667 mg g^{-1}	Zamani and Tabrizi (2015)
	Graphene oxide	Maximum absorption capability—714 mg g^{-1}	Yang et al. (2011)
	Rhamnolipid-functionalized graphene oxide	Adsorption isotherm order BET > Freundlich > Langmuir > Temkin	Wu et al. (2014)
Crystal violet	Magnetic nanocomposite hydrogels	Pseudo second order kinetics Langmuir isotherm, maximum sorption capability—769.23 mg g^{-1}	Pourjavadi et al. (2015)
	Chitosan-graphite oxide modified polyurethane foam	Fits Langmuir equation, maximum sorption capability—64.935 mg g^{-1}	Qin et al. (2015)
	Graphene oxide coated cotton wad	Thomas, Adams-Bohart, and Yoon-Nelson models were used on the experimental data.	Khaloo et al. (2016)
	Magnetic reduced graphene oxide	Langmuir monolayer sorption, pseudo second order kinetic model maximum sorption capability—64.93 mg g^{-1}	Sun et al. (2014)
	Magnetic bio-sorbent hydrogel beads	Langmuir isotherm model, maximum sorption capability—81.78, 69.67, 94.0, and 101.74 mg g^{-1} for Pb(II), Cu(II), crystal violet and congo red respectively	Sahraei et al. (2017)
Methyl violet	3D graphene oxide sponge	Sorption capability—397 and 467 mg g^{-1} activation energy—50.3 and 70.9 kJ mol^{-1} for methylene blue and methyl violet respectively	Liu et al. (2012)
	Magnetic water-soluble hyper branched polyol functionalized graphene oxide	Langmuir model pseudo second order kinetics	Hu et al. (2016)
	Amine-functionalized reduced graphene oxide-carbon nanotube	Langmuir model pseudo-second order kinetics maximum sorption capacities for methyl violet and methyl orange—298 and 294 mg g^{-1}, respectively,	Sarkar et al. (2014)
	Nano-graphite/Fe_3O_4 composite	Langmuir isotherm, pseudo-second-order kinetic model	Li et al. (2014a, b)
	Reduced graphene oxide	Pseudo second order kinetic model, Isotherm models follow the order Toth > Sips > Dubinin–Radushkevich (DR) > Scatchard > Langmuir > Temkin > Freundlich model for the pH range 4–6. DR > Scatchard > Toth > Sips > Langmuir > Temkin > Freundlich for pH 7–9	Sharma et al. (2014)

(continued)

Table 2 (continued)

Dye	Graphene based nano adsorbent	Details	Reference
Rhodamine B	Tannic acid functionalized graphene nanocomposite	Langmuir model, pseudo second order kinetic model maximum sorption capability—201 mg g^{-1}	Liu et al. (2015a, b)
	Graphene oxide/Beta zeolite composite	Langmuir and Freundlich isotherms maximum sorption capability—64.47 mg g^{-1}	Cheng et al. (2017)
	Graphene oxide-zeolite	Langmuir isotherm pseudo-second-order model maximum sorption capacities for zeolite, GO grafted zeolite (GO-zeolite) and benzene carboxylic acid derivatized GO-zeolite, was 50.25, 55.56 and 67.56 mg g^{-1} respectively	Yu et al. (2013)
	Magnetite/reduced graphene oxide nanocomposites	Freundlich isotherm pseudo second order kinetic model	Sun et al. (2011)
	reduced graphene oxide–Fe_3O_4 nanoparticles	Efficient regeneration of adsorbent	Geng et al. (2012)
Acridine Orange	GO-encapsulating alginate beads	Langmuir model	Sun and Fugetsu (2014)
	Graphene oxide	Maximum sorption capability of 1.4 g g^{-1} without situ reduction and with the in situ reduction maximum sorption capability was 3.3 g g^{-1}	Sun et al. (2012)
	Graphene oxide	Maximum sorption capability—229.8 mg g^{-1}	Wang et al. (2015a, b)
	Magnetic calcium silicate/graphene oxide	Freundlich isotherm model pseudo-second-order kinetics	Wang et al. (2016)
Neutral red	Graphene-SO_3H/Fe_3O_4	Langmuir and Freundlich sorption models pseudo-second-order maximum sorption capability—216.8 mg g^{-1}	Wang et al. (2013)
	L-Cysteine reduced graphene oxide	Maximum sorption capability for anionic indigo carmine (IC) and cationic neutral red (NR)—1005.7 mg g^{-1} and 1301.8 mg g^{-1}, respectively	Xiao et al. (2016)
	Magnetic Fe_3O_4 reduced Graphene Nanocomposite	Langmuir and Freundlich model pseudo second order	Wang et al. (2014)
	Chitin/graphene oxide	The maximum sorption capability for ramazol black and neutral red were 9.3×10^{-2} mmol g^{-1} and 57×10^{-2} mmol g^{-1}	González et al. (2015)
	Paramagnetic graphene oxide-Fe_3O_4 composite	Sorption capacities for methylene blue and neutral red were 167.2 and 171.3 mg g^{-1}, respectively.	Xie et al. (2012)

(continued)

Table 2 (continued)

Dye	Graphene based nano adsorbent	Details	Reference
Malachite green	Graphite oxide	Langmuir isotherm, maximum sorption capability—351 and 248 mg g^{-1} for methylene blue and malachite green, respectively	Bradder et al. (2010)
	Graphene oxide/cellulose bead	Langmuir isotherm pseudo-second-order kinetic model	Zhang et al. (2015)
	Magnetic β-cyclodextrin-graphene oxide nanocomposites	Langmuir isotherm pseudo-second-order kinetic model. Maximum sorption capability—990.10 mg g^{-1}	Wang et al. (2015a, b)
	Hematite-reduced graphene oxide composites	Langmuir isotherm pseudo-second-order kinetic model. Maximum sorption capability—438.8 mg g^{-1}	Liu et al. (2015a, b)
	Graphene oxide	Langmuir and Freundlich isotherm pseudo-second-order kinetic model. Maximum sorption capability—384.62 mg g^{-1}	Nuengmatcha et al. (2014)
Reactive black 5	Three-dimensional chitosan–graphene mesostructures	Maximum removal efficiency—97.5%	Cheng et al. (2012)
	Graphene oxide/chitosan	Langmuir, Freundlich and Langmuir–Freundlich model maximum sorption capability—277 mg g^{-1} @25 °C	Travlou et al. (2013a)
	Magnetic chitosan–graphene oxide	Langmuir, Freundlich, and Langmuir–Freundlich models, maximum sorption capacities at 25, 45, and 65 °C were 391, 401, and 425 mg/g, respectively	Travlou et al. (2013b)
	Magnetic graphene oxide	Maximum sorption capability—188 mg g^{-1}	Kyzas et al. (2013)
Congo red	Graphene oxide/chitosan/silica fibre	Fits Langmuir isotherm and pseudo first order model. The maximum sorption capability—294.12 mg g^{-1}	Du et al. (2014)
	Magnetic Fe_3O_4@Graphene	Second-order kinetic equation	Yao et al. (2012)
	pTSA-Pani@Graphene Oxide-CNT nanocomposite	Maximum sorption was observed in acidic conditions at 30 °C	Ansari et al. (2017)
	Graphene oxide/ferroferric oxide/polyethylenimine nanocomposites	Maximum removal capability—574.7 mg g^{-1}, pseudo-second-order model	Wang et al. (2017)
	Magnetic bio-sorbent hydrogel beads	Langmuir isotherm model with maximum sorption of 101.74 mg g^{-1} for congo red	Sahraei et al. (2017)

(continued)

Table 2 (continued)

Dye	Graphene based nano adsorbent	Details	Reference
Basic Red 12	Graphene/graphene oxide	Pseudo-second-order model	Moradi et al. (2015)
	Graphene oxide	Langmuir isotherm	Robati et al. (2016)
	Graphene surface	Pseudo second order model, maximum removal at pH 9	Shahryari-ghoshekandi and Sadegh (2014)
Orange G	Exfoliated graphene oxide/reduced graphene oxide	Effective sorption of cationic dyes is seen while anionic dyes exhibit negligible sorption due to the presence of oxygen containing functional groups such as epoxy, carboxyl, hydroxyl and ketone groups on its basal and edge planes	Ramesha et al. (2011)
	Magnetic graphene oxide	Pseudo second order model and Langmuir model. Maximum sorption capacity—20.85 mg g^{-1}	Deng et al. (2013)

6 Adsorption Mechanisms

The mechanism of adsorption is a necessary parameter for designing application. Adsorption capacity is dependent on total area per unit volume. If the surface is elongated, the active sites are highly open to adsorbate, hence enhancing the adsorption capacity (Mohanty et al. 2008). The graphene exhibits high surface area, but it is non-porous. Therefore, introduction of pores enhances the adsorption of graphene.

Another way to amplify the adsorption capacity of graphene may be by introducing different functional groups which bind the adsorbates. The functional groups promote formation of bonds between the adsorbent and adsorbate. The oxidation of graphene functionalises its surface with groups such as –CO, –COOH, and –OH and makes it hydrophilic (Fan et al. 2010).

Generally, graphene based materials were found to follow hydrogen bonding, electrostatic or π–π conjugation interactions with organic dyes (Chowdhury and Balasubramanian, 2014). These interactions take place between the sorbent and sorbate during adsorption.

Graphene oxide (GO) is an efficient dye adsorbent due to its unique conjugated, two-dimensional structure. It displays good adsorption efficacy for different dyes through π–π stacking interactions (Geim and Novoselov 2007; Li and Kaner 2008; Geim 2009; Huang et al. 2012a, b; Xu et al. 2009; Peigney et al. 2001). GO sheets have several oxygen-rich functional groups that creates negative charges. Due to the negative charges, there is strong electrostatic interactions with cationic dyes (Sharma and Das 2013; Huang et al. 2011; Huang et al. 2012a, b; Sui et al. 2011; Rao et al. 2009; Wu et al. 2013). Nevertheless, GO sheets are highly dispersible in water, which inhibits separation of dye-adsorbed GO sheets from aqueous phase. Therefore, various modified GO-based adsorbents have been synthesized to aid the separation of dye-adsorbed GO sheets from aqueous media (Geng et al. 2012; Wang et al. 2014; Liu et al. 2012)

There are various active sites in rGO namely (a) functional groups such as –COOH, and –OH, (b) free electrons present in the aromatic ring that are delocalized (c) surface groups consisting of oxygen (Pei and Cheng 2012). The main reason for stronger adsorption ability is the larger specific surface area of GO (Zhang et al. 2015). The adsorption mechanism over graphene oxide (GO)/cellulose bead (GOCB) composites follows three basic steps (i) the adsorbate drifted from the liquid to the surface of the sorbent (ii) the dye travelled inside the macro and mesopores of the sorbent (iii) sorption happened on the exterior of the sorbent. During physisorption the dyes exhibited affinity to be adsorbed by GO; this may be due to the electrostatic interactions amongst the π-electrons of GO and the cationic dye. Since MG is a planar molecule, it may be effortlessly adsorbed by π–π stacking interactions between the aromatic dyes and hexagonal framework of GO. Thus, graphene oxide (GO)/cellulose bead may be a promising adsorbent for organic contaminant adsorption. The functional groups in GO, is found to further enhance the adsorption (Zhang et al. 2015).

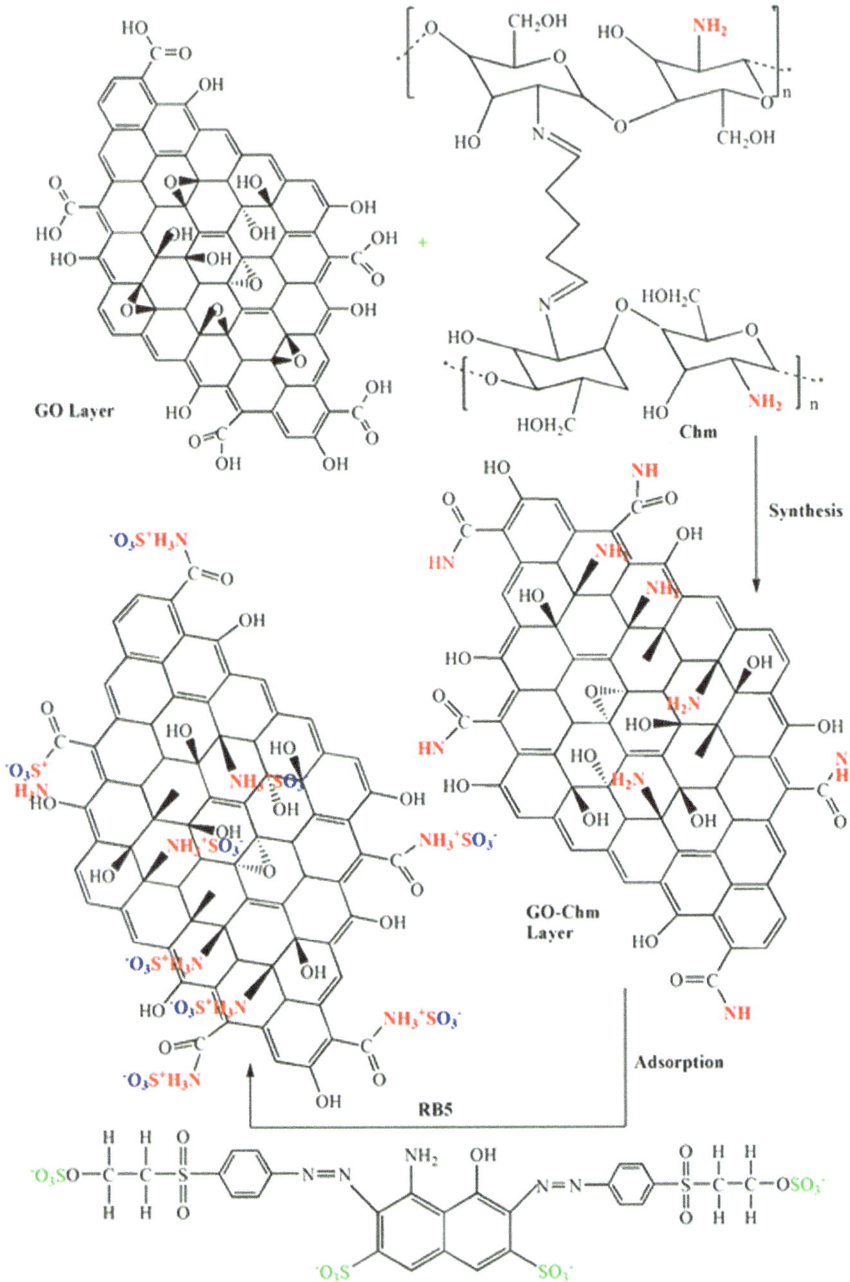

Fig. 12 Proposed mechanism for RB5 sorption onto the surface of synthesized GO-Chm. Adapted with permission from Travlou et al. (2013b) Copyright 2013 American Chemical Society

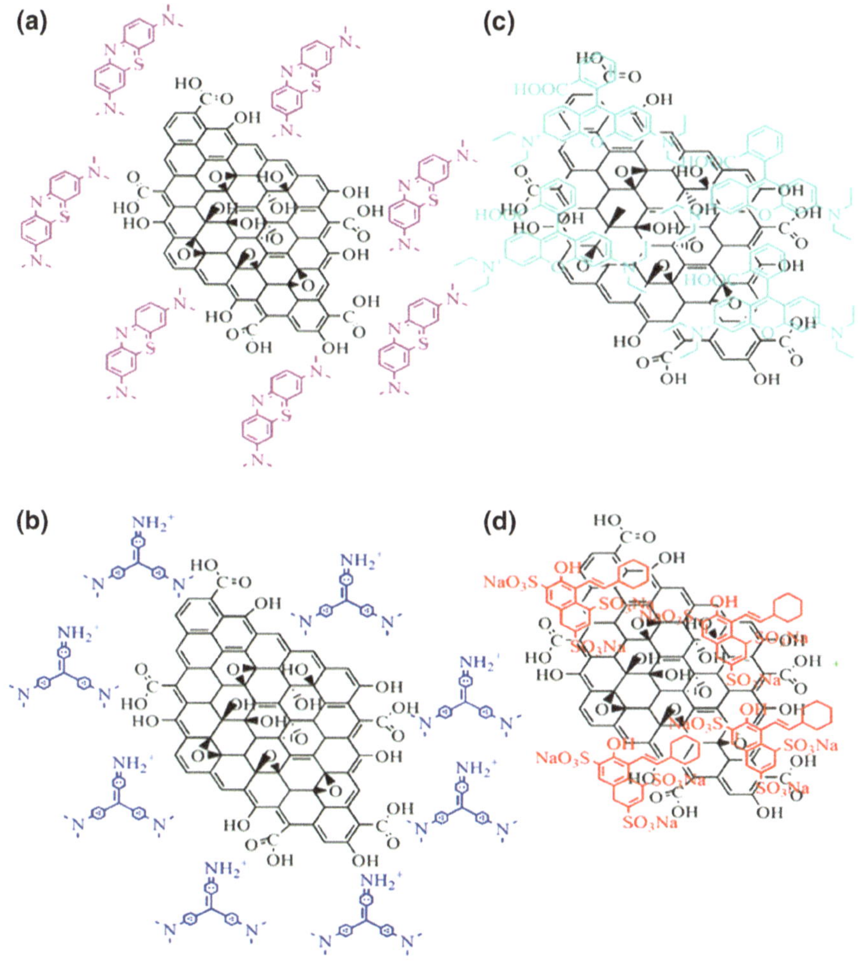

Fig. 13 Interactions of graphene based adsorbents with dyes, **a** exfoliated graphene oxide-methylene blue, **b** exfoliated graphene oxide-methyl violet, **c** exfoliated graphene oxide-rhodamine blue and **d** reduced graphene oxide-orange G

Travlou et al. (2013b) synthesised graphite oxide/magnetic chitosan's (GO-Chm) for the removal of Reactive Black 5. Characterization of the adsorbent showed that a considerable quantity of the amines is from the chitosan (i) were embedded in-between the GO layers and (ii) reacted with epoxy and carboxyl groups of GO. pH 3 was the best for the adsorption process. The presence of Fe_3O_4 nanoparticles resulted in an enhanced electroactive surface of adsorbents. The enhanced adsorption of GO–Chm composite may be due to the π–π interaction among the aromatic dye and GO basal planes. The binding of carboxyl groups to chitosan makes these interactions stronger. The process of adsorption of RB5 on GO-Chm is shown in Fig. 12.

Ramesha et al. (2011) suggested that between exfoliated graphene oxide and Rhodamine blue there was electrostatic interactions and weak van der Waals force, and amongst exfoliated graphene oxide and orange G, it may be only van der Waals interactions. Electrostatic interactions play a role in the case of exfoliated graphene oxide and methyl violet and exfoliated graphene oxide—methylene blue while in the case of exfoliated graphene oxide—orange G it may be van der Waals forces. The interactions amongst exfoliated graphene oxide and rhodamine blue may be electrostatic and van der Waals. This is diagrammatically presented in Fig. 13.

7 Conclusion and Future Perspectives

This chapter presents a small introduction to adsorption principles and adsorption isotherms. It explains the synthesis and uses of graphene and nano graphene materials in the removal of dyes. Dyes are used in textile, paper and printing industries. The wastewaters from these industries have high level of color, COD and are reported to be toxic. Mismanagement of these effluents can contaminate the soil and ground water. These effluents are resistant to biological degradation.

Large-scale treatment requires potentially superior adsorbents to exhibit enhanced dye remediation within short time and it need to be environment friendly. Proper separation of the adsorbents must be possible after adsorbing waste dyes. Recently, graphene and graphene nanocomposites have become popular as prospective dye sorbents due to their exceptional structure. Graphene exhibits high surface area, but it is non-porous. Therefore, introduction of pores improves the sorption capability of graphene. Sorption capability of graphene may be enhanced by introducing different functional groups which bind the adsorbates.

The functional groups promote formation of bonds between the adsorbent and adsorbate. Graphene exhibits extraordinarily good adsorption capacity for various dyes through van der Waals, electrostatic and π–π stacking interactions. This chapter consolidates the recent literature available for dye adsorption using graphene materials and its mechanism.

Acknowledgements The authors thank Clean Water—ESC0306; Project no 2B 3.3.5, and National Council for Scientific and Technological Development (CNPq, Brazil) for funding. The authors also thank the Director, CSIR-NEERI for his support.

Conflict of Interest The authors report no declarations of interest.

References

Adak A, Bandyopadhyay M, Pal A (2005) Removal of crystal violet dye from wastewater by surfactant-modified alumina. Sep Purif Technol 44(2):139–144

Amin MT, Alazba AA, Manzoor U (2014) A review of removal of pollutants from water or wastewater using different types of nano materials. Adv Mater Sci Eng. https://doi.org/10.1155/2014/825910 (Hindawi Publishing Corporation)

Anbia M, Samira S (2012) Removal of acid dyes from aqueous media by adsorption onto amino-functionalized nanoporous silica SBA-3. Dyes Pigm 94:1–9

Ansari MO, Kumar R, Ansari SA, Ansari SP, Barakat MA, Alshahrie A, Cho MH (2017) Anion selective pTSA doped polyaniline@ graphene oxide-multiwalled carbon nanotube composite for Cr (VI) and Congo red adsorption. J Colloid Interface Sci 496:407–415

Aoki H, Dresselhaus MS (2014) Physics of graphene. Nano science and technology. Springer International Publishing

Baek MH, Ijagbemi CO, Se-Jin O, Kim DS (2010) Removal of Malachite Green from aqueous solution using degreased coffee bean. J Hazard Mater 176(1):820–828

Bradder P, Ling SK, Wang S, Liu S (2010) Dye adsorption on layered graphite oxide. J Chem Eng Data 56(1):138–141

Cardoso NF, Lima EC, Royer B, Bach MV, Dotto GL, Pinto LAA, Calvete T (2012) Comparison of Spirulina platensis microalgae and commercial activated carbon as adsorbents for the removal of Reactive Red 120 dye from aqueous effluents. J Hazard Mater 241–242:146–153

Carneiro PA, Umbuzeiro GA, Oliveira DP, Zanoni MVB (2010) Assessment of water contamination caused by a mutagenic textile effluent/dyehouse effluent bearing disperse dyes. J Hazard Mater 174:694–699

Chen Y, Zhang B, Liu G, Zhuan X, Kang ET (2012) Graphene and its derivatives: switching ON and OFF. Chem Soc Rev 41:4688–4707

Cheng JS, Du J, Zhu W (2012) Facile synthesis of three-dimensional chitosan–graphene mesostructures for reactive black 5 removal. Carbohyd Polym 88(1):61–67

Cheng ZL, Li YX, Liu Z (2017) 0 Novel adsorption materials based on graphene oxide/Beta zeolite composite materials and their adsorption performance for rhodamine B. J Alloy Comp 708:255–263

Chowdhury S, Balasubramanian R (2014) Recent advances in the use of graphene-family nanoadsorbents for removal of toxic pollutants from wastewater. Adv Colloid Interface Sci 204:35–56

Clarke EA, Anliker R (1980) Organic dyes and pigments. Anthropogenic compounds. Springer, Berlin, pp 181–215

De Lima ROA, Bazo AP, Salvadori DMF, Rech CM, Oliveira DP, Umbuzeiro GA (2007) Mutagenic and carcinogenic potential of a textile azo dye processing plant effluent that impacts a drinking water source. Mutat Res Genet Toxicol Environ Mutagen 626:53–60

De Oliveira RAG, Zanoni TB, Bessegato GG, Oliveira DP, Umbuzeiro GA, Zanoni MVB (2014) The chemistry and toxicity of hair dyes. Quim Nova 37:1037–1046

Deng JH, Zhang XR, Zeng GM, Gong JL, Niu QY, Liang J (2013) Simultaneous removal of Cd (II) and ionic dyes from aqueous solution using magnetic graphene oxide nanocomposite as an adsorbent. Chem Eng J 226:189–200

Doğan M, Alkan M (2003) Removal of methyl violet from aqueous solution by perlite. J Colloid Interface Sci 267(1):32–41

Dreyer DR, Park S, Bielawski CW, Ruoff RS (2010) The chemistry of graphene oxide. Chem Soc Rev 39:228–240

Du Q, Sun J, Li Y, Yang X, Wang X, Wang Z, Xia L (2014) Highly enhanced adsorption of congo red onto graphene oxide/chitosan fibers by wet-chemical etching off silica nanoparticles. Chem Eng J 245:99–106

El Haddad M, Mamouni R, Saffaj N, Lazar S (2012) Removal of a cationic dye–Basic Red 12–from aqueous solution by adsorption onto animal bone meal. J Assoc Arab Univ Basic Appl Sci 12(1):48–54
Fan ZJ, Kai W, Yan J, Wei T, Zhi LJ, Feng J, Ren YM, Song LP, Wei F (2010) Facile synthesis of graphene nanosheets via Fe reduction of exfoliated graphite oxide. ACS Nano 5(1):191–198
Freundlich H (1906) Adsorption in solution. Phys Chem Soc 40:1361–1368
Geim AK (2009) Graphene: status and prospects. Science 324:1530–1534
Geim AK, Novoselov KS (2007) The rise of graphene. Nat Mater 6(3):183–191
Geng Z, Lin Y, Yu X, Shen Q, Ma L, Li Z, Pan N, Wang X (2012) Highly efficient dye adsorption and removal: a functional hybrid of reduced graphene oxide–Fe_3O_4 nanoparticles as an easily regenerative adsorbent. J Mater Chem 22(8):3527–3535
González JA, Villanueva ME, Piehl LL, Copello GJ (2015) Development of a chitin/graphene oxide hybrid composite for the removal of pollutant dyes: adsorption and desorption study. Chem Eng J 280:41–48
Graphene, complied by the class for physics of the royal Swedish academy of sciences, 2010. www.nobelprize.org/nobel_prizes/physics/laureates/2010/advanced-physicsprize2010.pdf
Gupta VK, Mittal A, Jain R, Megha M, Shalini S (2006) Adsorption of safranin-t from wastewater using waste materials—activated carbon and activated rice husks. J Colloid Interface Sci 303:80–86
Gupta SS, Sreeprasad TS, Maliyekkal SM, Das SK, Pradeep T (2012) Graphene from sugar and its application in water purification. Appl Mater Interfaces 4:4156–4163
Hamdaoui O (2006) Batch study of liquid-phase adsorption of methylene blue using cedar sawdust and crushed brick. J Hazard Mater 135:264–273
Han M, Yun J, Kim H (2012) Effect of surface modification of graphene oxide on photochemical stability of poly(vinyl alcohol)/graphene oxide composite. J Ind Eng Chem 18:752–756
Hosseinabadi-Farahani Z, Mahmoodi NM, Hosseini-Monfared H (2015) Preparation of surface functionalized graphene oxide nanosheet and its multicomponent dye removal ability from wastewater. Fiber Polym 16(5):1035–1047
Hu L, Li Y, Zhang X, Wang Y, Cui L, Wei Q, Ma H, Yan L, Du B (2016) Fabrication of magnetic water-soluble hyperbranched polyol functionalized graphene oxide for high-efficiency water remediation. Sci Rep 6
Huang Y, Chen X (2015) Carbon nanomaterial-based composites in wastewater purification. Nano Life 4:1441006-1–1441006-11
Huang X, Yin ZY, Wu SX, Qi XY, He QY, Zhang QC, Yan Q, Boey F, Zhang H (2011) Graphene-based materials: synthesis, characterization, properties, and applications. Small 7:1876–1902
Huang H, Lu S, Zhang X, Shao Z (2012a) Glucono-δ-lactone controlled assembly of graphene oxide hydrogels with selectively reversible gel-sol transition. Soft Matter 8:4609–4615
Huang X, Qi X, Boey F, Zhang H (2012b) Graphene-based composites. Chem Soc Rev 41:666–686
Huang R, Liu Q, Huo J, Yang B (2017) Adsorption of methyl orange onto protonated cross-linked chitosan. Arab J Chem 10:24–32
Hummers WS, Offeman RE (1958) Preparation of graphitic oxide. J Am Chem Soc 80:1339
Iram M, Guo C, Guan Y, Ishfaq A, Liu H (2010) Adsorption and magnetic removal of neutral red dye from aqueous solution using Fe_3O_4 hollow nanospheres. J Hazard Mater 181(1):1039–1050
Jauris IM, Matos CF, Saucier C, Lima EC, Zarbin AJG, Fagan SB, Machado FM, Zanela I (2016) Adsorption of sodium diclofenac on graphene: a combined experimental and theoretical study. Phys Chem Chem Phys 19:1526–1536
Kazim S, Ali V, Zulfequar M, Haq MM, Husain M (2007) Electrical transport properties of poly [2-methoxy-5-(2′-ethyl hexyloxy)-1, 4-phenylene vinylene] thin films doped with acridine orange dye. Physica B 393(1):310–315

Khaloo SS, Ahmadi Marzaleh M, Kavousian M, Bahramzadeh Gendeshmin S (2016) Graphene oxide coated wad as a new sorbent in fixed bed column for the removal of crystal violet from contaminated water. Sep Sci Technol 51(1):1–10

Kurade MB, Waghmode TR, Patil SM, Jeon BH, Govindwar SP (2017) Monitoring the gradual biodegradation of dyes in a simulated textile effluent and development of a novel triple layered fixed bed reactor using a bacterium-yeast consortium. Chem Eng J 307:1026–1036

Kurniawan TA, Lo WH (2009) Removal of refractory compounds from stabilized landfill leachate using an integrated H_2O_2 oxidation and granular activated carbon (GAC) adsorption treatment. Water Res 43:4079–4091

Kyzas GZ, Travlou NA, Kalogirou O, Deliyanni EA (2013) Magnetic graphene oxide: effect of preparation route on reactive black 5 adsorption. Mater 6(4):1360–1376

Kyzas GZ, Deliyanni EA, Matis KA (2014) Graphene oxide and its application as an adsorbent for wastewater treatment. J Chem Technol Biotechnol 89:196–205

Langmuir I (1918) The adsorption of gases on plane surfaces of glass, mica and platinum. J Am Chem Soc 40:1361–1403

Leechart P, Nakbanpote W, Thiravetyan P (2009) Application of 'waste' wood-shaving bottom ash for adsorption of azo reactive dye. J Environ Manag 90:912–920

Leme DM, de Oliveira GAR, Meireles G, dos Santos TC, Zanoni MVB, de Oliveira DP (2014) Genotoxicological assessment of two reactive dyes extracted from cotton fibres using artificial sweat. Toxicol In Vitro 28:31–38

Li D, Kaner RB (2008) Graphene-based materials. Science 320:1170–1171

Li C, Dong Y, Yang J, Li Y, Huang C (2014a) Modified nano-graphite/Fe_3O_4 composite as efficient adsorbent for the removal of methyl violet from aqueous solution. J Mol Liq 196: 348–356

Li L, Li X, Duan H, Wang X, Luo C (2014b) Removal of Congo Red by magnetic mesoporous titanium dioxide–graphene oxide core–shell microspheres for water purification. Dalton Trans 43(22):8431–8438

Lima EC, Adebayo MA, Machado FM (2015) Kinetic and equilibrium models of adsorption. In: Bergmann CP, Machado FM (eds) Carbon nanomaterials as adsorbents for environmental and biological applications. Springer, Berlin, pp 33–69

Liu Y, Xu H, Yang SF, Tay JH (2003) A general model for biosorption of Cd^{2+}, Cu^{2+} and Zn^{2+} by aerobic granules. J Biotechnol 102:233–239

Liu F, Chung S, Oh G, Seo TS (2012) Three-dimensional graphene oxide nanostructure for fast and efficient water-soluble dye removal. ACS Appl Mater Interfaces 4(2):922–927

Liu A, Zhou W, Shen K, Liu J, Zhang X (2015a) One-pot hydrothermal synthesis of hematite-reduced graphene oxide composites for efficient removal of malachite green from aqueous solution. RSC Adv 5(22):17336–17342

Liu K, Li H, Wang Y, Gou X, Duan Y (2015b) Adsorption and removal of rhodamine B from aqueous solution by tannic acid functionalized graphene. Colloids Surf, A 477:35–41

Lucas MS, Peres JA (2006) Decolorization of the azo dye reactive black 5 by Fenton and photo-Fenton oxidation. Dyes Pigm 71(3):236–244

Machado FM, Bergmann CP, Fernandes THM, Lima EC, Royer B, Calvete T, Fagan SB (2011) Adsorption of Reactive Red M-2BE dye from water solutions by multi-walled carbon nanotubes and activated carbon. J Hazard Mater 192:1122–1131

Machado FM, Bergmann CP, Lima EC, Royer B, de Souza FE, Jauris IM, Calvete T, Fagan SB (2012) Adsorption of Reactive Blue 4 dye from water solutions by carbon nanotubes: experiment and theory. Phys Chem Chem Phys 14:11139–11153

Mall ID, Srivastava VC, Agarwal NK, Mishra IM (2005) Adsorptive removal of malachite green dye from aqueous solution by bagasse fly ash and activated carbon-kinetic study and equilibrium isotherm analyses. Colloids Surf, A 264(1):17–28

Mohanty K, Das D, Biswas MN (2008) Treatment of phenolic wastewater in a novel multi-stage external loop airlift reactor using activated carbon. Sep Purif Technol 58(3):311–319

Moradi O, Gupta VK, Agarwal S, Tyagi I, Asif M, Makhlouf ASH, Sadegh H, Shahryari-ghoshekandi R (2015) Characteristics and electrical conductivity of graphene and graphene oxide for adsorption of cationic dyes from liquids: kinetic and thermodynamic study. J Ind Eng Chem 28:294–301

Murty MN, Kumar S (2011) Water pollution in India an economic appraisal. India infrastructure report, water: policy and performance for sustainable development. Infrastructure Development Finance Company. Oxford University Press

Namasivayam C, Kavitha D (2002) Removal of Congo Red from water by adsorption onto activated carbon prepared from coir pith, an agricultural solid waste. Dyes Pigm 54(1):47–58

Naushad M, Khan MR, ALOthman ZA et al (2015a) Removal of BrO_3^- from drinking water samples using newly developed agricultural waste-based activated carbon and its determination by ultra-performance liquid chromatography-mass spectrometry. Environ Sci Pollut Res 22:15853–15865. https://doi.org/10.1007/s11356-015-4786-y

Naushad M, ALOthman ZA, Alam MM et al (2015b) Synthesis of sodium dodecyl sulfate-supported nanocomposite cation exchanger: Removal and recovery of Cu^{2+} from synthetic, pharmaceutical and alloy samples. J Iran Chem Soc 12:1677–1686. https://doi.org/10.1007/s13738-015-0642-8

Naushad M, Mittal A, Rathore M, Gupta V (2015c) Ion-exchange kinetic studies for Cd(II), Co(II), Cu(II), and Pb(II) metal ions over a composite cation exchanger. Desalin Water Treat 54:2883–2890. https://doi.org/10.1080/19443994.2014.904823

Niculescu VC, Miricioiu MG, Ionete R, Zgavarogea R (2015) An overview of mesoporous adsorbents for food industry wastewater treatment. In: 15th International Multidisciplinary Scientific Geo Conference SGEM 2015. www.sgem.org. SGEM2015 Conference Proceedings, vol 1, pp 25–32

Nuengmatcha P, Mahachai R, Chanthai S (2014) Thermodynamic and kinetic study of the intrinsic adsorption capacity of graphene oxide for malachite green removal from aqueous solution. Orien J Chem 30(4):1463–1474

Park S, Ruoff RS (2009) Chemical methods for the production of graphenes. Nat Nanotechnol 4:217–224

Pathania D, Katwal R, Sharma G et al (2016) Novel guar gum/Al_2O_3 nanocomposite as an effective photocatalyst for the degradation of malachite green dye. Int J Biol Macromol 87:366–374. https://doi.org/10.1016/j.ijbiomac.2016.02.073

Pei SF, Cheng HM (2012) The reduction of graphene oxide. Carbon 50(9):3210–3228

Peigney A, Laurent C, Flahaut E, Bacsa RR, Rousset A (2001) Specific surface area of carbon nanotubes and bundles of carbon nanotubes. Carbon 39:507–514

Pourjavadi A, Nazari M, Hosseini SH (2015) Synthesis of magnetic graphene oxide-containing nanocomposite hydrogels for adsorption of crystal violet from aqueous solution. RSC Adv 5 (41):32263–32271

Prola LDT, Acayanka E, Lima EC, Umpierres CS, Vaghetti JCP, Santos WO, Laminsi S, Njifon PT (2013) Comparison of Jatropha curcas shells in natural form and treated by non-thermal plasma as biosorbents for removal of Reactive Red 120 textile dye from aqueous solution. Ind Crop Prod 46:328–340

Puchana-Rosero MJ, Adebayo MA, Lima EC, Machado FM, Thue PS, Vaghetti JCP, Umpierres CS, Gutterres M (2016) Microwave-assisted activated carbon obtained from the sludge of tannery-treatment effluent plant for removal of leather dyes. Colloids Surf A: Physicochem Eng Aspects 504:105–115

Qin J, Qiu F, Rong X, Yan J, Zhao H, Yang D (2015) Adsorption behavior of crystal violet from aqueous solutions with chitosan–graphite oxide modified polyurethane as an adsorbent. J Appl Poly Sci 132(17)

Radich JG, McGinn PJ, Kamat PV (2011) Graphene-based composites for electrochemical energy storage. Electrochem Soci Inter 20(1):63–66

Rajaram T, Das A (2008) Water pollution by industrial effluent in India: Discharge scenarios and case for participatory ecosystem specific local regulation. Futures 40:56–69

Rajeswari A, Vismaiya S, Pius A (2017) Preparation, characterization of nano ZnO-blended cellulose acetate-polyurethane membrane for photocatalytic degradation of dyes from water. Chem Eng J 313:928–937
Ramesha GK, Kumara AV, Muralidhara HB, Sampath S (2011) Graphene and graphene oxide as effective adsorbents toward anionic and cationic dyes. J Colloid Interface Sci 361(1):270–277
Ramphal (2013) An empirical study about water pollution in India. Int J Manag Soc Sci 1:52–58
Rao CNR, Sood AK, Subrahmanyam KS, Govindaraj A (2009) Graphene: the new two-dimensional nanomaterial. Angew Chem Int Ed 48:7752–7777
Rashed MN (2011) Acid dye removal from industrial wastewater by adsorption on treated sewage sludge. Int J Environ Waste Manag 7:175–191
Ren L, Qi X, Liu Y, Huang Z, Wei X, Li J, Yang L, Zhong J (2012) Upconversion-P25 graphene composite as an advanced sunlight driven photo catalytic hybrid material. J Mater Chem 22 (23):11765–11771
Ribas MC, Adebayo MA, Prola LDT, Lima EC, Cataluna R, Feris LA, Machado FM, Pavan FA, Calvete T (2014) Comparison of a homemade cocoa shell activated carbon with commercial activated carbon for the removal of reactive violet 5 dye from aqueous solutions. Chem Eng J 248:315–326
Robati D, Mirza B, Rajabi M, Moradi O, Tyagi I, Agarwal S, Gupta VK (2016) Removal of hazardous dyes-BR 12 and methyl orange using graphene oxide as an adsorbent from aqueous phase. Chem Eng J 284:687–697
Rocha OP, Cesila CA, Christovam EM, Barros SBDM, Zanoni MVB, de Oliveira DP (2017) Ecotoxicological risk assessment of the "Acid Black 210" dye. Toxicology 376:113–119
Sahraei R, Pour ZS, Ghaemy M (2017) Novel magnetic bio-sorbent hydrogel beads based on modified gum tragacanth/graphene oxide: Removal of heavy metals and dyes from water. J Clean Prod 142:2973–2984
Saito R, Dresselhaus G, Dresselhaus MS (1998) Physical properties of carbon nanotubes. Imperial College Press London
Sakuramoto S, Sasako M, Yamaguchi T, Kinoshita T, Fujii M, Nashimoto A, Furukawa H, Nakajima T, Ohashi Y, Imamura H, Higashino M (2007) Adjuvant chemotherapy for gastric cancer with S-1, an oral fluoropyrimidine. New Engl J Med 357(18):1810–1820
Sarkar C, Bora C, Dolui SK (2014) Selective dye adsorption by ph modulation on amine-functionalized reduced graphene oxide–carbon nanotube hybrid. Ind Eng Chem Res 53(42):16148–16155
Sengupta B (2004) Identification of hazardous waste streams, their characterization and waste reduction options in Dyes and dye intermediate sector. Central Pollution Control Board, Ministry of Environment and Forest, New Delhi
Shahryari-ghoshekandi R, Sadegh H (2014) Kinetic study of the adsorption of synthetic dyes on graphene surfaces. Jordan J Chem 9(4):267–278
Shamik C, Rajasekhar B (2014) Recent advances in the use of graphene-family nanoadsorbents for removal of toxic pollutants from wastewater. J Adv Colloid Interface Sci 204:35–56
Shaobin W, Sun H, Ang HM, Tade MO (2013) Adsorptive remediation of environmental pollutants using novel graphene-based nanomaterials. Chem Eng J 226:336–347
Sharif F, Gagnon LR, Mulmi S, Roberts EP (2017) Electrochemical regeneration of a reduced graphene oxide/magnetite composite adsorbent loaded with methylene blue. Water Res 114:237–245
Sharma P, Das M (2013) Removal of a cationic dye from aqueous solution using graphene oxide nanosheets: investigation of adsorption parameters. J Chem Eng Data 58:151–158
Sharma P, Saikia BK, Das MR (2014) Removal of methyl green dye molecule from aqueous system using reduced graphene oxide as an efficient adsorbent: kinetics, isotherm and thermodynamic parameters. Colloids Surf, A 457:125–133
Shirmardi M, Mahvi AH, Hashemzadeh B, Naeimabadi A, Hassani G, Niri MV (2013) The adsorption of malachite green (MG) as a cationic dye onto functionalized multi walled carbon nanotubes. Korean J Chem Eng 30:1603–1608
Sips R (1948) On the structure of a catalyst surface. J Chem Phys 16:490–495

Sui Z, Zhang X, Lei Y, Luo Y (2011) Easy and green synthesis of reduced graphite oxide-based hydrogels. Carbon 49:4314–4321

Sun L, Fugetsu B (2014) Graphene oxide captured for green use: influence on the structures of calcium alginate and macroporous alginic beads and their application to aqueous removal of acridine orange. Chem Eng J 240:565–573

Sun H, Cao L, Lu L (2011) Magnetite/reduced graphene oxide nanocomposites: one step solvothermal synthesis and use as a novel platform for removal of dye pollutants. Nano Res 4 (6):550–562

Sun L, Yu H, Fugetsu B (2012) Graphene oxide adsorption enhanced by in situ reduction with sodium hydrosulfite to remove acridine orange from aqueous solution. J Hazard Mater 203:101–110

Sun JZ, Liao ZH, Si RW, Kingori GP, Chang FX, Gao L, Shen Y, Xiao X, Wu XY, Yong YC (2014) Adsorption and removal of triphenylmethane dyes from water by magnetic reduced graphene oxide. Water Sci Technol 70(10):1663–1669

Swaminathan K, Sandhya S, Carmalin Sophia A, Pachhade K, Subrahmanyam YV (2003) Decolorization and degradation of H-acid and other dyes using ferrous–hydrogen peroxide system. Chemosphere 50:619–625

Travlou NA, Kyzas GZ, Lazaridis NK, Deliyanni EA (2013a) Graphite oxide/chitosan composite for reactive dye removal. Chem Eng J 217:256–265

Travlou NA, Kyzas GZ, Lazaridis NK, Deliyanni EA (2013b) Functionalization of graphite oxide with magnetic chitosan for the preparation of a nanocomposite dye adsorbent. Langmuir 29 (5):1657–1668

Tsai WT, Hsu HC, Su TY, Lin KY, Lin CM (2008) Removal of basic dye (methylene blue) from wastewaters utilizing beer brewery waste. J Hazard Mater 154:73–78

Waghmare SS, Arfin T (2015) Fluoride removal from water by various techniques: review. Int J Innov Sci Eng Technol 2(9):560–571

Wang S, Wei J, Lv S, Guo Z, Jiang F (2013) Removal of organic dyes in environmental water onto magnetic-sulfonic graphene nanocomposite. CLEAN–Soil Air Water 41(10):992–1001

Wang J, Ji S, Ma R, Wu Q, Wang C, Qiang J, Wang Z (2014) Synthesis of magnetic Fe_3O_4 modified graphene nanocomposite and its application on the adsorption of some dyes from aqueous solution. Sep Sci Technol 49(6):861–867

Wang D, Liu L, Jiang X, Yu J, Chen X (2015a) Adsorption and removal of malachite green from aqueous solution using magnetic β-cyclodextrin-graphene oxide nanocomposites as adsorbents. Colloids Surf, A 466:166–173

Wang J, Zhu H, Hurren C, Zhao J, Pakdel E, Li Z, Wang X (2015b) Degradation of organic dyes by P25-reduced graphene oxide: Influence of inorganic salts and surfactants. J Environ Chem Eng 648:1–7

Wang H, Chen Y, Wei Y (2016) A novel magnetic calcium silicate/graphene oxide composite material for selective adsorption of acridine orange from aqueous solutions. RSC Adv 6 (41):34770–34781

Wang L, Mao C, Sui N, Liu M, Yu WW (2017) Graphene oxide/ferroferric oxide/ polyethylenimine nanocomposites for Congo red adsorption from water. Environ Technol 38 (8):996–1004

Wolf EL (2004) Nanophysics and nanotechnology—an introduction to modern concepts in nanoscience. Wiley, New York

Wu T, Liu G, Zhao J, Hidaka H, Serpone N (1998) Photoassisted degradation of dye pollutants. V. Self-photosensitized oxidative transformation of rhodamine B under visible light irradiation in aqueous TiO_2 dispersions. J Phys Chem B 102(30):5845–5851

Wu S, He Q, Tan C, Wang Y, Zhang H (2013) Graphene-based electrochemical sensors. Small 9:1160–1172

Wu Z, Zhong H, Yuan X, Wang H, Wang L, Chen X, Zeng G, Wu Y (2014) Adsorptive removal of methylene blue by rhamnolipid-functionalized graphene oxide from wastewater. Water Res 67:330–344

Xiang Q, Yu J, Jaroniec M (2012) Graphene-based semiconductor photocatalysts. Chem Soc Rev 41:782–796

Xiao J, Lv W, Xie Z, Tan Y, Song Y, Zheng Q (2016) Environmentally friendly reduced graphene oxide as a broad-spectrum adsorbent for anionic and cationic dyes via π–π interactions. J Mater Chem A 4(31):12126–12135

Xie G, Xi P, Liu H, Chen F, Huang L, Shi Y, Hou F, Zeng Z, Shao C, Wang J (2012) A facile chemical method to produce superparamagnetic graphene oxide–Fe_3O_4 hybrid composite and its application in the removal of dyes from aqueous solution. J Mater Chem 22(3):1033–1039

Xu YX, Zhao L, Bai H, Hong WJ, Li C, Shi GQ (2009) Chemically converted graphene induced molecular flattening of 5,10,15,20-tetrakis(1-methyl-4-pyridinio) porphyrin and its application for optical detection of cadmium(II) ions. J Am Chem Soc 131:13490–13497

Yang ST, Chen S, Chang Y, Cao A, Liu Y, Wang H (2011) Removal of methylene blue from aqueous solution by graphene oxide. J Colloid Interface Sci 359(1):24–29

Yao Y, Miao S, Liu S, Ma LP, Sun H, Wang S (2012) Synthesis, characterization, and adsorption properties of magnetic Fe_3O_4@ graphene nanocomposite. Chem Eng J 184:326–332

Yu Y, Murthy BN, Shapter JG, Constantopoulos KT, Voelcker NH, Ellis AV (2013) 0 Benzene carboxylic acid derivatized graphene oxide nanosheets on natural zeolites as effective adsorbents for cationic dye removal. J Hazard Mater 260:330–338

Zamani S, Tabrizi NS (2015) Removal of methylene blue from water by graphene oxide aerogel: thermodynamic, kinetic, and equilibrium modeling. Res Chem Intermed 41(10):7945–7963

Zhang L, Li X, Huang Y, Ma Y, Wan X, Chen Y (2010) Controlled synthesis of few layered graphene sheets on a large scale using chemical exfoliation. Carbon 48:2367–2371

Zhang X, Yu H, Yang H, Wan Y, Hu H, Zhai Z, Qin J (2015) Graphene oxide caged in cellulose microbeads for removal of malachite green dye from aqueous solution. J Colloid Interface Sci 437:277–282

Zhuang Y, Yu F, Chen J, Ma J (2016) Batch and column adsorption of methylene blue by graphene/alginate nanocomposite: comparison of single-network and double-network hydrogels. J Environ Chem Eng 4(1):147–156

The manufacturer's authorised representative in the EU is Springer Nature Customer Service Centre GmbH, Europaplatz 3, 69115 Heidelberg, Germany. If you have any concerns regarding our products, please contact ProductSafety@springernature.com

Printed and bound by CPI Group (UK) Ltd, Croydon, CR0 4YY

30/09/2026

02225238-0015